Validation Standard Operating Procedures

Second Edition

A Step-by-Step Guide for Achieving Compliance in the Pharmaceutical, Medical Device, and Biotech Industries

Validation Standard Operating Procedures

Second Edition

A Step-by-Step Guide for Achieving Compliance in the Pharmaceutical, Medical Device, and Biotech Industries

Syed Imtiaz Haider, Ph.D.

CRC Press
Taylor & Francis Group
Boca Raton London New York

CRC Press is an imprint of the
Taylor & Francis Group, an **informa** business
A TAYLOR & FRANCIS BOOK

CRC Press
Taylor & Francis Group
6000 Broken Sound Parkway NW, Suite 300
Boca Raton, FL 33487-2742

First issued in paperback 2019

ISBN-13: 978-0-8493-9529-1 (hbk)
ISBN-13: 978-0-367-39077-8 (pbk)
Library of Congress Card Number 2005035145

Library of Congress Cataloging-in-Publication Data

Haider, Syed Imtiaz.
    Validation standard operating procedures : a step-by-step guide for achieving compliance in the pharmaceutical, medical device, and biotech industries / Syed Imtiaz Haider.-- 2nd ed.
       p. ; cm.
    Includes bibliographical references.
    ISBN-13: 978-0-8493-9529-1 (hardcover : alk. paper) ISBN-10: 0-8493-9529-1 (hardcover : alk. paper)
    1. Pharmaceutical technology--Quality control. 2. Pharmaceutical industry--Standards--United States. 3. Biotechnology industries--Standards--United States. 4. Medical instruments and apparatus industry--Standards--United States. [DNLM: 1. Equipment Safety--standards--United States. 2. Equipment and Supplies--standards--United States. 3. Biotechnology--standards--United States. 4. Drug Industry--standards--United States. 5. Quality Control--United States. 6. Technology, Pharmaceutical--United States. W 26 H149v 2006] I. Title: Step-by-step guide for achieving compliance in the pharmaceutical, medical device, and biotech industries. II. Title.

RS192.H353 2006
681'.761'021873--dc22
                                   2005035145

Visit the Taylor & Francis Web site at
http://www.taylorandfrancis.com

and the CRC Press Web site at
http://www.crcpress.com

CONTENTS

Section VAL 400.00

Section VAL 500.00

Section VAL 600.00

Section VAL 700.00

Section VAL 800.00

Section VAL 900.00

Section VAL 1000.00

Section VAL 1100.00

Section VAL 1200.00

Section VAL 1300.00

Section VAL 1400.00

Section VAL 1500.00

Section VAL 1600.00

Section VAL 1700.00

Section VAL 1800.00

Section VAL 1900.00

Section VAL 2000.00

Section VAL 2100.00

Section VAL 2200.00

Section VAL 2300.00

PREFACE

As the validation master plan execution program proceeds and the facility is integrated into regulatory guidelines of the FDA, current good manufacturing practices (cGMPs), good laboratory practices (GLPs), and the need for comprehensive and well-defined validation supporting standard operating procedures (SOPs) are required. As the validation program progresses and the systems are integrated into routine operation, there are fewer deviations and the SOPs become more precise and complete.

This book and CD-ROM provide an administrative solution for management. The execution of test functions defined in the validation master plan procedures is provided in the text and the electronic files. The validation standard operating procedure can help your company comply with GMP, GLP, and validation requirements imposed by the FDA.

The formats and style provided are generic and can be further amended. The contents of the SOPs are intended to build quality into the regulatory requirements. However, having a set of validation SOPs does not preclude adverse inspection findings because contents that satisfy one inspector may not satisfy another.

I strongly believe that the facility's technical management and staff should read the procedures to ensure that particular needs are addressed with reference to operational control within the organization and individual countries' regulatory requirements. It is, however, guaranteed to provide management with a tool to develop a set of validation SOPs in order to support the road map established for the on-time successful start-up of the facility operation in compliance with the GMP requirement.

The pharmaceutical industry and its top management are confronted today with a period in which the demands for validation, GMP, and GLP compliance have never been greater. The validation procedures in the present environment must be designed to ensure regulatory compliance in pharmaceutical operations as well as serve as a training source for undergraduates and graduates at the academic level.

The first edition comprised Val. Section 200 to Val. Section 1300, providing 75 validation SOPs covering design qualification, utilities qualification, facility qualification, equipment qualification, training needs, and major sterile and nonsterile operations-related equipment.

In Val. Section 1400 to Val. Section 2300, the second edition provides additional information needed to respond to increasing demand by official regulatory bodies worldwide. I have assembled a matrix that will provide a source for experienced and inexperienced practitioners in the pharmaceutical, biotech, and medical device industries.

In the second edition, 64 new validation SOPs are added to describe documentation required for sterility assurance, qualification and requalification template reports of major sterile equipments, critical applicable procedures, templates for certification guidelines, media fill procedures, environmental control guidelines, training, and critical environmental performance evaluation procedures.

The second volume provides 139 template procedures, protocols, and reports that can be downloaded and, after minor changes, adopted. The ready-to-use protocols allow end users to record all raw hard data, further enabling them to prepare final reports. The additional chapters included in this edition provide details on how to ensure sterility and prepare sterility assurance reports.

- Section Val. 1400 provides a unique template on how to prepare and submit sterility assurance and information data for submission to FDA in ANDA and NDA submissions.
- Section Val. 1500 provides 11 state-of-the-art validation protocols covering critical aseptic processing equipment.
- Section Val. 1600 describes 11 examples of how to prepare critical qualification and requalification reports related to aseptic operations and provides them as a proof of records to FDA for ANDA and ANDA file submissions.
- Section Val. 1700 provides 14 exclusive procedures describing critical monitoring operations to ensure sterility assurance.
- Comprehensive process simulation (media fill) test, media fill microbial examination, process simulation test protocol and template media fill run reports are provided in Section Val. 1800.
- Section Val. 1900 includes four aseptic processes associated with monitoring and qualification programs covering determination of components' bioburden before sterilization sterility test failure investigation, bacterial endotoxin determination in WFI, in-process finished product, and monitoring the bioburden, spore bioburden, and endotoxin present on stoppers and unprocessed vials.

- Section Val. 2000 describes the training needs of personnel involved in aseptic operations.
- Section Val. 2100 provides 14 critical environmental performance evaluation tests, including:
- Five discrete cleaning validation protocols related to aseptic manufacturing equipments, solution preparation tanks, mobile tanks, filtration assemblies, freeze dryers, and vial filling machine parts are provided in Section Val. 2200.
- Section Val. 2300 includes recommended reading.

Pharmaceutical, medical, and biotech industries are regulated worldwide to be in compliance with cGMP and GLP principles. Each company is required to create validation SOPs to qualify its equipment, utilities, buildings, and personnel. The template validation SOPs available enable end users to understand principles and elements of good manufacturing practice and provide documentation language ranging from generic to specific, depending on the detail level of the requirements.

Compliance to FDA regulations by the health care industry over the last decade has been a major goal, including those companies intending to export their product to the U.S. market. As a result, only a few companies are able to seek approval for exportation; one of the reasons behind this is the absence or inadequacy of validation SOPs.

The validation SOPs on the CD-ROM are valuable tools for companies in the process of developing or revising validation SOPs to achieve FDA, GMP, and GLP compliance. The documentation package is especially relevant to quality assurance personnel, engineers, utilities engineers, computer engineers, validation designers, internal and external auditors, or to anyone interested in developing a qualification documentation matrix.

Syed Imtiaz Haider

DEDICATION

This book is dedicated to my loving father Syed Mohsin Raza, and to my late mother, Syeda Kharsheed-un-Nissan, for their continuous motivation. I am also indebted to my wife, Syeda Shazia Fatima, my son, Syed Zeeshan Haider, and my daughter, Syeda Mehreen Fatima, for their patience while I compiled this book.

ACKNOWLEDGMENTS

I am thankful to Mr. Abdul Razzaq Yousef, the managing director of Gulf Pharmaceutical Industries, for always encouraging me on my professional achievements and continuously keeping me motivated.

I would also like to thank my friends and colleagues Ateeq Hashmi and Javed Zamir for their help and encouragement and for creating a professional environment. Special thanks to the staff of Taylor & Francis Books for their patience and diligence in the production of this book.

THE AUTHOR

**Syed Imtiaz Haider** earned his Ph.D. in chemistry and is a quality assurance and environmental specialist with over 15 years of experience in aseptic and nonaseptic pharmaceutical processes, equipment validation, and in-process control and auditing. Dr. Haider is the author and coauthor of more than 20 research publications in international refereed journals dealing with products of pharmaceutical interest, their isolation, and structure development. A professional technical writer, Dr. Haider has authored more than 2000 standard operating procedures based on FDA regulations, ISO 9001:2000, and ISO 14001:2004 standards. He is a certified QMS auditor of IRCA and registered associate environmental auditor of EARA. He has written more than ten quality system manuals for multidisciplinary industries and provided consultancy to the Ministry of Health, United Arab Emirates, Drug Control Laboratory, in developing a quality management system based on ISO 9003 and later transition to ISO 9001:2000.

Dr. Haider is working as a quality affairs director at Julphar, Gulf Pharmaceutical Industries, and is involved in the preparation of several abbreviated new drug application (ANDA) files, followed by successful FDA, EU GMP inspections leading to export of finished pharmaceutical products to the U.S. and European markets. He has also written *ISO 9001: 2000: Document Development Compliance Manual: A Complete Guide and CD-Rom* and *Pharmaceutical Validation Master Plan, The Ultimate Guide to FDA, GMP, and GLP Compliance.* Dr. Haider holds the intellectual copyright certificate of registration on an electronic documentation package on ISO 9000 and ISO 14001 from the Canadian Intellectual Property Office. He is also a contributing author of chapters on ISO 9001: 2000 and ISO 14001 in international publications.

THE AUTHOR

ABOUT THE BOOK

This book and CD-ROM take into account all major international regulations, such as FDA, EU GMP, cGMP, GLP, PDA technical monographs, PDA technical reports, PMA concepts, journals of PDA, GCP, and industry standard ISO 9000, to be in compliance with documentation guidelines. No other book in print deals exclusively with the key elements of validation procedure for pharmaceutical plants and provides hands-on templates to be tailored to achieve FDA compliance.

Validation SOPs are written to provide explicit instruction on how to achieve the standards for those responsible for writing and executing master validation plans for drug, drug–device combinations, diagnostics, pharmaceutical biotechnology, and bulk pharmaceutical chemical products. Included is the ready-to-use template so that one can immediately save time and expense without missing any critical elements.

The book provides instant answers to validation engineers, validation specialists, quality professionals, quality assurance auditors, and protocol writers about what to include in validation SOPs and how to enhance productivity.

Validation of aseptic pharmaceutical processes is specifically assembled in the second edition as a reference for use by managers, supervisors, and scientists in the pharmaceutical industry. The primary intent of this work is to guide design engineers, manufacturing personnel, research and development scientists, and quality control professionals in validating those processes needed for nonaseptic and aseptic pharmaceutical production.

The master validation SOPs on the CD-ROM are valuable tools for companies in a process of developing validation master plans and organizing SOPs to achieve FDA, GMP, and GLP compliance. The documentation package is especially relevant to quality assurance personnel, engineers, utilities engineers, computer engineers, validation designers, internal auditors, external auditors, or anyone interested in developing a qualification documentation matrix.

The second edition CD-ROM contains 139 validation SOPs; they are made available so that customers can input them into their computers and use their Microsoft Word programs to edit and print these documents. The contents of the procedures are written in simple and precise language to be in compliance with FDA regulation GMP and GLP requirements. The book ensures minimization of the number of documents to avoid the nightmare of the head of quality assurance at the time of FDA audit. The SOPs exclusively refer to the documents specially required for compliance; however, specific formats are not included to ensure that the electronic templates can be easily used by pharmaceutical, bulk pharmaceutical, medical device, and biotechnology industries.

The purpose of the second edition is to meet the need for a ready-to-use text on the validation of aseptic pharmaceutical production and to provide general information and guidelines. It is a compilation of various theories, sterilization variables, and engineering and microbial studies that can be used independently or in combination to validate equipment and processes. The concepts and methods presented in this edition are not intended to serve as a final rule. Reciprocal methods for achieving this purpose exist and should also be reviewed and consulted, if applicable.

The formats and style provided are generic and can be further amended. The contents of the SOPs are intended to build quality into the regulatory requirements. However, having a set of validation SOPs does not preclude adverse inspection findings because contents that satisfy one inspector may not satisfy another.

I strongly believe that the facility's technical management and staff should read the procedures to ensure that particular needs are addressed with reference to operational control within the organization and individual countries' regulatory requirements. It is, however, guaranteed to provide management with a tool to develop a set of validation SOPs in order to support the road map established for the on-time successful start-up of the facility operation in compliance with the GMP requirements.

Pharmaceutical, medical, and biotech industries are regulated worldwide to be in compliance with cGMP and GLP principles. Each company is required to create validation SOPs to qualify its equipment, utilities, buildings, and personnel. The template validation SOPs available enable end users to understand principles and elements of GMP and provide documentation language ranging from generic to specific, depending on the detail level of the requirements.

Compliance to FDA regulations by the health care industry over the last decade has been a major goal, including those companies intending to export their product to the U.S. market. As a result, the FDA inspects several companies around the world every year for their GMP and GLP compliance. Only a few companies are able to seek approval for exportation; one of the reasons behind this is the absence or inadequacy of validation SOPs. Key benefits involve but are not limited to:

- Successful facility operational start-up
- Minimized noncompliance
- Reduced reworks
- Reduced rejected lots
- Avoidance of recalled lots
- Help in new drug approval
- Satisfactory inspections
- Corporate image
- Financial gain
- Secure third-party contracts
- Corporate legal protection
- Utility cost reduction
- Minimized capital expenditures
- Fewer complaints
- Reduced testing
- Improved employee awareness

I believe that by following the broadly based example of these SOPs, new as well as experienced companies can benefit by enhancing their existing documentation to meet FDA and other regulatory requirements. Currently, no GMP document specifically describes the format of these validation SOPs.

The second volume and CD-ROM are designed for individuals specifically involved in writing and execution of master validation plans, development of protocols, and applicable procedures. This book provides a complete, single-source reference detailing conceptual design elements and 139 explicit procedures to provide sterility assurance.

INTRODUCTION

This book was designed and written for validation professionals responsible for writing and maintaining quality management systems for the successful operation of their companies. It provides a set of SOPs that can be used to manage and document critical validation and revalidation tasks in a pharmaceutical manufacturing facility.

The numbering of the sections and related SOPs begins with 200 and goes through 2300. In addition, the reader may add SOPs that are unique to his or her facility. The term "responsible person" is used extensively throughout the SOPs. The term refers to the person delegated as authority by management and deemed responsible for performing duties associated with validation tasks within the facility.

SOP Format

Information common to all SOPs is described next.

First Page

Company name — at the top of each SOP, a box is provided to enter your company's name.

SOP Number — each SOP is assigned a unique number that appears at the upper-left corner of each page.

Title — the title of each SOP appears at the top of the first two pages below the SOP number. The title describes the subject of the SOP.

Date — each SOP is assigned an effective date at the top of the page, to the right of the SOP number. The date describes the month, day, and year of implementation.

Author — each SOP is assigned a space to provide the author name, title, and department, along with signatures and dates.

Checked by — each SOP is assigned a space to provide the name, title, and department of the person responsible for checking the contents of the SOP requiring the signature and date.

Approved by — each page of the SOP provides a space for the signature of the quality assurance manager approving the SOP to prevent unauthorized changes.

Revisions — at the end of the first page is the revisions box. This box documents the revision number, section, pages, initials, and date.

Other Pages

Subject — each SOP begins with the subject to provide key description of the SOP.

Purpose — each SOP is supported with reasons describing the purpose.

Responsibility — the space for responsibility clearly identifies who must follow the procedures and who is responsible for the overall compliance with the SOP.

Procedure — following the purpose statement are the individual steps of the SOP, arranged in logical order to make the SOP easy to perform.

Reasons for Revision — at the end of each SOP, a space is provided to list the reasons why the SOP is changed, along with the date.

CD-ROM — an electronic copy of the generic validation SOPs is provided.

DISCLAIMER

Every effort has been made to ensure that the contents of the generic validation standard operating procedures are accurate and that recommendations are appropriate and made in good faith. The author accepts no responsibility for inaccuracies or actions taken by companies subsequent to these recommendations.

The similarity in the contents of the procedure with a particular reference to the test functions, acceptance criteria, qualification protocols, and checks may be incidental because of the similarity in principle and operations of pharmaceutical equipment.

SECTION

VAL 100.00

YOUR COMPANY
VALIDATION STANDARD OPERATING PROCEDURE

SOP No. Val. 100.10

Effective date: mm/dd/yyyy

Approved by:

TITLE: Introduction to Validation

AUTHOR:

_____

Name/Title/Department

_____

Signature/Date

CHECKED BY:

_____

Name/Title/Department

_____

Signature/Date

APPROVED BY:

_____

Name/Title/Department

_____

Signature/Date

REVISIONS:

No.	Section	Pages	Initials/Date

SOP No. Val. 100.10

Effective date: mm/dd/yyyy
Approved by:

SUBJECT: Introduction to Validation

PURPOSE

To describe the definition, types, and benefits of validation

RESPONSIBILITY

It is the responsibility of validation team members to follow the procedures. The quality assurance (QA) manager is responsible for SOP compliance.

PROCEDURE

1. Definition of Validation

Validation is a systematic approach to gathering and analyzing sufficient data which will give reasonable assurance (documented evidence), based upon scientific judgment, that a process, when operating within specified parameters, will consistently produce results within predetermined specifications.

2. Type of Validation

- Retrospective Validation
- Prospective Validation
- Concurrent Validation
- Revalidation

2.1 Retrospective Validation

Validation of a process for a product already in distribution, based on accumulated production, testing, and control dates. Summary of existing historical data.

2.2 Prospective Validation

Validation conducted prior to distribution either of a new product, or a product made under a revised manufacturing process. Validation is completed and the results are approved prior to any product release.

SOP No. Val. 100.10   **Effective date: mm/dd/yyyy**

                                                    Approved by:

2.3 Concurrent Validation

A combination of retrospective and prospective validation. Performed against an approved protocol but product is released on a lot-by-lot basis. Usually used on an existing product not previously validated or insufficiently validated.

2.4 Revalidation

To validate change in equipment, packaging, formulation operating procedure, or process that could impact product safety, efficacy, or potency. It is important to establish a revalidation program for critical equipment to maintain validity.

3. Importance of Validation

- Increased throughput
- Reduction in rejections and reworking
- Reduction in utility costs
- Avoidance of capital expenditures
- Fewer complaints about process-related failures
- Reduced testing in-process and in finished goods
- More rapid and reliable start-up of new equipment
- Easier scale-up from development work
- Easier maintenance of equipment
- Improved employee awareness of processes
- More rapid automation

REASONS FOR REVISION

Effective date: mm/dd/yyyy

- First time issued for your company, affiliates and contract manufacturers

SECTION

VAL 200.00

+---+
| **YOUR COMPANY** |
| **VALIDATION STANDARD OPERATING PROCEDURE** |
+---+

SOP No. Val. 200.10   Effective date: mm/dd/yyyy
                                       Approved by:

TITLE: Fundamentals of Validation SOPs

AUTHOR: _____
                             Name/Title/Department

                    _____
                                Signature/Date

CHECKED BY: _____
                             Name/Title/Department

                    _____
                                Signature/Date

APPROVED BY: _____
                             Name/Title/Department

                    _____
                                Signature/Date

REVISIONS:

No.	Section	Pages	Initials/Date

SOP No. Val. 200.10   Effective date: mm/dd/yyyy

                                             Approved by:

SUBJECT: Fundamentals of Validation SOPs

PURPOSE

The purpose of writing fundamentals of validation SOPs is to assist the parent company, affiliates, and contract manufacturers to understand and maintain the validation program to meet the GMP requirements.

RESPONSIBILITY

It is the responsibility of the quality assurance manager to develop and maintain the validation program. The departmental managers and contractors are responsible for the SOP compliance.

PROCEDURE

Introduction

1. Purpose and Scope of the SOP

The quality assurance department is responsible for providing support to the parent company, affiliates, and contract manufacturers in the development, upgrading, and maintenance of GMP requirements. Validation SOPs is required to give step-by-step direction in performing validation.

2. Definitions of Validation

- Action of proving, in accordance with the principles of good manufacturing practice, that any procedure, process, equipment, material, activity, or system actually leads to the expected result
- Documented evidence which provides a high degree of assurance that a specific process will consistently produce a product meeting its predetermined specifications and quality attributes and characteristics
- Obtaining and documenting evidence to demonstrate that a method can be relied upon to produce the intended result within defined limits

SOP No. Val. 200.10   Effective date: mm/dd/yyyy

                                       Approved by:

■ Action to verify that any process, procedure, activity, material, system, or equipment used in manufacture or control can, will, and does achieve the desired and intended results

3. Justification of Validation

Validation is basically good business practice. The objective is to achieve success in the first production of a new product.

■ Government regulations
  Current good manufacturing practices (GMPs) have been established all over the world. The GMPs basically serve as guidelines but do not provide step-by-step directions on how to achieve them. However, the validation master plan and associated SOPs exactly define responsibilities: who, when, where, and how much is sufficient to demonstrate.
■ Assurance of quality
  Validation provides confidence in the quality of products manufactured as the over quality of a particular process cannot be established due to the limited sample size. Validation leads to less troubleshooting within routine production. As a result, it reduces the number of customer complaints and drug recalls.
■ Cost reduction
  Processes running at marginal levels often cause costs because of necessary reinspection, retesting, rework, and rejection. Validation leads to the optimization of processes and results in minimization of those expenses.

4. Background of Validation

4.1 History

Since the mid-1970s validation has become an increasingly dominant influence in the manufacture and quality assurance of pharmaceutical products. In 1976 the FDA proposed a whole set of current GMP regulations which were revised several times.

Effective date: mm/dd/yyyy
Approved by:

4.2 Legal requirements

In several major countries GMP regulations are considered official law and noncompliance is prosecutable. Additional compliance policies, guides, and guidelines are not legally binding. However, the pharmaceutical industry follows them as a part of good management and business practice.

4.3 Market requirements

The demands in the health care industry are greater than ever because customers (government, physicians, pharmacists, patients, and health insurance companies) are more interested in product safety, efficacy, and potency and asking value for money.

Pharmaceutical products' quality must be consistent and meet the health and regulatory requirements. The pharmaceutical industry has the obligation to validate GMP to their process to be in compliance with GMP requirements.

4.4 Validation philosophy

All reputable companies have recognized the commitment for validation and laid down the respective policies in the quality assurance manual.

5. The Basic Concept of Process Validation

1. Requalification or revalidation
2. Calibration, verification, and maintenance of process equipment
3. Establishing specifications and performance characteristics
4. Selection of methods, process, and equipment to ensure the product meets specifications
5. Qualification or validation of process and equipment
6. Testing the final product, using validated analytical methods, in order to meet specifications
7. Challenging, auditing, monitoring, or sampling the recognized critical and key steps of the process

SOP No. Val. 200.10   Effective date: mm/dd/yyyy

                                                Approved by:

6. The Basic Concept of Equipment Validation

Equipment validation comprises installation qualification, operational qualification, and performance qualification. The intention is to demonstrate that equipment is qualified for processing.

7. The Basic Concept of Area and Facility Validation

The purpose of area and facility qualification is to demonstrate that the area and facility meet the design qualification requirements for temperature, humidity, viable, and nonviable count.

8. The Basic Concept of Utilities Validation

This validation demonstrates that the utilities required to support the process meet the desired standard for quality.

9. The Basic Concept of Cleaning Validation

The cleaning validation is required to demonstrate that, after cleaning, the equipment and surfaces are essentially free from product residues and traces of cleaning agents to prevent cross-contamination.

10. The Basic Concept of Computerized System Validation

Computer systems are used worldwide in the pharmaceutical industry and have direct bearing on product quality. The purpose of validation is to demonstrate that the intended product manufactured, packed, or distributed using a computerized controlled system will meet the safety, efficacy, and potency requirements per the individual monograph.

REASONS FOR REVISION

Effective date: mm/dd/yyyy

- First time issued for your company affiliates and contract manufacturers

YOUR COMPANY
VALIDATION STANDARD OPERATING PROCEDURE

SOP No. Val. 200.20   **Effective date: mm/dd/yyyy**

                                                          Approved by:

TITLE: **Validation Master Plan and Guideline for DQ, IQ, OQ, and PQ**

AUTHOR: _____

                                  Name/Title/Department

_____

                                      Signature/Date

CHECKED BY: _____

                                  Name/Title/Department

_____

                                      Signature/Date

APPROVED BY: _____

                                  Name/Title/Department

_____

                                      Signature/Date

REVISIONS:

No.	Section	Pages	Initials/Date

SOP No. Val. 200.20   **Effective date: mm/dd/yyyy**
                                        Approved by:

SUBJECT: Validation Master Plan and
Guideline for DQ, IQ, OQ, and PQ

PURPOSE

To provide the guideline for the preparation of validation master plan to meet the design qualification requirement

RESPONSIBILITY

It is the responsibility of all supervisors, validation officers, engineers, and managers to follow the procedure. The quality assurance manager is responsible for SOP compliance.

PROCEDURE

1. Policy

A validation master plan, or other equivalent document, must be prepared and approved. This must be initiated at the earliest practical point and must be reviewed and updated throughout the project. The validation master plan must address all the relevant stages of DQ, IQ, OQ, and PQ.

1.1 Introduction

The validation master plan is a summary document stating the intention and the methods to be used to establish the adequacy of the performance of the equipment, systems, controls, or process to be validated. It is approved by the quality assurance, validation, production, and engineering groups.

1.2 Validation master plan requirement

A validation master plan, or other equivalent document, will be prepared for all projects.

1.3 Validation master plan format

The physical format of the validation master plan is flexible.

SOP No. Val. 200.20   **Effective date: mm/dd/yyyy**

                                                        Approved by:

1.4 Validation master plan content

Content may include the following descriptions (but not be limited to):

1. Introduction
   1.1   Project Description
   1.2   What a Validation Master Plan Is
   1.3   Scope of Validation Master Plan
   1.4   Definition for the Term *Validation*
   1.5   Validation Team Member
   1.6   Validation Team Responsibility
2. Concept of Qualification or Validation
   2.1   Fundamentals
   2.2   Concept of a Validation Life Cycle
   2.3   Elements of Validation
   2.4   Documentation Format of Qualification Programs
   2.5   Numbering System
3. Revalidation
4. Facility Description
5. Description of Building
   5.1   Dry Production Facility
   5.2   Liquid and Semi-Solid Production Facility
   5.3   Parenterals Production Facility
6. Equipment Description
   6.1   Dry Production
   6.2   Liquid and Semi-Solid Production
   6.3   Parenterals Production
   6.4   Overprinting Area
   6.5   Quality Control
   6.6   Quality Assurance (In-Process)
   6.7   Product Development Laboratories
7. HVAC Description
   7.1   Dry Production Facility
   7.2   Liquid and Semi-Solid Production Facility
   7.3   Parenterals Production Facility
   7.4   Overprinting Area
   7.5   Quality Control
   7.6   Quality Assurance (In-Process)
   7.7   Product Development Laboratories

SOP No. Val. 200.20   **Effective date: mm/dd/yyyy**

                                           Approved by:

15.2 Process Flow, Variables, and Responses:
Powder for Suspension

15.3 Process Flow, Variables, and Responses:
Capsules

16. Process Description
Liquids and Semi Solid Products

16.1 Process Flow, Variables, and Responses
Syrups, Suspension, and Drops Products

16.2 Process Flow, Variables, and Responses
Cream, Ointment, and Suppository Products

17. Process Description
Parenterals Product

17.1 Process Flow, Variables, and Responses
Aseptic Fill Products

17.2 Process Flow, Variables, and Responses
Aseptic Fill Ready-to-Use Disposable Syringes

17.3 Process Flow, Variables, and Responses
Terminal Sterilization Products

17.4 Process Flow, Variables, and Responses
Lyophilized Products

18. Qualification of Process Equipment
Test Functions and Acceptance Criteria

18.1 Commuting Mull

18.2 Oven

18.3 V-Shell Blender

18.4 Tablet Compression

18.5 Capsulation

18.6 Powder Filling

18.7 Capsule Polisher

18.8 Tablet Coating

18.9 Syrup Manufacturing Vessel

18.10 Suspension Manufacturing Vessel

18.11 Drops Manufacturing Vessel

18.12 Mixer

18.13 Emulsifying Mixer

18.14 Filter Press

18.15 Cream, Ointment, and Suppository Manufacturing Vessel

18.16 Syrup, Suspension, and Drop Filling Machine

18.17 Cream and Ointment Filling Machine

SOP No. Val. 200.20   **Effective date: mm/dd/yyyy**

                                                    Approved by:

       18.18 Suppository Filling Machine
       18.19 Labeling Machine
       18.20 Capping Machine
       18.21 Cartonator
       18.22 Shrink Wrapping Machine
       18.23 Overprinting Machine
       18.24 Trays and Rack Washer
       18.25 Autoclave (Steam Sterilizer)
       18.26 Hot Air Tunnel (Dry Heat Sterilizer)
       18.27 Vial and Ampule Washing Machine
       18.28 Vial, Ampoule, and Syringe Filling Machine
       18.29 Freeze Dryer (Lyophilizer)
       18.30 Laminar Flow Unit
       18.31 Pass Through
   19. Validation of Support Processes Test Functions and Acceptance Criteria
       19.1 Washing of Components
       19.2 Sterilization of Components
       19.3 Depyrogenation of Components
       19.4 Aseptic Filling Validation (Media Fill Studies)
       19.5 Cross-Contamination Control
       19.6 Computerized Pharmaceutical System
   20. Quality Assurance/Control Laboratory Validation
       20.1 Laboratory Equipment Qualification
       20.2 Computer-Related Systems in QA and QC
   21. cGMP Procedures and Programs
       21.1 Engineering Change Control
       21.2 Calibration
       21.3 Preventive Maintenance Program
       21.4 Standard Operating Procedure (SOP)
       21.5 Facility Cleaning and Sanitization
       21.6 Environmental Monitoring Program
       21.7 HEPA Filter Integrity Testing
       21.8 Filter Integrity Testing
       21.9 Label Control Program
       21.10 cGMP Training
       21.11 Equipment Log Book, Status Tags, and Room Clearance
       21.12 Validation Files

22. Validation Schedule
23. Drawings and Layouts
    23.1 Dry Production Facility
    23.2 Liquid and Semisolid Production Facility
    23.3 Parenterals Production Facility
    23.4 Deionized Water System
    23.5 HVAC
    23.6 Water for Injection
    23.7 Steam Distribution
    23.8 Compressed Air Distribution
    23.9 Nitrogen Distribution
    23.10 Drainage System
    23.11 Personnel Flow
    23.12 Materials Flow
    23.13 Electrical Drawings
    23.14 Equipment Installation Drawings

1.5 Validation master plan review and update

The validation master plan is a dynamic document which will be reviewed, updated, and approved as required during the lifecycle of the project.

2. Guideline — Design Qualification

Design qualification is documented evidence that quality is built into the design of facilities and operations.

Policy: the company must prepare or adopt appropriate guidelines for design qualification (DQ), installation qualification (IQ), operational qualification (OQ), and performance qualification (PQ).

2.1 Background

The standard of any facility of operation is highly dependent upon the quality of the design and, therefore, the staff employed to undertake the work. A design qualification protocol, or report document, should detail and record the disciplined, structured approach followed. This will provide a useful lead into the installation qualification (IQ) stage.

SOP No. Val. 200.20   **Effective date: mm/dd/yyyy**

                                                            Approved by:

2.2. Typical document

The principlal information included should be as follows:

■ HAZOPs (hazard and operability studies)
■ Modular design, with drawings and specifications produced
■ Zone classification studies, etc.

Confirmation of the structured and rigorous approach to design, including comments on confirmation of design standards adopted referring to:

■ Key issues underwritten by QA
■ Company standards
■ Confirmation of the use of appropriately qualified staff
■ Confirmation of the attention paid to GMP issues (e.g., by GMP audits)
■ Key reference texts on cGMP issues
■ National and international codes
■ Confirmation that available technology transfer information has been used
■ Confirmation that change control systems have operated effectively

2.3 Design qualification report

Policy: upon completion of each stage (DQ, IQ, OQ, PQ), a document review or report detailing results compared against the requirement and recommendations for future work must be prepared and accepted by appropriate management before proceeding to the next stage of validation or implementing the process.

3. Guideline — Installation Qualification

This qualification is a documented demonstration that facilities and operations are installed as designed and specified and are correctly interfaced with systems.

SOP No. Val. 200.20 **Effective date: mm/dd/yyyy**

Approved by:

3.1 Installation qualification protocol

The IQ protocol should include a statement of the data required and acceptance criteria to be met for installation of the system or equipment to verify that the specification has been satisfied.

The protocol should include as applicable, but not be limited to:

- Engineering drawing and documents
- Building finishes
- Process and utilities (services) flow diagrams
- Piping and instrumentation diagrams
- Equipment and instrument specifications
- Manufacturers' drawing, equipment maintenance, and operating manuals
- Spare lists
- Maintenance schedules

The IQ protocol should also ensure that equipment and instrumentation is clearly described and suitably labeled as to vendor, model, capacity, materials, and other critical criteria.

The IQ protocol should ensure that instrumentation has been calibrated according to approved procedures and that the measurements are traceable to defined national or international standards. All such calibrations and detailed control parameters must be recorded and records securely kept.

It should also ensure that change control systems are in operation, and that all systems have been verified to operate under no load conditions.

3.2 Installation qualification report

- Results as compared against the protocols
- Recommendations for future work
- Must be prepared and accepted by appropriate management, as defined under 3.6 (responsibility), before proceeding to the next stage of validation or implementing the process

SOP No. Val. 200.20 Effective date: mm/dd/yyyy

Approved by:

4. Guideline — Operational Qualification

The operational qualification provides a documented demonstration that facilities an operation's function as specified.

4.1 Operational qualification (OQ) protocol

The OQ protocol should:

- Include a complete description of the purpose, methodology, and acceptance criteria for the operational tests to be performed
- Ensure that instrumentation is in current calibration
- Ensure that detailed control parameters have been established and recorded for each instrument loop
- Ensure change control systems are in operation
- Ensure that standard operating and maintenance procedures have been developed (drafted) for each system, to ensure continued operation under defined conditions
- Ensure that training modules and training sessions for production, engineering, and support personnel have been developed, conducted, and documented during this stage

Where appropriate and documented in the validation master plan, the IQ and OQ protocols may form a single document which clearly defines the acceptance for each test(s).

4.2 Operational qualification report

- Results as compared against the protocols
- Recommendations for future work

5. Guideline — Performance Qualification

The performance qualification is a documented program to demonstrate that an operation, when carried out within defined parameters, will consistently perform its intended function to meet predetermined acceptance criteria.

Effective date: mm/dd/yyyy

Approved by:

Operation, maintenance and calibration procedures: Before commencing performance qualification, authorized procedures must be in place for routine use of the facility or operation, training of operators, routine calibration and maintenance, notification and recording of problems, and for the definition of actions to be taken in the event of breakdown.

5.1 Performance qualification (PQ) protocol

The PQ protocol should:

- Include a complete description of the purpose, methodology, and acceptance criteria for the performance tests to be performed
- Ensure that change control systems are in operation
- Ensure that any outstanding actions (exceptions) from IQ or OQ are recorded and recommendations for remedial actions are justified and approved
- Ensure that maintenance and calibration routines are in operation
- Ensure that all SOPs have been finalized and approved at this stage
- Ensure that operating staff have been trained according to the approved SOPs
- Ensure that performance qualification testing has been carried out using the same personnel who will routinely operate the system or equipment
- Ensure that all deviations from the validation protocol are investigated and documented
- Ensure that sufficient lots or samples will be evaluated to demonstrate adequate process control
- Ensure that, before approval is given to allow PQ testing to proceed, all IQ and OQ results are reviewed and accepted

5.2 Performance qualification (PQ) report

- Results as compared against the protocols
- Recommendations for future work
- Must be prepared and accepted by appropriate management before proceeding to the next stage of validation or implementing the process

SOP No. Val. 200.20   **Effective date: mm/dd/yyyy**
                                       Approved by:

6. Guideline — Validation Review

A validation review procedure must be formally defined. Any validation review must be documented in detail and the results of any test should be compared to the original validation results. If the results are satisfactory, the facility or operation may continue to be used. If the results are not satisfactory, operations must be suspended. The facility or operation must be validated before further use.

6.1 Validation review procedure

The frequency of the validation review should be addressed in the final validation report and may be determined against elapsed time or the number of batches processed, anomalies in results of in-process and end-product testing, and questions arising from internal or external audits.

7. Guideline — Actions to Be Taken Following Test Failures

Any test failure during a validation exercise must be reviewed or analyzed to identify the origin of the failure. Action to be taken following test failure must be documented and authorized through a formal system.

7.1 Introduction

Test failures should be considered as *useful* events during validation exercises. The more failures detected during the validation, the more problems avoided during the future routine work.

7.2 Procedure

Appropriate qualified staff must study the results to identify the reason for failure, which may not be due to the facilities or operations under test, but instead due to:

- Inadequate test protocols
- Incomplete or substandard installation (at IQ and early stages of OQ)
- Badly calibrated off-line control apparatus
- Sampling errors

Effective date: mm/dd/yyyy

Approved by:

- Materials of the wrong specification
- Human factors

The failure may be justified as arising from the facilities or operations if:

- Required parameters cannot be achieved
- Equipment specification is incorrect
- Results are inconsistent or divergent
- Precision is poor

7.2.1 Actions to be taken once the source of the defect has been identified:

- For failures due to incomplete or substandard installation during IQ or early stages of OQ, authorized remedial action may be taken and the test repeated.
- For all other failures, an approved group will decide the correct course of action, for example:
  - To retest in the case of unreliable initial analytical tests, or resample according to a defined or authorized procedure
  - To introduce a change in order to correct or solve the problem (raw material, operation parameter, part of equipment, new step, the process, control system)
  - To apply more or less severe limits in the use of the process or equipment and update the SOP

7.3 Documentation

All remedial corrective action and preventive measures must be documented and records retained.

REASONS FOR REVISION

Effective date: mm/dd/yyyy

- First time issued for your company, affiliates, and contract manufacturers

YOUR COMPANY
VALIDATION STANDARD OPERATING PROCEDURE

SOP No. Val. 200.30

Effective date: mm/dd/yyyy

Approved by:

TITLE: Design Qualification Guideline for Minimizing the Risk of Product Cross-Contamination by Air Handling System

AUTHOR: _____

Name/Title/Department

_____

Signature/Date

CHECKED BY: _____

Name/Title/Department

_____

Signature/Date

APPROVED BY: _____

Name/Title/Department

_____

Signature/Date

REVISIONS:

No.	Section	Pages	Initials/Date

SOP No. Val. 200.30 Effective date: mm/dd/yyyy
Approved by:

SUBJECT: Design Qualification Guideline for Minimizing
the Risk of Product Cross-Contamination by
Air Handling System

PURPOSE

To provide the guideline to be followed for minimizing the risk of product cross-contamination with respect to the air handling system

RESPONSIBILITY

It is the responsibility of all key supervisors, validation officers, engineers, and managers to build guidelines into the plant architectural structure at the design phase, while placing orders, making purchase requisitions, and in the standard operating procedures to ensure prevention of cross-contamination. The quality assurance manager is responsible for SOP compliance.

PROCEDURE

This guide is intended to outline some of the considerations to be taken into account in the design of air handling systems in order to minimize the risk of product cross-contamination.

1. Potential Sources of Cross-Contamination

- Dispersal of product around manufacturing facilities through inadequately designed (non GMP) air handling system zoning, air locks, and room pressure differentials
- Spread of product around manufacturing facilities via environmental and process air handling systems
- Dispersal of product during the cleaning and maintenance of environmental and process air handling plant and equipment

SOP No. Val. 200.30   Effective date: mm/dd/yyyy

Approved by:

2. Containment of Product and Collection at Source

The practice of containment of product within the manufacturing process system should be maximized in order to minimize the potential sources of product cross-contamination outlined above.

Wherever product or material is exposed, adequately designed localized containment should be carefully considered, for example, local exhaust ventilation and containment booths, etc.

3. Zoning of Air Handling Systems

Specific air handling system consideration should be given to minimizing the dispersal of product by careful design of air handling zones including the following factors:

- Dedicated air handling system for specific manufacturing departments, for example, dry, liquid, and sterile product manufacture
- Separate air handling systems for manufacturing and non-manufacturing operations, for example, manufacturing and primary packing, secondary packing, QC laboratories such as in-process and administration facilities, etc.
- Dedicated air handling units for products containing specific active ingredients where possible
- Once-through air handling plants (i.e., no recirculation) or recirculation type systems (see Section 6)
- Air locks between air handling zones

4. Air Locks

Adequate consideration should be given to the design of air locks to minimize the spread of product between air handling zones.

- **Negative pressure air locks:** Account shall be taken of the risk of bringing product-contaminated air streams from two air handling zones closer together when using negative air locks.
- **Positive pressure air locks:** Careful consideration should be given to ensuring that air is not supplied to positive pressure air locks from a product-contaminated source.

SOP No. Val. 200.30　　　　　　　　　　**Effective date: mm/dd/yyyy**

　　　　　　　　　　　　　　　　　　　　Approved by:

5. Room Pressure Differentials

Room pressure differentials should be adequate to minimize the dispersal of product by ensuring air movement in a controlled and predetermined direction.

6. Recirculation vs. Once-Through Air Handling Systems

The following factors should be considered:

- Return air to air handling plant shall not be product contaminated.
- HEPA filters in air handling system give adequate protection against product cross-contamination. Select correct HEPA filter.
- Is air handling system serving other critical product manufacturing areas?

7. Room Air Distribution

The key function of an air handling system is to facilitate environmental conditions of temperature and humidity.

Provisions for reducing room particle count during manufacture operations should be provided through careful dilution of particle-contaminated room air with particle free supply.

8. The Use of HEPA Filters

The following are some of the factors which should be considered:

- Are terminal HEPA filters on room air supplies necessary to prevent risk of product migration from one manufacturing area to another through ductwork distribution systems when air handling plant is inoperative?
- Are terminal HEPA filters on room air returns necessary to prevent product contamination of air handling plant components *and* ductwork distribution system *or* risk of product migration from one manufacturing area to another through ductwork distribution systems?
- Are HEPA filters on main returns to air handling plants necessary to prevent product contamination of air handling plant components?
- Will production area, and hence return ductwork and air handling plant, be product contaminated? Is this acceptable?

SOP No. Val. 200.30   **Effective date: mm/dd/yyyy**

                                                  Approved by:

9. Air Handling Plants

Careful consideration should be given to the risk of product cross-contamination from one air handling plant to another during cleaning and maintenance procedures.

Precautions could include:

- The location of dust collection system air handling plants external to manufacturing building
- The use of safe-change filters where located in product contaminated air streams
- Segregation of air handling system plants

10. Fresh Air Intakes and Exhausts

The location and sign of fresh air intakes and discharge air exhausts to the atmosphere from air handling plants should be designed to eliminate the risk of product cross-contamination by short circuiting of air streams.

REASONS FOR REVISION

Effective date: mm/dd/yyyy

- First time issued for your company, affiliates, and contract manufacturers

YOUR COMPANY
VALIDATION STANDARD OPERATING PROCEDURE

SOP No. Val. 200.40

Effective date: mm/dd/yyyy

Approved by:

TITLE: **Design Qualification Guideline for Minimizing the Risk of Cross-Contamination of Facility, Equipment, and Process**

AUTHOR: _____

Name/Title/Department

_____

Signature/Date

CHECKED BY: _____

Name/Title/Department

_____

Signature/Date

APPROVED BY: _____

Name/Title/Department

_____

Signature/Date

REVISIONS:

No.	Section	Pages	Initials/Date

SOP No. Val. 200.40 Effective date: mm/dd/yyyy

Approved by:

SUBJECT: Design Qualification Guideline for Minimizing the Risk of Cross-Contamination of Facility, Equipment, and Process

PURPOSE

To provide the guideline to be followed for minimizing the risk of cross-contamination with respect to facility, equipment, and process design

RESPONSIBILITY

It is the responsibility of all key supervisors, validation officers, engineers, and managers to build the guidelines into the plant architectural structure at the design phase, while placing orders, making purchase requisitions, and in the standard operating procedures to ensure prevention of cross-contamination. The quality assurance manager is responsible for SOP compliance.

PROCEDURE

1. Good Design Practices

The design of facilities, equipment, and processes can minimize but not eliminate the risk of cross-contamination. The application of good design practices, if applied and monitored throughout the development of construction documents, will help prevent a high degree of cross-contamination. However, good design cannot overcome poor operating or quality control practices during operation of any facility.

The following points shall be given due consideration during facility design. Equipment which contains product contact surfaces or noncontacting potentially contaminating surfaces shall be reviewed for adequacy.

2. Process Design

- The product handling and transfer should be carried out in a closed system for minimizing cross-contamination from the environment and from personnel.

SOP No. Val. 200.40   **Effective date: mm/dd/yyyy**
                                           Approved by:

- Processes should also prevent contamination of the working environment and subsequent dispersal problems.
- Manual handling shall be minimal and only where desirable. No direct discharge from one vessel or system to another shall be allowed to maintain the process in a close environment. Where containment is not possible, contaminants should be collected at the source.
- Process lines and instrumentation are initially designed with the use of process flow diagrams (PFDs) and piping and instrument diagrams (P&IDs); these drawing are important in cross-contamination prevention. Piping should be designed to prevent mixing of gases, water, and other foreign substances. Check valves, backflow preventers, and other devices may be designed into processes to avoid mixing or backflow of fluids and gases.

3. Facility Design

Architectural finishes and room layout are important considerations. Personnel and product movements through the facility are key aspects in the minimization of cross-contamination. The directions of air flows, and air flow rates, are critical and should be considered during facility design.

4. Equipment Design

4.1 General considerations

Process equipment selection, design, and surface that contain product contact surfaces or potentially contaminating surfaces should be specified such that cross-contamination is minimized and clean ability is maximized. Design review should be conducted before the actual fabrication begins.

- Where possible, equipment should be dedicated to specific products.
- Equipment should be designed such that areas which would collect or retain drug residues are minimized. All surfaces should be smooth and free of pitting.
- The surface should not be reactive, additive, or absorptive as noted in 21CFR211.65, current good manufacturing practices: equipment construction.

SOP No. Val. 200.40   **Effective date: mm/dd/yyyy**

                                              Approved by:

- Where possible, incorporate cleaning apparatus into the equipment. This can be accomplished either with spray nozzles, spray wands, or spray ball diffusers. Drains should be carefully placed in the equipment so that rinse waters and solvents are allowed to fully drain.
- Sanitary fittings and valves should be used for product contact applications.

4.2 Clean in place (CIP)

Clean in place (CIP) is a system which involves circulating or once-through water rinses and chemical or sanitizing solutions which are discharged through plant and equipment while kept in an assembled state. The rinses and solutions are used such that all contaminated or soiled product contact surfaces are cleaned to an acceptably high and consistently reproducible state.

CIP systems are commonly used to eliminate environmental and personnel exposure to the contaminant. They are effective when the shutdown and disassembly of equipment in production would impact manufacturing efficiency. They are also used to improve the consistency and reproducibility of the cleaning process.

Computer control of CIP systems is common and programmed recipes can help control the consistency and quality of the cleaning procedure. Validated computer systems used for CIP should be considered.

4.3 Clean out of place (COP)

COP is used to describe the removal, disassembly, and opening of process equipment and systems for cleaning in other than its normal operating conditions. Care should be taken to minimize environmental exposure leading to cross-contamination and personnel interference.

4.4 Equipment Seals

Equipment seals on rotating shafts such as agitators, pumps, and compressors should avoid contact with products. Otherwise, seal lubricants should be food grade where permitted by the manufacturer.

4.5 Computer controls

The microprocessor-based systems should be programmed, challenged, and validated to eliminate the exposure of one product to another through control failure. Valves and actuation devices used to divert product flows should be programmed properly and validated for their adequate functionality.

5. Dust Collection Equipment

Adequate dust collection systems' "point of use" are recommended in areas where materials are handled. Where recovery of product is required from the dust collector systems, it is desirable that the system be dedicated to a single process or product line.

REASONS FOR REVISION

Effective date: mm/dd/yyyy

■ First time issued for your company, affiliates, and contract manufacturers

---

YOUR COMPANY
VALIDATION STANDARD OPERATING PROCEDURE

SOP No. Val. 200.50

Effective date: mm/dd/yyyy

Approved by:

TITLE: Design Qualification Guideline for HVAC System of a Pharmaceutical Plant

AUTHOR: _____

Name/Title/Department

_____

Signature/Date

CHECKED BY: _____

Name/Title/Department

_____

Signature/Date

APPROVED BY: _____

Name/Title/Department

_____

Signature/Date

REVISIONS:

No.	Section	Pages	Initials/Date

SOP No. Val. 200.50   **Effective date: mm/dd/yyyy**

                                                Approved by:

SUBJECT: Design Qualification Guideline for HVAC System of a Pharmaceutical Plant

PURPOSE

To provide the qualification design guide line for the HVAC design of pharmaceutical manufacturing plant

RESPONSIBILITY

It is the responsibility of all key supervisors, validation officers, engineers, and managers to build guidelines into the plant architectural structure at the design phase, while ordering materials and making purchase requisitions, and in the standard operating procedures to ensure prevention of cross-contamination. The quality assurance manager is responsible for SOP compliance.

PROCEDURE

1. Heating Ventilation and Air Conditioning

The HVAC assigned to the project should use the following guidelines to control the quality of the design from a cGMP perspective. The checklist that follows should be completed and used to control the quality of working drawings, revisions, and related documents.

2. Guideline For HVAC Design

As a predecessor to detailed design, the HVAC should meet with your company's basic project management. Parameters should be designed for the facility on an area-by-area basis. Meeting results should be documented and form the basis for design criteria. The approval of these criteria must be obtained prior to proceeding with design.

The purpose of the design criteria is to establish a basis for facility HVAC design and to provide your company with a document for FDA facility review.

SOP No. Val. 200.50   **Effective date: mm/dd/yyyy**
                                           Approved by:

The design criteria should address both outside and inside design criteria. Outside design criteria should be obtained from ASHRAE (American Society for Heating, Refrigeration, and Air Conditioning Engineers) data. Inside design criteria should include, but not be limited to, temperature, relative humidity, filtration level, minimum air change rate pressurization requirements, exhaust requirements, and cleanliness level.

Both temperature and humidity should be listed as design set points/minus tolerances. A listing of temperature and humidity ranges is subject to interpretation and should be avoided. Filtration should address final filter efficiency and location. Air change rates may be required for cooling load, or as required for a given area classification. Pressurization should be noted on the drawing in the form of actual room pressure levels from a common reference point, which should be noted on the design documents, along with any special exhaust requirements. The location of laminar flow hoods, biosafety hoods, and fume exhaust hoods should also be noted on these documents if the room is to be validated as a clean room or containment area.

2.1 Calculations

2.1.1 Airflow leakage rate calculations

In general, the calculation of the airflow leakage rate for a room should be based on the pressure differential established on the design criteria sheet. Assumed leakage rates based on a percentage of supply air are unacceptable. For leakage calculations, each wall should be considered separately regardless of whether the room is interior or exterior. In addition, it should be noted whether leakage is into or out of the room. Leakage calculation should contain a safety factor to compensate for less than ideal construction, deterioration of gasket sealing, and so on.

2.1.2 Cooling load calculations

Cooling load calculations should be done on a room-by-room basis. Careful attention should be given to process equipment loads because these can be significant heat generators. In calculating this load, consideration should be given to motor load, convection, and radiation from heated vessels and thinly insulated process and utility piping. Attention should also be given to potential heat gain from the air handling unit supply fan due to the high air change rates and multistage filtration. The actual operation of process equipment should be understood so that realistic assumptions on heat gain factors and diversity can be made. Process equipment heat loads with basis should be included in design criteria and reviewed and approved by responsible management and quality assurance manager.

2.1.3 Heating calculations

Heating is generally accomplished by a "reheat." Therefore, the heating calculation should include the energy required to raise the supply air temperature from its summertime design point to room temperature plus sufficient energy to offset winter heat loss.

2.1.4 Fan static pressure calculation

Fan static pressure should always be calculated, rather than relying on rules of thumb.

2.2 Airflow diagrams

Airflow diagrams should show the air handling unit components arranged in their proper sequence, flow measuring stations, reheat coils, location of each level of filtration, humidifiers, exhaust, and any other system components. Airflow diagrams should show supply, return, exhaust, infiltrations, and exfiltration airflows from each room. Obtaining engineering and quality assurance approval of airflow diagrams is required prior to starting ductwork layout.

2.3 HVAC P&IDs

P&IDs should be developed for each individual air handling system. The HVAC discipline should generate P&IDs for chilled water and hot water. Each of these systems should be independent from similar process-related systems and should be the responsibility of HVAC engineers, not process engineers. The HVAC department may also be required to produce P&IDs for other general utilities.

SOP No. Val. 200.50   Effective date: mm/dd/yyyy

Approved by:

2.4 HVAC ductwork and equipment arrangement drawings

Detailed HVAC drawings are an absolute necessity for a facility with the complexity common to pharmaceutical facilities. Single line drawings are unacceptable. All ductwork should be drawn as double-lined, fully dimensioned duct (including centerline elevation for round ducts and bottom of duct elevation for rectangular ducts). The location should be shown from the column lines. All ductwork should be coordinated with piping and electrical disciplines. Density may require multiple levels of duct drawings. Sections are required at any points where the plans do not completely and clearly define the design. A meeting must be convened to preplan special requirements for duct, pipe, and electrical prior to commending any design. Ductwork plans must clearly indicate duct material, insulation type, and pressure classification using duct construction tags.

When sizing return air ducts, keep in mind that calculated room leakage rates are empirical and will vary depending on construction tolerances; the acutal return air (R.A.) quantities may vary from theoretical design. Return ducts should be sized to handle at least 25% more than design airflow. Return air plenum designs should be avoided unless there is no potential for contamination problems.

2.5 General specifications of ducts

- Construction material *or* re ducts capital.
- The air duct must be constructed with galvanized steel.
- Other materials are authorized if specific uses (i.e., chemical resistance) are required.
- The perimeter of the duct must be constructed with the same material.
- The inside surface of the ducts must be smooth, except if specific equipments are required (noise attenuator, damper, etc.).
- The minimum thicknesses required are:

    Circular section:

    8/10 mm   for diameter ≤ 200 mm
    10/10 mm   for diameter ≤ 600 mm
    12/10 mm   for diameter ≥ 600 mm

SOP No. Val. 200.50

Effective date: mm/dd/yyyy

Approved by:

Rectangular section:

8/10 mm   for the greatest side ≤400 mm
10/10 mm   for the greatest side ≤600 mm
12/10 mm   for the greatest side ≤850 mm
15/10 mm   for the greatest side ≥850 mm

2.6 Design of ducts

- The use of ducts with circular section is preferred.
- The duct system must be airtight.
- The rectangular sections must be reinforced with folds or reinforcement to resist deformations and vibrations.
- The duct system, including elbows, change of section and other equipment, must be selected and designed to minimize the static pressure losses and the level of noise.
- The assembly of the rectangular ducts must be executed with riveted sleeves and retractable adhesive strips.
- The duct system must be designed to allow the cleaning of the ducts (access door, etc.).
- line before the opening). Obturable openings (Ø 25 mm) must be located in the ducts to allow the measurement of the air velocity (minimum 2 m of duct in straight
- All the external openings must be protected against small animals with a wire mesh (maximum opening: 10 mm).
- The connecting of the movable parts (filters) can be executed with semi-flexible metallic ducts. Their lengths must be as short as possible.
- The sections of the ducts must be designed to obtain the following air velocities:

≤5 m/s	in the blow ducts
≥4 m/s and ≤7 m/s	in the return or exhaust ducts
≥13 m/s	in the return or exhaust ducts of powder production rooms

2.7 Supports

The anchors and the hangings must be selected and positioned to allow the possible dismantling of the ducts (type "clamping").

SOP No. Val. 200.50   Effective date: mm/dd/yyyy

Approved by:

The number of the anchors must be selected to keep the bending of the ducts under L 300.

All the hangings must be cut to avoid the cutting parts.

The anchors, hangings, and other metallic parts must be protected against rust (galvanized or rust-preventing paint).

2.8 Piping for air conditioning

The piping installation for air conditioning must be executed with usual steel welded or wiredrawed.

They must be designed with a water velocity between 1.5 and 2 m/s.

This piping must be insulated following the characteristics of the air inside: $T° > 18°C$ should be insulated with armaflex thickness = 25 mm.

The insulation must be glued on the pipes and the joints between two strips of insulation must be airtight (gluing of an insulation strip on the joints) to prevent condensation on the pipes.

Pipes with $T° > 18°C$ should be insulated with rockwool or glasswool (thickness = 25 mm) recovered by a metallized vapor barrier.

2.9 Supports for piping

The supports of the piping must be galvanized steel. The contact between the pipes and the anchor must be insulated to prevent heat transfer (condensation on cold piping). Do not rigidly anchor tubing.

The distance between two supports must be at maximum 1 m for horizontal tubing (∅ ½"); 1.5 m for horizontal tubing (∅ ¾ to ⁴/₄"); 2 m for vertical tubing (→ ⁴/₄"). Place a support at every change of direction.

2.10 Control diagrams

Control diagrams should be drawn in P&ID format and should show all coil piping, including all valves, line numbers, and so on. A separate P&ID should be developed for each air-handling unit and should show all devices individually. Control diagrams should supplemented with sequences of operation either on the drawing or as part of the specifications. Control diagrams should include a control valve schedule that supplies the size and specifications for each valve.

3. Air Handling Units (AHUs)

3.1 General specifications

The air handling unit shall be made up of modular sections to suit each particular application. The sections shall be made with double skin isolated panels assembled on a galvanized steel base frame. The panels shall be constructed designed with two galvanized steel sheets (minimum thickness: 0.8 mm) and insulation (minimum thickness: 50 mm). The insulation must be non-flammable and conform to fire class A1 (according to DIN 4102). The panels must be reinforced (fold in the sheet or integrated reinforcement) to resist deformations and vibrations.

The assembly of the panels and the modular sections must be executed to obtain a perfectly smooth inside surface and a perfectly airtight volume. The panels must be fastened by stainless steel screws and must be removable. The sections shall be fitted with an access door for easy cleaning and maintenance of the equipment (fan, filters, heat exchanger, etc.) The access door must be locked with a mechanism able to press the door on an airtight seal; this must be continuous to guarantee the airtightness. The air handling unit must be designed without thermal bridges. The connecting of the ducts to the air handling unit must be done by flexible sleeves made with flame resistant materials (DIN 4102-A2).

3.2 Heat exchanger sections

- The heater must be designed with copper tubes, aluminium fins, and steel connections with male thread.
- The cooler must be designed with stainless tubes fines and frame. They must be installed in modular sections (description above) and must be laterally removable on guide rails (drawer). The equipment and the connections of the heat exchanger must be designed to allow the heat exchanger to be removed easily.
- The cleaning of the heat exchanger must be done easily (provide, if necessary, an empty section with access door).
- The cooling exchanger must be provided with a polipropylene droplets separator and a stainless steel tray to connect across a trap to the drain.

SOP No. Val. 200.50   **Effective date: mm/dd/yyyy**

                                              Approved by:

- The height of this trap and the diameter of the drain shall be calculated following the characteristics of the unit pressure and amount of water to eliminate.
- The heat exchanger must be calculated with an over capacity of 15%.
- The air velocity in the heat exchangers must be less than 3 m/sec.

3.3 Insulation

Air handling units should not contain internal insulation exposed to the airstream. Double wall construction sandwiching insulation between two metal panels of single wall construction with external insulation is a requirement.

3.4 Filters

Air handling units should be provided with prefilters, 2 in. thick at minimum, Farr 30/30s or equal, and 80 to 85% efficient filter. If HEPA filters are located within the air handling unit, they should be the last component in the direction of the airflow into the room.

The air filters must be made and used in accordance with ASHRAE 52-76 and EUROVENT 4/5 standards.

To assure filtration in the pharmaceutical industry, several types of filters must be used:

Sand trap:	to be placed on the fresh air inlet
Prefilter:	to be placed on the inlet of mixed air (fresh air and recycled air)
Filters:	to be placed directly after the fan section before the outlet of the air handling unit
Terminal filters:	to be placed directly after the fan section before the outlet of the air handling unit
Terminal filters:	to be placed in the room

The filters must be designed and fastened to allow an easy removal to change the filter and to keep a perfect peripheral airtightness. They must be provided with differential manometer to check them. (option: equipped with maximum contact point). An obturable opening (+/− 100 mm) must be placed before the fan section to inject the DOP. The number of terminal filters must be calculated to guarantee the same air velocity out of the

Effective date: mm/dd/yyyy

Approved by:

different filter of the same air handling unit. For each filter, a manometer should be connected upstream and downstream of the filter housing and pressure drop measured. When the pressure drop across the filter exceeds the recommended change point, the filter must be replaced.

Filter efficiency is defined by ASHRAE 52/76 or EUROVENT 4/5.

3.5 Fans

Air handling unit fans should contain a provision for modulating airflow to compensate for increased static pressure losses as the loading of filters increases. This is usually accomplished through the use of inlet vanes; other methods would be discharge dampers and variable speed drives.

Fan motors should be sized at 25% above the brake horsepower. Fans should be belt driven with variable pitch drives up to and including 25 HP and fixed pitch drives at about 25 HP. Fans must be radial and they must be balanced (static and dynamic) and mounted on anti-vibration supports. The connection of the fan outlet to the inside panels must be done by means of an internal flexible connector.

The transmission motor or fan is operated by drive belts and pulleys. The support frame of the motor must be designed to obtain the tension of the belts without modification of the alignment of the pulley and to adjust this alignment. The motor class must be calculated with an over capacity of 30% of the used power and must be connected through a safety switch placed on the air handling unit.

3.6 Dampers

Dampers must be designed with profiled damper blades operating in opposite directions, with an additional seal mounted on external, reinforced, polyamide, cogged wheels with slide bearings.

3.6.1 Coils

Air handling unit coils should have no more than eight fins per inch and be no more than six rows in depth to facilitate coil cleaning. Where more than four rows are required to obtain the cooling capacity, two coils should be placed in series. Spacing between the coils should be piped so that counterflow of air and water is achieved. Fins should be of the continuous, flat (noncorrugated) type.

SOP No. Val. 200.50   **Effective date: mm/dd/yyyy**

                                            Approved by:

Note: On large air handling unit systems, consideration should be given to bypassing the cooling coil with part of the return air to minimize the amount of reheat required. With coil bypass, pretreating the outside air for dehumidification may be required.

3.6.2 Notes

The HVAC specifications should state that before the activation of any air handling unit, all construction debris should be cleared away and the unit thoroughly vacuumed. Units should be vacuumed again before the installation of cartridge or bag filters. Prefilters should be installed before initial unit start-up. Cartridge or bag filters should be stored in a clean, dry place and should be installed after room finishes are complete. If HEPA filters are contained within the air handler, these should also be installed at this time. Comprehensive cleaning guidelines for duct and equipment must be provided as part of the construction specifications.

3.6.3 Sealing

All ductwork should be sealed in accordance with SMACNA Class A rating, which requires all seams, joints, fasteners, penetrations, and connections to be sealed. Sealant should be FDA acceptable for the application, and non-hydrocarbon based. Leakage rates as low as 1% total airflow are not uncommon. All ducts passing through a clean room wall or floor should be provided with stainless steel sheet metal collars and sealed at the opening. Details of sealing methods should be provided on the design documents.

3.6.4 Leak testing

Before ductwork is insulated, the installing contractor should leak test each duct system at 125% of the operating pressure. Leak testing should be witnessed and signed off to signify approval. Acceptable leakage rates and leak testing procedures and reports should be based upon the SMACNA HVAC Duct Leakage Test Manual.

3.6.5 Insulation for HVAC ductwork

All insulation should be in accordance with the "flame-spread" and "smoke-develop" ratings of NFPA Standard 255. Ductwork should be externally insulated. The use of internal duct liner is not acceptable. Some criteria will not allow the use of fiber glass insulation even on the exterior of the ductwork. Whenever ductwork is exposed in clean spaces, it should be insulated with a rigid board-type insulation and jacketed with either a washable metallic or PVC coated jacket. Jacketed ductwork should be of sufficient density to minimize the dimpling effect when the jacket is applied. Flexible blanket insulation is acceptable in concealed spaces. No insulation should be applied on the ductwork until the leak test has been performed, witnessed, and approved. Rigid duct insulation should also be used in mechanical rooms.

3.6.6 Insulation of the ducts

The ducts must be insulated following the characteristics of the air inside: $T° \leq 18°C$ should be insulated with armaflex thickness = 25 mm. The insulation must be glued on the ducts. The joints between two strips of insulation must be airtight (gluing of an insulation strip on the joints) to prevent condensation on the pipe. Return ducts, blow ducts $(T > 18°C)$ should be insulated with rockwool or glasswool (thickness = 25 mm recovered by a metallized vapor barrier).

3.6.7 Marking of the ducts

The ducts and other equipment must be marked after insulation by:

- Arrows to give the direction of the air flows
- A red strip on the blow ducts
- A blue strip on the return or exhaust ducts
- A green strip on the fresh air
- The number of the equipment (AHU, damper, sensors, etc.) corresponding to the P&I

3.6.8 Damper

The regulation dampers — air flow (blow, return, or exhaust) by room or by intake or return opening — must be manual with mechanical locking (minimal looseness allowed) located in the ducts. The general dampers, or those for most important sections, must be similar to the dampers for air handling unit; a suitable model should be selected.

SOP No. Val. 200.50   **Effective date: mm/dd/yyyy**

                                                      Approved by:

3.6.9 Sound attenuators

Sound attenuators should not be used in systems requiring sanitizing because the perforated face (interior) of the sound traps can collect dust and microorganisms.

3.6.10 Humidifier

Humidifiers serving clean areas and process areas should use only clean steam for humidification. Carbon steel piping and headers are not acceptable; 316L grade stainless steel should be used. All humidifier components (main body, valves, piping, manifold, etc.) should be made of 316L stainless steel.

3.6.11 Air distribution

Standard type (turbulent flow) diffusers should be used in class 100,000 areas. Terminal HEPA filters should be used in the ceilings of the areas that are class 10,000 or cleaner.

3.6.12 Return or exhaust air

Ceiling return or exhaust is acceptable in class 100,000 areas. Low wall returns should be used in class 10,000 or cleaner areas. All returns or exhausts must be louvered, removable types.

3.6.13 Intake and return openings

The number and the location of the openings must be selected to have a good distribution of the changing of air (cleanless class, t° uniformity).

    They must be designed following the requested air flows, velocities (at the opening and in contact with people), and level of noise.

    The openings must be constructed with stove enamelled steel or stainless steel.

    They must be mounted flush with the ceiling or the walls (or partitions) and sealed with paintable silicone.

    Models should be selected as appropriate.

Effective date: mm/dd/yyyy

Approved by:

3.6.14 Intake or return openings with filters

In the classified rooms 10,000, 100, 100 with LAF (following F.S. 209E), the intake openings must be fitted with absolute filters. The number and size must be selected to obtain the same air velocity (±10%) for all the terminal filters of the same air handling unit. The absolute filters' housing shall be airtight, mounted flush with the partitions, walls, or ceilings (maximum 2 mm of difference). The sealing gasket between the housing and the filter should be provided to prevent bypass of unfiltered filter. A manometer should be connected upstream and downstream of the filter housing and pressure drop measured. When the pressure drop across the filter exceeds the recommended change point, the filter must be replaced.

The recommended air velocity of a HEPA filter is 0.45 m/s. The HEPA filter and its housing must be tested mounted on site with the DOP. For return openings for powder production, the housing with HEPA filter must be fitted with a prefilter removable from the front. The recommendation is to use HEPA filter presealed in the housing located in a stainless steel frame.

REASONS FOR REVISION

Effective date: mm/dd/yyyy

- First time issued for your company, affiliates, and contract manufacturers

---

YOUR COMPANY
VALIDATION STANDARD OPERATING PROCEDURE

SOP No. Val. 200.60

Effective date: mm/dd/yyyy

Approved by:

TITLE: Design Qualification for the Prevention of Contamination of Non-Sterile Pharaceutical Products

AUTHOR: _____

Name/Title/Department

_____

Signature/Date

CHECKED BY: _____

Name/Title/Department

_____

Signature/Date

APPROVED BY: _____

Name/Title/Department

_____

Signature/Date

REVISIONS:

No.	Section	Pages	Initials/Date

SOP No. Val. 200.60   Effective date: mm/dd/yyyy

                                       Approved by:

SUBJECT: **Design Qualification for the Prevention of Contamination of Non-Sterile Pharmaceutical Products**

PURPOSE

To ensure that air supplied to manufacturing areas where product is exposed is not a source of significant contamination

RESPONSIBILITY

It is the responsibility of production manager, packaging manager, and technical service manager to follow the procedure. The quality assurance manager is responsible for SOP compliance.

PROCEDURE

This procedure applies to air supplied to all non-sterile secondary manufacturing and final stage primary manufacturing areas.

1. Policy

1.1 Air quality standards

The quality of air at or near the points of supply to manufacturing areas where product is exposed must comply with the following:

- Particles equal to or greater than 5 μm: 24,700 per m³
- Number of viable microorganisms permitted: 500 per m³

The quality of air at or near the points of supply of large volume forced air supplied coming into direct contact with products (e.g., fluid bed dryers, tablet coating machines, etc.) must comply with the following standards:

- Particles equal to or greater than 0.5 μm: 3,530 per m³
- Number of particles equal to or greater than 5 μm: 5 per m³
- Number of viable microorganisms: 5 per m³

SOP No. Val. 200.60   **Effective date: mm/dd/yyyy**

                                                Approved by:

1.2 Design of air systems

Air supply systems must be designed to avoid the introduction of contaminants into air flows or into manufacturing areas' local environmental conditions and associated risk factors. Air supply systems must be validated, operated, monitored, and controlled to deliver the required air quality.

1.3 Documentation and records

Documents must be compiled and kept up-to-date for each air supply system and must contain:

- Schematic as-built drawings
- Validation data
- Routine operating specifications
- Cleaning schedules
- Standard operating procedures
- A history of installation, changes, and modifications

The following records must be maintained:

- Monitoring, performance, and test data
- Engineering maintenance and calibration data
- Operating control data

REASONS FOR REVISION

Effective date: mm/dd/yyyy

- First time issued for your company, affiliates, and contract manufacturers

---

YOUR COMPANY
VALIDATION STANDARD OPERATING PROCEDURE

SOP No. Val. 200.70

Effective date: mm/dd/yyyy

Approved by:

TITLE: Design Qualification Guideline for
Cross-Contamination and Decontamination

AUTHOR: _____

Name/Title/Department

_____

Signature/Date

CHECKED BY: _____

Name/Title/Department

_____

Signature/Date

APPROVED BY: _____

Name/Title/Department

_____

Signature/Date

REVISIONS:

No.	Section	Pages	Initials/Date

SOP No. Val. 200.70   Effective date: mm/dd/yyyy

Approved by:

SUBJECT: Design Qualification Guideline for Cross-Contamination and Decontamination

PURPOSE

To provide the guideline to be followed for the prevention of cross-contamination and decontamination

RESPONSIBILITY

It is the responsibility of all key supervisors, validation officers, engineers, and managers to build the guidelines into the plant architectural structure at the design phase while placing orders, making purchase requisitions, and in the standard operating procedures to ensure prevention of cross-contamination. The quality assurance manager is responsible for SOP compliance.

PROCEDURE

1. Sources of Cross-Contamination

The following potential sources should be considered to assess the risk of cross-contamination.

1.1 Purchased product (semifinished) and raw material

Precontaminated due to poor control by the manufacturer

1.2 Retention of product

- In poorly designed equipment
- Poor cleaning practices

1.3 Product dispersion

- Around processing areas via working garments, due to inadequate changing or laundering procedures

- Via HVAC systems, fluid bed driers, film-coaters, vacuum systems, and product containers
- Poorly designed facilities or processes, or poor handling practices (e.g., during sampling or dispensing)

1.4 Product spillage

- Poorly designed processes, or due to poor handling during processing
- Poor handling during the cleaning of air conditioning plant air filter, dust extraction plant, or vacuum plant

2. Control of Cross-Contamination

The following are among the measures which may be necessary to control cross-contamination:

2.1 Containment

Closed manufacturing systems
Closed transfer systems
Clean-in-place systems
Containment boxes or isolators

2.2 Collection

Localized dust extraction
Containment booths

2.3 Other

Correct operating procedures
Good engineering practices in the design of equipment and facilities
Segregated plant
Safe-change filters
Product dedicated equipment
Disposable containers
Validated changing procedures
Validated laundering procedures

SOP No. Val. 200.70   **Effective date: mm/dd/yyyy**

                                       Approved by:

3. Monitoring Methods

3.1 Analytical methodology

All analytical methods should be validated and should have a detection limit at or below that necessary to show compliance with the product limit. Before each test, the recovery of the sampling method should be determined. Appropriate corrections should be made in the final calculation.

3.2 Start-up validation

Equipment — On introduction of each new piece of equipment or process, the cleaning method should be validated, using one or an appropriate combination of the following methods:

- Taking samples of rinse solvents from manufacturing equipment and measuring active residues
- Manufacturing a placebo batch after the manufacture of the active batch and measuring residues of the active in the placebo
- Taking several representative or "worst-case" swab samples from equipment and measuring active recovered from the swab

Facilities — Immediately after the start-up of a new or modified facility, potential for cross-contamination should be determined. Samples may be taken from the environment by air sampling, or from surfaces by swabbing.

3.3 Routine production

Visual checks for cleanliness before each use of a piece of manufacturing equipment should be made in addition to chemical testing. Air monitoring should be implemented to ensure that any contamination released during manufacturing operations is not being spread. Both product-contact and non-product-contact surfaces should be monitored to determine the level of contamination settling out from the environment. A product or placebo batch may also be tested to show the absence of residues from previous batches.

3.4 Validation review

A formal validation review should be carried out periodically to determine the necessity for revalidation.

SOP No. Val. 200.70   Effective date: mm/dd/yyyy
                                             Approved by:

3.5 Cross-Contamination limit

Limits for the monitoring program should be derived from product limits, which should be defined based on reasoned assessment of the risks and historical data.

4. Air handling systems

See SOP No. Val. 200.30, Guideline for Minimizing the Risk of Product Cross-Contamination Air Handling Systems.

5. Gowning

The gowning and laundry for each building shall be separate.

6. Cleaning

6.1 Introduction

The challenge to the cleaning procedure should be minimized by the use of dedicated equipment and facilities where possible (e.g., fluid bed drier filter bags). Segregation and transportation of used equipment should be considered. Areas of high risk should be recognized in cleaning procedures (e.g., dispensaries, sampling utensils, containers, pallets).

6.2 Procedures

Standard operating procedure should define cleaning of process equipment and facilities. Operators should be trained in cleaning SOPs.

6.3 Cleaning agents

Cleaning agents should be approved before use. Water used for cleaning and rinsing should comply with the requirements specified in the relevant SOP.

6.4 Validation

Removal of cleaning agents at the end of the cleaning procedure should be validated. Validated cleaning procedures should be used.

SOP No. Val. 200.70

Effective date: mm/dd/yyyy
Approved by:

6.5 Records

All cleaning should be recorded.

7. Facility, Equipment, and Process Design

See SOP No. Val. 200.40, Design Qualification Guideline for Minimizing the Risk of Cross-Contamination of Facility, Equipment and Process.

REASONS FOR REVISION

Effective date: mm/dd/yyyy

■ First time issued for your company, affiliates, and contract manufacturers

---

YOUR COMPANY
VALIDATION STANDARD OPERATING PROCEDURE

SOP No. Val. 200.80

Effective date: mm/dd/yyyy

Approved by:

TITLE: **Design Specifications for Process Water**

AUTHOR: _____

Name/Title/Department

_____

Signature/Date

CHECKED BY: _____

Name/Title/Department

_____

Signature/Date

APPROVED BY: _____

Name/Title/Department

_____

Signature/Date

REVISIONS:

No.	Section	Pages	Initials/Date

SOP No. Val. 200.80   **Effective date: mm/dd/yyyy**

                                      Approved by:

SUBJECT: Design Specifications for Process Water

PURPOSE

To provide guidelines for the selection of material to be used for the construction of equipment and supporting accessories to manufacture the process water

RESPONSIBILITY

It is the responsibility of the technical services manager and contractors to follow the procedure. The quality assurance manager is responsible for SOP compliance.

PROCEDURE

The following design specifications are establishing based on guidelines provided in USP 24. They can be further fine-tuned according to the individual company's requirement. For general guidelines refer to section D.

1. Construction Material

1.1 Stainless steel

All parts in contact with the process water, except gaskets and membrane of valves, must be AISI 316L stainless steel (distillator, tubing, heat exchangers, sensors, valves, fittings, tank, pump, and all instrumentation).

Finishing inside (contact with the process water): Ra < 0.8 μm and electropolished; outside: 150 grit finish.

1.2 Tubing

Tubing should not be welded longitudinally.

Tubing should be supplied with a material certification sheet for material construction and rugosity and be precleaned and capped. Handle material in such a way as to prevent introduction of contaminants into the piping system.

SOP No. Val. 200.80   **Effective date: mm/dd/yyyy**
                                       Approved by:

■ Tolerances: minimum tolerances to ensure proper alignment for automatic welding — ASTM A270
  Outside diameter: 0.5, 0.75, 1 in. + 0.002 over to 0.008 under in.
  1.5 in. + 0.002 over to 0.008 under in.
  2.3 in. + 0.002 over to 0.011 under in.
■ Wall thickness: +0.0065 in.
■ Ovality: +0.010 in.
■ Plane angularity: +0.005 degrees
■ Angle angularity: +0.005 degrees
  (fittings)
■ Squareness: face to tangent: +0.005 in.

1.3 Valves

All the valves in contact with the product must be diaphragm valves:

■ Weir type 316L stainless steel body
■ Interior finish as specified for tubing
■ Self draining
■ Weld connection with tubing or clamps
■ Diaphragm: Teflon (214S) + EPDM (325) (resistant to sterilization temperature)
■ Tolerances: inside dimensions as specified for tubing

1.4 Fittings

Material, wall thickness, and finishing should be as specified for tubing

1.5 Connections

The largest possible number of connections should be automatic TIG orbital welding without external filler wire (butt-welded). Manual welding or tri-clamp connections are authorized case–by–case, but have to be minimized and reserved; manual welding is needed for inaccessible orbital welding, tri-clamp for sensors connections, and moving equipment (pump, hose pipe, etc.).

SOP No. Val. 200.80   **Effective date: mm/dd/yyyy**

                                            Approved by:

1.6 Tri-Clamps

- **Ferrules** — Material, tolerances, wall thickness, and finishing as specified for tubings in welding type with one end for clamp connection
- **Gasket** — Teflon envelope with EPDM insert, steam resistant, inside diameters as specified for tubing
- **Clamps** — Type 304 stainless steel, three segments heavy-duty type with metal wing nut

1.7 Sampling valve

Type: Millipore, TC sampling valve

1.8 Distribution pumps

- Centrifugal type, sample stage nonoverloading, long coupled end-suction
  - Pump casing, back plate impeller pin, one type, back shrouded impeller and shaft shall be 316 stainless steel, 240 grit finish, electro-polished
  - Suction and discharge connections — sanitary clamp type
- Provide seal flush kit including valves, tubing and gauges as required for WFI Service
  - Working pressure: minimum 12 bar
  - Fully drainable
  - Steam sterilizable

1.9 Heat exchanger

- Welded double concentric tube or double tube sheet with tri-clamp sanitary fittings (tube side)
- Surface in contact with WFI: 316L electro-polished (Ra ≤ 8 μ)
- Fully drainable
- All necessary fittings to remove the cooling water from the exchanger to enable efficient steam sterilization

1.10   Storage tank

The inside surface should be 316L electro-polished (Ra ≤ 0.8 μ); the tank must be totally drainable. Storage tanks must be equipped with a spray ball on the circulation return line to sanitize all internal parts of the tank and to assure that the tank interior surfaces above the water level are continuously flushed. Tanks must be equipped with a vent filter (0.2 μm hydrophobic) installed in such a way as to prevent condensate from being trapped. The technical services manager and QA manager must approve the model.

Tanks with vent filters must be designed for pressures of −20 kPa and +100 kPa and must be equipped with a rupture disk. Tanks for hot loop and steam sterilizable line must be designed to withstand steam sterilization (121°C). A steam jacketed sterile vent filter must be used to avoid condensation in the filter and the vent filter housing temperature controlled. The tank for hot storage is steam jacketed and insulated for temperature maintenance. Minimum instrumentation shall include level indication, temperature recording controller, pressure gauge, and pressure relief valve.

1.11   Sensors

All required sensors (temperature, pressure, conductivity, flow meter, etc.) in contact with the process waters must be the sanitary type, connected by tri-clamp, and well mounted (temperature). The length of the electrical connections must be long enough to allow the calibration of the sensors.

1.12   Insulation

The insulation must be chloride free, minimum of 25 mm thickness, and protected by a jacket. The jacket must be stainless steel for the technical rooms and fully closed in the classified rooms (PVC pre-molted type, PVC sealed).

2.   Calculation Criteria

The installation (pump, loop, tubing) should be designed to meet the following parameters:

SOP No. Val. 200.80   **Effective date: mm/dd/yyyy**

                                                 Approved by:

- No dead leg ≥6 of the internal diameter of the unused pipe (measured from the axis of the pipe in use)
- For tubing larger than 3 cm, no dead leg ≥3 internal diameter of the unused pipe
- Water flow velocity ≥1.5 m/sec without any sampling or during sampling (long times)
- Water flow velocity ≥0.9 m/sec during sampling (short time)

3. Execution

Tubing must be provided precleaned and capped. The prefabricated parts must be stocked cleaned and capped to prevent introduction of contaminants into the piping system. The bending of tubes is not allowed.

3.1 Drainage

The supports in contact with the tubing must be stainless steel 302:

- 1 m for horizontal tubing (⌀ 0.5 in.)
- 1.5 m for horizontal tubing (⌀ 0.75 to 1 in.)
- 2 m for vertical tubing (→ 1 in.)
- 3 m for tubing (>1 in.)

Do not rigidly anchor tubing. Place a support at every change of direction.

3.2 Welded stainless steel connections

3.2.1 Certification

Welders shall be certified to a qualified welding procedure for the applicable material in accordance with ASME Section IX. Welders shall be certified in the use of the specific equipment and material used in the welding process.

3.2.2 Procedure

To perform automatic TIG orbital welding, the use of a machine with a printout of the welding parameters is preferable. Develop a set of acceptable parameters for the tubing that will be welded, subject to approval. The welding machine supplier shall provide a written procedure for the calibration of the welding controller. Equipment shall be set up by a designated welding supervisor or inspector who shall check the settings, comply with the master program, and sign and date the welding log to that effect. Check inert gas bottles at each setting change and at suitable intervals throughout the day.

Use bench welding whenever possible (automatic). Field welding (manual) shall be kept to a minimum. Hand welding will be permitted only with approval on a case–by–case level. Do not use ferrous material, tools, or equipment (carbon steel cutting tools) in the fabrication or installation of systems.

3.2.3   Purging

Purge oxygen from the weld area before commencing welding, using an inert gas back-up. Back-up gas shall utilize full line purge. Use 99.996% minimum purity argon gas for the back-up purge and for the torch gas. Provide certificates.

3.2.4   Alignment and tacking

Accomplish tack welding in a manner that will not cause any deleterious effect on completed welds. Make tacks as light as possible to reduce excessive heat that may cause structural changes to the granular composition of the system components. Do not allow tacks to penetrate the inner surface of components. Use back-up purge gas for tack welding. Provide four tacks per weld maximum. Cracks or improper tack welds will be rejected. When welding, center the electrodes over the butt weld joint by the use of setting gauge.

3.2.5   Inspection

Examine each external weld visually to ensure there are no surface defects, and record. Examine each interior weld and adjacent areas, both visually and by the use of a boroscope. Welds not accessible with the boroscope must be examined by γ-ray with photo. 3 shot by weld two perpendicular to the tube axis (moved at 90°) and one oblique compared to the tube axis.

Effective date: mm/dd/yyyy

                                            Approved by:

Document and log each weld as follows:

- Provide the "as built" isometric view with the number, location, and type of welds (manual, automatic)
- Engrave each weld with a number
- Location and number of weld
- Name of welder
- Name of inspector
- Name of boroscope user
- Videotape of 10% of automatic welds and 100% of manual welds; One "test" weld per day in automatic mode
- Date

3.2.6   Weld defects

Welds will be rejected and removed (including the heat affected zone) under the following conditions:

- Internal concavity of the weld: none allowed
- Internal convexity of the weld: ≤25% of the wall thickness
- Full penetration of the entire interior weld joint periphery: required
- Lack-of-fusion: none allowed
- Crevices: none allowed
- Slag-or-inclusions: none allowed
- Cracks: none allowed
- Burn through: none allowed
- Sugaring: none allowed
- M.S. alignment (high/low): maximum allowable is 0.02 in.

Rewelding of defective welds is not permitted.

3.3 Pressure tests

The pressure tests must be done before each cleaning and passivation. A first test must be executed during 1 hour with an inert gas. The second test must be hydraulic with DI water, during 6 hours, without loss of the initial pressure. The test pressure must be 1.5 times the used pressure.

Effective date: mm/dd/yyyy

Approved by:

3.4 Cleaning and passivation

- **Cleaning 1** — Cleaning medium shall be compatible with electro-polished tubing and finish. Cleaning is mandatory. Operate according to a written preapproved procedure. Record and log cleaning dates and steps.
- **Passivation** — Passivation is mandatory. Use a 15 to 20% weight/weight nitric acid solution in water. Circulate at least 1 hour. Operate according to a written preapproved procedure. Do not use fluorhydric acid for passivation. Record and log passivation dates and steps. Drain the system completely.
- **Rinsing** — Fill the system with DI water. Circulate for 15 min, then flush each use point outlet and equipment connection until the pH of discharge water is balanced with the inlet pH. Record and log cleaning dates and steps.

3.5 Identification

The whole equipment (pump, tank, sensor, valves, heat exchangers, etc.) must be tagged with an engraved stainless steel tag. The tag number must follow the number of the P&I.

4. Installation and Material of Construction and Component Selection

4.1 General

Adequate consideration should be given to installation techniques; they can adversely affect the mechanical, corrosive, and sanitary integrity of the system. Following are the considerations which shall be built in the system:

- Valve installation attitude should promote gravity drainage.
- Pipe supports should provide appropriate slopes for drainage and should be designed to support the piping adequately under worst-case thermal conditions.
- Methods of connecting system components, including units of operation, tanks, and distribution piping, should preclude problems.
- Stainless steel welds should provide reliable joints that are internally smooth and corrosion free. Low carbon stainless steel, compatible

SOP No. Val. 200.80   **Effective date: mm/dd/yyyy**

                                                       Approved by:

wire filler where necessary, inert gas, automatic welding machines, and regular inspection and documentation help to ensure acceptable weld quality.

■ Final cleaning and passivation shall be performed for removing contamination and corrosion products and to reestablish the passive corrosion resistant surface.

■ Plastic materials can be fused (welded) in some cases and also require smooth, uniform internal surfaces. Adhesives should be avoided due to the potential for voids and chemical reaction.

■ Mechanical methods of joining, such as flange fittings, shall be performed carefully to avoid the creation of offsets, gaps, penetrations, and voids.

■ Control measures shall include good alignment, properly sized gaskets, appropriate spacing, uniform sealing force, and avoidance of threaded fittings.

4.2 Materials of construction

■ Materials of construction selected should be compatible with control measures such as sanitizing, cleaning, and passivation.

■ Materials selected should be able to handle elevated operating, sanitization temperature, and chemicals or additives to be used to clean, control, or sanitize the system.

■ Materials should be capable of handling turbulent flow and elevated velocities without wear on the corrosive barrier impact such as the passivation-related chromium oxide surface of stainless steel.

■ The finish on metallic materials such as stainless steel, whether it is a refined mill finish, polished to specific grit, or an electropolished treatment, should complement system design and provide satisfactory corrosion and microbial activity resistance.

■ Auxiliary equipment and fittings that require seals, gaskets, diaphragms, filter media, and membranes should exclude materials that permit the possibility of extractable, shedding, and microbial activity.

■ Insulating materials exposed to stainless steel surfaces should be free of chlorides to avoid the phenomenon of stress corrosion cracking that can lead to system contamination and the destruction of tanks and critical system components.

SOP No. Val. 200.80   **Effective date: mm/dd/yyyy**
Approved by:

- Specifications are important to ensure proper selection of materials and serve as a reference for system qualification and maintenance.
- Information such as mill reports for stainless steel and reports of composition, ratings, and material handling capabilities for non-metallic substances should be reviewed for suitability and retained for reference.
- Component (auxiliary equipment) selection should be made with the assurance that it does not create a source for contamination intrusion.
- Heat exchangers should be double tube sheet or concentric tube design.
- Heat exchangers should include differential pressure monitoring or utilize a heat transfer medium of equal or better quality to avoid problems if leaks develop.
- Pumps should be of sanitary design with seals that prevent contamination of the water.
- Valves should have smooth internal surfaces with the seat and closing device exposed to the flushing action of water, such as occurs in diaphragm valves.
- Valves with pocket areas or closing devices (e.g., ball, plug, gate, and globe) that move into and out of flow area should be avoided.

4.3 Sanitization

- Microbial control in water systems should be achieved primarily through sanitization practices. Systems should be sanitized using either thermal or chemical means. In-line ultraviolet light at a wavelength of 254 mm can also be used to sanitize water in the system continuously.
- Thermal approaches to system sanitization shall include periodically or continuously circulating hot water and the use of steam.
- These techniques are limited to systems that are compatible with the higher temperatures needed to achieve sanitization, such as stainless steel and some polymer formulations. Although thermal methods control biofilm development, they are not effective in removing established biofilms.
- Chemical methods, where compatible, shall be used on a wider variety of construction materials. These methods typically employ oxidizing agents such as halogenated compounds, hydrogen peroxide, ozone,

SOP No. Val. 200.80   **Effective date: mm/dd/yyyy**
                                             Approved by:

or peracetic acid. Halogenated compounds are effective sanitizers but are difficult to flush from the system and tend to leave biofilms intact. Compounds such as hydrogen peroxide, ozone, and peracetic acid oxidize bacteria and biofilms by forming reactive peroxides and free radicals (notably hydroxyl radicals). The short half-life of these compounds, particularly ozone, may require that they be added continuously during the sanitization process. Hydrogen peroxide and ozone rapidly degrade to water and oxygen; peracetic acid degrades to acetic acid in the presence of ultraviolet light.

■ Ultraviolet light impacts the development of biofilms by reducing the rate of new microbial colonization in the system; however, it is only partially effective against planktonic microorganisms. Alone, ultraviolet light is not an effective tool because it does not eliminate existing biofilm. However, when coupled with conventional thermal or chemical sanitization technologies, it is most effective and can prolong the interval between system sanitizations. The use of ultraviolet light also facilitates the degradation of hydrogen peroxide ozone.

■ Sanitization steps should be validated to demonstrate the capability of reducing and holding microbial contamination at acceptable levels.

■ Validation of thermal methods should include a heat distribution study to demonstrate that sanitization temperatures are achieved throughout the system.

■ Validation of chemical methods should demonstrate adequate chemical concentrations throughout the system. In addition, when the sanitization process is completed, effective removal of chemical residues must be demonstrated.

■ The frequency of sanitization derived from the trend analysis of the microbiological data should be used as the alert mechanism for maintenance. The frequency of sanitization should be established such that the system operates in a state of microbiological control and does not exceed alert levels.

REASONS FOR REVISION

Effective date: mm/dd/yyyy

■ First time issued for your company, affiliates, and contract manufacturers

---

YOUR COMPANY
VALIDATION STANDARD OPERATING PROCEDURE

SOP No. Val. 200.90

Effective date: mm/dd/yyyy

Approved by:

TITLE: **Design Specifications for Water for Injection Production and Distribution**

AUTHOR: _____

Name/Title/Department

_____

Signature/Date

CHECKED BY: _____

Name/Title/Department

_____

Signature/Date

APPROVED BY: _____

Name/Title/Department

_____

Signature/Date

REVISIONS:

No.	Section	Pages	Initials/Date

SOP No. Val. 200.90

Effective date: mm/dd/yyyy

Approved by:

SUBJECT: Design Specifications for Water for Injection Production and Distribution

PURPOSE

To provide guidelines for the selection of material to be used for the construction of equipment and supporting accessories to manufacture the process water

RESPONSIBILITY

It is the responsibility of the technical services manager and contractors to follow the procedure. The quality assurance manager is responsible for SOP compliance.

PROCEDURE

The following guidelines and technical specifications must be followed:

1. Principle

The water for injection is produced by distillation of purified water. The distillator filling the storage tank shall be located at a suitable location per approved layout. Connection shall be provided per plan, e.g., 1 for the CIP of the mfg. tank and 1 for the distribution in building and to allow filling of the tanks.

Some points of use are cold points and work with a cold exchanger followed after the point of use by a heat exchanger. From the two secondary tanks, start two small loops with cooling and heating. The main storage tank is automatically filled by the distillator (level switch) and must be full at the beginning of a production day. The secondary storage tanks are filled during the night and can only be filled during the day in case of low level.

2. Main Characteristics

- Obtained from: purified water
- Quality: following the USP 24
      microbiology: <10 CFU/100 ml
- Production: distillator 4 effect
      capacity: 500 kg/h
- Temperature: ≥80°C at each point of the loop
- Pressure: ≥2 bars
- Temperature keeping: obtained by heating the jacket of
      the tank; if necessary, installation of
      a heat exchanger in the loop.

3. Sterilization

The tanks and the loops must be steam sterilized; each pump must be separately sterilized. During the sterilization, the cooling or heating waters of the exchangers must be drained; the Fo. value must be ≥15 at each point of the loops.

4. Consumption Points

The consumption shall be evaluated for full time during the production day (e.g., washing machine and freeze drier).

5. Monitoring and Control

The following characteristics must be monitored: conductivity, C°T at the return of the loop, and flow at the return of the loop. A sampling valve must be located near each main point of use.

6. Secondary Loops — Cold Point

For the cold points of use, a secondary loop is used. The velocity of the water must follow the recommended values in each loop (primary and secondary). A regulation valve shall be provided to balance the loss of pressure between primary and secondary.

SOP No. Val. 200.90   **Effective date: mm/dd/yyyy**

                                                  Approved by:

When requested by the user located at the point of use, the first exchanger on the secondary loop starts to cool the water. This one is automatically heated at minimum 85°C before the return in the primary loop. When the temperature asked for the cold point of use is reached, the sampling value is automatically opened. The users ask to stop the sampling. A safety timer must be installed to prevent the draining of the installation. Safety timing is adjustable (maximum 1 h); temperature of cold point: 30°C ± 5°C.

7. Federal Regulation

USP XXIV specifies the limits and method of testing for chemical and pyrogenic contaminants for various compendial classifications of water, such as purified water and water for injection.

The U.S. Food and Drug Administration has established various good manufacturing practices for pharmaceutical products. Selected excerpts of these that have impact on water quality are reproduced on the following pages.

Note: The underlined words have been added by the author for emphasis.

§210.3 Definitions

a) The following definitions of terms apply to Parts 210 through 229 of this chapter.
b) The terms are as follows:
    (3) "Component" means any ingredient intended for use in the manufacture of drug product, including those that may not appear in such drug product.

                              * * *

    (5) "Fiber" means any particle with a length of at least three times greater than its width.
    (6) "Non-fiber releasing filter" means any filter, which after any appropriate pretreatment such as washing or flushing, will not release fibers into the component or drug product that is being filtered. All filters composed of asbestos or glass fibers are deemed to be fiber-releasing filters.

                              * * *

(2) "Batch" means a <u>specific quantity</u> of a drug that has uniform character and quality, within specified limits, and is produced according to a single manufacturing order during the same cycle of manufacture.

* * *

(10) "Lot" means a batch, or a specific identified portions of a batch, having uniform character and quality specified limits; or, in the case of a drug product produced by <u>continuous process,</u> it is a specific identified amount produced in a <u>unit of time</u> or quantity in a manner that assures its having <u>uniform character and quality within specified limits.</u>

§211.48 Plumbing

(a) Potable water shall be supplied under <u>continuous positive pressure</u> in a plumbing system free of defects that could <u>cause continuous contamination to any drug product.</u> Potable water shall meet the standards prescribed in the Public Health Service Drinking Water Standards set forth in Subpart J of 42 CFR Part 72. Water not meeting such standards shall not be permitted in the plumbing system.

§211.72 Filters

(a) Filters used in the manufacture, processing, or packing of injectable drug products intended for human use shall not release fibers into such products. Fiber-releasing filters may not be used in the manufacture, processing, or packing of these drug products unless it is not possible to manufacture such drug products without the use of such a filter.

(b) If use of a fiber-releasing filter is necessary, an additional non-fiber-releasing filter of 0.22 micron maximum mean porosity (0.45 micron if the manufacturing conditions so dictate) shall subsequently be used to reduce the content of particles in the drug product. Use of an asbestos containing filter, with or without subsequent use of a specific non-fiber-releasing filter, is permissible only upon submission of proof to the appropriate bureau of the Food and Drug Administration that use of a non-fiber-releasing filter will or is likely to, compromise the safety or effectiveness of the drug product.

SOP No. Val. 200.90 **Effective date: mm/dd/yyyy**

Approved by:

§212.3 Definitions

* * *

(11) "Static line" means any pipe containing liquid that is not emptied or circulated at least <u>once every 24 hours.</u>

Subpart B — Organization and Personnel
§212.22 Responsibilities of quality control unit

(a) The <u>quality control unit</u> shall have the responsibility and authority to test and accept or reject the <u>design, engineering, and physical facilities</u> of the plant, the equipment, and the manufacturing process and control procedures to be used in the manufacture, processing, packing, and holding of each large volume parenteral drug product. The quality control unit shall reject any such plant, equipment, process, or procedure if it does not comply with the provisions of this part or if, in the opinion of the quality control unit, it is not suitable or adequate to assure that the drug product has the characteristics it purports or is represented to possess.

* * *

(c) The <u>quality control unit</u> shall have the responsibility and authority to test and approve or reject any <u>changes in previously approved plant, equipment, processes, procedures</u>, and container-closures and delivery systems before utilization in the manufacture, processing, packing and holding of a large volume parenteral drug product.

Subpart C — Buildings and Facilities
§212.42 Design and construction features

* * *

(c) There shall not be <u>horizontal fixed pipes</u> or conduits over <u>exposed components,</u> in-process materials, drug products, and drug product contact surfaces, including drug product containers and closures after the final rinse.

(d) In each <u>physically separated area, pipes or conduits</u> for air or <u>liquids</u> shall be <u>identified</u> as to their contents. Such identification shall be by name, color code, or other suitable means.

SOP No. Val. 200.90   **Effective date: mm/dd/yyyy**
                                            Approved by:

§212.49 Water and other liquid-handling systems

(a) Filters may not be used at any point in the water for manufacturing or final rinse piping system.

(b) Backflow of liquids shall be prevented at points of interconnection of different systems.

(c) Pipelines for the transmission of water for manufacturing or final rinse and other liquid components shall:

(1) Be constructed of welded stainless steel (nonrusting grade) equipped for sterilization with steam, except that sanitary stainless steel lines with fittings capable of disassembly may be immediately adjacent to the equipment of valves that must be removed from the lines for servicing and replacement.

(2) Be sloped to provide for complete drainage.

(3) Not have an unused portion greater in length than six diameters of the unused pipe measured from the axis of the pipe in use.

§212.67 Equipment cleaning and maintenance

The following requirements shall be included in written procedures and cleaning schedules:

(a) All equipment and surfaces that contact components, in-process materials, drug products or drug product contact surfaces such as containers and closures shall be cleaned and rinsed with water meeting the quality requirements stated in §212.224.

(b) Immediately prior to such contact, equipment and surfaces specified in paragraph (a) of this section shall be given a final rinse with water meeting the quality requirements stated in §212.225.

(c) Steam used to sterilize liquid-handling systems or equipment shall be free of additives used for boiler control.

§212.68 Equipment calibration

(a) Procedures shall be written and followed designating schedules and assigning responsibility for testing or monitoring the performances or accuracy of automatic or continuously operating equipment, devices, apparatus, or mechanisms, such as, but not limited to, the following:

SOP No. Val. 200.90   **Effective date: mm/dd/yyyy**
                                                 Approved by:

(1) <u>Alarms and controls on sterilizing equipment</u>
(2) Temperature-recording devices on sterilizers.
(3) <u>Pressure gauges</u>
(4) Mechanisms for maintaining sterilizing medium uniformity
(5) Chain speed recorder
(6) <u>Heat exchanger pressure differential monitor</u>
(7) Mercury-in-glass thermometer
(8) <u>Written records of such calibrations, checks, examinations, or inspections</u> shall be maintained, as specified in §212.183.

§212.72 Filters

(a) The integrity of all <u>air filters shall be verified upon installation and maintained through use. A written testing program adequate to monitor integrity of filters</u> shall be established and followed. Results shall be recorded and maintained as specified in §212.183.

§212.76 Heat exchangers

<u>Heat exchangers</u>, other than the welded double-concentric-tube type or double-tube sheet type, must employ a pressure differential and a means for monitoring the differential. The pressure differential shall be such that the fluid requiring a higher microbial quality shall be that with the greater pressure. Written records of the pressure differential monitoring shall be maintained as required in §212.183

§212.78 Air vents

<u>All stills and tanks holding liquid requiring microbial control</u> shall have air vents with non-fiber-releasing <u>sterilizable filters capable of preventing microbial contamination of the contents.</u> Such filters shall be designed and installed so that they do not become wet. <u>Filters shall be sterilized and installed aseptically.</u> Tanks requiring air vents with filters include those holding water for manufacturing or final rinsing, water for cooling the drug product after sterilization, liquid components, and in-process solutions.

§212.100 Written procedures, deviations

* * *

(b) <u>Written procedures</u> shall be established, and shall be followed. Such procedures shall:

    (1) Ensure that all <u>static lines are flushed</u> prior to use. Such procedures shall require that flushing produce a turbulent flow for 5 minutes and that all valves on the line are opened and closed repeatedly to flush the valve interior.

§212.182 <u>Equipment cleaning and use log</u>

(a) <u>Written records</u> of the corrective action taken pursuant to §212.24 (a) and (c), and §212.225 (a) and (b), including validation of the effectiveness of the action, shall be maintained.

(b) <u>Written records</u> of <u>equipment usage</u> shall include documentation of the length of time the equipment was in use as indicated in §212.111.

(c) <u>Written records</u> demonstrating a <u>positive pressure differential,</u> as described in and required by §212.76, shall be maintained.

* * *

(e) For <u>filtration equipment,</u> or devices, <u>written records documenting the installation, replacement, and sterilization</u> (where appropriate) of filters such as those indicated in §§212.72, 212.77(b) and (c), 212.78, and 212.222(a) shall be maintained.

§212.183 Equipment calibration and monitoring records

<u>Written records of calibration and monitoring tests</u> and readings performed shall be maintained for at least 2 years after the expiration date of each batch of drug product produced by the equipment.

(a) Calibration records shall include:

    (1) A description of the equipment

    (2) The date the equipment was purchased

    (3) The operating limits of the equipment

    (4) The date, time, and type of each test

    (5) The results of each test

    (6) The signature of each person performing a test

    (7) The date the equipment was installed

(b) Monitoring records shall include:
  (1) A description of the equipment
  (2) The date the equipment was installed
  (3) The date the equipment was last calibrated, if appropriate
  (4) The operating limits of the equipment
  (5) The date and time of the recording
  (6) The reading
  (7) The signature of each person performing the monitoring
(c) Corrective measures employed to bring the equipment into compliance with its operating specification shall be:
  (1) Recorded in the appropriate equipment log
  (2) Noted in the calibration and/or monitoring record
  (3) Immediately followed by testing to assure that the corrective measures were adequate to restore the required operating characteristics

§212.183 Batch production and control records

These records shall include the following information where appropriate:

(1) Verification that static lines were flushed prior to use according to established written procedures in §212.100(b).

§212.188 Batch production and control records

The review and approval of production and control records by the quality control unit shall extend to those records not directly related to the manufacture, processing, packing, or holding of a specific batch of large volume parenteral drug product but which have a bearing on the quality of batches being produced. Such indirectly related records shall include:

(a) Those dealing with equipment calibration or standardization
(c) Those demonstrating the quality of water produced by various processing systems.
(d) Those demonstrating the quality of air produced by various systems

§212.190 Air and water monitoring records

Written records of the air and water monitoring test results, readings, and corrective measures taken shall be maintained for at least 2 years after the expiration date of each batch of drug product produced in the area being monitored or containing the water as a component.

The record shall include at a minimum, the following information:

(a) Identity of the material being monitored
(b) Each characteristic being monitored
(c) Each specification limit
(d) Each testing method used
(e) Site sampled or monitored
(f) The date and time of each monitoring or testing
(g) The result of each test or monitoring reading
(h) Batch number and expiration date of the drug product being processed in the area or equipment, or to which the component is being added at the time of monitoring or sampling
(i) Corrective measures employed to bring the area, component or product into compliance with specifications
(j) Retesting results to verify the adequacy of the corrective measures

Subpart L — Air and Water Quality
§212.220 General requirements

(a) Air or water as described in this part may not be used until the plant, processes, and procedures used in producing and distributing it have been tested and approved by the quality control unit as capable of consistently producing air or water meeting the requirements set forth in this subpart.
(b) In addition to the requirements of this subpart, air and water quality shall be monitored as specified in Subpart J.
(c) The results of all testing and data generated shall be recorded and maintained as required by §212.180.
(d) Procedures designating schedules, assigning responsibility, and describing in detail the action to be taken to assure that the systems produce and deliver air and water that conform to the requirements set forth in this subpart shall be written. Such procedures shall also specify the corrective action to be taken when testing reveals that the established standards are not being met. Records of corrective actions shall be maintained, as specified in §212.190.

SOP No. Val. 200.90   **Effective date: mm/dd/yyyy**
                                           Approved by:

§212.224 Water for cleaning or initial rinsing

Water used to cleanse or initially rinse drug product contact surfaces such as containers, closures, and equipment shall:

> Drinking Water Standards set forth in Subpart J of 42 CFR Part 72;
> (b) Be subjected to a process such as chlorination for control of microbial population;
> (c) Contain not more than 50 microorganisms per 100 millimeters in three consecutive samples from the sampling site when tested by the method specified in §212.225(b) after neutralizing bacteriocidal agents, if present.

§212.225 Water for manufacturing or final rinsing

Water used as a component or as a final rinse for equipment or product contact surfaces shall:

(a) Conform to the specifications in the U.S.P. for "Water for Injection":
(b) Contain not more than 10 microorganisms per 100 millimeters in three consecutive samples from the same site when samples of 250 millimeters or more are tested for total aerobic count by the plate method set forth in Microbial Limit Tests in the current revision of the U.S.P. Alternate methodology may be used provided that data are available to demonstrate that the alternate method is equivalent to the official method. When the microbial quality falls below that specified in this section, use of such water shall cease, and corrective action shall be taken to clean and sterilize the system so that the water conforms to the limit.
(c) Be stored in a suitable vessel or system including a piping network for distribution to points of use:
   (1) At a temperature of at least 80°C under continuous circulation, or
   (2) At ambient or lower temperature for not longer than 24 hours, after which time such water shall be discarded to drain.

§212.226 Water for drug product cooling

Water used in the sterilizer as a drug product cooling medium shall:

(a) Be <u>treated</u> to eliminate microorganisms:
(b) Contain not more than <u>one microorganism per 100 millimeters</u> in three consecutive samples from the same sampling site when one liter or more are tested for total aerobic count by a membrane filtration method and placing each membrane filter on appropriate nutrient media after neutralizing any bacteriocidal agents present in the water samples.

§212.227 <u>Boiler feed water</u>

Feed water for boilers supplying steam that <u>contacts components,</u> in-process materials, drug products, and <u>drug product contact surfaces</u> shall not contain <u>volatile additives</u> such as <u>amines or hydrazines.</u>

§212.233 <u>Water quality program</u> design

(a) <u>Water quality monitoring</u> shall include:
    (1) <u>Sampling and testing</u> of water for manufacturing or final rinsing at least once a day. All sampling ports or points of use in the distribution system shall be sampled at least weekly.
    (2) Sampling water for drug product cooling at a point just before entry into the sterilizer at least once each sterilizer cycle and testing by the method described in §212.226.
    (3) Sampling and testing water for cleaning or initial rinsing at least once a week. All sampling points or points of use in the distribution system shall be sampled at least monthly.
(b) Boiler feed water shall be sampled and tested periodically for the presence of volatile additives.
(c) <u>If three consecutive samples</u> of drug product cooling water <u>exceed microbial limits,</u> the sterilizer loads shall be rejected and shall not be reprocessed.

§212.231 <u>Monitoring</u> of air and <u>water quality</u>

(a) <u>After the plant, equipment, manufacturing processes,</u> and control procedures have been <u>tested and approved</u> by the <u>quality control unit,</u> there shall be performed in accordance with <u>written procedures</u> and schedules a <u>sampling and testing program</u> that is designed to monitor the microbial flora of the plant and its environment. The design of the sampling and testing program shall include

SOP No. Val. 200.90   **Effective date: mm/dd/yyyy**

                                                      Approved by:

      monitoring of air and water quality in accordance with require-
ments set forth in this subpart and taking corrective action when
such requirements are not met.

(b) If the results of any <u>one sample</u> of air or water <u>exceed the quality
limits</u> specified in this subpart, <u>more frequent</u> sampling and testing
shall be required to determine the need for corrective action.

(c) Representative colonies of microorganisms found by the monitoring
required in this section shall be identified by genus. The colonies
shall be quantified.

(d) <u>Written records of all test findings</u> and <u>any resultant corrective
measures taken</u> shall be <u>maintained,</u> as specified in §212.190.

REASONS FOR REVISION

Effective date: mm/dd/yyyy

■ First time issued for your company, affiliates, and contract manufacturers

YOUR COMPANY
VALIDATION STANDARD OPERATING PROCEDURE

SOP No. Val. 200.100

Effective date: mm/dd/yyyy

Approved by:

TITLE: Design Specifications for Purified Water (DIW)
Production and Distribution

AUTHOR: _____

Name/Title/Department

_____

Signature/Date

CHECKED BY: _____

Name/Title/Department

_____

Signature/Date

APPROVED BY: _____

Name/Title/Department

_____

Signature/Date

REVISIONS:

No.	Section	Pages	Initials/Date

SOP No. Val. 200.100   Effective date: mm/dd/yyyy

Approved by:

SUBJECT: Design Specifications for Purified Water (DIW) Production and Distribution

PURPOSE

To provide guidelines for the selection of material to be used for the production and distribution of DI water

RESPONSIBILITY

It is the responsibility of the technical services manager, validation manager, and contractors to follow the procedure. The quality assurance manager is responsible for SOP compliance.

PROCEDURE

The following guidelines and technical specifications must be followed:

1. Principle

Purified water is produced by the water treatment of available drinking water. The water treatment unit filling the storage tank shall be located at a suitable place per approved layout. From this tank, start hot loops, e.g., one for the recirculation in the water treatment unit (if required) and one for distribution in the building, to allow filling of the tanks and to feed the distillator and the steam generator. From the two secondary tanks start two loops to provide several points of use in the buildings (as required).

2. Main Characteristics

- Obtained from: available drinking water analysis report
- Quality: following USP 24
                       microbiology: <20 CFU/ml
- Production: to propose by the supplier
- Temperature: ≤20°C ± 5°C at each point of the loop
- Pressure: ≥2 bars
- Temperature keeping: obtained by a cold exchanger located at the return of the loop

3. Sanitization

The tanks and the loops must be sanitized NLT 80°C. The DIW is heated at 80°C and maintained at this temperature for 1 h once a day, during the night. After 1 h at 80°C, the loop must be cooled at 20°C by the cold exchanger located at the return of the loops.

4. Consumption Points

Provide the points as necessary (distillator, steam generator, tanks DIW, etc.).

5. Monitoring and Control

The following characteristics must be monitored: conductivity, °T at the return of the loop and between the two heat exchangers, and flow at the return of the loop. A sampling valve must be located near each main point of use.

REASONS FOR REVISION

Effective date: mm/dd/yyyy

- First time issued for your company, affiliates, and contract manufacturers

```
┌─────────────────────────────────────────────────────────┐
│                  YOUR COMPANY                           │
│     VALIDATION STANDARD OPERATING PROCEDURE             │
└─────────────────────────────────────────────────────────┘
```

SOP No. Val. 200.110   Effective date: mm/dd/yyyy

                                        Approved by:

TITLE: Design Specifications for Pure Steam
                Production and Distribution

AUTHOR: _____
                           Name/Title/Department

                _____
                           Signature/Date

CHECKED BY: _____
                           Name/Title/Department

                _____
                           Signature/Date

APPROVED BY: _____
                           Name/Title/Department

                _____
                           Signature/Date

REVISIONS:

No.	Section	Pages	Initials/Date

SOP No. Val. 200.110   Effective date: mm/dd/yyyy

Approved by:

SUBJECT: Design Specifications for Pure Steam Production and Distribution

PURPOSE

To provide guidelines for the selection of material and techniques to be used for the construction of equipment and supporting accessories used to manufacture the process water

RESPONSIBILITY

It is the responsibility of the technical services manager and contractors to follow the procedure. The quality assurance manager is responsible for SOP compliance.

PROCEDURE

The following guidelines and technical specifications must be followed:

1. Principle

The pure steam is produced by the steam generator of the purified water. The steam generator shall be located at a suitable place per approved layout. From this, start the tubing to the different points of use. A system to collect the condensate must be provided for the most important point of use. The collected condensate could be used to feed the industrial steam generator.

2. Main Characteristics

- Obtained from: purified water
- Quality: the condensate must have the qualities required for water for injection
- Production: to propose by the supplier power: industrial steam (6 bars)
- Pressure available: 4 bars (relative)

SOP No. Val. 200.110   **Effective date: mm/dd/yyyy**

                                               Approved by:

3. Consumption Points

These points shall be established considering the requirements, they can be used simultaneously during a production day.

- Autoclave
- The SIP of the freeze drier

The sterilization of the WFI installation must be done beyond the workings hours.

REASONS FOR REVISION

Effective date: mm/dd/yyyy

- First time issued for your company, affiliates, and contract manufacturers

SECTION

VAL 300.00

| YOUR COMPANY |
| VALIDATION STANDARD OPERATING PROCEDURE |

SOP No. Val. 300.10

Effective date: mm/dd/yyyy
Approved by:

TITLE: **Validation Glossary**

AUTHOR:

_____
Name/Title/Department

_____
Signature/Date

CHECKED BY:

_____
Name/Title/Department

_____
Signature/Date

APPROVED BY:

_____
Name/Title/Department

_____
Signature/Date

REVISIONS:

No.	Section	Pages	Initials/Date

SOP No. Val. 300.10

Effective date: mm/dd/yyyy
Approved by:

SUBJECT: Validation Glossary

PURPOSE

To provide the glossary of terms used in the SOPs with a particular reference to the validation and cGMP

RESPONSIBILITY

It is the responsibility of the validation team to understand and implement the terms defined in the validation lexicon. The QA manager is responsible for the SOP compliance.

PROCEDURE

Batchwise Control

The use of validated in-process sampling and testing methods in such a way that results prove that the process has done what it purports to do for the specific batch concerned, thus assuring that control parameters have been appropriately respected.

Calibration

Demonstrating that a measuring device produces results within specified limits of those produced by a reference standard device over an appropriate range of measurements. This process results in corrections that may be applied if maximum accuracy is required.

Calibration Program

An element of quality assurance ensuring that all tests and measurements used to control and monitor the process or to test the product are capable of producing results that are accurate and precise to the extent dictated by importance of the measurement.

SOP No. Val. 300.10   Effective date: mm/dd/yyyy
                                        Approved by:

Certification

Documented testimony by qualified authorities that a system qualification, calibration, validation, or revalidation has been performed appropriately and that the results are acceptable.

Control Parameters

Those operating variables that can be assigned values that are used as control levels.

Control Parameter Range

Range of values for a given control parameter that lies between its two outer limits or control levels.

Edge of Failure

A control parameter value that, if exceeded, means adverse effects on state of control or fitness of use for the product.

Installation Qualification

Documented verification that all key aspects of the installation adhere to appropriate codes and approved design intentions and that manufacturers' recommendations are suitably considered.

Operating Variables

All factors, including control parameters, that may potentially affect process state of control or fitness for use of the end product.

Operational Qualification

Documented verification that the system or subsystem performs as intended throughout all anticipated operating ranges.

Effective date: mm/dd/yyyy

Approved by:

Process Development

Establishing evidence that all process control parameters and all control parameter ranges are validated and optimized.

Process Validation

Establishing documented evidence that a process does what it purports to do.

Prospective Validation

Establishing documented evidence that a system does what it purports to do based on a preplanned protocol.

Proven Acceptable Range

All values of a given control parameter that fall between proven high and low worst case conditions.

Quality Assurance

The activity of providing, to all concerned, the evidence needed to establish confidence that the quality function is being performed adequately.

Quality Control

The regulatory process through which industry measures actual quality performance, compresses it with standards, and acts on the difference.

Retrospective Validation

Establishing documented evidence that a system does what it purports to do based on review and analysis of historic information.

Revalidation

Repetition of the validation process or a specific portion of it.

SOP No. Val. 300.10   **Effective date: mm/dd/yyyy**

                                              Approved by:

State of Control

A condition in which all operating variables that can affect performance remain within ranges that the system or process performs consistently and as intended.

Sterilization Process

A treatment process from which probability of any microorganism survival is less than 10^{-6}, or one in a million.

The Quality Function

The entire collection of activities from which industry achieves fitness for use, no matter where these activities are performed.

Validation

Establishing documented evidence that a system does what it purports to do.

Validation Change Control

A formal monitoring system by which qualified representatives of appropriate disciplines review proposed or actual changes that might affect validated status and cause corrective action to be taken so that the system retains its validated state of control.

Validation Protocol

A document that spells out what tests are to be performed, how the tests are to be performed, what data are to be collected, and what the acceptance criteria are.

Validation Report

A scientific report of the results derived from executing a validation protocol.

SOP No. Val. 300.10

Effective date: mm/dd/yyyy

Approved by:

Worst Case

The highest or lowest value of a given control parameter actually evaluated in a validation exercise.

REASONS FOR REVISION

Effective date: mm/dd/yyyy

- ■ First time issued for your company, affiliates, and contract manufacturers

---

YOUR COMPANY
VALIDATION STANDARD OPERATING PROCEDURE

SOP No. Val. 300.20

Effective date: mm/dd/yyyy

Approved by:

TITLE: Organization for Validation

AUTHOR: _____

Name/Title/Department

_____

Signature/Date

CHECKED BY: _____

Name/Title/Department

_____

Signature/Date

APPROVED BY: _____

Name/Title/Department

_____

Signature/Date

REVISIONS:

No.	Section	Pages	Initials/Date

SOP No. Val. 300.20

Effective date: mm/dd/yyyy

Approved by:

SUBJECT: Organization for Validation

PURPOSE

To describe the functions and responsibilities of the validation team to meet the cGMP compliance

RESPONSIBILITY

It is the responsibility of all concerned departments to follow the procedure. The QA manager is responsible for SOP compliance.

PROCEDURE

1. Validation Coordinator

All validation activities through the different progress steps should be coordinated by one person, preferably the quality assurance manager.

2. Validation Task Force/Certification Team

The team should consist of managers of the departments involved in the validation and outside vendors (if applicable); for example:

- Quality assurance manager
- Production manager
- Technical services manager
- Product development manager
- Calibration manager
- Quality control manager
- Approved vendors (outside)

2.1 Responsibilities

- Scope of validation
- Validation priorities
- Acceptance criteria

SOP No. Val. 300.20   **Effective date: mm/dd/yyyy**

                                                Approved by:

- Approving of validation protocols and reports
- Validation change control

3. Validation Working Groups

The executive part of the validation work should be delegated to dedicated personnel:

- A member of the validation task force
- Representatives from relevant departments
- A representative from quality assurance
- A representative from technical services
- A representative from product development laboratory
- A representative from quality control
- A representative from the vendor (outside)

4. Validation Planning and Scheduling

- Manpower resources
- Document preparation
- Filed execution
- Calibration
- Lab support
- Test and balance/filter certification
- Start-up and commissioning

REASONS FOR REVISION

Effective date: mm/dd/yyyy

- First time issued for your company, affiliates, and contract manufacturers

---

| YOUR COMPANY |
| VALIDATION STANDARD OPERATING PROCEDURE |

SOP No. Val. 300.30   Effective date: mm/dd/yyyy

                                           Approved by:

TITLE: Revalidation

AUTHOR: _____

                          Name/Title/Department

                _____

                          Signature/Date

CHECKED BY: _____

                          Name/Title/Department

                _____

                          Signature/Date

APPROVED BY: _____

                          Name/Title/Department

                _____

                          Signature/Date

REVISIONS:

No.	Section	Pages	Initials/Date

SOP No. Val. 300.30   **Effective date: mm/dd/yyyy**

                                               Approved by:

SUBJECT: Revalidation

PURPOSE

To describe the necessity and reasons for revalidation

RESPONSIBILITY

It is the responsibility of the validation team members to follow the procedures. The quality assurance manager is responsible for SOP compliance.

PROCEDURE

1. Evaluation of Revalidation Necessity

Revalidation provides a guarantee of consistent system, process, or equipment usage. It assures that monitoring controls are sensitive enough to identify major problems or drifts in quality and that process or equipment variations have no adverse effect on quality.

2. Reasons for Revalidation

A revalidation has to be performed in cases of:

- ■ Change of (or in):Formula of the product
  - Process
  - Equipment
  - Facility (influencing process)
  - Control methods
  - Batch size
  - Hardware and software
  - Cleaning agents
  - Material changes
  - Supplier change
- ■ Deviations in results of in-process and final controls
- ■ Extensive maintenance or repairs of equipment

Note: Prior to starting the revalidation, it should be evaluated whether the whole or only a part of the system, process, or equipment has to be revalidated.

Effective date: mm/dd/yyyy

Approved by:

3. Revalidation Cycle

In the event that no major changes are brought into the systems, process, and facilities, retrospective validation shall be conducted every 3 years.

REASONS FOR REVISION

Effective date: mm/dd/yyyy

■ First time issued for your company, affiliates, and third-party manufacturers

| YOUR COMPANY |
| VALIDATION STANDARD OPERATING PROCEDURE |

SOP No. Val. 300.40

Effective date: mm/dd/yyyy

Approved by:

TITLE: **Retrospective Validation**

AUTHOR: _____

Name/Title/Department

_____

Signature/Date

CHECKED BY: _____

Name/Title/Department

_____

Signature/Date

APPROVED BY: _____

Name/Title/Department

_____

Signature/Date

REVISIONS:

No.	Section	Pages	Initials/Date

SOP No. Val. 300.40

Effective date: mm/dd/yyyy
Approved by:

SUBJECT: Retrospective Validation

PURPOSE

To describe the requirements and criteria for retrospective validation

RESPONSIBILITY

The validation team is responsible for following the procedure. The quality assurance manager is responsible for SOP compliance.

PROCEDURE

1. Definition

Retrospective validation is the most pertinent for use by most pharmaceutical companies, establishing documented evidence that a system does what it purports to do based on review and analysis of historical information.

2. Objective

To demonstrate that the process has performed satisfactorily and consistently over time and, therefore, can be relied upon to deliver the same product quality in the future on a continuous basis.

3. Product Selection Criteria for Retrospective Validation

Following are criteria to establish the fact that the process can be categorized for retrospective validation:

- Used analytical test methods and results adequately specific
- Unchanged process
- Personnel procedures consistent in performance
- Available data from process history and testing clearly identified
- Unchanged suppliers of materials

A minimum of 20 batches should be evaluated over a specified time interval.

SOP No. Val. 300.40   **Effective date: mm/dd/yyyy**
                                             Approved by:

4. Retrospective Validation

- Setting up of process equipment
- Identification of measuring equipment
- Discussion of the influence possibilities on critical process parameters
- Identification of process equipment
- Selection of process and product parameters to be considered
- Fixing of requirements for process and product parameters
- Fixing of setting up the statistical evaluation
- Average value
- Minimum value
- Maximum value
- Standard deviation
- Correlation coefficient

5. Assessment of the Investigation

After review of all statistical data per approved standard operating procedure, assessment should be made to suggest important steps for future production or further investigations, if necessary.

REASONS FOR REVISION

Effective date: mm/dd/yyyy

- First time issued to your company, affiliates, and contract manufacturers

---

YOUR COMPANY
VALIDATION STANDARD OPERATING PROCEDURE

SOP No. Val. 300.50

Effective date: mm/dd/yyyy

Approved by:

TITLE: **Validation Change Control**

AUTHOR: _____

Name/Title/Department

_____

Signature/Date

CHECKED BY: _____

Name/Title/Department

_____

Signature/Date

APPROVED BY: _____

Name/Title/Department

_____

Signature/Date

REVISIONS:

No.	Section	Pages	Initials/Date

SOP No. Val. 300.50   Effective date: mm/dd/yyyy

                                                  Approved by:

SUBJECT: Validation Change Control

PURPOSE

To describe the procedure to prevent uncontrolled changes in validated equipment and processes

RESPONSIBILITY

All concerned departmental managers are responsible for following the procedures. The quality assurance manager is responsible for SOP compliance.

PROCEDURE

1. Definitions

Change: Any subsequent departure from the approved flow chart and activities, or a modification to documentation, equipment, packaging, utilities, facilities, formulations, processes, or computer systems.

Change control: A formal monitoring system by which qualified representatives of the appropriate discipline review actual changes that might affect a validated status to determine the need for corrective action ensure that the system retains its validated state.

2. Process Change Request (PCR)

■ Changes will be needed or at least desired for a variety of reasons, such as new or improved functions, errors not detected earlier, alterations to accept new equipment in the system, etc.

■ The changes may involve, but are not restricted to:
  a) Equipment (e.g., fluid bed instead of lytzen drier, etc.)
  b) Procedure and process (e.g., addition, deletion, or revision of an existing procedure, changing any parameters, etc.)
  c) Material (e.g., excess quantity to be added, etc.)

- The departmental manager will initiate a **process change request** form prior to any change in the approved and authorized procedure.
- The form will be forwarded to the quality assurance manager to review and approve.
- The approved document will be sent back to the initiating department.
- A copy of the approved form will be sent to the initiator, with a copy to the QA manager.
- The change may be approved on a case–by–case basis, or may be suggested to be incorporated in the current document as a permanent change.
- In case of dispute, a management review committee or material review board meeting will be called to resolve the problem.

3. Engineering Change Control

- To make any changes in existing equipment configuration, parts, or software or in utilities (HVAC), facility systems, etc., the engineering change control form will be raised.
- The purpose is to monitor and ensure that a validated system remains validated by recognizing and addressing the potential impact of the change of the existing system.
- Prior to the change, the concerned engineer will raise the engineering change control form signed by the departmental engineer, clearly mentioning the proposed change (and, if applicable, attaching the drawing of current and proposed systems).
- The form will be forwarded to the QA manager and for approval and a decision regarding the requirement for validation.
- The approved form will be sent back to the initiating department.
- Copies will be marked to validation and department file.

4. Change Request Control in Technical Documents

- The departmental manager (initiator) will initiate the change request form prior to implementing any change in the approved and authorized technical document.

SOP No. Val. 300.50   **Effective date: mm/dd/yyyy**

                                                       Approved by:

- The change may involve:
  - a) Standard test methods (STMs)
  - b) Standard control procedure (SCPs)
  - c) Raw material purchase specifications (RMPSs)
  - d) In-process packaging specifications (IPSs)
  - e) In-process manufacturing specifications (IMSs)
  - f) Finished product specifications (FPSs)
  - g) Packaging material specifications (PMSs)
  - h) Master packaging instructions (MPIs)
  - i) Manufacturing formula and method (MFM)
- The form will be sent to the QA manager to approve the change in specification.
- A copy of the approved form will be sent to the initiator, and a copy to all other concerned departmental managers, if required, by the systems manager.
- The changes may be approved on a case–by–case basis or they may be suggested to be incorporated in the current document as a permanent change.

5. Packaging Materials (New and Existing) Design Change Control

- The packaging materials involved are labels, leaflets, and boxes. The procedure describes how to manage change control in new and existing (in-use) specifications and to place orders for purchase.
  - New packaging materials specifications: products for which the labels, leaflets, and box specifications are designed, developed, and approved for the first time.
  - Existing packaging materials specifications: products for which packaging material specifications (leaflets, labels, and boxes) are already in use.
- The changes in the test and color or design of leaflets, labels, and boxes of (in-use) packaging materials specifications are followed as stated in the change control approval matrix of the company.
- The changes in the existing packaging materials may be initiated by the following departments marketing (promotional require-ments), registration (regulatory or registration requirement), or the drug information department (to update any information and imple-ment changes required by health authorities).

SOP No. Val. 300.50

Effective date: mm/dd/yyyy

Approved by:

REASONS FOR REVISIONS

Effective date: mm/dd/yyyy

■ First time issued for your company, affiliates, and contract manufacturers

SECTION

VAL 400.00

| YOUR COMPANY |
| VALIDATION STANDARD OPERATING PROCEDURE |

SOP No. Val. 400.10   Effective date: mm/dd/yyyy

                                       Approved by:

TITLE: Calibration of Instruments

AUTHOR: _____

                    Name/Title/Department

            _____

                    Signature/Date

CHECKED BY: _____

                    Name/Title/Department

            _____

                    Signature/Date

APPROVED BY: _____

                    Name/Title/Department

            _____

                    Signature/Date

REVISIONS:

No.	Section	Pages	Initials/Date

SOP No. Val. 400.10

Effective date: mm/dd/yyyy

Approved by:

SUBJECT: Calibration of Instruments

PURPOSE

This procedure describes conditions pertaining to an adequate and organized calibration system of measuring devices.

RESPONSIBILITY

The calibration manager is responsible for establishing an adequate and organized calibration program. Supervision of the whole program should be done by the quality assurance manager.

PROCEDURE

- Upon receipt, all equipment should be reviewed by the calibration lab manager and the user department to establish if the equipment has to be kept on calibration program.
- Inclusion and exclusion lists should be maintained with reasons, location, and final approval by the QA manager.
- Classification of equipment as critical or noncritical with regard to calibration necessities should be maintained.
- The equipment variables affecting the product quality shall be identified as critical for calibration.
- Determination of accuracy requirements of the instrument should be in consent with the quality assurance and calibration departments.
- Calibration of the instrument should be on time prior to expire of calibration date by informing the calibration department.
- New, changed, or repaired instrument should be prior to use.

1. Tasks of Calibration Department

- Decide which instruments can be calibrated internally and which have to be calibrated by subcontractors.

SOP No. Val. 400.10   Effective date: mm/dd/yyyy

                                       Approved by:

■ Purchase and control measuring equipment and reference material necessary for internal calibration certification of calibrated instruments. Label them with the date when last calibrated and also date when the next calibration is due. Write a calibration report and, if necessary, an incident report or a calibration variance report.

2. Reference Standards

The reference standards used for calibration of instruments should be checked for accuracy annually by an authorized measuring institution (e.g., Office of Weights and Measures or Bureau of Standards). The reference standard employed should have an uncertainty of measurement which is 1/10 to 1/5 the uncertainty required in the measurement equipment.

3. Cumulative Effect of Errors

The uncertainties of all equipment used in the calibration procedure and the method of combining them should be shown on the calibration certificate.

4. Training of Personnel Performing the Calibration Work

Only trained personnel should be employed for performing the calibration work. The above requirements are applicable for subcontractors performing calibration work on behalf of your company. Outside calibration laboratories should be inspected.

5. Intervals of Calibration

All critical items should be calibrated every 6 months. Noncritical items should be calibrated every 12 months. If necessary, recalibration frequency should be reviewed and revised.

6. Periodic Review of the Calibration Program

Periodic review of the calibration system should happen once a year using an established inspection checklist.

7. FDA Regulations

Some specific GMP calibration regulations that address the pharmaceutical and device industries are:

CFR 21 Part 58 — Good Laboratory Practice for Non-clinical Laboratory Studies

Section 58.63 — Maintenance and Calibration of Equipment

(a) Equipment shall be adequately inspected, cleaned, and maintained. Equipment used for the generation, measurement, or assessment of data shall be adequately tested, calibrated, and/or standard.

(b) The written standard operating procedures required under Section 58.81(b) (11) shall set forth in sufficient details the methods, materials, and schedules to be used in the routine inspection, cleaning, maintenance, testing, calibration and/or standardization of equipment, <u>and shall specify remedial action to be taken in the event of failure or malfunction of equipment.</u> The written standard operating procedures shall designate the person responsible for the performance of each operation, and copies of the standard operating procedures shall be made available to laboratory personnel.

(c) Written records shall be maintained of all inspection, maintenance, testing, calibrating and/or standardizing operations. These records, containing the date of the operation, shall describe whether the maintenance operations were routine and followed the written standard operating procedures. Written records shall be kept of non routine repairs performed on equipment as a result of failure and malfunction. Such records shall document the nature of the defect, how and when the defect was discovered and any remedial action taken in response to the defect.

* * *

Current Good Manufacturing Practices for Finished Pharmaceuticals

Section 211.25 — Personnel Qualifications

(a) Each person engaged in the manufacture, processing, packing, or holding of drug product shall have the education, training, and experience, or any combination thereof, to enable that person to perform the assigned functions. Training shall be in the particular

SOP No. Val. 400.10   **Effective date: mm/dd/yyyy**
                                           Approved by:

operations that the employee performs and in Current Good Manufacturing Practice (including the Current Good Manufacturing
Practice regulations in this chapter and written procedures required
by these regulations) as they relate to the employee's functions.
Training in Current Good Manufacturing Practice shall be conducted by qualified individuals on a continuing basis and with
sufficient frequency to assure that employees remain familiar with
cGMP requirements applicable to them.

Section 211.68 — Automatic, Mechanical, and Electronic Equipment

(a) Automatic, mechanical, or electronic equipment of other types of
    equipment, including computers, or related systems that will per-
    form a function satisfactorily, may be used in the manufacture,
    processing, packing, or holding of a drug product. If such equip-
    ment is so used, it shall be routinely calibrated, inspected, or
    checked according to a written program designed to assure proper
    performance. Written records of those calibration checks and
    inspections shall be maintained.
(b) Appropriate controls shall be exercised over computer or related
    systems to assure that changes in master production and control
    records or other records are instituted only by authorized person-
    nel. Input to and output from the computer or related system of
    formulas or other records or data shall be checked for accuracy.
    A backup file of data entered into the computer or related system
    shall be maintained except where certain data, such as calculations
    performed in connection with laboratory analysis, are eliminated
    by computerization, or other automated processes. In such
    instances, a written record of the program shall be maintained
    along with appropriate validation data. Hard copy or alternative
    systems, such as duplicates, tapes, or microfilm, designed to assure
    that backup data are exact and complete and that it is secure from
    alteration, inadvertent erasures, or loss shall be maintained.

Section 211.160 — General Requirements

(a) The establishment of any specification, standards, sampling plans,
    test procedures, or other laboratory control mechanisms required by
    this subpart, including any change in such specifications, standards,

sampling plans, test procedures, or other laboratory control mech-anisms, shall be drafted by the appropriate organizational unit and reviewed and approved by the quality control unit. The require-ments in this subpart shall be followed and shall be documented at the time of performance. Any deviation from the written spec-ifications, standards, sampling plans, test procedures, or other laboratory control mechanisms shall be recorded and justified.

(b) Laboratory controls shall include the establishment of scientifically sound and appropriate specifications, standards, sampling plans, and test procedures designed to assure that components, drug product containers, closures, in-process materials, labeling and drug products conform to appropriate standards of identity, strength, quality and purity. Laboratory controls shall include:

(4) The calibration of instruments, apparatus, gauges and recording devices at suitable intervals in accordance with an established written program containing specific directions, schedules, limits for accuracy and precision, and provisions for remedial action in the event accuracy and/or precision limits are not met. Instruments, apparatus, gauges, and recording devices not meeting established specifications shall not be used.

Section 820.61 — Measurement Equipment

All production and quality assurance measurement equipment, such as mechanical, automated, or electronic equipment, shall be suitable for its intended purposes and shall be capable of producing valid results. Such equipment shall be routinely calibrated, inspected, and checked according to written procedures. Records documenting these activities shall be main-tained. When computers are used as part of an automated production or quality assurance system, the computer software programs shall be vali-dated by adequate and documented testing. All program changes shall be made by a designated individual(s) through a formal approval procedure.

(a) Calibration

Calibration procedures shall include specific directions and limits for accuracy and precision. There shall be provisions for remedial action when accuracy and precision limits are not met. Calibration shall be performed by personnel having the necessary education, training, background, and experience.

(b) Calibration Standards
     Where practical, the calibration standards used for production and
     quality assurance measurement equipment shall be traceable to
     the national standards of the National Bureau of Standards, Depart-
     ment of Commerce. If national standards are not practical for the
     parameter being measured, an independent reproducible standard
     shall be used. If no applicable standard exists, an in-house standard
     shall be developed and used.
(c) Calibration Records
     The calibration date, the calibrator, and the next calibration date
     shall be recorded and displayed, or records containing such
     information shall be readily available for each piece of equipment
     requiring calibration. A designated individual(s) shall maintain a
     record of calibration dates and of the individual performing each
     calibration.

[The last GMP included in this listing is the original document that
identified calibration program requirements but was never formally issued.]

CFR 21 Part 212 — Good Manufacturing Practices for Drugs

Section 212.68 — Equipment Calibration

(a) Procedures shall be written and followed designating schedules
     and assigning responsibility for testing or monitoring the perfor-
     mance or accuracy of automatic or continuously operating equip-
     ment, devices, apparatus, or mechanisms such as, but not limited
     to, the following:
     (1) Temperature-recording devices on sterilizing equipment
     (2) Temperature-recording devices on sterilizers
     (3) Pressure gauges
     (4) Mechanisms for maintaining sterilizing medium uniformity
     (5) Chain speed recorder
     (6) Heat exchanger pressure differential monitor
     (7) Mercury–in–glass thermometer
(b) Written records of such calibrations, checks, examinations, or
     inspections shall be maintained, as specified in Section 212.183.

Section 212.183 — Equipment Calibration and Monitoring Logs

Written records of calibration and monitoring tests and readings performed shall be maintained for at least 2 years after the expiration date of each batch of drug product produced by the equipment.

(a) Calibration records shall include
    (1) A description of the equipment
    (2) The date the equipment was purchased
    (3) The operating limits of the equipment
    (4) The date, time, and type of each test
    (5) The results of each test
    (6) The signature of each person performing a test
(b) Monitoring records shall include:
    (1) A description of the equipment
    (2) The date the equipment was installed
    (3) The date the equipment was last calibrated, if appropriate
    (4) The operating limits of the equipment
    (5) The date and time of the recording
    (6) The reading
    (7) The signature of each person performing the monitoring
(c) Corrective measures employed to bring the equipment into compliance with its operating specifications shall be:
    (1) Recorded in the appropriate equipment log
    (2) Noted in the calibration and/or monitoring record
    (3) Immediately followed by testing to assure that the corrective measures were adequate to restore the required operating characteristics.

Section 212.192 — Production Record Review

The review and approval of production and control records by the quality control unit shall extend to those records not directly related to the manufacture, processing, packing, or holding of a specific batch of large volume parenteral drug product but which have a bearing on the quality of batches being produced. Such indirectly related records shall include:

(a) Those dealing with equipment calibration or standardization.

SOP No. Val. 400.10

Effective date: mm/dd/yyyy

Approved by:

REASONS FOR REVISION

Effective date: mm/dd/yyyy

- First time issued for your company, affiliates, and contract manufacturers

| YOUR COMPANY |
| VALIDATION STANDARD OPERATING PROCEDURE |

SOP No. Val. 400.20   Effective date: mm/dd/yyyy
                                                       Approved by:

TITLE: Periodic Review of the Calibration Program

AUTHOR: _____
                              Name/Title/Department

                   _____
                              Signature/Date

CHECKED BY: _____
                              Name/Title/Department

                   _____
                              Signature/Date

APPROVED BY: _____
                              Name/Title/Department

                   _____
                              Signature/Date

REVISIONS:

No.	Section	Pages	Initials/Date

SOP No. Val. 400.20

Effective date: mm/dd/yyyy

Approved by:

SUBJECT: Periodic Review of the Calibration Program

PURPOSE

To ensure that the approved procedures for calibration of all reference standards are effective and meet the cGMP requirements

RESPONSIBILITY

It is the responsibility of the calibration manager to follow the procedure. The quality assurance manager is responsible for SOP compliance.

PROCEDURE

1. Calibration System

- Are there approved procedures for the calibration of all reference standards and measuring equipment?
- Are the management responsibilities defined in SOPs promptly detected?
- Are there deficiencies within the system to prevent subsequent inaccuracies?
- Is there a procedure to ensure corrective action and preventive measures?
- Are limits of calibration uncertainty properly defined?

2. Periodic Review of the Calibration System

- Is there an SOP to review the measuring and calibration system?
- Are the reviews conducted periodically?
- Are the records of the reviews maintained and do they provide objective evidence of the effectiveness of the system?
- Is management informed about the results of the review and is corrective action taken?

SOP No. Val. 400.20   **Effective date: mm/dd/yyyy**

                                                 Approved by:

3. Planning

- Is there a system to establish and plan needs of calibration and measurement before starting new projects?
- Are the necessary reference standards and measuring equipment determined?
- Have the skills and training required by the calibration and measuring personnel been established?
- Are controlled environments provided where necessary (temperature, humidity, vibration, etc.)?

4. Calibration Limits

- Does the calibration system identify the source and magnitude of uncertainties associated with calibration and product characteristics?

5. Documented Calibration Procedures

- Are the approved procedures for controlling the calibration of reference standards and measuring equipment available and used for product verification?
- Where no in-house procedures are available, are there appropriate and identified published standard practices or manufacturer's written instructions available?
- Is there a system in place to conduct the inspection of the existing system to ensure adherence to existing procedures?
- Are the outside calibration labs audited and certified?

6. Records

- Do the records include details of calibration controls, environmental data, designated error limits, and information necessary to establish traceability?
- Does the system include the retention of calibration certificates or data used in support of all calibration of measuring equipment?
- Does the record system allow for calling forward, at the appropriate interval, equipment requiring calibration?

SOP No. Val. 400.20   Effective date: mm/dd/yyyy

                                       Approved by:

- Do the records indicate that the equipment is capable of performing measurements within the designated limits?
- Are all the records required maintenance?

7. Calibration Labeling

Is there a procedure of labeling that identifies the calibration status of reference standards and measuring equipment?

8. Sealing for Integrity

Where necessary, is a sealing provided to prevent access to the equipment?

9. Intervals of Calibration

- Have calibration intervals been established for all reference standards and measuring equipment based on the equipment manufacturer's recommendations or knowledge of equipment stability, purpose, and level of usage?
- Are trend data reviewed from previous calibration records to adjust the calibration intervals?

10. Invalidation of Calibration

- Does the SOP ensure the immediate removal from use, or conspicuous identification, of any reference standard or measuring equipment that has not been calibrated in accordance with the established time schedule, has failed in operation in any measurement parameter, or shows evidence of physical damage?
- Does the SOP provide for immediate notification of equipment failures or damage likely to have compromised product quality?

11. Subcontractors

- Do procedures ensure that a subcontractor employs a measurement and calibration system that complies with the company's requirements?

SOP No. Val. 400.20

Effective date: mm/dd/yyyy
Approved by:

- Is responsibility accepted for ensuring that the procedures employed by a subcontractor for calibration and measurement work are suitable and properly documented?
- Is there a procedure for periodically evaluating subcontractors for adherence to company requirements for traceability, suitability of method, and documentation practices?

12. Traceability

Can all calibrations performed in-house or by subcontractors be traced to a national or international reference standard?

13. Environmental Control

Do the procedures for calibration and measurement of the equipment indicate the environment required to ensure accuracy and precision?

14. Training

Do personnel performing calibration functions have appropriate experience or training applicable to the type of calibration work undertaken?

REASONS FOR REVISIONS

Effective date: mm/dd/yyyy

- First time issued for your company, affiliates, and contract manufacturers

---

YOUR COMPANY
VALIDATION STANDARD OPERATING PROCEDURE

SOP No. Val. 400.30

Effective date: mm/dd/yyyy

Approved by:

TITLE: **Calibration and Validation Equipment**

AUTHOR: _____

Name/Title/Department

_____

Signature/Date

CHECKED BY: _____

Name/Title/Department

_____

Signature/Date

APPROVED BY: _____

Name/Title/Department

_____

Signature/Date

REVISIONS:

No.	Section	Pages	Initials/Date

SOP No. Val. 400.30   Effective date: mm/dd/yyyy
                                                   Approved by:

SUBJECT: Calibration and Validation Equipment

PURPOSE

To provide the basic instrument list required for validation. This list should be regarded as a collection of examples; several alternatives are possible.

RESPONSIBILITY

It is the responsibility of the calibration manager to follow the procedure. The quality assurance manager is responsible for SOP compliance.

PROCEDURE

1. Instrument Selection

After the selection of instruments required to perform calibration per SOP No. Val. 400.10, the calibration manager will procure the instruments.

2. Equipment Procurement

The basic instruments required for validation are listed in the following table as a guideline; several alternatives are possible.

Sl. No.	Equipment Name	Model	Manufacturer
1.	Kaye Validator	X1310CE	Kaye Instruments
2.	Aerosol Particle Counter	CI – 500	Climet Instruments
3.	Liquid Particle Sensor	RLLD 1-100H	Climet Instruments
4.	Micro Monometer	AXD – 530	Alnor
5.	Aerosol Photometer	TDA – 2G	Air Techniques (ATI)
6.	Temp. & Humidity Recorder	THDx	Dickson
7.	Hand Held Digital Thermometer	51	Fluke
8.	Bead Probe	80 PK – 1	Fluke
9.	Surface Probe	80 PK – 3A	Fluke
10.	Immersion Probe	80 PK – 2A	Fluke
11.	Exposed Junction Probe	80 PK – 6A	Fluke
12.	Air Probe	80 PK – 4A	Fluke
13.	Tachometer	93-412 E	Connemara Electronics

SOP No. Val. 400.30

Effective date: mm/dd/yyyy

Approved by:

Sl. No.	Equipment Name	Model	Manufacturer
14.	Digital Lux Meter	93-1065 L	Connemara Electronics
15.	Temp. & Humidity Meter	HMI/HM 31	Vaisala Helsinki
16.	Air Velocity Meter	TESTO – 440	Testoterm Inc.
17.	Air Contamination	—	
18.	Measurement Device	RCS 94001	Reuter
19.	Conductivity Meter	LF 530	WTW
20.	pH Meter	764	Knick
21.	Analogue Digital Converter	Helios II	Fluke
22.	Digital Multimeter	M 2005	AVO
23.	Oscilliscope	V – 212	Hitachi
24.	Digital Stopwatch	331 – 382	RS
25.	Digital Manometer	P 200 AH	Digitron
26.	Digital Micromanometer	MP –6 KSR	Air Neotronics
27.	Dew Point Meter	SADP-5	Shaw
28.	Thermocouple	KM 2067	Kane May
29.	Temperature Calibrator	TRU	Technosyn
30.	Pressure Calibrator	—	Beamex, Digitron
31.	Current Source	1021	Time Electronics
32.	Millivolt Pot Source	Momocal	Ero Electronics
33.	Resistance Box	1051	Time Electronics
34.	Anemometer	8350	TSI

REASONS FOR REVISION

Effective date: mm/dd/yyyy

- First time issued for your company, affiliates, and contract manu-facturers

SECTION

VAL 500.00

YOUR COMPANY
VALIDATION STANDARD OPERATING PROCEDURE

SOP No. Val. 500.10   **Effective date: mm/dd/yyyy**

                                           Approved by:

TITLE: **Training on the Job**

AUTHOR: _____

                              Name/Title/Department

             _____

                                  Signature/Date

CHECKED BY: _____

                              Name/Title/Department

             _____

                                  Signature/Date

APPROVED BY: _____

                              Name/Title/Department

             _____

                                  Signature/Date

REVISIONS:

No.	Section	Pages	Initials/Date

SUBJECT: Training on the Job

PURPOSE

To describe the key elements to be considered part of the training to meet the cGMP requirements

RESPONSIBILITY

The departmental managers are responsible for developing and maintaining the training program. The quality assurance manager is responsible for SOP compliance.

PROCEDURE

Besides the normal cGMP training, personnel should be trained on a regular basis with a particular reference to their daily work. The on-job-training program should ensure the following elements:

- Entry-level training for those who never worked in a pharmaceutical company
- List of all equipment
- Description of relevant equipment
- List of applicable standard operating procedures
- List of applicable drawings of the equipment
- Operational procedures
- Materials flow procedures
- Personnel flow
- Operational sequence of the production process in the equipment
- Identification of critical parameters
- Cleaning procedures
- Operational SOPs read by all operations at least once a year

1. Training Documentation

Training records shall be maintained for all personnel. A list of personnel authorized to work in each department, to operate each piece of equipment, and to carry out each process should be maintained.

Effective date: mm/dd/yyyy
Approved by:

2. Training of Personnel

Training should take into consideration, as appropriate, control laboratories (including technical, maintenance, calibration, and cleaning personnel), and other personnel whose activities could affect the quality of the product.

The effectiveness of the training should be assessed periodically. Training records should be maintained and reviewed.

REASONS FOR REVISIONS

Effective date: mm/dd/yyyy

■ First time issued for your company, affiliates, and contract manufacturers

---

| YOUR COMPANY |
| VALIDATION STANDARD OPERATING PROCEDURE |

SOP No. Val. 500.20

Effective date: mm/dd/yyyy

Approved by:

TITLE: **Good Manufacturing Practices**

AUTHOR: _____

Name/Title/Department

_____

Signature/Date

CHECKED BY: _____

Name/Title/Department

_____

Signature/Date

APPROVED BY: _____

Name/Title/Department

_____

Signature/Date

REVISIONS:

No.	Section	Pages	Initials/Date

SOP No. Val. 500.20   Effective date: mm/dd/yyyy

                                            Approved by:

SUBJECT: Good Manufacturing Practices

PURPOSE

To describe and implement the elements of current good manufacturing practice to produce high quality products with optimum potency, efficacy, and safety

RESPONSIBILITY

It is the responsibility of departmental managers to follow the procedure. The quality assurance manager is responsible for SOP compliance.

PROCEDURE

1. Introduction

Training new and old employees on cGMP guidelines is an essential requirement of the pharmaceutical industry in order to maintain and produce quality pharmaceutical products on a continuous basis, and to assure patients that the products used meet safety, efficacy, and potency requirements.

Following is the generic audit checklist of which the personnel working in the pharmaceutical industry must be aware:

2. General

Check the following:

1. Workers coming into direct contact with the product who have been medically examined and cleared are allowed in the area.
2. Workers in the area are wearing clean uniforms and caps.
3. All areas are cleaned per respective SOPs.
4. No eating, smoking, or unhygienic practices are allowed in the manufacturing area.
5. Facility is neat and clean, and lockers provided in the facility are clean. SOP of cleaning is followed.
6. All manufacturing operations are carried out in separate areas intended for such purposes.

7. All apparatus and equipment to be used in the operation have been cleaned per their respective SOPs.
8. The contents of all vessels and containers used in manufacturing and storage between manufacturing stages are identified by clearly legible labels having all the required information.
9. All machines and equipment bear legible clearance labels having all the required information.
10. All respective SOPs are followed.
11. Insecticutors on every entrance are available as appropriate.
12. Under-construction or maintenance areas should be well segregated from the manufacturing area.
13. All the specific areas have required temperature, relative humidity (Rh), and air pressures.
14. Clogs are in use instead of street shoes.
15. Gauges and measuring equipment have valid calibration tags.

3. Stores and Weighing Area

1. Weighing and measuring equipment are of appropriate accuracy and are calibrated.
2. All incoming supplies of raw material and packaging commodities are stored per SOP.
3. Stores' receiving labels are fixed on all containers and sample labels are affixed as appropriate.
4. Release labels are fixed on approved items and rejected labels on rejected items.
5. All rejected items are kept segregated from other materials.
6. Cleaning in the area is done according to schedule and is effective.
7. Workers coming into direct contact with product are wearing clean masks.
8. Curtains of the weighing booth are drawn during dispensing to avoid cross-contamination.
9. Dispensed materials are kept in locked, clean cages with proper labels.
10. Log books of the weighing area and calibration records of weighing balances are updated.
11. Raw material card entries are on time, complete, and correct.

SOP No. Val. 500.20   **Effective date: mm/dd/yyyy**

                                                    Approved by:

12. Temperature and Rh are within requirements and limits, and the records are maintained.
13. Storage of raw materials, packaging materials, bulk products and finished product is appropriate.

4. Packaging Section

1. All packed units after completion of batch are kept in finished goods area in stores.
2. All coded and uncoded labels and boxes are kept in cages with identification labels.
3. Workers coming into direct contact with products are wearing masks.
4. Operators wear disposable gloves while handling the product.
5. There is proper physical separation or segregation between the lines.
6. Temperature and humidity records of the dry and soft packaging area are updated and maintained.

5. Soft and Dry Production

1. All employees are attired appropriately according to the specific garmenting SOP for soft and dry production.
2. Operators are wearing masks and gloves and using beard covers (head cover for covering beard) when necessary.
3. Relevant cleaning procedures for area and equipment should be followed.
4. Cleaning schedules should be followed.
5. Work areas are clearly labeled.
6. All vessels and utensils are labeled as to their cleanliness status.
7. Manufacturing instructions are at hand during processes and are approved and accurately followed.
8. Valid calibration labels are on all equipment.
9. Vessels should be labeled with product lot number and stage of processing.
10. Sampling tools are dedicated for each building and should be used accordingly.
11. Maintenance tools should also be dedicated for each building.

12. Air inlets and outlets are functional.
13. Materials holding containers are intact and can be closed completely.
14. Materials and personnel flow are dedicated and being followed.
15. Temperature and humidity are under the limits.
16. Dispensing and all processing are carried out in the prescribed and dedicated areas.
17. Line clearance procedure should be strictly followed and counter-checked to avoid mix-ups.
18. Devices for the presence of labels and overprinting are challenged prior to initiation of work.
19. Effective segregation of printed materials from not printed and rejected materials is in place.
20. Utilities lines are clearly marked and labeled.
21. Storage of product (bulk and finished) is under their labeled or prescribed storage condition.
22. All records must be completed at the time of action.
23. Employees have undergone training in GMPs, SOPs, and sterile area techniques and respective SOPs are followed
24. Employees have full knowledge about their job functions.
25. The department is well maintained for cleanliness and spacious enough for equipment and operations.
26. Area and equipment are clean at the end of the day's work.
27. Specific procedure for the cleaning of major equipment is followed.
28. Daily calibration records of balances are complete.
29. Correction to writing errors are made by crossing out, initialing, dating, and writing the reason (if necessary).
30. All the products have status labels.
31. All the log books (equipment and area usage records) are updated.

6. Good Documentation Skill

1. The correct ink color and type of proper writing instrument are used (pencils should not be used).
2. Additional documents (e.g., charts, printout) are properly included.
3. Printing and writing can be easily read.
4. Errors are properly corrected with one line through the original entry, initialed, dated, and explained (if necessary).
5. Calculations are reviewed and verified.

6. Numbers are properly rounded off; the correct significant figures are used.
7. Spelling of the product names and other words is correct.
8. Lot numbers and product ID codes are doublechecked and are correct.
9. Information is recorded as it is obtained.
10. Proper formats for date and time are used.
11. All abbreviations are approved and standardized.
12. Blank spaces are properly handled.
13. The original entry is made into the official record.

In short, good documentation skill should produce proper records with the following characteristics:

- Permanent
- Accurate
- Prompt
- Clear
- Consistent
- Complete
- Direct
- Truthful

7. Parenteral Facility

1. In clean room operations, employees should be correctly attired according to relevant SOP.
2. A fresh set of garments should be used on each entry into the clean room.
3. Sets of garments ready for use should be sealed and labeled.
4. Sterile production area should be in a good state of repair and neat.
5. Relevant cleaning procedures for area and equipment should be followed.
6. Cleaning schedules should be followed.
7. Cleaning and sanitization agents per the SOP should be labeled with the expiration date.
8. Records should be maintained for the preparation of cleaning and sanitization agents.

9. No simultaneous opening of doors on the clean and lesser clean class.
10. All alarm systems are functional.
11. Work areas are clearly labeled.
12. Product and product components are exposed only where protected by LAF stream providing an air quality of 100 or better.
13. Handling and all working practices should be such to avoid contamination and generation of particles.
14. All vessels and utensils are labeled as to their cleanliness status.
15. Manufacturing instructions are at hand during processes and are approved and accurately followed.
16. Vessels should be labeled with the product lot number and stage of processing.
17. Relevant sterilization and other charts and printouts are fully labeled, verified, and approved.
18. Valid calibration labels are on all equipment.
19. Air inlets and outlets are functional.
20. Material-holding containers are intact and can be closed completely.
21. Materials and personnel flow is dedicated and being followed.
22. Air differentials are maintained between the different areas of processing per their levels of cleanliness.
23. Temperature and humidity are under the limits.
24. Dispensing and all processing are carried out in the prescribed and dedicated areas.
25. Line clearance procedures should be strictly followed and counterchecked to avoid mix-ups.
26. Devices for the presence of labels and overprinting should be challenged prior to initiation of work.
27. Effective segregation of printed materials from not printed and rejected.
28. Utilities lines are clearly marked and labeled.
29. Storage of product (bulk and finished) is under its labeled or prescribed storage condition.
30. All records must be completed at the time of action.
31. Watches, bracelets, jewelry, or rings should not be worn during work.
32. Persons ill or having opened or bandaged wounds should inform the supervisor and should not be allowed in the clean room areas.
33. Monitor that no adhesive tapes are used in the sterile area in future.

8. Dos and Don'ts to Be Followed

- Do wear identification card all the time at work.
- Do follow clothing instructions before entering the plant.
- Do wash hands before entering and leaving the personal facility.
- Do read, understand, and follow standard operating procedures related to work.
- Do follow good manufacturing practices at work.
- Do observe personal hygiene at work.
- Do keep nails clean, and cut them every week.
- Do get a haircut every month.
- Do trim beard every day before coming to work.
- Do not take food and drinks inside the plant, and especially do not chew gum.
- Do not enter the plant in personal clothing, street shoes, or with unwashed hands.
- Do not go to the bathroom or outside the plant in working uniform, head cover, and clogs.
- Do not enter the plant if suffering with some contagious disease or a skin lesion.
- Do not carry anti-dust face mask, disposable hand gloves, or other working tools to lockers.
- Do not keep medicines inside the locker, except as prescribed for personal use by a doctor.
- Do not carry batch record and support documents in pockets or to the lockers.
- Do not remove medicines from the line for personal use.
- Do not taste and touch raw materials for personal experience during dispensing and processing.
- Do not change labels from the container, unless authorized.
- Do not remove items from the pallets, unless authorized.
- Do not cough inside the plant without a handkerchief.
- Do not use cosmetics and jewelry at work.
- Do not keep used sanitary towels inside the lockers.
- Do not sign batch records unless authorized.
- Do not use sampling and other tools dedicated for building A in building B, or vice versa.

SOP No. Val. 500.20

Effective date: mm/dd/yyyy

Approved by:

REASONS FOR REVISION

Effective date: mm/dd/yyyy

- ■ First time issued for your company, affiliates, and contract manufacturers

SECTION

VAL 600.00

YOUR COMPANY
VALIDATION STANDARD OPERATING PROCEDURE

SOP No. Val. 600.10

Effective date: mm/dd/yyyy

Approved by:

TITLE: Guidelines for Area Classification and Air Handling

AUTHOR: _____

Name/Title/Department

_____

Signature/Date

CHECKED BY: _____

Name/Title/Department

_____

Signature/Date

APPROVED BY: _____

Name/Title/Department

_____

Signature/Date

REVISIONS:

No.	Section	Pages	Initials/Date

SOP No. Val. 600.10

Effective date: mm/dd/yyyy

Approved by:

SUBJECT: Guidelines for Area Classification and Air Handling

PURPOSE

To provide the guideline for classification of areas to prevent mix–ups and cross-contaminatio

RESPONSIBILITY

It is the responsibility of all departmental managers to follow the procedure. The quality assurance manager is responsible for SOP compliance.

PROCEDURE

The different areas within a pharmaceutical plant must be identified. According to the classification, the following requirements should be fulfilled:

1. Nonclassified Areas

1.1 Black area

Nonmanufacturing areas, such as warehouse, administration, and workshop facilities, comprise the black area. Control of the warehouse storage conditions (temperature and relative humidity) is recommended, where necessary.

1.2 Dark gray area

These are areas where primary packaged products are handled (e.g. visual inspection of ampules, vials, blisters, and final packaging operations). Depending on the product mix manufactured, the control of temperature and relative humidity may be necessary.

SOP No. Val. 600.10   Effective date: mm/dd/yyyy
                                            Approved by:

2. Classified Areas

2.1 Gray area

These are areas where personnel could come into direct contact with open materials (starting materials, intermediate product, bulk product, and open primary packaging materials), for example, all operations (sampling, compounding, and producing) for manufacturing of solids, semisolids, and liquids for oral and topical use.

2.2 Solids

Classification:	no classification
Air supply:	filter efficiency ≥95%
Air exhaust:	filter efficiency ≥95%
Air pressure:	Generally negative pressure with respect to adjacent areas
	Material and personnel airlocks should have over-pressure with respect to adjacent areas (e.g., corridors)
	Continuous monitoring of pressure differences (about 10 to 15 pa)
Air change:	The following number of air changes should be met in working areas
	Low material potency: at least 10/h
	High material potency: between 10 and 20/h
	Recirculation of air is only allowed when working with low potency materials; when working with high potency materials, recirculation is not recommended. In case of built-in HEPAs, recirculation is possible while working with low potency materials
Spot exhaustion:	Local extraction at dust generating points by a separate system leading into a cyclone is recommended
Temperature:	To be monitored
Relative humidity:	To be monitored

2.3 Clean area

These are areas where materials of low viable count are handled. The area should be constructed and used in such a way as to reduce the

Effective date: mm/dd/yyyy
                                    Approved by:

introduction, generation, and retention of contaminants. The operations can include a sterilizing process.

2.3.1 Semisolids and liquids for oral and topical use

Classification:	Class 100,000 (unmanned)
Air supply:	Filter efficiency ≥95% for particles ≥0.5 µm
Air pressure:	Positive pressure with respect to less clean areas
	Continuous monitoring of pressure differential with respect to less clean areas
Air changes:	At least 20/h
Temperature:	To be monitored
Relative humidity:	To be monitored

2.3.2 Preparation of parenterals

Classification:	Minimally class 10,000 area (unmanned)
	Minimally class 100, 000 area (manned)
Air supply:	HEPA filters ≥99.97% efficiency for particles ≥0.5 µm
	Partly recirculation air is allowed
Air pressure:	Positive pressure with respect to less clean areas
	Continuous monitoring of pressure differential with respect to less clean areas (about 10 to 15 Ps)
Air changes:	At least 20/h
Temperature:	20 to 25°C recommended, to be monitored
Relative humidity:	40 to 60% recommended, to be monitored

3. Aseptic Area

These are areas within a clean area designed and constructed, serviced, and used with the intention to protect sterile products from microbiological and particulate contamination.

Classification:	Minimally class 100 (unmanned)
	Minimally class 10,000 (manned)
	Particle load to be monitored continuously
Air supply:	HEPA filters ≥99.995% efficiency for particles ≥0.5 µm
	Partial recirculation of air is possible

SOP No. Val. 600.10   **Effective date: mm/dd/yyyy**

                                               Approved by:

Air exhaust:	Low level air extraction
Air pressure:	Positive pressure with respect to less clean area
	Continuous monitoring of pressure differential with respect to less clean area (about 10 to 15 Pa)
Air changes:	More than 20/h
Temperature:	20 to 25°C recommended, to be monitored
Relative humidity:	40 to 60% recommended, to be monitored

4. Critical Area

These are areas where the product, prepared under aseptic conditions, as well as open sterile containers and closures, is exposed to the environment.

Classification:	Class 100 (laminar flow installations)
Air velocity:	Flow of 0.30 m/s (vertical) or 0.45 m/s (horizontal)
Air supply:	Filter efficiency 99.997% for particles ≥ 0.5 μm

5. General Recommendations

At the cross point between nonclassified and classified areas, as well as between the different classified areas, separate airlocks for materials and personnel should be installed. Additionally, airlocks to the clean or to the aseptic areas should be equipped with interlocking doors. Only authorized personnel should be allowed to enter the classified areas.

REASONS FOR REVISION

Effective date: mm/dd/yyyy

- First time issued for your company, affiliates, and contract manufacturers

YOUR COMPANY
VALIDATION STANDARD OPERATING PROCEDURE

SOP No. Val. 600.20

Effective date: mm/dd/yyyy

Approved by:

TITLE: **Guideline for Area Validation: Clean Area**

AUTHOR:

_____
Name/Title/Department

_____
Signature/Date

CHECKED BY:

_____
Name/Title/Department

_____
Signature/Date

APPROVED BY:

_____
Name/Title/Department

_____
Signature/Date

REVISIONS:

No.	Section	Pages	Initials/Date

SOP No. Val. 600.20

Effective date: mm/dd/yyyy

Approved by:

SUBJECT: Guideline for Area Validation: Clean Area

PURPOSE

To describe the procedure for the area validation of clean area

RESPONSIBILITY

Concerned departmental managers are responsible for following the procedure. The quality assurance manager is responsible for SOP compliance.

PROCEDURE

1. Measurements in Unmanned Condition

1.1 Integrity testing of HEPA filters

Procedure

According to SOP No. Val. 600.30

Requirements

According to SOP No. Val. 600.30

Frequency

Initial validation: once per HEPA filter unit
Revalidation: annually

1.2 Air pressure situation

Procedure

The overpressure situation should be measured with a pressure gauge over 24 h. Also refer to SOP No. Val. 600.30.

Effective date: mm/dd/yyyy

Approved by:

Requirements

With regard to less clean adjacent areas, the average overpressure should be at least 10 Pa with a maximum deviation of ±2Pa with regard to the average value.

Frequency

Initial validation: once for the clean area
Revalidation: annually

1.3 Particulate matter in air

Procedure

The particulate matter in air should be measured with a particle counter

Sampling procedure

The measurements should be performed at a height of 1.5 m above floor level. The minimum volume of air required per sample can be calculated using the following formula:

Volume = 20 particles/class limit (particles/volume)

Each sample of air tested at each location shall be of sufficient volume such that at least 20 particles would be detected.

The minimum number of sampling points required for measuring the particulate matter in a clean area can be read from Table 1.

Table 1

Square feet	Square meter	Sampling points
100	9.2	4
200	18.4	8
400	36.8	16
1,000	92.0	40
2,000	184.0	80
4,000	368.0	160
10,000	920.0	400

This means that not less than two locations should be sampled for a clean area. At each location at least 5 measurements of a volume calculated as described in 1.3 should be performed. Afterwards, the average of the obtained values for each sampling location should be calculated. The average value of the measurements from each sampling location should comply with the requirements.

Requirements

$\leq$10,000 particles $\geq$0.5 μm/cft or $\leq$353,000 particles $\geq$0.5 μm/m^3
$\leq$70 particles $\geq$5 μm/cft or $\leq$2,470 particles $\geq$5 μm/m^3

Frequency

Initial validation: once
Revalidation: annually

1.4 Temperature

Procedure

Temperature sensors should be placed on different locations within the clean area. The investigation should be performed over at least 24 hours.

Requirements

The temperature should be between 20 and 25°C.

Frequency

Initial validation: once
Revalidation: annually

1.5 Relative humidity

Procedure

The measurements should be performed with a hygrometer. The investigation should be done over at least 24 h.

SOP No. Val. 600.20

Effective date: mm/dd/yyyy
Approved by:

Requirements

Within the clean area a relative humidity of 40 to 60% should be maintained.

Frequency

Initial validation: once
Revalidation: annually

1.6 Microbiological quality of air

Procedure

To determine the airborn microbial contamination level, refer to SOP No. Val. 600.30.

Requirements

Class 100   <3 cfu/m^3 of air
Class 10,000   <20 cfu/m^3 of air
Class 100,000   <100 cfu/m^3 of air

Frequency

Initial validation: once
Revalidation: refer to SOP No. Val. 600.40

1.7 Air Changes

Procedure

Determine the number of air changes per hour in the clean area by calculation from the inlet air flow and the area volume.

Requirements

At least 20 air changes/h should be achieved.

Frequency

Initial validation: once
Revalidation: annually

SOP No. Val. 600.20

Effective date: mm/dd/yyyy

Approved by:

2. Measurements in Manned Condition

2.1 Particulate matter in air

Procedure

The particulate matter in air should be measured with a particle counter.

Sampling procedure

The measurements should be performed at a height of 1.5 m above floor level. The minimum volume of air required sample can be calculated from step 1.3. The minimum number of sampling points required for verifying the particulate matter in a clean area can be read from Table 2 above.

This means that, for a clean area, not less than two locations should be sampled. At each location at least 5 measurements should be performed. Afterward, the average of the obtained values for each sampling location should be calculated.

The average value of the measurements from each sampling location should comply with the requirements:

$\leq$100,000 $\geq$0.5 μm/cft or $\leq$3,530,000 $\geq$0.5 μm/m^3
$\leq$700 $\geq$5 μm/cft or $\leq$24,700 $\geq$5 μm/m^3

2.2 Restoration time

The clean area is classified as class 10,000 in unmanned condition and as class 100,000 in manned condition. The time lapse to comply again with class 10,000 requirements after finishing work and after personnel have left the clean area has to be determined.

Procedure

The particulate matter in air should be measured with a particle counter.

Requirements

The class 10,000 conditions should be reached within 20 min after personnel have left the area.

Effective date: mm/dd/yyyy
                                       Approved by:

Frequency

Initial validation: once
Revalidation: annually

2.3 Air pressure situation

For procedure requirements and frequency, see Section 1.3.

2.4 Temperature

For procedure, requirements, and frequency, see Section 1.4.

2.5 Relative humidity

For procedure, requirements, and frequency, see Section 1.5.

2.6 Surface and floor contamination

Procedure

The measurements should be performed in accordance with SOP No. Val. 600.30.

Frequency

Initial validation: once
Revalidation: annually
   In addition to these investigations with regard to validation requirements, a monitoring program, as described in SOP No. Val. 600.30. Guidelines for area validation aseptic areas should be implemented.

2.7 Microbiological quality of air

Procedure

Measurements should be performed in accordance with SOP No. Val. 600.30.

Requirements

According to SOP No. Val. 600.30

SOP No. Val. 600.20   **Effective date: mm/dd/yyyy**

                                            Approved by:

2.8 Air changes

For procedure, requirements, and frequency see Section 1.7.

REASONS FOR REVISION

Effective date: mm/dd/yyyy

- First time issued for your company, affiliates, and contract manufacturers

---

YOUR COMPANY
VALIDATION STANDARD OPERATING PROCEDURE

SOP No. Val. 600.30

Effective date: mm/dd/yyyy

Approved by:

TITLE: Aseptic Area Validation Procedures

AUTHOR: _____

Name/Title/Department

_____

Signature/Date

CHECKED BY: _____

Name/Title/Department

_____

Signature/Date

APPROVED BY: _____

Name/Title/Department

_____

Signature/Date

REVISIONS:

No.	Section	Pages	Initials/Date

SOP No. Val. 600.30   Effective date: mm/dd/yyyy

                                                Approved by:

SUBJECT: Aseptic Area Validation Procedures

PURPOSE

To describe the procedure for the validation of aseptic area to prevent cross-contamination and demonstrate environmental control

RESPONSIBILITY

All concerned departmental managers and validation engineers are responsible for following the procedure. The quality assurance manager is responsible for SOP compliance.

PROCEDURE

1. Testing

The controlled areas will be subjected to the following set of performance tests:

1. Air flow, volume, and distribution (also designated as post-balancing verification)
2. HEPA filter/leak (DOP)
3. Temperature control
4. Humidity control
5. Air flow and uniformity
6. Pressure control
7. Airborn particle count
8. Induction leak
9. Air flow patterns
10. Recovery
11. Particle dispersion
12. Airborn microbial
13. Surface bioburden
14. Lighting level

The instruments used for these tests should be calibrated and included with the report. The tests will be performed at rest, dynamic, and stress conditions.

1.1 At-rest testing

Performance tests executed under at-rest (complete and with production equipment installed and operating, but without personnel within the facility) conditions will serve as baseline information, and are needed to determine simulated fully operational conditions. After this analysis certain procedures, equipment, methods, etc., are changed.

1.2 Dynamic testing

Dynamic testing occurs when production equipment is installed and in normal operation with all services functioning, and with personnel present and performing their normal functions in the facility. The tests performed will serve to obtain a clear representation of the prevailing environmental conditions.

1.3 Stress testing

Testing at stress conditions will be performed to determine the ability of the system to remain stable at all times during operation conditions defined as continuity. Stress testing will be used to determine the ability of the system to recover after an unacceptable limit has been reached.

2. Reporting Forms

The protocols define procedures to be used to verify the performance of qualified equipment. As part of validation, the results obtained should be carefully recorded and compared with the design conditions. Deviations or diversions contrary to the specified levels determine the suitability of the controlled environment, so the reporting form represents the document for certification or acceptance of the system. The reporting form should show the following information:

> Date, start, and finish time
> Name of person performing the test
> Location of the test
> List of testing equipment with serial numbers
> Calibration dates

SOP No. Val. 600.30 **Effective date: mm/dd/yyyy**

Approved by:

Temperature (when applicable)
Humidity (when applicable)
Air velocity (when applicable)
Design conditions
Actual conditions
Signatures of those involved in the test
Diagrams showing test locations

The aseptic area validation matrix is summarized in the following table.

Table 1   Aseptic Area Validation Matrix

Tests	At Rest Initial	At Dynamic	Stress
Air Flow, Volume, and Uniformity Test	X	—	NR
HEPA filter/Leak Test (DOP)	X	NR	NR
Temperature Control Test	X	X	X
Humidity Control Test	X	X	X
Air Flow and Uniformity Test	X	X	X
Air Pressure Control Test	X	X	X
Airborn Particle Count Test	X	X	X
Induction Leak Test	X	—	—
Air Flows Patterns Test	X	X	NR
Recovery Test	X	X	—
Particle Dispersion Test	X	—	NR
Airborn Microbial Test	X	X	X
Surface Bio-burden Test	X	X	X

—: Optional

NR: Not Required

3. Air Flow Velocity, Volume, and Uniformity Tests

Purpose

These test procedures are performed to determine average air flow velocity and uniformity of velocity within a cleanroom, clean zone, or unidirectional flow work zone, as well as to determine air supply volume uniformity.

SOP No. Val. 600.30

Effective date: mm/dd/yyyy

Approved by:

Typically, either airflow velocity or air flow volume testing will be performed. Total volume may in turn be used to determine the air exchange rate (room air volume changes per hour) for the clean room.

Equipment

- Air flow velocity: electronic microanemometer with tube array thermal anemometer, vane-type anemometer, or equivalent
- Air flow volume: electronic micromanometer with appropriate air flow hood, or equivalent
- Suitable support stand where necessary

Method

- Air flow velocity (unidirectional) test: The work zone for which air flow velocities are measured is the unidirectional air flow volume designated for clean work, characterized by an entrance plane normal to the air flow. The entrance plane is typically no more than 30 cm (12 in.) from the supply source. Alternative spacing may be selected, especially for horizontal flow conditions, provided that air flow remains unobstructed in any manner that would significantly affect test results. The unidirectional air flow velocity test should be performed as follows:
  1. Divide the work zone entrance plane into a grid of equal areas. Individual areas should not exceed 0.37 m² (4 ft²).
  2. Support the anemometer sensor probe with a suitable stand. The use of a stand will prevent errors resulting from disruption of the unidirectional air flow that can be caused by the body or arm if the probe is hand-held. Orient the probe perpendicular to the velocity flow vector to be measured. Probe positions for air flow velocity testing are the designated grid test locations, at the work zone entrance plane. All test positions should be within unobstructed, unidirectional air flow.
  3. Measure the air flow velocity at each test position. Allow at least five seconds for each measurement and record the average reading during that period.

SOP No. Val. 600.30   **Effective date: mm/dd/yyyy**

                                                Approved by:

■ Air flow velocity (nonunidirectional) test: In a nonunidirectional clean room or clean zone, air flow velocity measurements should be made for each terminal HEPA filter (or supply air diffuser, if applicable); there is no entrance plane as such.
Note: The measurement of air flow volume is usually preferable to measurement of air flow velocity and is a more representative test of the final filter air supply for nonunidirectional air flow clean rooms.

The air flow velocity test in a nonunidirectional clean room should be performed as follows:

1. Support the anemometer sensor probe with a suitable stand so that optimum control of test positions can be maintained. Orient the probe perpendicular to the velocity flow vector being measured.
2. Measure and record the velocity at the approximate center of each filter area of 0.37 m² (4 ft²). The probe should be positioned at a distance of no more than 15 cm (6 in.) from the filter face. The effect of nonuniform velocity across the filter face can be minimized by taking more readings per unit area or by using a tube array sensor.
   ■ Air flow volume test: The supply air flow volume is measured by using a flow hood in a manner that includes all of the air issuing from each terminal filter or supply diffuser. The air flow volume test should be performed as follows:
   1. Place the flow hood opening completely over the filter or diffuser, seating the face of the hood against a flat surface to prevent air bypass and inaccurate readings.
   2. Measure and record the volume flow rate in l/sec (ft³/min) for each filter or diffuser.

Acceptance criteria

■ The average air flow velocity, or the average or total air flow volume, for the clean room or clean zone should be within ±5% of the value specified for the clean room or clean zone, or within other standardized tolerance limits.
Note: Air flow volumes must be normalized, if filters or diffusers differ in size, before the average airflow volume can be calculated.
■ The relative standard deviation should not exceed 15%.

SOP No. Val. 600.30   Effective date: mm/dd/yyyy

                                                 Approved by:

4. HEPA Filter/Leak Test (DOP)

Purpose

To ensure against HEPA filter failure due to damage during installation or operation

Equipment

DOP polydisperse aerosol generated by blowing air through liquid dioctyl phthalate (DOP) at room temperature. The approximate light-scattering mean droplet size distribution of the aerosol is:

$$99\% + \text{less than } 3.0 \ \mu m$$
$$95\% + \text{less than } 1.5 \ \mu m$$
$$92\% + \text{less than } 1.0 \ \mu m$$
$$50\% + \text{less than } 0.72 \ \mu m$$
$$25\% + \text{less than } 0.45 \ \mu m$$
$$11\% + \text{less than } 0.35 \ \mu m$$

- DOP aerosol generator — compressed-air operated, equipped with Laskin-type nozzles.
- Aerosol photometer — light-scattering type with a threshold sensitivity of at least 10^{-3} mg/l. Capable of measuring concentrations in the range of 80 to 120 mg/l, and with air sample flow rate of 1 ft^3 + 10%/min.

Method

- This test shall be performed only by certified or previously trained personnel.
- Introduce DOP aerosol upstream of the filter through a test port and search for leaks downstream with an aerosol photometer.
- Filter testing shall be performed after operational air velocities have been verified and adjusted where necessary.
- Align the aerosol photometer.
- Position the smoke generator so the DOP aerosol will be introduced into the air stream ahead of the HEPA filters.

SOP No. Val. 600.30   Effective date: mm/dd/yyyy

                                            Approved by:

- Open the appropriate number of nozzles until a DOP challenge concentration of 100 mg/liter of air is reached. This challenge concentration is measured upstream of the HEPA filter and is evidenced by a reading between 4 and 5 on the logarithmic scale of the aerosol photometer.
- Scan each filter by holding the photometer probe approximately 1 in. from the filter face and passing the probe in slightly overlapping strokes, at a traverse rate of not more than 10 ft/min, so that the entire face is sampled.
- Make separate passes with the photometer probe around the entire periphery of the filter, along the bond between the filter medium and the frame, and along all other joints in the installation through which leakage might bypass the filter medium

Acceptance criteria

- HEPA filters 99.99% (MIL-STD 282, Eurovent 4/4, UNI 7833) and more: challenge aerosol penetration is lower or equal to 0.01% of the upstream concentration (= filter class EU13 and EU14 [BS 3928, BSI Document 90/73834 and draft DIN 24.184] and class H13, H14, U15, U16 and U17 [draft CEN/TC/195])
- HEPA filters 99.97% (MIL-STD 282, Eurovent 4/4, UNI 7833): challenge aerosol penetration is lower or equal to 0.03% of the upstream concentration (= filter class EU12 [BS 3928, BSI Document 90/73834 and draft DIN 24.184] and class minimum H12 [draft CEN/TC/195])
- HEPA filters 95% (MIL-STD 282, Eurovent 4/4, UNI 7833): challenge aerosol penetration is lower or equal to 5% of the upstream concentration. (= filter class EU10 and EU 11 [BS 3928, BSI Document 90/73834 and draft DIN 24.184] and class minimum H10 and H11 [draft CENT/TC/195])

Repairs

No repair is authorized without the acceptance of the contractor. If the contractor agrees to repair the filter, the medium to repair the filter should be agreed upon (usually silicones; do not use silicones when interfering

SOP No. Val. 600.30

Effective date: mm/dd/yyyy
Approved by:

with product, e.g., clean rooms for painting purposes). Each repair must be properly documented on the worksheet.

In any case, the maximum surface to be repaired is less than 5% of the visible surface of the filter and any dimension of any repair is maximum 4 cm. Other criteria for repair can be agreed upon with the contractor.

Frequency

Initial validation: once
Revalidation: every six months for class 100 areas; every year for class 10,000 and 100,000 areas respectively

5. Temperature Control Test

Purpose

To demonstrate the ability of the air handling system to control temperature at designed standards all year

Equipment

■ Calibrated DICKSON temperature and humidity recorder or equivalent

Method

■ Air conditioning systems are to be in continuous operation for at least 24 h prior to performing these tests. All lights in the sterile core are to be on during the testing as well as during the 24-h preconditioning period.
■ Measure and record temperatures at 1–h intervals for a period of 8 h (at least 3 days) at each of the indicated locations for each room.
■ The test should be repeated for at-rest and dynamic conditions.

Acceptance criteria

The system shall be capable of maintaining temperature and humidity as per designed standards all year round, with the specified occupancy and heat generation design levels. As a guideline, in practice, the temperature should be to U.S Federal Standards 209E: 22.2 ± 2.8°C

Frequency

Initial validation: once
Revalidation: annually

6. Humidity Control Test

Purpose

To demonstrate the capability of the air handling system to control humidity at the specified level for each room

Equipment

- Calibrated DICKSON humidity and temperature recorder or equivalent

Method

- Execute this test after all the balancing procedures have been concluded.
- Measure and record humidities for the conditions and locations specified for every room under at-rest and dynamic conditions.
- Operate the system for at least 6 h prior to the start of the test.
- Measure and record the RH at 1–h intervals for the period of 24 h (at least 3 days) at each of the indicated locations for each room.

Acceptance criteria

The relative humidity at each grid point shall be equal to the specified levels and tolerance limits indicated on each recording form. If these levels are attained, the system is accepted.

Requirements

U.S Federal Standard 209E: 30 to 45% relative humidity

SOP No. Val. 600.30   Effective date: mm/dd/yyyy

Approved by:

Frequency

Initial validation: once
Revalidation: annually

7. Air Flow and Uniformity Test

Purpose

- To demonstrate that the air system is balanced and capable of delivering sufficient air volumes to maintain a minimum cross-sectional velocity under the absolute terminal/filter modules of at least 90 fpm measured 6 in. downstream of the filters
- To verify air velocities before the air encounters an obstruction; this should be conducted when new filters are installed in the system
- To verify horizontal and vertical air velocity components at the point at which the clean air reaches an obstacle or a surface 40 in. above the floor, whichever occurs first

Equipment

Hot-wire anemometer and stand

Method

- These tests are to be executed in every room where an absolute terminal filter module is installed.
- Draw a grid on the floor as indicated in the room diagram.
- Measure and record the velocity at the center of each grid at the specified heights.
- Allow no objects within 10 ft of the anemometer, except for built-in equipment. Minimize the number of people during the at-rest testing.
- Measurements should be taken for a minimum of 15 sec.
- Record the pressure readings (in inches) from the manometer connected to the module's plenum.

SOP No. Val. 600.30   **Effective date: mm/dd/yyyy**
                                           Approved by:

Acceptance criteria

Average measured clean air velocity shall be according to designed standard at 6 in. downstream from the filter face. Velocity differences within the same plenum should be no more than 25%.

Frequency

Initial validation: once
Revalidation: annually

8. Pressure Control Test

Purpose

To demonstrate the capacity of the system to control pressure levels within the specified limits

Equipment

Inclined pressure gauge with resolution of 0.01 in. of water

Method

- All HVAC and laminar flow systems are to be in continuous operation when performing these tests.
- To avoid unexpected changes in pressure and to establish a baseline, all doors in the sterile facility must be closed and no traffic is to be allowed through the facility during the test.
- Pressure readings are taken with the high and low pressure tubing at the specified following locations. The following stress conditions (Table 2) should be simulated while monitoring pressure.

Table 2

	Test 1	Test 2	Test 3	Test 4
Gowning	Closed	Open	Closed	Open
Ancillary	Closed	Open	Closed	Open
Primary room	Closed	Open	Closed	Open
Corridor	Closed	Open	Closed	Open

SOP No. Val. 600.30   **Effective date: mm/dd/yyyy**
                                             Approved by:

Acceptance criteria

- Pressure differentials should be as indicated in the design conditions at all times under static conditions.
- Pressure differentials should be maintained as indicated in the design conditions under standard simulated operating conditions.
- Pressure differentials should be above 0.02 at the primary environments when stress conditions occur.
- The system will not be acceptable if, at any time during normal dynamic, static, or stress conditions, the pressure in the primary environments becomes less than zero or negative.

Frequency

Initial validation: once
Revalidation: annually

9. Airborn Particle Count Test

Purpose

To establish that, at critical work locations within clean rooms, a count of less than 100 particles per cubic foot of air, 0.5 μm in diameter or larger, is maintained. The other critical areas of class 10,000 and class 100,000 are also maintained.

Equipment

CI-500 laser particulate counter with printer

Method

- The measurement should be performed at a height of 1.5 m above the floor level.
- These tests are performed after the HEPA filter leak tests and air velocity tests are completed.

SOP No. Val. 600.30　　　　　　　　　**Effective date: mm/dd/yyyy**

　　　　　　　　　　　　　　　　　　　　Approved by:

- To obtain baseline data with the room in static conditions, perform the following tests with operational personnel absent and the equipment at rest:
  1. Using the particle analyzer, count particles greater than or equal to 0.5 μm in diameter at heights of 40 in. in the center of each grid.
  2. If the particle count in the 0.5 μm range is less than 50 per cubic foot of air, four additional counts at this location are taken to place these particle counts within a 50% confidence interval.
- After completion of these tests, if the absolute air filtration modules are operating within accepted limits, repeat steps 1 and 2 with operational personnel present and the fill equipment running. If at any time there is a deviation from accepted parameters, the various components of the systems in operation are reviewed.

The minimum volume of air to be sampled can be read from Table 3:

Table 3

Testing for particles	Air volume required
≥ 0.1 μm	0.1 cft or 2.8 l
≥ 0.3 μm	0.1 cft or 2.8 l
≥ 0.5 μm	0.2 cft or 5.6 l
≥ 5 μm	0.3 cft or 8.4 l

The minimum number of sampling points can be read from the Table 4:

Table 4

Square feet	Square meter	Sampling points
100	9.2	4
200	18.4	8
400	36.8	16
1,000	92	40
2,000	184	80
4,000	368	160
10,000	920	400

SOP No. Val. 600.30   Effective date: mm/dd/yyyy

                                             Approved by:

Table 5 shows room air classification based on Federal Standard 209 E:

Table 5   Federal Standard 209 E class limits in particles per cubic foot of size equal to or greater than particle size shown

Class	*Measured Particles size (μm)*				
	0.1	*0.2*	*0.3*	*0.5*	*5.0*
1	35	7.5	3	1	NA
10	350	75	30	10	NA
100	NA	750	300	100	NA
1,000	NA	NA	NA	1,000	7
10,000	NA	NA	NA	10,000	70
100,000	NA	NA	NA	100,000	700

Acceptance criteria

- The air system can be considered validated when the results of three consecutive sets of tests are within accepted operational parameters.
- At any of the designated critical locations (where any sterilized product or material is exposed to the working environment), the particulate count shall not exceed 100 particles 0.5 μm in diameter and larger per cubic foot of air.
- The same test should be repeated at ancillary environments. Ancillary environments shall not exceed a particle count of 100,000 particles 0.5 μm in diameter and larger per cubic foot of air, in order to be considered acceptable by current regulations. It is common practice to design and operate ancillary environments at levels not exceeding 10,000 particles of 0.5 μm per cubic foot of air, in order to provide additional protection to the final product while it is processed at the critical area.

SOP No. Val. 600.30

Effective date: mm/dd/yyyy

Approved by:

Frequency

Initial validation: once
Revalidation: annually

10. Induction Leak Test

Purpose

- To determine if there is intrusion of unfiltered air into the clean work areas from outside the clean room enclosure through joints and cracks in the walls, ceiling, etc., other than from the pressurized air supply system
- To determine unfiltered air intrusion into the clean room through open entrance doorways

Equipment

Optical particle counter

Method

- Measure the concentration outside the clean room enclosure immediately adjacent to the surface or doorway to be evaluated. This concentration should be at least 100,000 particles per cubic foot of a size equal to greater then 0.5 μm. If the concentration is less, generate an aerosol to increase the concentration.
- To check for construction joint leakage, scan all joint areas from a distance of 6 in. at a speed of approximately 10 in./min.
- To check for intrusion of open doorways, measure the concentration inside the enclosure at 10 in. from the open door.
- Repeat the same test in front of any openings, i.e., pass-through, electrical outlets, and any openings connecting with the outside.
- Repeat this test while opening and closing clean room entrance doors.

Acceptance criteria

No construction joint leaks or intrusion through open doors should exceed 0.1% of the measured external concentration. See Tables 6 and 7.

Table 6   Air Cleanliness Level: Definition of Classes

	Maximum number of particles			
	0.5 µm		5 µm	
Class	Per cubic foot	Per cubic meter	Per cubic foot	Per cubic meter
100	100	3.5	—	—
10,000	10,000	350	65	2.3
100,000	100,000	3500	700	25

Table 7   Guidelines for Cleanliness Levels Required during Manufacturing of a Parenteral Drug

Operation	Class	Cleanliness level (particles 0.5 µm and larger)
Warehousing	—	—
Preparation	100,000	No more than 100,000
Filtration	100,000	No more than 100,000
Filling area	100,000 or better	No more than 100,000
Filling line (point of use)	100	No more than 100

11. Air Flow Patterns Test

Purpose

- To determine air flow interaction with machinery and equipment in a critical area protected with a laminar flow clean air system
- To determine the air flow patterns during fill line operations
- To select and improve the flow pattern that generates the minimum turbulence and best washing capabilities

SOP No. Val. 600.30   **Effective date: mm/dd/yyyy**
                                              Approved by:

Equipment

White visible or yellow smoke generator, anemometer, 35-mm camera or videotape recorder

Method

- Verify that the laminar flow devices in the sterile core are operational.
- Check air velocities at 6 in. from the filter face to ensure that the device is operating within the specified laminar flow velocity (90 ft/min or more).
- Verify that the ventilation and air conditioning systems are operating and balanced.
- If the system operates according to the specified operating parameter, begin to generate white visible smoke at the critical locations. A critical location is defined as any area where sterilized product or material is exposed to the working environment.
- Generate white smoke inside and over each component that forms part of the line (to avoid damage to the materials or equipment, cover them tightly with plastic). Film the smoke as it travels through each critical area of the machine.
- Smoke should flow through these critical areas. If the air returns (back-flows) due to turbulence, the system cannot be accepted and must be rebalanced or adjusted. Slight turbulence, due to equipment configuration, is not significant as long as the air does not return to the critical areas.
- If the air does not back-flow, continue to film. If the smoke back-flows to the critical working area at any point during this operation, procedures must be established to prevent cross-contamination and reentry into these areas. If the unit passes, proceed.
- Determine if the generated turbulence can carry contaminants from other areas to critical points of the line. If so, adjust the air flow to ensure a minimum of turbulence and rapid cleaning. If the turbulence cannot be stopped, a different aerodynamic pattern must be found (covers and diffusers can be used over the filling equipment). If turbulence carries contaminants from any area to the critical areas, the system should be reevaluated and analyzed in terms of the filling, capping, and laminar-flow equipment.

SOP No. Val. 600.30

Effective date: mm/dd/yyyy

Approved by:

Acceptance criteria

- If the results of the test are unsatisfactory, the laminar flow system cannot be validated and the rest of the validation tests should not be carried out until a satisfactory operation has been reached. Otherwise, the system is valid and can be certified.
- Should corrective changes be necessary, the changes are made and recorded, and the validation process repeated.
- A turbulent air flow should not be present at each area location.

Frequency

Initial validation: once
Revalidation: annually

12. Recovery Test

Purpose

To determine the capabilities of the system to recover from internally generated contamination

Equipment

Visual smoke generator, particles counter, and hot wire anemometer

Method

- With smoke generation output tube located at a predesignated location, generate smoke for 1 to 2 min and shut off.
- Wait 2 min and then advance the sample tube of the particle counter to a point directly under the smoke source and at the level of the work zone. Record the particle count. If it is not 100 per cubic foot or less, repeat the test with the wait interval increased in increments of 0.5 min until counts are less than 100 per cubic foot.
- Repeat for all grid areas, recording recovery time for each area.

Acceptance criteria

The recovery time should be not more than 2 min.

SOP No. Val. 600.30   **Effective date: mm/dd/yyyy**
                                             Approved by:

Frequency

Initial validation: once
Revalidation: annually

13. Particle Dispersion Test

Purpose

To verify the parallelism of air flow through the work zone and the
capability of the clean room to limit the dispersion

Equipment

Visual smoke generator, particle counter, and hot wire anemometer

Method

- Perform this test after completion of the air velocity uniformity tests.
- Divide the work zone into 2 × 2 ft grids of equal area.
- Set up the smoke generator, with outlet tube pointing in the
  direction of air flow and located at the center of a grid area at the
  work zone entrance plane.
- If smoke is introduced with air pressure, adjust it to provide a
  smoke outlet velocity equal to the room air velocity at that point.
- Operate the particle counter with the sample tube at the normal
  work level and at a point remote from the smoke source. Verify
  that the counter indicates particle concentrations less than 200
  particles of 0.5 μm or greater.
- Move the sample tube in toward the smoke source from all
  directions at this level to the point where particle counts show a
  sudden and rapid rise to high levels (10^6 per cubic foot). This
  defines the envelope of dispersion away from the smoke source
  and demonstrates the airflow parallelism control of the room.
- Repeat for all grid areas. Prepare a diagram showing grid areas
  and corresponding dispersion envelopes.

Acceptance criteria

The degree to which dispersion away from the smoke source is confined and the regularity of the pattern (indication of directional drift in one direction) is a matter of the configuration of the line. It is recommended that dispersion should not extend beyond 2 ft rapidly from the point of smoke source, i.e., at 2 ft from the generation point, the particle count should be less than 100 per cubic foot of the 0.5 μm size and larger.

Frequency

Initial validation: once
Revalidation: annually

14. Airborn Microbial Test

Purpose

To determine the airborn microbial contamination level

Equipment

Solid surface impactor with a rotating collection surface or staged plates (Anderson-Slite)

Method

After proceeding with calibration and indications given in the operating manual, proceed as follows:

- Aseptically prepared collection plates are placed in the sampling apparatus. Petri dishes used must be sterilized prior to filling. Verify the adequacy of petri dish dimensions so that the operational characteristics are maintained according to the manufacturers specifications. Plastic petri dishes are not recommended because static changes are likely to be present in plastic that will reduce the collection efficiency.
- Any general purpose, solid bacteriological medium, such as trypticase soy agar or blood agar, can be used. Selective media are

SOP No. Val. 600.30   **Effective date: mm/dd/yyyy**
                                           Approved by:

not recommended, since they inhibit the repair and growth of injured or stressed cells.

■ Verify the air sample rate, time, and location of the plate before starting the sampling. Sampling time should be 20 min at every location. After the sampling is complete, remove the collection plates, cover, and identify them. Identification should include date, sampling instrument number, location, and plate number.

■ Plates are then taken to an incubator and maintained inverted, to prevent condensation drop for a period of 18 to 24 h at 35°C.

■ After incubation, the number of colonies on each plate is counted, using a standard bacterial colony counter.

Acceptance criteria

The following table provides the Federal Standard 209 E for cleanliness level:

Table 8   Air Cleanliness Guidelines in Colony-Forming Units (cfu) in Controlled Environments (Using a Slite-to-Agar Sampler or Equivalent)

Class			
SI	U.S. Customary	cfu per cubic meter of air	cfu per cubic feet of air
M3.5	100	Less than 3	Less than 0.1
M5.5	10,000	Less than 20	Less than 0.5
M6.5	100,000	Less than 100	Less than 2.5

Establishment of sampling plan and sites

■ During initial start-up or commissioning of a clean room or other controlled environment, specific locations for air and surface sampling should be determined. Consideration should be given to the proximity to the product and whether air and surfaces might be in contact with a product or sensitive surfaces of the container-closure system. Such areas should be considered critical areas requiring more monitoring than nonproduct-contact areas. In a parenteral vial filling operation, areas of operation would typically include the container-closure supply paths of opened containers and other inanimate objects (e.g., fomites) that personnel routinely handle.

■ The frequency of sampling will depend on the criticality of specified sites and the subsequent treatment received by the product after it has been aseptically processed. Table 9 shows suggested frequencies of sampling in decreasing order of frequency of sampling and in relation to the criticality of the area of the controlled environment sampled.

Table 9   Suggested Frequency of Sampling on the Basis of Criticality of Controlled Environment

Sampling Area	Frequency of Sampling
Class 100 or better room designations	Each operating shift
Supporting areas immediately adjacent to Class 100 (e.g., class 10,000)	Each operating shift
Other support areas (class 100,000)	Twice/week
Potential product/container contact areas	Twice/week
Other support areas to aseptic processing areas but nonproduct contact (class 100,000 or lower)	Once/week

Frequency

Initial validation: once
Revalidation: annually

15. Surface Bioburden Test

Purpose

To determine the microbial contamination level on surfaces

Equipment

Cotton swabs or RODAC plate (nutrient agar culture medium)

SOP No. Val. 600.30   **Effective date: mm/dd/yyyy**
                                           Approved by:

Method

Take the swab stick from the tube and gently swab 25 cm^2 of area (walls, floor, equipment, etc.) and place back in tube containing 5 mL sterile buffer and test or incubate per official procedure.

This technique is to be used after decontamination procedures. Agar media left on the surface could represent a problem. Therefore, immediate decontamination procedures should follow sampling. Identify every plate, indicating the exact location where the sample was taken. Room landmarks should be noted for present and future reference.

Acceptance criteria

The maximum number of colonies per square foot should not exceed the limits in Tables 10 and 11.

Table 10   Surface Cleanliness Guidelines of Equipment and Facilities in cfu Controlled Environment

Class		cfu per Contact Plate*
S.I	*U.S Customary*	*Gloves*
M3.5	100	3
M5.5	10,000	5
		10 (floor)

* Contact plate areas vary from 24 to 30 cm^2. When swabbing is used in sampling, the area covered should be greater than or equal to 24 cm^2 but no larger than 30 cm^2.

Table 11   Surface Cleanliness Guidelines in Controlled Environments of Operating Personnel Gear in cfu

Class		cfu per Contact Plate*	
S.I.	*U.S. Customary*	*Gloves*	*Personnel clothing and garb*
M3.5	100	3	5
M5.5	10,000	10	20

* See Table 10 above (*).

SOP No. Val. 600.30   Effective date: mm/dd/yyyy
                                             Approved by:

Frequency

Initial validation: once
Revalidate: annually

16. Air Pressure Control Test

Purpose

To demonstrate that air pressure is maintained in critical and adjacent areas as specified in the HVAC design according to the specification limits

Equipment

Inclines pressure gauge with resolution of 0.01 in. water

Method

- All HVAC and laminar flow systems are to be in continuous operation when performing these tests.
- To avoid unexpected changes in pressure and to establish a baseline, all doors in the sterile facility must be closed and no traffic is to be allowed through the facility during the test.
- Pressure readings are taken with the high and low pressure tubing at the following locations: (refer to each room's diagram).

Acceptance criteria

- The minimum positive pressure differential between the room and any adjacent area of less clean requirement should be 0.05 in. water (12 Pa), with all entryways closed. When the entryways are open, the blower capacity should be adequate to maintain an outward flow of air to minimize contamination migrating into the room.
- A filtered air supply should maintain a positive pressure relative to surrounding areas under all operational conditions and flush the area effectively. Final filtration should be at or as close as possible to the point of input to the area. A warning system should be included to indicate failure in the air supply and an indicator

of pressure difference should be fitted between areas where the difference is critical.

■ Aseptic processing areas should have a positive pressure differential relative to adjacent less than clean areas. A pressure differential of 0.05 in. water (12 Pa) is acceptable.

■ The sterile filling and sealing rooms are kept under higher air pressure than adjacent rooms, while the remaining clean rooms (in which containers, raw materials, or other parts of the finished product are processed) are under greater air pressure than the surrounding nonsterile environment. Such pressure differentials help minimize the inward flow of airborn contaminants.

17. Air Changes

Purpose

To demonstrate that an adequate number of air changes are maintained within the classified rooms per HVAC design

Method

Determine the number of air changes per hour in the aseptic area by calculation from the inlet air flow and the area volume.

Requirements

More than 20 air changes per h should be achieved.

Frequency

Initial validation: once
Revalidation: annually

REASONS FOR REVISION

Effective date: mm/dd/yy

■ First time issued for your company, affiliates, and contract manufacturers

| YOUR COMPANY |
| VALIDATION STANDARD OPERATING PROCEDURE |

SOP No. Val. 600.40

Effective date: mm/dd/yyyy

Approved by:

TITLE: Microbiological Monitoring of Areas Used for Production of Solids, Semi-Solids, and Liquids

AUTHOR: _____

Name/Title/Department

_____

Signature/Date

CHECKED BY: _____

Name/Title/Department

_____

Signature/Date

APPROVED BY: _____

Name/Title/Department

_____

Signature/Date

REVISIONS:

No.	Section	Pages	Initials/Date

SOP No. Val. 600.40 Effective date: mm/dd/yyyy

Approved by:

SUBJECT: Microbiological Monitoring of Areas Used for Production of Solids, Semi-Solids, and Liquids

PURPOSE

To define the procedure for the monitoring of surfaces

RESPONSIBILITY

The quality control manager is responsible for following the procedure. The quality assurance manager is responsible for SOP compliance.

PROCEDURE

1. Monitoring Surfaces

Surfaces in areas of production of solids and liquid for oral use are tested periodically to indicate the adequacy of cleaning and sanitizing procedures, and to detect contamination caused by personnel. The samples are taken routinely under normal working conditions. The techniques used are as follows:

Contact plates

The test is performed with RODAC plates (replicate organisms detecting and counting). These plates contain a solid culture medium in a specially designed petri dish (diameter 65 mm). The convex surface extends above the walls of the plate. The surface of the medium is pressed against the test surface, transferring the microorganisms to the nutrient agar. Incubation colonies are formed wherever microorganisms were present on the tested surface. Disinfect the area of contact after sampling.

Swabbing

Swabs (sterile, moistened, alginate wool tips) are used for sampling discrete surfaces areas or difficult-to-reach locations. Subsequent direct inoculation on solid media or membrane filtration of the swab rinsing fluid can be used to cultivate the micro-organisms.

The use of contact plates is the recommended technique. The RODAC plate method is the simplest, but it is useful only for flat surfaces. Swab samples afford examination of corners, crevices, and other area inaccessible to RODAC plates.

Materials

RODAC dish (Falcon) code No. 1034, containing a culture medium made to the formula as indicated in the USP should be used. This medium contains approximately 0.7 g/1 lecithin and 5.0 g/1 polysorbate 80, two commonly used neutralizers, to inactivate residual disinfectants on the spot where the sample is collected.

Normal petri dishes are not suitable because the lid will make contact with the convex surface of the agar, which for this test must protrude above the brim of the dish.

Alcohol 70%, sterile (0.2 μm filtered) should be used.

Technique

- All handlings should be performed in a way as aseptic as possible.
- Prepare the culture medium according to the USP and fill the medium into sterile RODAC dishes to the brim to obtain a convex surface. Extreme care should be taken to prevent the formation of air bubbles and to prevent the medium from overflowing (if either occurs, these plates should be discarded). Allow the plates to solidify. Preincubate the plates at 35°C for 24 h and then at 25°C for 24 to 48 h.
- Carefully introduce the RODAC dish at the area to be tested and remove the lid. Carefully but firmly press the plate with a slight rolling movement onto the surface being examined and hold for a few seconds without moving it. Remove the dish and replace the lid immediately. The tested area is then to be cleaned and disinfected with 70% alcohol.
- First, incubate the plates at 35°C for 48 h and subsequently at 25°C for another 5 days. Count the formed colonies.

Sampling locations and number of samples: per room, 1 to 3 representative samples from the walls in a height of 1 to 2 m.

SOP No. Val. 600.40   **Effective date: mm/dd/yyyy**

                                           Approved by:

Interpretation of results

The contamination rate of the area is stated as the number of viable microorganisms per 100 cm^2. The area of a standard RODAC dish is 25 cm^2.

Requirements

≤200 cfu/100 cm^2 under working conditions

Actions in case of defect

An evaluation of the reasons for the contamination and the performance of the cleaning procedure should take place. Furthermore, an additional cleaning may be necessary.

Frequency

Twice a year

1.1 Monitoring equipment surfaces

For the procedure, requirements, actions in case of defect, and frequency, see Section 1.1. Sampling locations and sample number: the locations have to be established in-house, three samples per equipment.

1.2 Monitoring floors

For procedure, action in case of defect, and frequency, see Section 1. Sampling locations and number of samples: per room, one to three representative samples scattered over the total floor space.

Requirements

≤500 cfu/100 cm^2 under working conditions

2. Monitoring Air

Procedure

Although the microbiological quality of the air can be estimated by various methods, the RCS method is recommended. The biotest RCS (centrifugal

SOP No. Val. 600.40   **Effective date: mm/dd/yyyy**

                                                Approved by:

air sampler) collects viable airborn particles in combination with the principle of agar impaction.

As air is drawn towards the impeller blades, it is subjected to centrifugal acceleration. Viable microorganisms in the sampled air are impacted at high velocity onto the surface of the agar strip. After sampling, the strip containing the nutrient is incubated at 30 to 35°C for days. The colonies can be enumerated by visual examination and the results should be recorded. Sampling time with the RCS should be 2 min (= 80 1 air).

Sampling locations and number of samples: per room, one sample taken in the middle of the room in a height of between 1 and 2 m.

Requirements

$\leq$200 cfu/m^3 under rest conditions
$\leq$400 cfu/m^3 under working conditions

Actions in case of defect

An evaluation of the reasons for the contamination should occur. A discussion of the results with the technical services department should take place.

Frequency

Twice a year

REASONS FOR REVISION

Effective date: mm/dd/yyyy

■ First time issued for your company, affiliates, and contract manufacturers

YOUR COMPANY
VALIDATION STANDARD OPERATING PROCEDURE

SOP No. Val. 600.50

Effective date: mm/dd/yyyy
Approved by:

TITLE: Efficiency Testing for Disinfectants

AUTHOR: _____
Name/Title/Department

_____
Signature/Date

CHECKED BY: _____
Name/Title/Department

_____
Signature/Date

APPROVED BY: _____
Name/Title/Department

_____
Signature/Date

REVISIONS:

No.	Section	Pages	Initials/Date

SOP No. Val. 600.50   Effective date: mm/dd/yyyy
Approved by:

SUBJECT: Efficiency Testing for Disinfectants

PURPOSE

To ensure that disinfectants used to clean the facility are effective

RESPONSIBILITY

It is the responsibility of the quality control manager to follow the procedure. The quality assurance manager is responsible for SOP compliance.

Definitions

Sanitizer is specifically defined as any chemical that kills microbial contamination in the form of vegetative cells.

Frequency

The *membrane filtration technique* is used once, prior to the introduction of a new disinfectant within the production department.

The *surface testing technique* is used prior to any changes in the recommended procedure for evaluating its effectiveness on surfaces to be treated and demonstrating activity against contaminated for various contact times.

Equipment

Incubator 23°C and 32°C,
  Standard loop
  Membrane filtration unit
  Vortex
  Cultures
  ■ *P. aeruginosa* ATCC 9027
  ■ *E. coli* ATCC 8739
  ■ *S. aureus* ATCC 6538
  ■ *C. albicans* ATCC 10231
  ■ *B. subtilis* var. *niger* ATCC 9372
  ■ *A. niger* ATCC 16404
  ■ Organisms recovered from plant environment

SOP No. Val. 600.50   Effective date: mm/dd/yyyy

                                                         Approved by:

Sterile distilled water
Sterile screw-cap test tubes
Sterile buffer (dilution blanks)
Poured, sterile soybean casein digest agar (SCDA) petri plates
Poured, sterile potato dextrose agar (PDA)
Sterile pipettes, 25 ml, 10 ml, and 1 ml sizes
Sterile forceps
Sterile membrane filtration unit with 47 mm diameter
0.45 μm pore size
Solvent resistant membrane
Sterile, empty petri plates
Sterile, empty containers of ≈200 ml volume
Sterile Ca alignate swabs
Brushed stainless steel strips

PROCEDURE

1. Preparation of Challenge Inocula

Except for *B. subtilis*, streak slants (one or more) of appropriate agar (PDA for *C. albicans* and *A. niger*, SCDA for all others) with specific microorganisms from stock culture. Incubate at 30 ± 1°C for 48 to 72 h, except *A. niger*, which may need extended incubation for good sporulation.

Harvest cells by withdrawing ≈3 ml of buffer from a 10-ml tube and pipetting onto a slant. Using a pipette, gently scrape the slant to suspend the cells. Withdraw the suspension and transfer it back into the tube for buffer.

Do a microbial plate count of suspensions as necessary and dilute with appropriate buffer to obtain final working suspensions of 10^4 to 10^5 CFU/ml.

1.1 Sample preparation

Dilute the tested sanitizer to the use dilution (according to the recommendations of the manufacturer). Prepare further dilutions of the use dilution (10^{-1} or 10^{-2}). Adjust pH to 6.8 to 7 and attempted at 30°C.

Membrane filtration technique

In duplicate, pipette 10 ml of each of the dilutions into separate sterile test tubes (changing pipettes after each transfer). Provide 12 tubes for each dilution and time interval, as there are six challenge organisms. Using

SOP No. Val. 600.50

Effective date: mm/dd/yyyy

Approved by:

a calibrated pipette, inoculate each separate complement of dilution tubes with different challenge microorganisms, using an inoculum volume of 0.1 ml. Shake well and let stand at 30°C. For positive controls, inoculate appropriate duplicate dilution blanks with challenge microorganisms.

Assay at 5, 10, and 15 min intervals. Perform the entire operation in a laminar-flow hood. At the conclusion of each challenge time interval, pass the contents of each tube through a separate membrane filter unit. Wash each membrane with 3 × 100 ml portions of neutralizer solution. Remove each membrane from its filter unit and place face up on the surface of appropriate poured agar plate. Positive controls must be tested last.

Incubate all plates at 30 ± 1°C for 48 to 72 h. Examine each day for signs of growth of the inocula. Count and record.

Growth promotion controls: control membranes must show confluent growth of each of the challenge microorganisms.

Surface testing technique

Clean each polished stainless steel strip (or any surface to be treated) with detergent and rinse well with distilled water. Spot-inoculate four steel strips with a known volume of a culture for each dilution of sanitizer. Each strip should contain approximately 100 organisms. Allow drying for 30 min at room temperature. After drying, immerse two strips in sanitizer solution to be tested. Allow the contact for 5, 10, or 15 min.

Repeat this for each test organism. For positive control use sterile DW instead of sanitizer solution. For negative control use sterile strips. Swab each strip thoroughly with a premoistened Ca alginate swab and place in *Lecithin broth* (or other suitable neutralizer). Vortex for 30 sec, then perform the usual pour plate method.

Acceptance criteria

The recommended disinfectant solution must be able to establish a 5-log reduction of each of the inoculated microorganisms within 5 min.

REASONS FOR REVISION

Effective date: mm/dd/yyyy

- First time issued for your company, affiliates, and contract manufacturers

YOUR COMPANY
VALIDATION STANDARD OPERATING PROCEDURE

SOP No. Val. 600.60   **Effective date: mm/dd/yyyy**

                                            Approved by:

TITLE: Drinking Water

AUTHOR: _____

                          Name/Title/Department

_____

                              Signature/Date

CHECKED BY: _____

                          Name/Title/Department

_____

                              Signature/Date

APPROVED BY: _____

                          Name/Title/Department

_____

                              Signature/Date

REVISIONS:

No.	Section	Pages	Initials/Date

SOP No. Val. 600.60   **Effective date: mm/dd/yyyy**

       Approved by:

SUBJECT: Drinking Water

PURPOSE

To describe the acceptable standard of drinking water

RESPONSIBILITY

It is the responsibility of the validation manager and concerned departmental managers to maintain consistent quality of drinking water. The QA manager is responsible for SOP compliance.

PROCEDURE

Drinking water serves as the starting material from most forms of water covered by compendial monographs and plays an important role. The source of drinking water is either municipality or privately drilled wells. According to the USP, "Drinking water may be used in the preparation of USP drug substances but not in the preparation of dosage forms, or in the preparation of reagents or test solution." The major concern is the microbial content of drinking water.

1. Microbiological Investigation

1.1 Total aerobic viable count

Procedure

Method: Pour plate
Minimum sample size: 1 mL
Culture media: Plate count agar
Incubation time: 47 to 72 h
Incubation temp.:   30 to 35°C

Requirements

Total aerobic microbial count: ≤500 cfu/ml (grown at 30°C)

SOP No. Val. 600.60

Effective date: mm/dd/yyyy
Approved by:

1.2 E. coli *and coliforms*

Test for the presence of *E. coli* and coliforms:

Sample volume:	At least 100 ml
Test method:	Membrane filtration, place the membrane filter into 50 ml of lactose broth 1%
Incubation temperature:	35 to 37°C
Incubation time:	20 to 28 h
Observation time:	40 to 48 h
Provisional inspection:	Suspension for *E. coli* and coliforms if gas or acid formation is observed.
Further identification:	Transfer to Endo agar
Incubation temperature:	35 to 37°C
Incubation time:	20 to 28 h

Requirements

Total *E. coli*	Not detectable in 100 ml
Total coliforms	Not detectable in 100 ml

Frequency of microbiological investigation

The drinking water quality shall be monitored and the results should be evaluated to establish a routine monitoring program.

REASONS FOR REVISION

Effective date: mm/dd/yyyy

■ First time issued for your company, affiliates, and contract manufacturers

YOUR COMPANY VALIDATION STANDARD OPERATING PROCEDURE

SOP No. Val. 600.70

Effective date: mm/dd/yyyy

Approved by:

TITLE: Purified Water

AUTHOR: _____

Name/Title/Department

_____

Signature/Date

CHECKED BY: _____

Name/Title/Department

_____

Signature/Date

APPROVED BY: _____

Name/Title/Department

_____

Signature/Date

REVISIONS:

No.	Section	Pages	Initials/Date

SOP No. Val. 600.70   **Effective date: mm/dd/yyyy**

                                        Approved by:

SUBJECT: Purified Water

PURPOSE

To describe the acceptable standard for purified water USP

PROCEDURE

1. Definition

Purified water is water obtained by a suitable process. It is prepared from water complying with the U.S. Environmental Protection Agency National Primary Drinking Water Regulations or comparable regulations of the European Union or Japan. It contains no added substances.

2. Complying with Monograph (e.g., USP XXIV) Requirements

Since no added substances are allowed, if chloride and ozone are applied in the preparation of purified water, the following should be ensured: tests for total organic carbon and conductivity apply to purified water produced on site for use in manufacturing. Purified water packaged in bulk for commercial use elsewhere shall meet requirements of all tests under sterile purified water, except labeling and sterility, per USP XXIV.

Frequency

In continuous production facilities (minimum 10 working days), water should be sampled at least weekly.

     From noncontinuous production facilities water must be sampled from each production batch.

Acceptance criteria

Conductivity: 1.3 ms/cm at 25°C
TOC: <500 ppb

SOP No. Val. 600.70 Effective date: mm/dd/yyyy
Approved by:

3. Microbiological Investigations

Requirements

Total aerobic viable count: Establish in-house requirements
Action limit: 100 cfu/ml
Total *E. coli*:   Not detectable
Total coliforms: Not detectable

Frequency

Samples for microbiological investigations must be tested within 6 hours of sample collection. For continuous production facilities (minimum 10 working days) water should be sampled at least weekly. For noncontinuous production facilities, water must be sampled prior to each production batch.

Acceptance criteria

The requirements of USP XXIV are met.

REASONS FOR REVISION

Effective date: mm/dd/yyyy

■ First time issued for your company, affiliates, and contract manufacturers

---

YOUR COMPANY
VALIDATION STANDARD OPERATING PROCEDURE

SOP No. Val. 600.80   Effective date: mm/dd/yyyy

                                        Approved by:

TITLE: Water for Injection

AUTHOR: _____
                              Name/Title/Department

                  _____
                              Signature/Date

CHECKED BY: _____
                              Name/Title/Department

                  _____
                              Signature/Date

APPROVED BY: _____
                              Name/Title/Department

                  _____
                              Signature/Date

REVISIONS:

No.	Section	Pages	Initials/Date

SOP No. Val. 600.80   Effective date: mm/dd/yyyy
                                                 Approved by:

SUBJECT: Water for Injection

PURPOSE

To provide the guideline for the validation of water for injection (WFI) to be maintained in compliance with USP 24 monograph

RESPONSIBILITY

It is the responsibility of the quality control, validation, and technical services managers to follow the procedure. The quality assurance manager is responsible for SOP compliance.

PROCEDURE

1. Definition

The USP 24 monograph states, "Water for injection is water purified by distillation or reverse osmosis. It contains no added substance."

The British Pharmacopia states, "Water for injections is sterilized, distilled water, free from pyrogens. It is obtained by distilling potable water, purified water or distilled water...."

Although the USP allows generation of water for injections by use of reverse osmosis, this is not in line with the FDA opinion. Therefore, water for injections should always be generated by distillation.

To fulfill the monograph requirements with respect to the maximum allowed pyrogen content of about 0.25 endotoxin units per ml, the water used for distillation must be of very high microbial quality.

Prerequirement for water used for generation of water for injection is that it must comply with the microbiological requirements stated under Drinking Water, SOP No. Val. 600.60.

2. Chemical Investigation

Requirements

Complying with the monograph requirements (USP 24)

Frequency

Once per week

SOP No. Val. 600.80   **Effective date: mm/dd/yyyy**

                                              Approved by:

Acceptance criteria

Conductivity: <1.3 ms/cm at 25°C
TOC: <500 ppb

3. Microbiological Investigation

3.1 Pyrogens

Requirements

Endotoxin content: ≤0.25 EU/ml

3.2 Total aerobic viable count

Requirements

- ≤10 cfu/100 ml, plus a complete absence of pseudomonas and *E. coli*. Frequency of microbiological investigations. Samples for endotoxins and total aerobic viable count should be investigated within 8 hours after collection.
- Water for injection used for other purposes (e.g., washing and rinsing product components and equipment) should be tested for pyrogens and aerobic viable count at least weekly.
- Water for injection used for product compounding should be tested each day for pyrogens and aerobic viable count.

4. Storage of Water for Injection

Water for injection is not allowed to be retained in the system for longer than 24 h unless it is maintained at a minimum temperature of 85°C in a continuously circulating loop.

REASONS FOR REVISION

Effective date: mm/dd/yyyy

- First time issued for your company, affiliates, and contract manu-facturers

YOUR COMPANY
VALIDATION STANDARD OPERATING PROCEDURE

SOP No. Val. 600.90

Effective date: mm/dd/yyyy

Approved by:

TITLE: Validation of a Water System

AUTHOR: _____

Name/Title/Department

_____

Signature/Date

CHECKED BY: _____

Name/Title/Department

_____

Signature/Date

APPROVED BY: _____

Name/Title/Department

_____

Signature/Date

REVISIONS:

No.	Section	Pages	Initials/Date

SOP No. Val. 600.90   **Effective date: mm/dd/yyyy**

                                         Approved by:

SUBJECT: Validation of a Water System

PURPOSE

To provide the guideline to validate the water system for the pharmaceutical industry

PROCEDURE

The following types of water systems are preferable for use in the pharmaceutical industry:

Purified Water — Purified water is described in the USP 24 as "water obtained by distillation, ion-exchange treatment, reverse osmosis, or other suitable process. It is prepared from water complying with the regulations of the Federal Environmental Protection Agency with respect to drinking water. It contains no added substances."

Water for Injection — Water for injection is described in the USP 24 as "water purified by distillation or reverse osmosis. It contains no added substances." See Tables 1 and 2.

Table 1   USP 24 Chemical Requirements for Purified Water and Water for Injection

Component	Purified Water	Water for Injection
Conductivity	<1.3 µS/cm at 25°C	<1.3 µS/cm at 25°C
TOC	<500 ppb	<500 ppb
Pyrogens (EU/ml by LAL)	—	<0.25 EU/ml

Table 2   Microbiological Levels for Compendial Waters in cfu/ml

	Purified Water	Water for Injection
Pyrogen	—	<0.25 EU/ml
Sterility	<100 cfu/ml	<10 cfu/100 ml
Pathogens	Absent	Absent

SOP No. Val. 600.90   Effective date: mm/dd/yyyy

                                       Approved by:

The validation process is divided into the following stages:

- Prevalidation of the full system
- Construction validation
- Start-up validation
  - Functional operation
  - Procedures verification
  - Quality limits
- System qualification
- Approval of the system for use

After finishing the above stages, a monitoring program should be established.

1. Prevalidation of the Total System Design

- Flow schematics for the designed (layouts) water system showing all of the instrumentation valves, controls, and monitors, numbered serially
- Complete description of the features and function of the system
- Specifications for the equipment (storage tanks, heat exchangers, pumps, valves and piping components) to be used for water treatment and pretreatment
- Detailed specifications for sanitary system controls
- Procedures for cleaning the system, both after construction and ongoing
- Sampling procedures to monitor water quality and the operation of the equipment

2. Construction Validation

Construction validation shall be conducted to avoid irreparable damage due to the use of unsuitable techniques.

- System components and construction materials
  Major equipment, such as distillation unit and WFI storage tanks, should be inspected before it is shipped from the supplier to verify operational function and compliance with specifications.

SOP No. Val. 600.90   **Effective date: mm/dd/yyyy**

                                        Approved by:

Equipment should be examined immediately upon arrival.
- Verification of construction procedures
List of procedures should be established and reviewed.
- Construction completion
  - As-built drawing completed and approved
  - Checking of dead legs
  - Checking for proper slope for draining
- Pressure testing of the system
During construction it is impossible to avoid contamination of the piping with airborn ferrous particles from installation of structural steel and carbon steel piping components. If the stainless steel is kept dry, this may not be a problem. If the stainless steel piping is allowed to become wet, e.g., from a hydrostatic pressure test, the system should be tested with dry, oil-free air. If water is used, then provision must be made to thoroughly clean the system immediately after the hydrostatic pressure test.
- Post-construction cleaning
Flush the system to remove dust and major debris. Recirculate detergent or alkali cleaner at elevated temperatures to remove grease or oil. Flush and recirculate an acid at elevated temperatures to dissolve any ferrous particles in the system. Flush with water of the same quality as will be used in service. The cleaning procedure shall be validated by making chemical analysis of surface residues.
- System functional checkout
The instrumentation and controls should be adjusted and calibrated to ensure proper monitoring and control of functions.

3. Start-Up Validation

- Functional operation:
- Verify consistency of operation of equipment and controls by repeated cycles of start-up and shutdown of all equipment and controls. Simulate manual, automatic, and emergency conditions. Verify suitability of design under all conditions.
- Establish preliminary monitoring program to ensure validation conditions, specificity, and calibration maintenance.
- Equipment logs, filter logs, and monitoring records must be properly documented.

4. Quality Limits

- Verify that the water produced by the system meets all the pre-defined chemical, microbiological, and pyrogenecity specifications.
- Verify sanitization temperature and pressure by steam.
- Establish target and alert limits for chemical and microbial quality.

5. System Qualification

Once the validation report is completed and approved, a qualification run should be made with the system to verify that validations will be duplicated in normal operation.

6. Approval for Use

If all requirements have been satisfied and all validation documents have been approved, the quality assurance managers may release the water system for production use.

7. Microbiological Investigations: Water for Injection

7.1 Viable count

Procedure

Take 100 ml sample from each tap point, one before the start of the production, and the second at the middle of the working day. An equivalent sample taken from the last sample point of the generating device and from the first sample point of the returning water should be used as a reference.

Requirements

Viable count ≤10 cfu/100 ml for each sample

SOP No. Val. 600.90   **Effective date: mm/dd/yyyy**
                                           Approved by:

Frequency

Preproduction and concurrent validation: three working days a week (e.g., Monday, Wednesday, Friday)
Ongoing validation: each tap point once a week
Reference sample each working day

7.2 Pseudomonads and coliform

Procedure

Take 300 ml sample from the last sample point of the generating device and from the first sample point of the returning water.

Requirements

Pseudomonads/Coliform: 0 cfu/100 ml for each sample

Frequency

Each last day in the work week.

7.3 Pyrogens

Procedure

Take 10 ml sample from each tap point. An equivalent sample taken from the last sample point of the generating device and from the first sample point of the returning water should be used as a reference.

Requirements

≤0.25 EU/ml for each sample

Frequency

Preproduction and concurrent validation: each tap point, once a week
Ongoing validation: each tap point, once a month

Effective date: mm/dd/yyyy
Approved by:

8. Chemical Investigation

8.1 Investigation according to the pharmocopias (USP 24)

Procedure

From the last sample point of the generating system and from the first sample point of the returning water a sample of 1 l from each must be taken, at the same time, at the beginning of the working week and at the moment when water is first pumped through the distribution network circuit.

Requirements

Fulfilling of the pharmcopia requirements (USP 24)

Frequency

Prevalidation and concurrent validation: once a week
Ongoing validation: once a month

8.2 Quick limit-test for an indicator ion

Chemical investigations are rather time consuming. Therefore an indicative test, in which the indicator ion is the chloride ion (Cl^-), to get information about the status of the water system quickly.

Procedure

After 3 min of prerinsing, a sample of 50 ml is taken from the last sample point of the generating device and from the first sample point of the returning water.

Limit Test

To 10 ml of the water sample, 1 ml of dilute nitric acid and 0.2 ml of silver nitrate solution are added.

SOP No. Val. 600.90 **Effective date: mm/dd/yyyy**
Approved by:

Used reagents: Dilute nitric acid
Contains about 12.5% M/V of HNO_3
Silver nitrate solution
A 1.7% M/V solution

Requirements

The test solution should show no change in appearance for at least 15 min.

Frequency

Concurrent validation: Daily
Ongoing validation: At the first working day of the week

9. Physical Investigation

9.1 Conductivity

For monitoring system performance, a conductivity meter should be present or built in the water system for continuous measurement of the conductivity. If this device is not available, samples must be taken manually and measured respectively.

Procedure

After prerinsing for 3 min, samples of about 50 ml should be taken.

Requirements

Normally, for pharmaceuticals, the requirement is ≤1 µS/cm, but this depends on the local terms of reference.

Frequency

Concurrent validation: at the start and in the middle of each working day
Ongoing validation: at the start of each working day

SOP No. Val. 600.90 **Effective date: mm/dd/yyyy**

Approved by:

9.2 pH

In purified water it is hardly possible to measure the pH.

Procedure

After prerinsing for 5 min, samples of about 50 ml should be taken from the sample point of the returning water, after which a few drops of a saturated potassium chloride solution should be added. Then pH can be measured.

Requirements

The pH for each sample should be between 5 and 7 (for information only).

Frequency

Concurrent validation: at the start and in the middle of each working day
Ongoing validation: at the start of each working day

9.3 Temperature

The temperature within the system plays an important role from a micro-biological point of view.

Requirements

For water prepared by distillation and to be used as water for injection, the requirement is >80°C. Systems with built-in reverse osmosis modules and ultrafiltration devices must comply with the supplier specification.

9.4 Pressure testing

The pressure within the system is an essential factor for functioning of the water generating and distribution system. Therefore pressure prior to and after subunits of the system should be registered continuously.

SOP No. Val. 600.90　　　　　　　　　　**Effective date: mm/dd/yyyy**

　　　　　　　　　　　　　　　　　　　　Approved by:

Requirements

Should meet supplier specification

REASONS FOR REVISION

Effective date: mm/dd/yyyy

- First time issued for your company, affiliates, and contract manufacturers

---

**YOUR COMPANY
VALIDATION STANDARD OPERATING PROCEDURE**

SOP No. Val. 600.100

Effective date: mm/dd/yyyy

Approved by:

TITLE: Oil-Free Compressed Air System

AUTHOR: _____

Name/Title/Department

_____

Signature/Date

CHECKED BY: _____

Name/Title/Department

_____

Signature/Date

APPROVED BY: _____

Name/Title/Department

_____

Signature/Date

REVISIONS:

No.	Section	Pages	Initials/Date

SOP No. Val. 600.100   Effective date: mm/dd/yyyy

Approved by:

SUBJECT: Oil-Free Compressed Air System

PURPOSE

To describe guideline for the validation of the oil-free compressed air system

RESPONSIBILITY

It is the responsibility of the technical service manager to follow the procedure. The quality assurance manager is responsible for SOP compliance.

PROCEDURE

1. Types of Compressed Air Systems

Two types of compressed air systems are found in an aseptic manufacturing facility:

- An instrument air system normally consists of conventional oil-lubricated compressors and is used for operating instruments and machinery where no contact with the product or product environment exists.
- An oil-free compressed air system is normally used in aseptic areas and often may be involved with product contact.

The system consists of an oil-free compressor, drier, storage tank, and distribution system. The validation process consists of installation qualification, operational qualifications, and actual validation testing of the operational system.

2. Installation Qualification of Oil-Free Compressor

- Verify and document specifications on purchase order against actual delivery.
- Check and document that no oil or other lubricant is used in the compressor.
- Verify and document that all required utilities are connected properly.
- Verify prestartup procedures.
- Document calibration performed.

3. Installation Qualification of Compressed Air Storage Tank

- Check and document that the materials of construction are as specified.
- Check storage tank for adequate capacity.
- Perform and document pressure hold test to determine that the leak rate is within specification.
- Perform and document the cleaning procedures after installation.
- Check and document all pressure ratings.
- Calibrate all critical pressure gauges and control sensors on the storage tank.

4. Installation Qualification Distribution System

- Check and document that the materials of construction are as specified.
- Follow the drawings of the system to trace the actual constructed system and make an "as built" drawing.
- Pressure test the system and document.
- Clean the system with detergent or solvent and document the procedures.
- Label all piping and components.

5. Operational qualification

5.1 Chemical investigation

Sampling procedure

Materials: Gas bag, 3.8 liter capacity (rubber bladder), with stopcock
            Rubber tubing of appropriate size
            Aluminium foil squares (10 × 10 cm)
Sampling: Use method as recommended by the manufacturer with all safety precautions.

6. Identification

Procedure

Use gas chromatograph for the identification of compressed air. For comparison, an air standard should be used.

SOP No. Val. 600.100

Effective date: mm/dd/yyyy

Approved by:

Requirements

The identity test for oil-free compressed air must show a chromatogram with no additional peaks other than those obtained with the air standard.

Frequency

Initial validation: once at all critical supply points
Revalidation: once at all critical supply points

7. Moisture Content

Procedure

Use dew-point meter to determine moisture content from critical supply point

Requirements

Moisture content measurements at supply point should not be greater than in-house specification

Frequency

Initial validation: one test per day for the first 30 days of the operation of the system from a different location each day. The test program should cover all critical supply points.
Revalidation: one test from each critical supply point per month.

8. Oil Content

Procedure

Use oil indicators.

Requirements

Oil content of oil-free compressed air should be not more than 0.01 ppm.

Frequency

Initial validation: one test per day for the first 30 days of the operation of the system from all critical supply point locations each day.

SOP No. Val. 600.100

Effective date: mm/dd/yyyy

Approved by:

Revalidation: one test from each critical supply point

9. Nonviable Particle Count

Procedure

The outlet of the supply point is opened and purged for 5 min. Adjust to a volume flow of about 30 l/min. The particle counter is connected to the outlet; at the maintained flow a minimum volume of 90 liters is monitored. Each supply point should be investigated in the same way.

Requirements

No requirements; for information only

Frequency

Initial validation: once for each supply point.
Revalidation: every 3 months

9.1 System supply reliability test

Document the system pressure twice a day over a period of about 30 working days. The data generated should be compared with the specifications of the system.

10. Certification

The system can be certified after successful execution and documentation of above tests.

REASONS FOR REVISION

Effective date: mm/dd/yyyy

- ▪ First time issued for your company, affiliates, and contract manufacturers

YOUR COMPANY
VALIDATION STANDARD OPERATING PROCEDURE

SOP No. Val. 600.110   Effective date: mm/dd/yyyy

        Approved by:

TITLE: Nitrogen Distribution System

AUTHOR: _____

                   Name/Title/Department

                   _____

                   Signature/Date

CHECKED BY: _____

                   Name/Title/Department

                   _____

                   Signature/Date

APPROVED BY: _____

                   Name/Title/Department

                   _____

                   Signature/Date

REVISIONS:

No.	Section	Pages	Initials/Date

SOP No. Val. 600.110   Effective date: mm/dd/yyyy
                                        Approved by:

SUBJECT: Nitrogen Distribution System

PURPOSE

To describe the procedure for validation of the nitrogen distribution system to be installed and operated per specification

RESPONSIBILITY

The technical services manager is responsible for following the procedure. The quality assurance manager is responsible for SOP compliance.

PROCEDURE

1. Installation Qualification of Gas Generator

- Check availability of electrical supply (voltage, amperage, frequency).
- Chilled water (supply volume, temperature)
- Compressed air (pressure, cubic feet per minute, oil-free)

2. Pre-startup Procedures

- Clean out system to remove construction debris, lubricants, and residue from manufacture.
- Fill all lubricant reservoirs.
- Check all utility connections.
- Test safety device (automatic shutdown and general safety equipment) for proper operation.

3. Calibration of All Critical Gauges and Instrumentation

- Identify and write standard operating procedures for calibration.
- Perform calibration of critical gauges.

SOP No. Val. 600.110   **Effective date: mm/dd/yyyy**

Approved by:

4. Operational Qualification

- Verify alarm control.
- Perform calibration requirements identified in the manual or established by the validation team.
- Operate the equipment at low, medium, and high speed per operations manual to verify the operating control.
- Verify all switched and push buttons are functioning properly.
- Establish procedures for operation, maintenance, and calibration.
- Establish training program for the relevant staff.

5. Storage Tank

- Check the storage tank for capacity per specification.
- Check that material and construction conform to specification.
- Conduct a pressure hold test and determine that the leak rate is within specifications.
- Document cleaning procedures done on the tank after installation.
- Calibrate and document all pressure gauges and sensors, both monitoring and controlling.

6. Distribution

- Confirm that the materials of construction and design parameters are per specification.
- Verify the as-built drawings.
- The system should be pressure tested and documented to confirm its integrity.

7. Chemical Investigation

For sampling procedure see SOP No. Val. 600.100, step 5.1, of the guideline for an oil-free compressed air system.

7.1 Compliance with pharmacopial monograph

Procedure

Samples for analysis should be taken from each critical supply point, and meet the official monographic requirement.

Requirements

Results should be in compliance with pharmacopial monograph

USP: 99% pure (nitrogen)
      0.001 or less (carbon monoxide content)

Frequency

Initial validation and revalidation: one test from each supply point
Revalidation: every 3 months

7.2 Moisture content

Procedure

Determine the moisture content at critical points. Use dew-point meter.

Requirements

No requirements; for information only (or per in-house specification)

Frequency

Initial validation and revalidation: one test from each critical supply point

8. Physical Investigation

8.1 Nonviable particle count

Procedure

Purge the outlet of the supply point for 5 min. Adjust to a volume flow of about 30 l/min. Connect particle counter to the outlet for sampling. All supply points should be investigated in the same way.

SOP No. Val. 600.110   **Effective date: mm/dd/yyyy**
                                              Approved by:

Requirements

No requirements; for information only (per in-house specification)

Frequency

Initial validation: once for each supply point
Revalidation: every 3 months

System supply reliability test

Verify the system pressure twice a day over a period of about 30 working days.

Acceptance criteria

System should meet the specification requirement.

9. Microbiological Investigation

9.1 Viable particle count

Procedure

Purge the outlet of the supply point for 5 min. Adjust to a flow volume of about 30 l/min. Pass the nitrogen through an air sampler provided with a 0.22 μm filter. Sampling time is about 5 min. Perform the procedure twice. One filter should be incubated for anaerobic, the other for aerobic viable count.

Filters used in the system (commonly at point of use) should be included in a routine filter integrity testing program. Establish the life cycle of the filters to maintain the system.

Requirements

No requirements; for information only (per in-house requirements)

Frequency

Initial validation and revalidation: once for each supply point

SOP No. Val. 600.110

Effective date: mm/dd/yyyy

Approved by:

REASONS FOR REVISION

Effective date: mm/dd/yyyy

■ First time issued for your company, affiliates, and contract manufacturers

| YOUR COMPANY |
| VALIDATION STANDARD OPERATING PROCEDURE |

SOP No. Val. 600.120   Effective date: mm/dd/yyyy

                                        Approved by:

TITLE: Clean Steam

AUTHOR: _____
                            Name/Title/Department

               _____
                            Signature/Date

CHECKED BY: _____
                            Name/Title/Department

               _____
                            Signature/Date

APPROVED BY: _____
                            Name/Title/Department

               _____
                            Signature/Date

REVISIONS:

No.	Section	Pages	Initials/Date

SOP No. Val. 600.120

Effective date: mm/dd/yyyy
Approved by:

SUBJECT: Clean Steam

PURPOSE

To describe the validation guideline for clean steam

RESPONSIBILITY

It is the responsibility of the utilities manager to follow the procedure. The quality assurance manager is responsible for SOP compliance.

PROCEDURE

1. Steam System

The following types of steam systems are normally found in modern aseptic manufacturing facilities:

- "House steam" consists of a steam generator and distribution system made of either iron or steel, which are both subject to rusting. This system is normally used in applications where contact with product or product contact surfaces does not occur.
- "Clean steam" systems are normally constructed of nonrusting (stainless steel) materials and typically use either distilled or deionized water as feed water; no additives are allowed to be used due to contact with products.

2. Types of Steam Systems

	Plant steam	Clean steam
Feed water	Portable, softened, or deionized water	Water for injection (distilled, reverse osmosis) or purified water
Material system construction	Iron or steel or stainless steel	Stainless steel
Use of additives	Yes — hydrazines, amines, etc.	None
Condensate	Commonly reused	May or may not be reused

SOP No. Val. 600.120   Effective date: mm/dd/yyyy
                                        Approved by:

3. Major Pieces of Equipment

■ Deionizer
■ Distillation equipment (optional)
■ Holding tank
■ Steam generator
■ Distribution system

3.1 Installation qualification deionizer

■ Check unit for conformance to purchase specifications.
■ Connect unit to required utilities and verify the correctness of utilities connected.
■ Check and document all plumbing connections.
■ Calibrate and document that all instrumentation is operating correctly.

3.2 Operational qualification deionizer

■ Test that water of appropriate quality (conductivity) is produced by the system using written standard operating procedures.
■ Verify that the regeneration system works satisfactorily.
■ Verify the adequacy of the ultraviolet light device used for maintaining deionized systems in a sanitary condition.

3.3 Installation qualification distillation equipment

■ Document that the distillation equipment received conforms to the purchase specifications.
■ Verify the correctness of utilities connections.
■ Complete all required prestart-up maintenance procedures (including cleaning).
■ Calibrate, check, and document all critical process instrumentation.

3.4 Operational qualification distillation equipment

■ Using the written standard operating procedure, start up and run the distillation equipment.

■ Check and verify that the water produced by the still conforms to specifications (quality and quantity).

3.5 Installation qualification holding tanks

■ Verify and document that the tank conforms to purchase specifications.
■ Pressure test the vessel to determine that the leak rate conforms to specifications, then document.
■ Ensure that the pressure rating of the vessel conforms to specifications.
■ Perform all required cleanout procedures for start-up.
■ Calibrate, check, and document all instrumentation systems.

3.6 Operation qualification holding tanks

■ Check all instrumentation systems during actual operation and document.
■ Check heating system (control) for correct operation.
■ Fill tank with distilled water and hold for typical production cycle to determine that the water quality does not change adversely during storage.

3.7 Installation qualification clean steam generator

■ Verify and document that the steam generator conforms to purchase specifications.
■ Connect the generator to the required utilities; verify and document that they are correctly connected.
■ Passivate generator after installation.
■ Calibrate, check, and document all critical process instrumentation.

3.8 Operational qualification clean steam generator

■ Determine the normal operating parameters of the system.
■ Verify and document that all instrumentation and alarms are working correctly.
■ Check and document that the steam produced meets quantitative and qualitative specifications. (The steam output should be condensed and then tested against current USP WFI).

SOP No. Val. 600.120

Effective date: mm/dd/yyyy

Approved by:

3.9 Installation qualification distribution system

- Check and document that materials of construction conform to specifications.
- Using design drawings, verify distribution system to determine that specifications have been met.
- Complete and document the cleaning of the system prior to start-up.
- Pressure test the system under actual production conditions and document the results.

3.10 Operational qualification distribution system

- Test and verify all use points of the system for adequate supply of steam under maximum load or production conditions.
- Steam quality should be tested at use points by condensing steam and condensing current USP, WFI, on the condensate.
- Use points should also be checked to determine that excess condensate is not present under operating conditions.

4. General Sampling Procedure

A condenser is connected to the clean steam supply point. Allow the clean steam to flow through the condenser without coolant flowing through the condenser for approximately 30 min. This will sterilize the condenser and allow it to be cleaned. The sampling bottles, condenser tubing, and fittings must be depyrogenated prior to use. After turning on the coolant, sampling from the lowest point of the condenser can be started. Caution should be observed during the sampling of clean steam, as it is dangerous.

5. Chemical Investigation

Procedure

Chemical investigation is performed according to the pharmacopial monograph, Water for Injection. Take a minimum of three condensate samples (sample volume 1 l each) from each supply point.

SOP No. Val. 600.120   **Effective date: mm/dd/yyyy**
                                            Approved by:

Requirements

The acceptance criteria of USP monographs should be met.

6. Microbiological Investigation

6.1 Viable count

Procedure

Take a minimum of three condensate samples (sample volume 300 ml each) from each supply point.

Requirements

≤10 cfu/100 ml

6.2 Pyrogen test

Procedure

Endotoxin determination is performed according to the bacterial endotoxins test of the USP. Take a minimum of three condensate samples (sample volume 100 ml each) from each supply point.

Requirements

No endotoxins detectable

Frequency

The investigations should be performed at the initial validation and should be repeated after maintenance and repair work on the clean steam system.

REASONS FOR REVISION

Effective date: mm/dd/yyyy

- ■ First time issued for your company, affiliates and contract manufacturers

YOUR COMPANY
VALIDATION STANDARD OPERATING PROCEDURE

SOP No. Val. 600.130

Effective date: mm/dd/yyyy

Approved by:

TITLE: Vacuum System

AUTHOR: _____

Name/Title/Department

_____

Signature/Date

CHECKED BY: _____

Name/Title/Department

_____

Signature/Date

APPROVED BY: _____

Name/Title/Department

_____

Signature/Date

REVISIONS:

No.	Section	Pages	Initials/Date

SOP No. Val. 600.130   Effective date: mm/dd/yyyy

Approved by:

SUBJECT: Vacuum System

PURPOSE

To describe the procedure for validation of the vacuum system

RESPONSIBILITY

It is the responsibility of the utilities manager to follow the procedure. The quality assurance manager is responsible for SOP compliance

PROCEDURE

1. Vacuum Systems

Three vacuum systems are commonly used in modern aseptic manufacturing facilities: (1) house vacuum systems, (2) vacuum systems dedicated to lyophilization equipment, and (3) vacuum systems dedicated to autoclaves or other sterilization equipment.

2. Installation Qualification of Vacuum Pump

2.1 Documentation

As-built drawings of the vacuum system within the plant should be available. The vacuum system should be installed in accordance with the set specifications. Records about maintenance repairs and modifications should be filed.

Check and document that the pumps conform to purchase specifications. Connect the pumps to the required utilities and document that the utilities are correct. Tighten flanges and mounts. Fill pumps with oil (if required). Check shock mountings and remove shipping restraints. Calibrate. Check and document all critical process instrumentation.

3. Operational Qualification of Vacuum Pumps

The following tests should be executed with disconnected vacuum-consuming equipment

SOP No. Val. 600.130   Effective date: mm/dd/yyyy
                                         Approved by:

3.1 Maximum obtainable vacuum

Procedure

Evacuate the system and determine the maximum obtainable vacuum.

Requirements

No requirements; for information only

Frequency

Initial validation and revalidation once. The test should be repeated after maintenance, repairs, and modifications of the system.

3.2 Vacuum hold test

Procedure

Evacuate the system to its maximum obtainable vacuum and then isolate the pump from the system. Monitor the vacuum over a period of 45 min.

Requirements

A closed system should not lose more than 3 to 7 KPa within 45 min of testing. The actual performance of the system will vary depending on the length of the system and the number of valves. Establish acceptance criteria for each individual system under test.

Frequency

Initial validation and revalidation once. The test should be repeated after maintenance, repairs, and modifications of the system.

3.3 Time required for reaching the maximum obtainable vacuum

Procedure

Evacuate the system to its maximum obtainable vacuum and record the time needed for this.

SOP No. Val. 600.130   Effective date: mm/dd/yyyy
                                          Approved by:

Requirements

No requirements; for information only

Frequency

Initial validation and revalidation once. The test should be repeated after maintenance, repairs, and modifications of the system. Following trials should be performed with connected vacuum-consuming equipment.

3.4   Determination of the worst case load

Procedure

Testing should also be performed using maximum demands on the system to determine the worst case load.

Requirements

No requirements; for information only

Frequency

Initial validation and revalidation once. The test should be repeated when new, additional equipment is connected to the vacuum system.

4.   Installation Qualification of Reservoir Tank

4.1   Documentation

Document that the reservoir conforms to purchase specification and invoice. Verify and document that the vessel meets or exceeds the pressure rating (vacuum) specified in the purchase specifications. Perform vacuum hold tests on the tank and document. Acceptance ID tests will vary with the size of the system. A positive pressure test is often done in order to find leaks. Perform and document cleaning procedures used prior to placing the vessel in service. This completes the normal testing done on the tank prior to joining it to the vacuum system.

SOP No. Val. 600.130   **Effective date: mm/dd/yyyy**
                                               Approved by:

5. Installation Qualification Distribution System

Check and document that distribution system materials of construction conform to specifications. Using design drawings, determine dimensions and all design features such as filters, strainers, check valves, etc. An as-built drawing should then be created to document the system. All branches of the system should be labeled. Pressure test the system using positive and negative pressure testing (pressure and vacuum hold tests). Complete and document cleaning procedures prior to system start-up. Calibrate all critical gauges, alarms, and automatic controllers.

6. Vacuum System Certification Testing

Test the system under production conditions to determine that it can reproducibly reach the vacuum required within the normal time constraints. The results of the testing should then be documented for future reference.

REASONS FOR REVISION

Effective date: mm/dd/yyyy

- First time issued for your company, affiliates, and contract manufacturers

YOUR COMPANY
VALIDATION STANDARD OPERATING PROCEDURE

SOP No. Val. 600.140   Effective date: mm/dd/yyyy

Approved by:

TITLE: Validation of an HVAC System

AUTHOR: _____

Name/Title/Department

_____

Signature/Date

CHECKED BY: _____

Name/Title/Department

_____

Signature/Date

APPROVED BY: _____

Name/Title/Department

_____

Signature/Date

REVISIONS:

No.	Section	Pages	Initials/Date

SOP No. Val. 600.140   Effective date: mm/dd/yyyy
Approved by:

SUBJECT: Validation of an HVAC System

PURPOSE

To provide the guideline for validation of the HVAC system to meet the design qualification requirement and to be capable of operating within established limits and tolerances

RESPONSIBILITY

It is the responsibility of the technical service manager and contractors to follow the procedure. The quality assurance manager is responsible for SOP compliance.

PROCEDURE

1. Documentation

Check the quality of materials received against purchasing specifications.

1.1 Scope

The HVAC system, in general, comprises the following elements:

- Design
- Calculation
- Airflow
- HVAC P and ID
- HVAC ductwork and equipment arrangement
- Control P and ID
- Air handling units
- Reheat coils
- Ductwork
- Insulation for HVAC ductwork
- Sound attenuators
- Humidifiers
- Air distribution
- Supply air grills

SOP No. Val. 600.140 Effective date: mm/dd/yyyy

Approved by:

- Exhaust air grills
- Sand trap
- Motorized dampers
- Round duct dampers
- Diffusers
- Programmable logic controller panel with control unit
- Chillers
- Recirculation pumps
- Valves
- Pressurization unit

The controls involved are generally as follows:

- Air velocity sensor
- Room pressure sensor
- Duct-mounted temperature sensor
- Duct-mounted combined temperature and RH sensor
- Differential pressure switch
- Airflow switch
- On/Off motorized damper actuator
- Two-way modulating valve
- Three-way modulating valve
- Actuator for modulating valve
- Damper actuator

1.2 Documentation

Check that the approved layout is available for reference; verify the intended purpose of the rooms against the standard requirement. For technical information (rooms, activity cleanness class, pressure, temperature, RH, number of air changes and AHUs), prepare a table for reference.

2. Characteristics Checks Performed

The following checks may be performed to assure that the HVAC system meets the requirements specified in the contract.

SOP No. Val. 600.140   Effective date: mm/dd/yyyy

Approved by:

2.1 General

Verify that formal design criteria have been developed and approved and that the document being checked conforms in all respects to the approved design criteria.

2.2 Specific

Design

1. Duct hangers and supports, in clean areas, are designed to prevent dust collection.
2. Outside intake and exhaust points are filtered with bird screens.
3. Access doors are provided at all in-line devices requiring access (coils, vanes, etc.).
4. Duct work materials are in accordance with design criteria.
5. Air intakes are located as far away as possible from any upwind air exhaust.
6. Process equipment heat loads have been established and reviewed.
7. Locations of laminar flow hoods, biosafety hoods, and exhaust hoods are noted.
8. Directional airflow based on pressure differential is noted along with special exhaust requirements.
9. Pressurization is noted in the form of actual room pressure levels from a common reference point. Reference point is not outdoors (affected by wind).
10. Filtration addresses final filter efficiency and location.
11. Temperature and humidity are listed as design set points with plus and minus tolerances.
12. Inside design criteria include, but are not limited to, temperature, relative humidity, filtration level, minimum air change rate, and pressurization requirements.

Calculations

1. Heating calculations include heat required to raise supply air temperature from summer design point to room temperature plus sufficient heat to offset winter heat loss.

2. Cooling load calculations have been performed on a room-by-room basis. Considerations include motor load, radiation from heated vessels, radiation from thinly insulated process and utility piping and equipment, electrical loads, and AHU supply fan motors for air handling units.

3. The calculation for the airflow leakage rate for each room is based on the pressure differential established on the design criteria sheet and not on a percentage of supply air.

4. Fan static pressure is calculated.

Air flow diagrams

1. Supply, return, exhaust, infiltration, and exfiltration airflow from each room are shown on airflow diagrams

2. Air flow diagrams for air handling unit components show, in proper sequence, flow measuring stations, reheat coils, location of each filtration level, humidifiers, exhaust fans and other system components.

HVAC P and IDs

1. HVAC, chilled water, and hot water P and IDs are generated independently of process-related systems and have been checked using a P and ID checklist.

HVAC ductwork and equipment arrangement drawings

1. All ducts identify materials of construction, insulation type, and pressure classification.

2. Duct center-line locations are shown from column lines.

3. Sections are provided at any design point not clearly defined.

4. Ductwork is coordinated with other disciplines, especially piping and electrical.

5. Duct work is drawn as double-lined and fully dimensioned duct (including center-line elevation for round duct and bottom-of-duct elevation for rectangular duct).

6. HVAC drawings are detailed.

SOP No. Val. 600.140

Effective date: mm/dd/yyyy
Approved by:

Control P and IDs

1. Separate control diagrams have been generated for each air handling unit, showing all control devices individually.
2. Control diagrams have been supplemented with sequences of operation and a control valve schedule showing the size and specification for each valve.
3. Control diagrams show coil piping, including valves and line number.

Air handling units

1. Insulation
   - Double-wall construction sandwiching insulation between two metal panels or single-wall construction with external insulation is used.
   - Insulation has not been placed on the inside of the air supply ducts.
   - Check compliance with all relevant drawings.
   - Check physical appearance.
   - Check positioning.
   - Check leveling.
   - Check piping connection with pump and valve.
   - Check bolt tightness in section joints.
   - Check denting on coil fin.
   - Check free rotation of impeller.
   - Check that all components of AHU have been installed per manufacturer instruction.
   - Check belt drive alignment.
2. Filters
   - Air handling unit has been provided with prefilters two inches thick at a minimum and medium efficiency cartridge or bag filters.
   - Check compliance with all relevant drawings and check physical appearance.
   - Check complete cleanliness of filter casing.
   - Check gasket and clamp.
   - Check correct air flow direction.
   - Check proper mounting.
   - Check the serial number and relevant test certificate.

3. Fans
   - Fans are belt drive with variable-pitch drives for 25 HP or less and fixed-pitch drives for drives above 25 HP.
   - Air handling unit fans are able to modulate air flow by using inlet vanes, discharge dampers, or variable speed drives.
   - Fan motors are sized 25% above brake horsepower.
4. Sand traps
   If the sand traps are used due to the climatic conditions of the area, the following checks shall be performed on fan:
   - Check overall cleanliness.
   - Check all components, bolts, and fixing, and secure.
   - Check physical damage to casings, impeller, drives.
   - Check alignment of pulleys and couplings.
   - Check belt tension and match.
   - Check inlet guide vanes over full range of movement.
   - Check anti-vibration mountings, free to vibrate.
   - Check impeller, free to rotate.
   - Check for abnormal noise in bearings on free rotation.
   - Check for grease nipples on fan and motor bearings.
   - Check blanked off temperature nipple on bearings.
   - Check drive guards with speed measuring openings.
   - Check inlet protection guard.
   - Check flexible connections.
   - Reinstall drive guard.
   - Check motor type to specification.
   - Check power wiring of motor.
   - Check air flow direction.
   - Check tag number and manufacturer's ID plate.
5. Coils
   - Handling unit coils have no more than eight fins per in. and are no more than six rows in depth.
   - Coils are properly oriented.
   - Fins are continuous and flat (noncorrugated).
   - Coils are piped to achieve air and water counterflow.
   - Coils are spaced a minimum of 24 in. apart.
   - If cooling capacity requires more than six rows, two coils are to be placed in series.
   - Water connections and manifolds are mounted.
   - Updated FID is available.

SOP No. Val. 600.140   **Effective date: mm/dd/yyyy**

                                                    Approved by:

- Check visual damage on casings.
- Check damage on fins and tubes.
- Check bolting air side.
- Check bolting flanges or threaded connection.
- Check clearance.
- Check vents.
- Check drains and drain pipes.
- Check water traps, siphons, and condensate pipe.
- Verify water trap height against design under pressure on cooling coils.
- Check stress-free connection to manifolds.
- Check for venting possibilities of pipe work to coil.
- Check for draining possibilities of pipe work to coil.
- Check air flow direction through coil.
- Check water flow direction through coil.
- Check painting quality of steel parts.
- Check configuration of control station, pumps, and valves.
- Check adequate supporting and hangers.
- Check for correct tag number and manufacturer's ID plate.

6. Reheat coils
   - Maximum fin spacing on reheat coils does not exceed eight fins per in. (more than two rows are rarely required).
   - Smaller reheat coils are not sized for a large water-side temperature difference.
   - Reheat coils are electric (multistage or SCR controlled) or hot water.

7. Duct work
   - Materials
     Duct work should be constructed of lock forming quality aluminum, galvanized steel, or stainless steel.
   - Specifications
     Specific tables are generated for each pressure classification and each duct system pressure class is clearly defined.
   - Flexible ductwork
     - Round, flexible duct contains aluminum foil liner in lieu of a vinyl liner.
     - The helix of flexible ductwork has formed on the outside of duct's surface.
   - Flexible ductwork complies with NEPA 90A and 90B.
   - Check compliance with all relevant drawings.

Effective date: mm/dd/yyyy
Approved by:

- Check physical appearance.
- Check cleaning of ducts.
- Check flange joints.
- Check fixings of damper.
- Check continuity of insulation and vapor sealing.
- Check fixing of inspection door.
- Elbows
  - All vaned elbows have an excess door located nearby for cleaning the vanes.
  - Either elbows of the radius type and without turning vanes or square mitered elbows equipped with single-thickness, extended-edge turning vanes are used.
- Dampers
  - Balancing dampers have a self-locking regulator suitable for securing the damper at the desired setting and making adjustments.
  - Manual balancing dampers are installed where required and shown clearly on design drawings.
  - End switches and wiring are completed. Perform the following checks on motorized damages:
    - Check overall cleanliness.
    - Check damages on casing, blades, spindels, and seats.
    - Check clearance.
    - Check free movement of blades.
    - Check relative position of blades in multileaf dampers.
    - Check pinning to damper spindles.
    - Check position of blades to quadrant indication.
    - Check fixation of damper bearings.
    - Check control linkages for alignment, rigidity free movement without slop
    - Check position of end switches and adjust.
    - Check wiring of motor and switches.
    - Check coupling of damper with other dampers and mode of operation.
    - Check fail position, e.g., fail open, fail closed, fail as is.
    - Check direction of low (if relevant).
    - Check open/closed indication.
    - Check manufacturer's ID plate.

SOP No. Val. 600.140   **Effective date: mm/dd/yyyy**

                                                     Approved by:

- System shut off checks for motorized dampers
  - Check position of damper and system shutoff.
  - Check fail position at loss of power (simulate).
  - Check working of end switches.
  - Check free movement of blades.
  - Check controlling movement of damper.
  - Check stressless operation of control linkages.
  - Check mode of operation of linked dampers.
  - Check and adjust course of servomotor.
  - Check control linkages for alignment, rigidity-free movement without slop
- System running checks for motorized dampers
  - Check the position of the damper with the system on.
  - Check fail position at loss of power (simulate).
  - Check working of end switches.
  - Check free movement of blades.
  - Check controlling movement of damper.
  - Check stressless operation of control linkages.
  - Check mode of operation of linked dampers.
  - Check and adjust course of servomotor.
  - Check control linkages for alignment, rigidity-free movement without slop
  - Verify deformation, due to (under) pressure.
8. Access doors
   - Hinged access doors have not been used.
   - Access doors are large enough for a person to clean obstructions from the interior of the duct and are provided with extension collars and sash locks.
   - Access doors are clearly indicated on the drawings.
   - Access doors are located at reheat coils and at any other surface that could collect material within the duct.
9. Hangers and supports
   Exposed duct work has smooth rod hangers threaded only at the end.
10. Diffusers
    The following checks are recommended on diffusers:
    - Check the cleanliness of the diffuser and connection box.
    - Check for any damage.
    - Verify correct type in correct location.

- Check connection to box and duct.
- Check rigid suspension.
- Check open position of regulating damper.
- Check and adjust deflection vanes.
- Check correct sealing to ceiling.
11. Insulation for HVAC ductwork
    - Rigid duct insulation is used in mechanical rooms.
    - Jacketed ductwork is dense enough to minimize dimpling.
    - Exposed duct work in clean spaces is insulated with rigid board-type insulation and jacketed with a washable metallic or PVC jacket
    - No internal duct insulation has been used.
    - Check complete mounting of units, correctly fitted in link.
    - Check air handling units.
    - Check supply and exhaust fans.
    - Check supply, return, and exhaust air duct work.
    - Check motorized, regulating, and nonreturn dampers.
    - Check fire and smoke exhaust dampers.
    - Check filter units.
    - Check diffusers.
    - Check control instruments and apparatus.
    - Check electrical and pneumatic connections to apparatus.
    - Check clearance for fans, coils, and filters.
    - Check that air leakage test is complete.
    - Check that duct work is cleaned and flushed.
    - Check that terminal units are cleaned.
    - Check that fire dampers are in open position.
    - Check that smoke exhaust dampers are in closed position.
    - Check that measurement point is identified and holes drilled.
    - Check that access to measurement is points free.
    - Check that prefilters are installed.
    - Check that HEPA filters are available and installed.
12. Sound attenuators
    - Sound attenuators are not used in systems requiring periodic sanitizing.
13. Electrical power and control system
    - Power supply to electrical panels is cut off.
    - Transit package is removed from equipment.
    - Panels and switch gears are clean and undamaged.

- Connections are tight on busbars and wiring.
- Fuse rating is correct.
- Starter overloads are correctly set according to circuit diagrams.
- Thermal cut-outs are correctly set.
- Internal links on starters are correct.
- All contractors, etc. are correctly fixed.
- Dash pots are correctly charged; time adjustments and levels are identified.
- There is no loose wiring.
- All cover plates are fixed.
- Power and control wiring is complete and in accordance with circuit diagram.

14. Humidifiers
   - Stainless steel (316L grade) has been used for piping, headers, and all humidifier components.
   - Humidifiers use lean steam for humidification.
   - Check physical appearance.
   - Check location.
   - Check leveling.
   - Denting on condenser fins should be removed, if any.
   - Remove transit fixture from compressor floating mounting.

15. Air distribution
   - All returns and exhaust are the louvered, removable core type.
   - Class 100 areas are protected with plastic curtain sheets.
   - Class 10,000 or cleaner areas have low wall returns.
   - Class 100,000 areas have ceiling returns.
   - Return and exhaust air
   - Class 10,000 or cleaner areas have terminal HEPA filters in the ceiling.

16. Terminal filters (HEPA) internal
   - Preinstallation condition
     - Check compliance with all relative drawings.
     - Check physical appearance.
   - Inspection during installation
     - Check complete compliance of filter casing.
     - Check gasket and clamps.
     - Check correct airflow direction.
     - Check proper mounting.
     - Check the serial number of relevant test certificate.

SOP No. Val. 600.140   Effective date: mm/dd/yyyy

                                        Approved by:

17. Chiller
    ■ Check compliance with all relevant drawings.
    ■ Check physical appearance.
    ■ Check foundation.
    ■ Check positioning.
    ■ Check leveling.
    ■ Check bolt tightness.
    ■ Check free rotation of impeller.
    ■ Check that flow directions are correct.
    ■ Check that pump has been installed per manufacturer instruction.
    a. Visual inspection, observations, and adjustments:
       ■ Check visual damages and external cleanliness.
       ■ Check bolting and tightness.
       ■ Belt drive and coupling are securely aligned and tensioned.
       ■ Belt drive and coupling are a matched set.
       ■ Bellows are correctly mounted.
       ■ Drive guards are installed.
       ■ Tachometer access is available.
       ■ Positioning of pump shaft is correct.
       ■ Measuring gauges are installed.
       ■ Antivibration mounting is in concrete base.
       ■ Impeller is free to rotate.
       ■ Flow direction is correct.
       ■ Glands are packed and adjusted for correct drip rate.
       ■ Gland drains are fitted and free of dirt.
       ■ Pumps and motor are correctly lubricated.
       ■ Tag number and manufacturer's ID plate are correct.
       ■ Power supply and controls are connected and correct.
       ■ Starter overload and fuse ratings are correctly set.
    b. Pretest condition checks for chiller pumps:
       ■ Visual inspection report is available.
       ■ System is pressurized and completely vented.
       ■ All normally open valves are fully open.
       ■ All bypass and normally closed valves are closed.
       ■ All thermostatically controlled valves are (blocked) open.
       ■ Motorized valves are set to manual override.
       ■ Automatic control valves are set to full flow.
       ■ Standby pumps are isolated.

SOP No. Val. 600.140   **Effective date: mm/dd/yyyy**
                                              Approved by:

- Pump casing is vented.
- Direction and rotation speed is correct.
- Motor, pump, and drive are free of vibration.
- Motor, pump, and drive are free of undue noise.
- Motor starting current is correct.
- Motor running current is equal between phases.
- Motor and bearings are not overheating.
- No water sea page of lubrication from housing is detected.
- Pressure in the system keeps stable.

c. Preinstallation condition chiller piping:
  - Check compliance with all relevant drawings.
  - Check physical appearance.

d. Inspection during installation:
  - Check finishing of joints.
  - Fix valves and accessories per drawing.
  - Supporting
  - Hydraulic test at 1.5 times working pressure (1 hour)
  - Check continuity of insulation.

e. Pressure testing (hydraulic) procedure for water piping system:
   Procedure No: 210029/PTA/PIPINGS, REV O
   After completion of all hot jobs like welding and grinding, the pipeline is subject to pressure tests to ensure leak tightness of the system. Before starting the pressure test, it must be ensured that all instrument mounting sockets and connections are in suitable arrangement. At the top-most point of the piping system, two numbers of pressure gauge must be fixed with suitable isolation valves. Water should be filled with a hose connection and air should be vented from all high points of the system. When all air is vented out, water will come out from all vent points, then the vent point should be closed properly to prevent any leakage of water. Now, by means of a small hand pump, the system pressure should be increased by means of introducing a suitable quantum of water. The test pressure should be 1.25 to 1.5 times the working pressure of the system.

   When the desired pressure is developed, isolate the system by closing the isolation valve and ensure no physical leakage takes place from all welding joints as well as bolted joints. If any leakage is identified, then release the pressure. After rectification of the leakage, repeat the same procedure. The duration of the test shall be dependent upon the volume of the system and should not be more than 24 h. While starting and completing the test, record the ambient temperature and system pressure.

f. Gauge reading

Pressure and temperature gauge (dial type) can be read from the same elevation of the gauge to prevent paradox error.

g. Acceptable criteria

There should not be any visible leakage from the piping system during the testing. This could be ensured by physical verification of joints (welded and bolted) and from the pressure gauge reading. However, the gauge reading can vary due to differences in ambient temperature. In that case, physical verification of joints will be the acceptance criteria.

h. Pretest conditions expansion and pressurization system

- End of erection
- Electrical wiring completed

i. Visual inspections, observations, and adjustments

- Check cleanliness.
- Check for any damage.
- Check piping connections, drains, and safety relief.
- Verify electrical connections.
- Check rigid fixation of compressor and switchboard.
- Check (automatic) water filling system.
- Check tag number and ID plate.
-     Verify, set, and note:
  System pressure:
  (min = height of piping above tank + prepressure)
  Pressure of safety relief valve
  Min. alarm pressure
  Max. alarm pressure
- Check overall cleanliness.
- Check for any damage on casing, piping, or fans.
- Check rigidity and leveling base frame.
- Check antivibration mountings.
- Check free rotation of condenser fans.
- Check undamaged insulation of evaporator.
- Check correct insulation of Freon tubing.

SOP No. Val. 600.140   **Effective date: mm/dd/yyyy**
                                             Approved by:

- Check for stress-free connection of water piping.
- Check that leak test Freon tubing is executed.
- Check that final Freon charge is at working pressure.
- Check that electrical power is connected.
- Check that fuses are fitted of correct rating.
- Check that overloads and timers are set.
- Check that freeze protection systems are installed.
- Check that glycol and brine are added and correct concentration.
- Check that instrumentation is installed at correct location.
- Check that pipeline tappings for temperature and pressure measurements are fitted on chilled water and cooling water circuits.
- Check for tag number and manufacturer's ID plate.

18. Documentation
    Verify that the following manuals are available for all the components of the system form:
    - Manufacturers
    - Specifications
    - Operations
    - Maintenance
    - Calibration procedures
    - Design room list
    - List of drawings
    - Test certificates for terminal filters and send trap louvers
    - Certificates for AHU, filters for air handling units, chillers, controls, and electronical panels

19. Construction drawings
    The following drawings have been inspected and accurately represent the system as built:
    - P&ID
    - Isometics
    - Orthographic

20. Operating procedures
    The following procedures have been inspected and accurately represent the system as built:
    - Operation
    - Maintenance
    - Calibration

SOP No. Val. 600.140   Effective date: mm/dd/yyyy

                                                          Approved by:

21. Environmental performance
   Controlled environments should generally be subjected to the following set of performance tests related to environmental control:
   - Air flow, volume, and distribution (also designated as post-balancing verification)
   - HEPA filter/leak (DOP)
   - Temperature
   - Humidity
   - Air pressure control
   - Airborn particle count
   - Induction leak
   - Air flow patterns
   - Recovery
   - Particle dispersion
   - Airborn microbial
   - Surface bioburden
   - Air changes
   - Lighting level
   - Sound level

These tests are required for the final certification and validation of the environment. The instruments used for them should be in perfect condition, and current calibration reports should be included with the report.

REASONS FOR REVISION

Effective date: mm/dd/yyyy

- First time issued for your company, affiliates, and contract manufacturers

SECTION

VAL 700.00

YOUR COMPANY
VALIDATION STANDARD OPERATING PROCEDURE

SOP No. Val. 700.10   Effective date: mm/dd/yyyy

                                           Approved by:

TITLE: Validation of a Steam Sterilizer

AUTHOR: _____

                        Name/Title/Department

            _____

                        Signature/Date

CHECKED BY: _____

                        Name/Title/Department

            _____

                        Signature/Date

APPROVED BY: _____

                        Name/Title/Department

            _____

                        Signature/Date

REVISIONS:

No.	Section	Pages	Initials/Date

SOP No. Val. 700.10

Effective date: mm/dd/yyyy
Approved by:

SUBJECT: Validation of a Steam Sterilizer

PURPOSE

To provide a written procedure to be used as a guideline for the certification/validation of a steam sterilizer

RESPONSIBILITY

It is the responsibility of the production manager, validation manager, and concerned departmental managers to follow the procedure. The QA manager is responsible for SOP compliance.

PROCEDURE

Different types of sterilizers are used in the pharmaceutical industry. However, the following criteria are generally considered common.

- A thermostatic steam trap to efficiently remove condensate from the chamber. This is open when cool and closed when in contact with steam. As condensate collects, the trap opens due to the slight temperature reduction, and the condensate is discharged. There is also a trap to remove condensate from the steam jacket.
- A safety door mechanism to prevent opening while the unit is under pressure. The locking device may be actuated directly by internal pressure or indirectly through an automatic switch. The door itself may be the swing-out or sliding type.
- A pressure vessel constructed according to the American Society of Mechanical Engineers (ASME) code to withstand the required internal steam pressures
- A steam jacket and insulation to conserve energy designed primarily to heat the metal mass of the vessel and to limit heat loss from within the vessel. Some laboratory and small special–use sterilizers are unjacketed.
- A chamber pressure indicator
- A microbial retentive vent filter (optional)
- A cycle timer and (usually) a sequencing controller

SOP No. Val. 700.10   Effective date: mm/dd/yyyy
                                       Approved by:

■ A temperature control system. Although operating under pressure, temperature is the controlling factor in steam sterilization. The modern temperature controller is made up of several key elements to sense, record, and react. These are discussed in a later section.

1. Prevalidation Protocol

The documentation of prevalidation protocols should be as follows:

■ A brief description or scope. This is a basic but complete explanation of the sterilizer and its ancillary equipment, including physical characteristics and function.
■ Detailed specifications list
■ A list of pertinent drawings
■ System installation check sheet
■ Calibration records of all instrumentation
■ Listing of key devices and brief description of each
■ Operational record that compares actual operating parameters to specifications

2. Preparing for Validation

A critical part of the validation study is the temperature measurement. Several items will be required to measure and record temperature effectively.

Type T (copper-constant) thermocouples are most applicable in steam sterilizer validation work. Their working temperature range is wide and they are resistant to corrosion in moist environments. A high grade of thermocouple wire should be chosen. Premium grades of wire accurate to as close as 0.1°C at 121°C are recommended. These must then be calibrated against a temperature standard traceable to the National Bureau of Standards (NBS).

The acceptable error should be no greater than the sum of the thermocouple wire accuracy (e.g., +0.1 to 0.3°C) and the degree of traceability of the NBS reference instrument (i.e., ±0.2°C). Thermocouples that do not meet this criterion should be replaced.

Calibration of thermocouples should be carried out at two temperatures. One of these is an ice-point reference at 0.0°C. The other should be a hot point slightly higher than the expected sterilization temperature. Correction factors are applied at both temperatures and the response of

SOP No. Val. 700.10   Effective date: mm/dd/yyyy
                                                 Approved by:

the thermocouple over the temperature range can be linearized. The corrected temperature measurements are used to calculate F_0.

3. Validation Protocol

The documentation (installation qualification, operational qualification, etc.) established prior to initiating validation studies provides the foundation for the subsequent validation. A comprehensive steam sterilization protocol should include the following items:

- Description of objectives of the validation study
- Responsibilities of validation personnel
- Identification and description of the sterilizer and its process controls
- Identification of standard operating procedures
- Description of or SOP for instrument calibration procedures
- Identification of calibration procedures for temperature-monitoring equipment (thermocouples, data loggers, etc.)
- Description of the studies to be conducted as under
  - Bioburden determination
  - Microbiological challenge
  - Empty chamber heat distribution
  - Loaded chamber heat penetration
  - Container mapping
  - Evaluation of drug product cooling water (where applicable)
  - Integrity testing of vent filter membranes associated with the sterilizer
- Process parameter acceptance criteria

The review process may initiate supportive changes in the experimental design resulting in protocol revision. Once the protocol is approved, the validation study may begin.

4. Empty Chamber Testing

The initial testing is performed on an empty chamber to measure temperature distribution. The thermodynamic characteristics of the empty sterilizer are depicted in a temperature distribution profile, which will serve to locate hot or cold areas in the sterilizer by mapping the temperatures at various points in the chamber.

SOP No. Val. 700.10

Effective date: mm/dd/yyyy

Approved by:

The temperature profile is obtained by placing at least 10 thermocouples distributed in the empty tunnel or batch sterilizer in such a way as to determine heat profiles. In the flames sterilizer the thermocouples should be placed at the level of the ampules. The thermocouple tips should be suspended to avoid contacting any solid surfaces (wall, ceiling, support rods, etc.). A good profile should demonstrate uniform temperatures across the sterilizer.

The temperature range must conform to the protocol requirements. All environmental factors should closely represent actual manufacturing conditions (relative humidity, room temperatures, static air pressure, and balance). All control settings are recorded, including any variable that will affect the cycle (key process variables such as temperature set points, heating elements settings, cycle-timer set point, belt speed, etc.). The cycle timer (batch), belt speed (tunnel or flame), controller operating temperature span, and production charts can be verified by a multipoint temperature recorder with an integral timer.

5. Container Mapping and Container Cool Point

Prior to initiating loaded chamber heat penetration studies, a container mapping study should be conducted. The intent of this study is to determine the coolest point within a liquid-filled container.

In general, the smaller the container volume, the less likely the detection of a discernible cold spot. Nevertheless, temperature mapping should be conducted on all the different container types, sizes, and fill volumes that will be subject to validation.

The number of thermocouples positioned within the container will be dependent on the container volume. A sufficient number of thermocouples should be positioned in areas representing the upper, middle, and lower portions of the container. Error in cold spot determinations may be introduced by employing an excessive number of thermocouples within the container. The error may be attributed to thermocouple mass and the resulting baffling effects may influence the normal convection currents of the liquid.

It is also possible to use a single thermocouple at different positions in multiple runs. This requires careful control of autoclave temperature to reduce error caused by run-to-run variation. Repeat studies are required to establish reproducible cold points and temperature profiles of the liquid in the container. The profile point having the lowest temperature or lowest F_0 is designated as the cold spot.

SOP No. Val. 700.10

Effective date: mm/dd/yyyy
Approved by:

In subsequent loaded chamber heat penetration studies, penetration thermocouples should be positioned within the container at the previously determined cold spot. The temperature profile of the container should remain constant among different sterilizing chambers, utilizing steam heat as the sterilizing medium.

6. Heat Distribution Studies

The intent of this study is to demonstrate the temperature uniformity and stability of the sterilizing medium throughout the sterilizer. Temperature distribution studies should be conducted on both empty and loaded chambers with maximum and minimum load configurations. Temperature uniformity may be influenced by the type, size, design, and installation of the sterilizer. The manufacturer of the vessel, based on the variables mentioned, should determine a satisfactory empty chamber temperature uniformity.

A narrow range is required and is generally acceptable if the variation is less than ±10°C (±2°F) of the mean chamber temperature. Significant temperature deviations greater than ±2.5°C (~±4.5°F) of the mean chamber temperature may indicate equipment malfunction. Stratified or entrapped air may also cause significant temperature variations within the sterilizer chamber. Initially, a temperature distribution profile should be established from studies conducted on the empty chamber. Confidence may be gained through repetition, and therefore empty chamber studies should be conducted in triplicate in order to obtain satisfactory assurance of consistent results.

Subsequent to the empty chamber studies, maximum load temperature distribution studies should be conducted to determine if the load configuration influences the temperature distribution profile obtained from the empty chamber studies. The thermocouples utilized in the heat distribution studies are distributed geometrically in representative horizontal and vertical planes throughout the sterilizer. The geometric center and corners of the sterilizer should also be represented. An additional thermocouple should be placed in the exhaust drain adjacent to the sensor that controls vessel temperature, if possible.

The number of thermocouples utilized in the heat distribution study will be dependent on sterilizer size. In a production-size sterilizer, 15 to 20 thermocouples should be adequate. The thermocouples utilized for loaded chamber heat distribution studies should be positioned in the same locations used for empty chamber heat distribution studies. The

SOP No. Val. 700.10   Effective date: mm/dd/yyyy

                                                Approved by:

uniformity and stability of the sterilizing medium are monitored in the distribution studies; consequently, the temperature probes should be suspended to avoid contacting solid surfaces and should not be placed within any containers. Temperatures must be obtained at regular intervals (e.g., each minute) throughout the time duration specified for a normal production cycle.

7. Loaded Chamber Heat Penetration Study — Load Cool Point

The intent of this study is to determine the coolest points within a specified load and configuration. Cool points originate because of the varied rate of heat transfer throughout the load. It is therefore imperative that heat penetration studies be conducted to determine cool points within a loading pattern and ensure that they are consistently exposed to sufficient heat lethality.

Load cool points are dependent on load configurations and the types of items that comprise the load (liquid-filled containers, process equipment, etc.). Prior to conducting heat penetration studies, maximum and minimum load configurations must be established. The penetration thermocouples are positioned within liquid-filled containers at the cool point previously determined by container mapping studies. The probed containers should be distributed uniformly throughout the load. When the load consists of multiple layers or pallets, a sufficient number of thermocouple-probed containers should be employed to provide an equal representation among layers.

Heat penetration studies conducted on maximum and minimum loads should be repeated until temperature data are obtained for all representative areas of the load. It may be necessary to reposition thermocouples in order to study different areas. Several runs, usually three of each thermocouple configuration, will provide confidence in the repeatability of the temperature profile.

A heat penetration study defining load cool points is not limited to load configurations composed of liquid-filled vials. The same principles can be applied to process equipment loads (filters, hoses, etc.) subject to steam sterilization. Penetration thermocouples are positioned at points within the process equipment suspected to be the most difficult for steam heat penetration. Temperature data are obtained from representative maximum and minimum loads in order to establish temperature profiles

SOP No. Val. 700.10   **Effective date: mm/dd/yyyy**

Approved by:

depicting load cool points. Equipment load configurations may be designed to allow reasonable flexibility for the operating department by permitting the use of partial loads. In this case partial loads would be defined as a portion of the established maximum validating load.

Heat penetration studies are also employed to determine points within a load configuration that achieve higher temperatures and consequently greater F_0 values. The temperature data obtained may be significant when heatable products are involved in the sterilization process and the potential for product degradation exists. The cool points established for a specified load and configuration will eventually be utilized to control the exposure time in subsequent routine production runs. The temperature sensors that control sterilization–cycle–exposure time at process temperature may be positioned within the load at the previously detected cool point. Consequently the entire load is exposed to sufficient heat lethality and achieves the desired F_0 value.

Lethal rates can be determined from the temperature data obtained from the heat penetration studies. The temperature data are converted by the following formula:

$$L = \log^{-1} \frac{T_0 - T_b}{Z} = 10(T_0 - T_b)/Z$$

where
T_0 = temperature within the container
T_b = process temperature (121°C)
Z = temperature required to change the D value by a factor of 10
L = lethality.
F_0 is then determined by integrating the lethal rates throughout the heating process:

$$F_0 = 10^{(T-121)}/10_{dt}$$

or

$$F_0 = \sum 10^{(T-121)}/10_{\Delta t}$$

SOP No. Val. 700.10

Effective date: mm/dd/yyyy

Approved by:

where

Δt = time interval between temperature measurements

T = product temperature at time t in °C.

When the sterilization process temperature deviates from 121°C, the amount of time providing equivalent lethality can be determined by the following formula:

$$F_t^Z = \frac{F_{121}^Z}{L}$$

where

F_t^Z = the equivalent time at temperature T delivered to a container for the purpose of sterilization with a specific value of Z

F_{121}^Z = the equivalent time at 121°C delivered to a container for the purpose of sterilization with a specific value of Z (if Z = 10°C, then $F_{121}^Z = F_0$).

8. Microbiological Challenge Studies

Biological castles are employed during heat penetration situations in order to demonstrate the degree of process lethality provided by the sterilization cycle. Calibrated biological indicators utilized for this purpose function as bioburden models providing data that can be utilized to calculate F_0 or substantiate and supplement physical temperature measurements obtained from thermocouples.

The most frequently utilized to challenge moist heat sterilization cycles are Bacillus stearothermophilus and Clostridium sporogenes, spore-forming bacteria are selected because of their relatively high heat resistance. In addition to the selection of an appropriate organism for use as a biological indicator, the concentration and resistance of the indigenous microbial population is established.

The biological indicator can be prepared to adequately challenge a sterilization cycle designed to provide a 10:6 probability of microbial survival with respect to indigenous bioburden. The concentration of spores utilized as the biological indicator can be determined from the following formula:

SOP No. Val. 700.10   **Effective date: mm/dd/yyyy**
                                                     Approved by:

$$D_S \ (\log N_i + 6) = D_{bi} \ (\log N_0 + 1)$$

where

N_i = the load of microorganisms on the product to be sterilized
D_S = D value of the most resistant isolate
N_0 = number of organisms on the biological indicator
D_{bi} = D value of biological indicator.

9. The Validation Report

Record keeping is a prime requirement of current good manufacturing practices. The records required for a validated steam sterilization cycle are listed below:

- Qualification reference documents (specifications, drawings, and calibration records)
- Operational qualification protocol and record
- Approved validation protocol
- Raw calibration and validation data
- Approved validation report

Frequency

Initial validation: three times
Revalidation: once in a year

REASONS FOR REVISION

Effective date: mm/dd/yyyy

- First time issued for your company, affiliates, and contract manufacturers

YOUR COMPANY
VALIDATION STANDARD OPERATING PROCEDURE

SOP No. Val. 700.20

Effective date: mm/dd/yyyy

Approved by:

TITLE: Hot Air Sterilization Tunnel Certification and Validation Guideline

AUTHOR: _____
Name/Title/Department

_____
Signature/Date

CHECKED BY: _____
Name/Title/Department

_____
Signature/Date

APPROVED BY: _____
Name/Title/Department

_____
Signature/Date

REVISIONS:

No.	Section	Pages	Initials/Date

SOP No. Val. 700.20   Effective date: mm/dd/yyyy

                                                      Approved by:

SUBJECT: Hot Air Sterilization Tunnel Certification and Validation Guideline

PURPOSE

To provide a written procedure to be used as a guideline for the certification and validation of a dry heat sterilizer

RESPONSIBILITY

It is the responsibility of production manager, validation manager and concerned departmental managers to follow the procedure. The quality assurance manager is responsible for SOP compliance.

Introduction

Laminar flow sterilization tunnels are widely used in high-speed aseptic manufacturing. Typically, laminar flow tunnels contain three sections: (1) preheating, (2) heating, and (3) cooling.

Sterilization occurs at temperatures higher than 300°C in the heating section. After sterilization, cooling is necessary before container filling. It is therefore very important to keep conditions sterile in the cooling section (up to the filling station) by keeping the cooling section at a slight positive pressure towards the tunnel room (2 to 3 Pa). A higher overpressure would result in cooling the heating section with cooling air, decreasing the sterilization efficiency of the heating section.

The certification activities include a series of process documentation and qualification studies that start with the initial installation of a sterilization system and continue as process engineering changes or new or revised product introductions are required. Qualification activities comprise installation, operational, change, and performance phases.

SOP No. Val. 700.20   **Effective date: mm/dd/yyyy**

                                               Approved by:

PROCEDURE

1. Installation Qualification (IQ)

The initial IQ hot air sterilizer tunnel certification shall consist of the development of the following information package:

- Hot air sterilizer tunnel dimensions
- Product carrier description
- Utility support system description
- Sterilizer equipment description
- Equipment control system description

2. Process Description

- Description of the sterilization medium employed
- Description of the cycle steps and process functions initiated during the sterilization process
- Type of process control employed, i.e., time and temperature or product container control
- System operating procedures and system flow diagrams

3. Product Safety

To confirm that product safety considerations have been addressed, review of tunnel construction and operation materials for product contact potential or suitability shall be documented. Tunnel construction materials, which contact the sterilization medium, shall be identified. This would include:

- Product carriers
- All exposed potential medium contact surfaces including heating and cooling sections
- Heat generating, cooling, and conveying system

Equipment lubricants with potential product contact implications must be verified as not jeopardizing product integrity. Lubricants should be identified.

4. Critical Process Instrumentation List

- Temperature control and monitoring systems
- Pressure control and monitoring systems
- Carrier drive monitoring systems
- Critical system alarms

The following equipment installation qualification checks shall be performed:

4.1 DOP tests of HEPA filters

Test objective

To demonstrate that HEPA filters are properly installed by verifying the absence of bypass leakage and other defects such as tears and pinhole leaks

Test method

This test is performed only by certified or previously trained personnel who introduce DOP aerosol upstream of the filter through a test port and search for leaks downstream with an aerosol photometer. Filter testing is performed after operational air velocities have been verified and adjusted where necessary.

Align the aerosol photometer as follows:

1. Position the smoke generator so the DOP aerosol will be introduced into the air stream ahead of the HEPA filters.
2. Open the appropriate number of nozzles until a DOP challenge concentration of 100 mg/l of air is reached. This challenge concentration is measured upstream of the HEPA filter, and is evidenced by a reading of between 4 and 5 on the logarithmic scale of the aerosol photometer.
3. Scan each filter by holding the photometer probe approximately 1 in. from the filter face and passing the probe in slightly overlapping strokes at a traverse rate of not more than 10 ft/min, so that the entire face is sampled.
4. Make separate passes with the photometer probe around the entire periphery of the filter, along the bond between the filter medium and the frame, and along all other joints in the installation through which leakage might bypass the filter medium.

SOP No. Val. 700.20   Effective date: mm/dd/yyyy
                                       Approved by:

Acceptance criteria

- Entrance section filters and cooling section filters
- Local DOP penetration ≤0.01% of the upstream concentration
- Heating section filters
- Local DOP penetration ≤0.1% of the upstream concentration provided that local results of hot tunnel particle counting air cleanliness classification are within specifications (particle counts at any location of the heating section ≤100 particles ≥0.5 $\mu m/ft^3$ and zero particle ≥5 $\mu m/ft^3$)

Equipment

DOP polydisperse aerosol is generated by blowing air through liquid dioctylphthalate (DOP) at room temperature. The approximate light scattering mean droplet size distribution of the aerosol is 99% + less than 3.0 μm and 95% + less than 1.5 μm.

The DOP aerosol generator is compressed-air operated, equipped with Laskin–type nozzles. The aerosol photometer is a light-scattering type with a threshold sensitivity of at least 10^{-3} mg/l, capable of measuring concentrations in the range of 80 to 120 mg/l, and with air sample flow rate of 1 ft^3 + 10% per min. This instrument is to be calibrated per manufacturer recommendation.

4.2 Air velocity and homogeneity at the exit of HEPA filters

Test objective

To demonstrate that air speed is homogeneous in each section of the tunnel (entrance, heating, cooling). The air speed values and homogeneity are important for uniform heating (sterilization) and uniform cooling of glass containers.

Test method

- Draw a grid on the floor of tunnel.
- Measure and record the velocity at the center of each grid at the specified heights.

- Allow no objects near the anemometer, except for built-in equipment.
- Measurements should be taken for a minimum of 15 sec.
- Record the pressure readings (in in.) from the manometer connected to the module's plenum.

Equipment

Hot-wire anemometer and stand

Acceptance criteria

From left to right, speed variation should not be more than 30% around the mean. From each filter the speed uniformity must be greater than ±20% relative to the mean, per filter not more than one location out of this limit.

4.3 Air velocity and uniformity on the tunnel conveyor

Test objective

To demonstrate that air flow is continuous from top to bottom along the whole surface of the conveyor (air flow from bottom to top would contaminate glassware with particles from the conveyor and machinery)

Test method

- Draw a grid on the floor of tunnel.
- Measure and record the velocity at the center of each grid at the specified heights.
- Allow no objects near the anemometer, except for built-in equipment.
- Measurements should be taken for a minimum of 15 sec.
- Record the pressure readings (in in.) from the manometer connected to the module's plenum.

Equipment

Hot-wire anemometer and stand

SOP No. Val. 700.20

Effective date: mm/dd/yyyy
Approved by:

Acceptance criteria

There should be no measured speed at any point from the bottom to the top (anemometer). No opposite flow should be visualized with the Drager tube and pump. It is preferable that left to right speed variation be lower than 30% around the mean. The number of points out of this limit is to be minimized.

4.4 Hot tunnel particle countings

Test objective

To demonstrate that the air handling system of the tunnel (hardware and software) is able to produce at the level of the top of the container a class 100 air on all the surface of the conveyor.

Test method

- These tests are performed after the HEPA filter leak tests and air velocity tests are completed.
- To obtain baseline data with the room in static conditions, perform the following tests with operational personnel absent and the equipment at rest:
  1. Using the particle analyzer, count particles greater than or equal to 0.5 μm in diameter at heights of 40 in. in the center of each grid.
  2. If the particle count in the 0.5 μm range is less than 50 per ft^3 of air, four additional counts at this location are taken to place these particle counts within a 50% confidence interval.
- After completion of these tests, if the absolute air filtration modules are operating within accepted limits, repeat steps 1 and 2 with operational personnel present and the fill equipment running. If at any time there is a deviation from accepted parameters, the various components of the systems in operation are reviewed, repaired, or adjusted until the desired conditions are achieved.

Equipment

Laser particulate counter

SOP No. Val. 700.20

Effective date: mm/dd/yyyy
Approved by:

Acceptance criteria

All locations with particle counts:

≤ 100 particle ≥ 0.5 $\mu m/ft^3$
zero particle ≥ 5 $\mu m/ft^3$

4.5 Empty tunnel heat distribution

Test objective

The objective of the empty distribution runs will be to evaluate:

- Heating characteristics of the sterilizer, product carrier system, and the sterilization medium employed
- Ability of the sterilizer to hold the required sterilization parameters
- Ability of the sterilization cycle control mechanisms to operate as intended

Test method

A review of all sterilization specifications assigned to the sterilizer under consideration shall be made, with the specifications cycle requiring the maximum peak dwell temperature and heating rate to be selected for the empty sterilizer heat distribution runs. During the empty sterilizer heat distribution runs, sterilizer parameters and equipment component status shall be visually monitored to confirm applicable control operations.

Technical criteria

- Fixed thermocouples shall be located at key sterilizer positions, as justified by the sterilizer operation and control characteristics (i.e., at exhaust or vent line, in recirculation heating medium line, next to controller sensor, as applicable).
- Distribution thermocouples for sterilizer shall be located throughout the chamber per plan and traceable location diagram. Sufficient functional thermocouples shall be used during distribution runs conducted in sterilizer to assure adequate distribution determination.

SOP No. Val. 700.20   **Effective date: mm/dd/yyyy**
Approved by:

- Traveling temperature sensors for continuous sterilizer shall be located throughout the conveyor system per plan and traceable placement diagram. The temperature sensors shall be placed in various locations within each distribution run.
- Heat distribution data shall include evaluation of the coldest and hottest sterilizer zones, the mean distribution temperature observed, the range of distribution temperatures observed, and the heat-up and cool-down times obtained.

Acceptance criteria

- The distribution runs must meet the time and temperature requirements of the corresponding specifications or operating procedures.
- All function initiations required during the operating modes must have occurred as specified.

4.6 System alarm and safeguard checks

Test objective

To confirm that all alarm feature input and output loops function as intended

Test method

All alarm features available on the sterilizer system under consideration, both program controlled or separately wired, shall be challenged to confirm appropriate functionality.

Technical criteria

- Where possible, each alarm and safeguard feature should be challenged by simulation of actual alarm conditions within the sterilizer equipment system. Where simulation of physical alarm conditions would be impractical, alarm circuitry may be challenged by use of electrical input signals.
- The following alarm and safeguard systems should be checked, as applicable:

- Power or electrical system interruption alarms
- Chamber door-open alarms
- Cycle sequence alarms
- Timer system alarms
- High or low temperature alarms
- Chain speed and R.P.M. alarms
- Fan-on alarms
- Computer or controller data entry safeguard system alarms

Acceptance criteria

All alarm and safeguard features shall respond to their corresponding system condition signal as specified.

Support documentation

The following documents shall be included in the IQ certification package:

- Sterilizer engineering drawings
- Sterilizer operation procedure
- Sterilizer sanitization procedure
- Sterilizer maintenance procedures
- Sterilizer specification utilization list
- Distribution thermocouple location diagrams
- Temperature sensing unit location diagram (continuous sterilizers)
- Sterilizer process log sheets
- Empty chamber heat distribution test data summaries
- Copy of appropriate specifications used
- Test data summary sheets for each function evaluation
- Test and equipment pre- and postcalibration status listings

5. Operational Qualification (OQ)

5.1 Background

The intent of sterilizer OQ studies will be to:

- Confirm that sterilizers are capable of processing at established time and temperature ranges that assure conformance with respective specification requirements

SOP No. Val. 700.20   **Effective date: mm/dd/yyyy**

                                               Approved by:

■   Confirm that established sterilization cycles deliver a uniform and reproducible heat input to products assigned to each cycle

The sterilization test functions required to qualify or validate the sterilizer will include process heat distribution, process heat penetration, and process microbial and depyrogenation validation, as applicable.

The OQ phase of sterilizer validation shall consist of the development of an information package fulfilling the documentation requirements of the generic equipment operational qualification. The following sterilizer-specific documentation shall be incorporated in the operation qualification:

■   Brief sterilizer equipment or process description shall be included for initial certification.

■   The products utilized for testing subsection shall include a listing of the items used for OQ test function runs and items utilized for sterilizer bulking during test function runs.

■   The sterilizer utilization list subsection shall include a listing of the sterilization cycles being validated and the corresponding product list numbers assigned to each specification.

The sterilizer utilization list and the following OQ test requirements summary will be utilized to determine the products assigned to the sterilizer that shall be subjected to the type and number of test function runs required to establish overall sterilizer qualification or validation. The test function subsections shall include test objectives, test methods and acceptance criteria, as follows.

5.2   Heat distribution

Test objective

To evaluate the heating characteristics of the tunnel, carrier system, and sterilization medium employed under loaded conditions

Test method

■   Distribution thermocouples shall be located in each run as described in IQ empty tunnel heat distribution runs.

■ All distribution runs shall be performed, monitored, and documented in accordance with the respective sterilizer operating procedure.

Acceptance criteria

All distribution runs must meet the parameter requirements of the corresponding specification and established production sterilization cycle.

5.3 Heat penetration

Test objective

■ To evaluate the heating characteristics of items within the tunnel when subjected to the sterilization medium
■ To evaluate the relative heating characteristics of items and reference thermocouples where applicable
■ To establish production work order sterilization parameters

Test method

■ Heat penetration runs may be conducted in conjunction with required heat distribution runs.
■ Thermocouple or temperature sensor probes shall be placed within the penetration test containers in accordance with established written container preparation procedures. Test containers may be trays, pans, commodities, etc., depending upon the testing required.
  ■ Thermocouple and temperature sensor probe placement within the containers shall be documented.
  ■ Where applicable, thermocouple placement shall be in the container cold zone, as determined from generated container mapping studies.
■ Each heat penetration run shall include thermocouple temperature sensor probe containers distributed throughout the tunnel, per planned and traceable location diagram.
■ The number of heat penetration test containers per run shall agree with that required for heat distribution thermocouples.
■ If previous empty tunnel heat distribution test runs have identified hot or cold zones, at least one of the penetration test containers must be placed in each of these zones per run.

SOP No. Val. 700.20

Effective date: mm/dd/yyyy
Approved by:

■ The heat penetration sample containers shall be loaded into the tunnel carrier in an orientation consistent with planned production run loading and the corresponding container heat mapping study loading method.

Acceptance criteria

■ All heat penetration data collected during each run must meet the requirements for the corresponding specification.

■ The production operating ranges and windows established from the heat penetration runs must assure all products in the test runs will meet the calculated requirements for the corresponding specification. If a satisfactory operating range is not established using minimum and maximum loading parameters, intermediate loading conditions must be tested.

■ Where tunnel peak dwell temperature and time are to be used for routine production cycle control, or as back-up control, correlation of sterilizer peak dwell time and temperature with the hottest and coldest profile container must be shown for each run, where applicable.

5.4 Componentry microbial or pyrogen challenges

Test objective

To confirm the biological relationship between parametrically determined process lethalities, by demonstrating the ability of the sterilizer to effectively reduce the challenge material to an acceptable level

Test method

1. Componentry microbial challenges
   ■ The number of challenge containers, preparation methods, spore crop type, and inoculation levels described in the appropriate documentation shall be followed in the production of test items for each run.
   ■ Heat penetration test containers of corresponding container size and type and fill volume shall be placed adjacent to the challenge test items in each run.

- At least one set of microbial challenge or penetration test containers shall be placed in the sterilizer cold zone per run (where applicable).
- Test container placement shall be defined per planned and traceable location diagrams.
- Maximum sterilizer loading configurations shall be used when conducting challenge test runs.
- The corresponding production cycle time and temperature control parameters that deliver subminimal specification conditions shall be utilized when conducting the challenge runs.
- If absolute minimum time and minimal temperature parameters are not used during the componentry challenge runs, manufacturing order parameter limits must reflect the parameters used during these runs.

2. Componentry pyrogen challenges
   - The number of challenge containers, preparation methods, endotoxin identification, and inoculation levels described in the appropriate documentation shall be followed in the production of test items for each run.
   - Heat penetration test containers of corresponding container size and type and fill volume shall be placed adjacent to the challenge test items in each run.
   - At least one set of pyrogen challenge or penetration test containers shall be placed in the sterilizer cold zone per run where applicable.
   - Test container placement shall be defined per planned and traceable location diagrams.
   - Maximum sterilizer loading configuration shall be used when conducting challenge test runs in all other sterilizers.
   - The corresponding production cycle time and temperature control parameters that deliver subminimal specification conditions shall be utilized when conducting the challenge runs.

Acceptance criteria

- A minimum microbial challenge spore log reduction of equal to or greater than six must be shown for each run.
- A minimum pyrogen challenge must be equal to or greater than three log reductions for each run.

SOP No. Val. 700.20   Effective date: mm/dd/yyyy
                                            Approved by:

- At least 15% of the required functional container time and temperature values must show subminimal process conditions, per run.
- Cold zone temperature correction must be used where applicable.

5.5 Support documentation

The following documents shall be included in each operational qualification certification package, as applicable:

- Sterilizer operating procedure
- Current sterilizer utilization list
- Copies of the specifications used during the OQ function test runs
- Copies of all penetration distribution and challenge thermocouple placement diagrams
- Description of the bulking items used, where applicable
- Copies of all microbial challenge protocols, which should include identification of the types, crop numbers, and D values for the biological indicators used
- Copies of all pyrogen challenge protocols, which should include materials used, lot number, sensitivity, and inoculation levels
- Key test and equipment instrumentation pre- and postcalibration status supporting each function
- Test data summaries

The temperature data collected during each operational qualification run shall be summarized so that the following information, as applicable, can be readily determined for each run:

- Sterilizer heat-up time
- Duration of sterilizer peak dwell
- Minimum and maximum sterilizer temperatures during peak dwell
- Sterilizer cool-down time
- Commodity or component heat-up time
- Peak dwell residence time or carrier speed
- Minimum and maximum item temperature during peak dwell
- Item cool-down time

Effective date: mm/dd/yyyy

Approved by:

Process Engineering Change Qualification

Modifications to sterilizer equipment systems shall be accompanied by initiation and completion of a formal engineering change request and authorization documentation package. Plant engineering, manufacturing and quality assurance shall be responsible for determining whether a change impacts the certified functions of the sterilizer. Changes involving the modification of the sterilizer, carrier design, sterilization medium supply or distribution systems, or sterilizer operation or control mode will require the performance of heat distribution and penetration runs with items bracketing thermal mass characteristics represented by the sterilizer utilization list.

All packages shall address the IQ and OQ documentation requirements affected by the change and shall include:

■ Description of the proposed change
■ Documented reason or rationale for the proposed change
■ Description of the test functions required to qualify and validate the sterilization process after the change is made, as applicable
■ Confirmation that documentation affected has been updated after the change takes place

Frequency

Initial validation: three times
Revalidation: twice per year

REASONS FOR REVISION

Effective date: mm/dd/yy

■ First time issued for your company, affiliates, and contract manufacturers

YOUR COMPANY
VALIDATION STANDARD OPERATING PROCEDURE

SOP No. Val. 700.30

Effective date: mm/dd/yyyy

Approved by:

TITLE: Freeze Drier

AUTHOR: _____
Name/Title/Department

_____
Signature/Date

CHECKED BY: _____
Name/Title/Department

_____
Signature/Date

APPROVED BY: _____
Name/Title/Department

_____
Signature/Date

REVISIONS:

No.	Section	Pages	Initials/Date

SOP No. Val. 700.30

Effective date: mm/dd/yyyy
Approved by:

SUBJECT: Freeze Drier

PURPOSE

To describe the procedure for validation of the freeze drier to ensure it meets the installation, operational, and performance qualification criteria

RESPONSIBILITY

It is the responsibility of the production manager, quality control manager, and technical services manager to follow the procedure. The quality assurance manager is responsible for SOP compliance.

PROCEDURE

1. Vacuum Leak Testing

Procedure

Perform the vacuum leak test with an empty freeze drier and condenser. After reaching the maximum available vacuum, the vacuum pump is switched off after a delay time and the valves are closed. Monitor the vacuum decrease for a period of 24 h. Let the condenser cool to avoid water evaporation and vacuum decrease.

Requirements

Maximum available vacuum: meet manufacturing specification
Vacuum decrease: meet specification requirements

2. Shelf Temperature Study

Uniformity of temperature distribution across the shelves is important in order to maintain product uniformity.

SOP No. Val. 700.30

Effective date: mm/dd/yyyy
Approved by:

Procedure

To study the temperature distribution and differences compared with the set values of each shelf, a temperature evaluation study of each shelf in a complete cooling-down and warming-up cycle should be performed as follows:

Distribute ≥15 thermocouples over the shelf surface, and at the refrigerant fluid inlet and outlet.

Run following cycle:

Start point	End Point	Time period
1) +20°C	+50°C	minimum time
2) –50°C	–50°C	hold for 30 min
3) –50°C	+50°C	minimum time
4) +50°C	+50°C	hold for 30 min

Requirements

Average refrigerant temperature should be ±3°C during the holding period compared with set temperature Cooling down velocity, warming up velocity, temperature distribution among the different shelves, and average shelf temperature should be in compliance with the manufacturer specifications.

Frequency

Initial validation: once
Revalidation: once per year

3. Temperature Distribution Study

Simulate the freeze drying process to perform an empty chamber temperature distribution study.

SOP No. Val. 700.30

Effective date: mm/dd/yyyy
Approved by:

Procedure

Establish the minimum quantity of thermocouples to be used for temperature distribution study. The thermocouples should preferably be placed in recognized cold spots during the shelf temperature study.

Requirements

The temperature distribution within the different shelves compared with the average temperature measured within the stable phase of the process and the average shelf temperature must comply with supplier specifications.

Frequency

Initial validation: once
Revalidation: once per year

4. Pressure Testing

Procedure

During simulated lyophilization processes, the pressure inside the freeze drier should be measured and compared with the set values.

Requirements

Average pressure should be no more than ±1 Pa compared with the set values.

Frequency

Initial validation: three times
Revalidation: once per year

5. Sterilization

Steam sterilization technique is commonly used for this purpose.

SOP No. Val. 700.30   Effective date: mm/dd/yyyy
                                             Approved by:

Procedure

Perform temperature distribution study. Determine the quantity of thermo-couples to be used during the investigation using the following formula:

Number of thermocouples = usable chamber volume (liters)/100 liters + 1

The minimum number of thermocouples recommended to conduct the study is 5. Additional thermocouples shall be placed in these condensate drains and condenser of the freeze drying chamber, of the condenser and of the steam inlet.

The F_0 value of the cold curve should be calculated from all run processes. Following these data the worst case can be recognized.

Requirements

$F_0 \geq 12$ min

Frequency

Initial validation: once for the whole chamber and afterward three times
    repeated for recognized cold spots
Revalidation: once per year

6. Microbiological Investigation

Procedure

Use spore strips for microbiological investigation:
    Bacillus stearothermophilus
    D-value ≥ 1.0 min
    Population between 10^4 and 10^7
    The cold spots recognized during the temperature investigation (worst case) are investigated again by placing a spore strip near each thermo-couple location which has been proved to be a cold spot. Besides, the two spore strips shall be used as a positive control.

Frequency

Initial validation: two times at the identified cold spot
Revalidation: once per year

SOP No. Val. 700.30   Effective date: mm/dd/yyyy

Approved by:

Requirements

Sterilized spore strips: no growth

Growth of the positive control strips: show growth

7. Pressure Testing during Sterilization Process

Procedure

The pressure profile during the sterilization cycle should be monitored and compared with the theoretical pressure calculated from sterilization temperatures.

Requirements

Measured pressure = theoretical pressure ±5 kPa (refer to specification)

Measured temperature = theoretical temperature ±3°C (refer to specification)

Frequency

Initial validation: three times

Revalidation: once per year

8. Media Fills

The verification of aseptic processes entails the use of media to assess the suitability of the handling procedures.

Procedure

Media-filled vials (soybean casein digest broth or fluid thioglycollate medium) are filled on the filling line, transported to the freeze drier, loaded into the chamber, subjected to a simulated lyophilization process, stoppered, sealed, and incubated for 14 days at 25°C. The minimum number of vials to be used is 5000.

Requirements

Infection rate ≤0.1% (PDA)

SOP No. Val. 700.30   Effective date: mm/dd/yyyy
                                       Approved by:

Frequency

Initial validation: three times
Revalidation: every 6 months

9. Oil Contamination

To assess the possibility of contamination of the freeze drying chamber with oil during the final drying process by diffusion of oil out of the vacuum pump.

Procedure

Place silica gel thin layer chromatogram plates on the shelves and run the normal process cycle. The TLCs are analyzed on their content of pump oil and refrigerant oil.

Requirements

Content pump oil: <1.00 µg/cm^2 × h

Frequency

Initial validation: three times with different locations of the TLCs during the three runs
Revalidation: once per year

10. Cross-contamination Test after Cleaning the Device

Cross-contamination test must be performed for freezers not dedicated to one process.

Procedure

Prepare vials containing an inert material (e.g., 5% w/v mannitol solution) and run the freeze drying process after the actual product run. Analyze the vials with inert material for the active content of the previous product.

SOP No. Val. 700.30   Effective date: mm/dd/yyyy
Approved by:

Requirements

The cross-contamination rate should meet the in-house criteria.

Frequency

For the product with the most serious active or the highest content of active: once
Revalidation: once every 5 years

REASONS FOR REVISION

Effective date: mm/dd/yyyy

- ▪ First time issued for your company, affiliates, and contract manufacturers

---

YOUR COMPANY
VALIDATION STANDARD OPERATING PROCEDURE

SOP No. Val. 700.40

Effective date: mm/dd/yyyy

Approved by:

TITLE: Ampule and Vial Washing Machine

AUTHOR: _____

Name/Title/Department

_____

Signature/Date

CHECKED BY: _____

Name/Title/Department

_____

Signature/Date

APPROVED BY: _____

Name/Title/Department

_____

Signature/Date

REVISIONS:

No.	Section	Pages	Initials/Date

SOP No. Val. 700.40   Effective date: mm/dd/yyyy

                                                Approved by:

SUBJECT: Ampule and Vial Washing Machine

PURPOSE

To describe the procedure for validation of an ampule and vial washing machine to maintain microbiological quality

RESPONSIBILITY

It is the responsibility of the production manager and technical service manager to follow the procedure. The quality assurance manager is responsible for SOP compliance.

PROCEDURE

1. Installation Qualification

- Verify approved purchase order.
- Verify invoice.
- Check manufacturer and supplier.
- Verify model number and serial number.
- Check for any physical damage.
- Confirm location and installation requirements per recommendation of manufacturers.
- Verify that the utilities required are available.
- Installation shall be conducted per the instructions provided in the manual.
- Ensure that all relevant documentation is received:
  - User manual
  - Maintenance manual
  - List of change parts
  - Electrical drawings
  - Medical drawings

SOP No. Val. 700.40

Effective date: mm/dd/yyyy
Approved by:

2. Operational Qualification

- Verify alarm control.
- Perform calibration requirements, identified in the manual or established by the validation team.
- Operate the equipment at low, medium, and high speed per operations manual to verify the operating control.
- Verify that all switches and push buttons are functioning properly.
- Establish procedures for operation, maintenance, and calibration.
- Establish training program for relevant staff.

3. Cleaning Process Efficiency

Procedure

Wash ampule and vials using different available washing programs. Determine the efficiency of the process by comparing the quantity of present sticking particles (particle size classes: >50 μm and 2 to 50 μm) prior to and after washing the glassware.

Fill washed and dried glassware with sterile filtered water, sonicate for 5 minutes and test with a particle counter.

Requirements

Sticking particles: 90% reduction

4. Microbiological Quality of Water Used for Glassware Washing

4.1 Recirculating water

Test the water used for prewashing in ampule and vial washing machine for viable count.

Procedure

Perform hourly sampling for a 3-day period.

4.2 Purified water supply system

Procedure

Perform sampling for a period of 2 weeks each day from each needle outlet and test for microbial count.

Requirements

Total aerobic microbial count: <100 cfu/ml

5. Water Filter Integrity

Procedure

Perform pressure hold test or a forward flow test.

Requirements

The filters should meet the integrity test as proposed by the filter supplier. The allowed usage time of a filter must be fixed (e.g., maximum pressure differences between water inlet and outlet).

Frequency

Initial validation: at the time of installation
Revalidation: each filter change

REASONS FOR REVISIONS

Effective date: mm/dd/yyyy

■ First time issued for your company, affiliates, and contract manufacturers

YOUR COMPANY
VALIDATION STANDARD OPERATING PROCEDURE

SOP No. Val. 700.50

Effective date: mm/dd/yyyy
Approved by:

TITLE: Washing, Sterilizing, and Drying Machine
for Stoppers

AUTHOR: _____
Name/Title/Department

_____
Signature/Date

CHECKED BY: _____
Name/Title/Department

_____
Signature/Date

APPROVED BY: _____
Name/Title/Department

_____
Signature/Date

REVISIONS:

No.	Section	Pages	Initials/Date

SOP No. Val. 700.50

Effective date: mm/dd/yyyy

Approved by:

SUBJECT: Washing, Sterilizing, and Drying Machine for Stoppers

PURPOSE

To provide the guideline for validation of the washing, sterilizing and drying machine for stoppers to meet the cGMP requirement

RESPONSIBILITY

It is the responsibility of the technical service and quality control managers to follow the procedure. The quality assurance manager is responsible for SOP compliance.

PROCEDURE

1. Installation Qualification

- Verify approved purchase order.
- Verify invoice.
- Check manufacturer and supplier.
- Verify model number and serial number.
- Check for any physical damage.
- Confirm location and installation requirements per recommendation of manufacturers.
- Verify that the utilities required are available.
- Installation to be conducted per instructions provided in the manual.
- Ensure all relevant documentation is received:
  - User manual
  - Maintenance manual
  - List of change parts
  - Electrical drawings
  - Medical drawings

SOP No. Val. 700.50   Effective date: mm/dd/yyyy
                                       Approved by:

2. Operational Qualification

- Verify alarm control.
- Perform calibration requirements, identified in the manual or established by the validation team.
- Operate the equipment at low, medium, and high speed per operations manual to verify the operating control.
- Verify that all switches and push buttons are functioning properly.
- Establish procedures for operation, maintenance, and calibration.
- Establish training program for relevant staff.

3. Performance Qualification

3.1 Cleaning efficiency

3.1.1 Reduction of sticking visible particles

Procedure

The number of stoppers should be chosen according to a total stopper-surface of 200 cm². Place the stoppers to be tested into a particle-free container with a pair of tweezers. Shake the stoppers for about 30 sec with a filtered and dilute detergent solution. Filter the obtained solution through a 0.45 μm filter, as reference. Repeat the same procedure without stoppers.

Perform the test and reference test three times. Inspect the filter afterwards visually, using microscope using lateral incident light with an enlargement of 40x.

Frequency

Initial validation: each stopper type should be tested once prior to and once after running the standard process with a loading of 100%.
Revalidation: once every 3 years

Requirements

Reduction of particles should be 80% compared with unwashed stoppers.

3.1.2 Reduction of sticking subvisible particles

Procedure

Transfer the stoppers into a particle-free container with a pair of tweezers. Fill eight stoppers in each container, add 10 ml of sterile filtered water, sonicate for 5 minutes.

Measure the particle load with a particle counter in the measuring ranges (92 to 5, 5 to 10, 10 to 25, 25 to 50, and >50 μm). From each container perform 20 measurements in the different ranges. As a reference test, perform the same procedure without stoppers.

Frequency

Initial validation: each stopper type should be tested once prior to and after running the standard process with a loading of 100%.
Revalidation: once every 3 years.

3.1.3 Reduction of pyrogen content

Procedure

Prepare five stoppers covered with 1000 endotoxin units each. Place the stoppers in a net and place them in the process vessel during a standard run with a 100% loading. Determine the endotoxin quantity present on the five stoppers.

Frequency

Perform the test three times.

Requirements

A 3-log reduction

3.1.4 Content of silicone

Procedure

After processing with 50 and 100% loading, the silicone content on the stoppers should be determined. The unprocessed stoppers' silicone content shall also be determined as a reference value.

SOP No. Val. 700.50

Effective date: mm/dd/yyyy
Approved by:

Frequency

Validation: initial for each stopper type and loading rate one time
Revalidation: once every 3 years

Requirements

Content processing of the stoppers
Silicone content: 5 to 20 $\mu m/cm^2$

3.1.5 Sterilization

3.1.6 Determination of the F_0

Procedure

With loading rates of 0, 50, and 100% using each stopper type, place thermocouples in the process vessel with predetermined location.

Frequency

Validation: each loading rate one time
The worst case loading identified from the initial testing should be repeated three times.
Revalidation: every 6 months

Requirements

$F_0 \geq 12$ min

3.1.7 Microbiological investigation

Procedure

- Bacillus stearothermophilus
- D–value ≥ 1 min
- Viable count $\geq 10^3$ bacteria/ampule

The ampules should be sterilized together with the stoppers; three ampules must be used.

Frequency

Three times at the worst case loading rate as determined above.

Requirements

No growth of the test organism after the sterilization

3.1.8 Drying process

Procedure

Run the standard process with 100% loading rate. Collect the dried stoppers in predried glass vessels. The stoppers are afterward cut into pieces of ±80 mg and analyzed using the Karl Fischer moisture determination method. The moisture out of the stoppers is obtained at increased temperature and blown over with nitrogen into the reaction vessel.

Frequency

Once for each stopper type

Requirements

Moisture content equal to or lower than that of unprocessed stoppers

3.1.9 Feedwater

Procedure

Monitor pyrogen and microbial content of feed water as the cleaning and depyrogenation effects are dependent from the water quality used during processing of the stoppers.

Requirements

pyrogen content ≤0.25 EU/ml
viable count ≤10 cfu/ml

SOP No. Val. 700.50

Effective date: mm/dd/yyyy

Approved by:

3.1.10 Rework procedure

If the reprocessing of stoppers that already passed the whole process is unavoidable, the whole process should be revalidated.

REASONS FOR REVISION

Effective date: mm/dd/yyyy

- ■ First time issued for your company, affiliates, and contract manufacturers

| YOUR COMPANY |
| VALIDATION STANDARD OPERATING PROCEDURE |

SOP No. Val. 700.60

Effective date: mm/dd/yyyy

Approved by:

TITLE: Ampule and Vial Filling Machine

AUTHOR: _____

Name/Title/Department

_____

Signature/Date

CHECKED BY: _____

Name/Title/Department

_____

Signature/Date

APPROVED BY: _____

Name/Title/Department

_____

Signature/Date

REVISIONS:

No.	Section	Pages	Initials/Date

SOP No. Val. 700.60

Effective date: mm/dd/yyyy

Approved by:

SUBJECT: Ampule and Vial Filling Machine

PURPOSE

To describe the validation guideline for the ampule and vial filling machine to be free from contamination during the filling cycle

RESPONSIBILITY

It is the responsibility of the production manager and technical service manager to follow the procedure. The quality assurance manager is responsible for SOP compliance.

PROCEDURE

1. Installation Qualification

- Verify approved purchase order.
- Verify invoice.
- Check manufacturer and supplier.
- Verify model number and serial number.
- Check for any physical damage.
- Confirm location and installation requirements per recommendation of manufacturers.
- Verify that the utilities required are available.
- Installation shall be conducted per instructions provided in the manual.
- Ensure all relevant documentation is received:
  - User manual
  - Maintenance manual
  - List of change parts
  - Electrical drawings
  - Medical drawings

SOP No. Val. 700.60

Effective date: mm/dd/yyyy

Approved by:

2. Operational Qualification

- Verify alarm control.
- Perform calibration requirements identified in the manual or established by the validation team.
- Operate the equipment at low, medium, and high speed per operations manual to verify the operating control.
- Verify that all switches and push buttons are functioning properly.
- Establish procedures for operation, maintenance, and calibration.
- Establish training program for relevant staff.

3. Classification at Filling Point

Procedure

The particle load should be examined at the location near the filling points with a particle counter.

Requirements

- ≤100 particles ≥0.5 μm/cft or ≤3000 particles ≥0.5 μm/m^3
- 0 particles ≥5 μm/cft or m^3

4. Particle Contamination of Ampules and Vials during Filling Procedure

Ampules and vials should be filled with water for injection and afterward be inspected on the contamination with particles (particle classes: ≤10 μm and ≤25 μm). The inspection can be performed with a particle counter.

Requirements

The USP 24 requirements for "particulate matter in injections, small-volume injections" must be fulfilled. The contamination with particles during the filling step should be equivalent at all available machines.

SOP No. Val. 700.60   **Effective date: mm/dd/yyyy**
                                       Approved by:

5. Filling Trials (Media Fill Run)

Procedure and requirements

Refer to SOP No. Val. 700.70.

6. Filling Volume Accuracy

The filling accuracy should be within ±% of the adjusted and desired filling volume in accordance with the machine specification, etc.
Attention limit: ±1%
Action limit: ±2%

REASONS FOR REVISION

Effective date: mm/dd/yyyy

■ First time issued for your company, affiliates, and contract manufacturers

YOUR COMPANY
VALIDATION STANDARD OPERATING PROCEDURE

SOP No. Val. 700.70   Effective date: mm/dd/yyyy

                                                       Approved by:

TITLE: Media Fill Run

AUTHOR: _____
                          Name/Title/Department

              _____
                          Signature/Date

CHECKED BY: _____
                          Name/Title/Department

              _____
                          Signature/Date

APPROVED BY: _____
                          Name/Title/Department

              _____
                          Signature/Date

REVISIONS:

No.	Section	Pages	Initials/Date

SOP No. Val. 700.70   **Effective date: mm/dd/yyyy**

                                           Approved by:

SUBJECT: Media Fill Run

PURPOSE

To describe the media fill procedure to qualify the aseptic following lines

RESPONSIBILITY

It is the responsibility of the production manager and technical service manager to follow the procedure. The quality assurance manager is responsible for SOP compliance.

PROCEDURE

1. Documentation

The document should include at least the following:

- Identification of the process to be simulated and a copy of the batch record to be used
- Identification of the rooms to be used
- Identification of the filling line and equipment to be used
- Type of container and closure to be used
- Line speeds (low, normal, and high)
- Number of units to be filled
- Number and type of interventions to be included in the test
- Number of personnel participating
- Media or placebo materials to be used
- Volume of medium to be filled into the containers
- Incubator identification, and incubation time and temperature for the filled units
- Environmental monitoring to be performed
- Growth promotion test
- Box and tray number and time, especially of any positive units
- Verification of medium sterility

Procedure

All the process should be videotaped or observed to gain further insight into problem resolution. The test should be performed under worst case approach.

SOP No. Val. 700.70

Effective date: mm/dd/yyyy

Approved by:

Units to be filled

The probability of detection of nonsterility in a media fill $= 1 - (1 - x)^n$, where x = acceptable contamination rate and n = the number of vials filled. It should be the same as a normal production run, but not less than 3000 units.

2. Solutions

2.1 Media to be used

Soybean casein digest broth shall be used for normal media runs but thioglycolate broth shall be used for the detection of anaerobic organisms, especially when a filtered nitrogen purge is used to ensure anaerobic conditions.

2.2 Compounding operations

A suitable holding tank is sterilized in place by steam. The following sterilized parts are aseptically connected to the tank: a blender valve, a safety relief valve, and a vent filter assembly containing a 0.2 μm filter. Prepare the media according to manufacturer instruction. Adjust pH.

The medium is sterilized by autoclaving it to 121°C for 30 min, after which it is rapidly cooled to 25 to 30°C. Growth media is handled in a manner similar to the production process being simulated. The medium is passed through the run as though it were an actual product batch, and all routine procedures used in manufacture of a batch are performed (i.e., filter integrity testing, sampling, etc.).

Once the medium has been processed, it is held for a period of time at least equal to that for aseptically produced materials. Any aseptic manipulations performed during and at the end of the hold period should be simulated as well (i.e., sampling, refiltration, hold times, and product recalculation).

2.3 Verification of medium sterility

Perform sterility test for the bulk media.

SOP No. Val. 700.70

Effective date: mm/dd/yyyy

Approved by:

2.4 Filling operations

The containers and closures, if necessary, are cleaned and sterilized using SOPs. It is preferable to use materials, components, and closures that have remained in the aseptic processing area for extended periods.

The filling machine is operated at the predetermined fill rate for the container size utilized, as well as at the fastest speed (handling difficulty) and the slowest speed (maximizing).

The containers are sealed and the medium-filled units are collected in sequentially numbered trays or boxes (notified to the filling time). It is preferable to use materials, components, and closures which have remained in the aseptic processing area for extended periods.

The filled units should be briefly inverted and swirled after filling to assure closure contact with the medium. Increase the size of filling crew to more than the number necessary to fill the batch. All routine activities which take place on the filling line should be a part of the test (i.e., weight adjustments, replenishment of containers, addition of components, change of filling pump, change filter, etc.). Increase the size of filling crew to more than the number necessary to fill the batch.

3. Lyophilized products

3.1 Compounding and filling operations

3.2 Lyophilization operations

The method employed for lyophilization process simulation testing generally is similar to those used for solution fills with the addition of the transport and freeze-drying steps. However, it should focus on loading and sealing activities, which are presumed to be the greatest source of potential contamination.

Containers are filled with medium, and stoppers are partially inserted. The containers are loaded into the lyophilizer. A partial vacuum is drawn on the chamber and this level is held for a predetermined time. The vacuum must not be so low as to permit the medium in the containers to boil out.

The chamber is then vented and the stoppers are seated within the chamber. The stopper units are removed from the aseptic processing area and sealed.

SOP No. Val. 700.70   Effective date: mm/dd/yyyy

                                              Approved by:

An anaerobic condition exists if there is need for sterile inert gas to break the vacuum on the chamber and remain in the container after sealing. The use of anaerobic medium (e.g., alternative fluid thioglycolate medium) would be appropriate where the presence of anaerobic organisms has been confirmed in either environmental monitoring or, more likely, during end product sterility testing.

Where anaerobic organisms have not been detected in the environmental monitoring or sterility testing, lyophilizer process simulation tests should utilize TSB and air.

Alternatively, an appropriate number of glass vials are filled to the proper level with sterilized WFI, following which the filled bottles are subjected to the lyophilization process. The processed bottles are then filled with a known volume of a sterile liquid medium, sealed, and incubated as described above.

4. Powders

Selection of placebo powder

The chosen material must be easily sterilizable, dispersible, or dissolvable in the chosen medium. The principal sterile placebo materials (irradiated in a final container) are lactose, mannitol, polyethylene glycol 6000, and sodium chloride. The material should pass solubility testing at the desired concentration with suitable amount and time of agitation.

4.1 Compounding operations

A quantity of an appropriate sterilized placebo powder is blended with sterile excipients prior to filling (if needed) in a manner similar to the production process being simulated. The medium is passed through the run as though it were an actual product batch, and all routine procedures used in manufacture of a batch are performed. Once the medium has been processed, it is held for a period of time at least equal to that for aseptically produced materials. Any aseptic manipulations performed during and at the end of the hold period should be simulated hold times and product recalculation.

SOP No. Val. 700.70   Effective date: mm/dd/yyyy
                                              Approved by:

4.2 Filling operations

The containers and closures are cleaned and sterilized using SOPs. The filling machine is operated at the predetermined fill rate for the container size utilized, as well as at the fastest speed (handling difficulty) and the slowest speed (maximizing). The containers are sealed and the medium-filled units are collected in sequentially numbered trays or boxes (notified to the filling time).

All routine activities which take place on the filling line should be a part of the test (i.e., weight adjustments, replenishment of containers, addition of components, change of filling pump, change filter, etc.). Increase the size of filling crew to more than the number necessary to fill the batch.

4.3 Powder reconstitution

Before incubation of the vials, powder should be reconstituted with adapted media (TSB or thioglycolate broth) using aseptic technique under laminar flow. The reconstitution volume is according to the volume described in the original formula. Strict environmental monitoring should be followed through this step.

Incubation Conditions

The incubation period should be not less than 14 days per procedure. An incubation temperature in the range of 20 to 35°C may be used depending upon information gained from the environmental monitoring (during routine production, sterility testing, and media filling).

Inspection

It may be advisable to inspect containers midway through the incubation period. The filled units should be briefly inverted and swirled after filling to assure closure contact with the medium. Personnel who have had specific training in the visual inspection of media-filled units should perform these incubator checks.

SOP No. Val. 700.70   Effective date: mm/dd/yyyy

Approved by:

Growth promotion test

The medium in the final containers should be tested for growth promotion method according to the in-house or official monograph. The units used for growth testing must subject to the same processing steps (e.g., cleaning, depyrogenation, sterilization, filtration, filling, lyophilization, reconstitution) up to the point at which they are placed into incubation.

Container inspection

The containers should be inspected for any breach of integrity which may have gone undiscovered during release inspection prior to incubation, or could have occurred during post-inspection handling (e.g., transport to incubator, microbiological inspections).

Damaged containers should not be considered in the evaluation (acceptance) of the aseptic processing.

An identification of the organism may be performed, but the information will most likely be of little value for damaged containers.

Acceptance criteria

Number of vials with microbial growth × 100

The % of contamination =

$$\frac{\text{Number of vials filled} - \text{Number of damaged vials}}{}$$

The acceptable percentage of contamination is ≤0.1%

Failure investigation and corrective action

A contaminated container should be examined carefully for any breach in the container system. All positives (from integral containers) should be identified to at least genus, and to species whenever possible. The identification of contaminant should be compared to the database of the organisms recently identified.

The biochemical profile of the contaminant can then be compared to that of microorganisms obtained from the sterility tests and bioburden and environmental monitoring programs, in order to help identify the potential sources of the contaminant.

SOP No. Val. 700.70   Effective date: mm/dd/yyyy
                                       Approved by:

■ If media fill contaminant is the same as sterility test contaminant:
■ Increase media fill vial quantities and routine filling environmental monitoring to identify the source of contamination.
■ Review environmental data obtained during line setup.
■ If media fill contaminant is the same as media fills environmental contaminant:
■ Increase routine environmental monitoring to determine if the contamination potential exists during routine filling operation.
■ If media fill contaminant is the same as routine environmental contaminant:
■ Increase media fill environmental monitoring (in the same location) to confirm the contaminant source.
■ If sterility test contaminant is the same as media fill environmental contaminant:
■ Increase routine environmental monitoring (in the same location) and number of media fill vial to conform.
■ If sterility test contaminant is the same as routine environmental contaminant:
■ The sterility test is voided.
■ Investigate sterility test procedures and room sanitation and sterilization methods to eliminate cause.
■ If media fill environmental contaminant is the same as routine environmental contaminant:
■ Increase the number of media fills vial in media fill in order to determine the product risk potential.
■ Review monitoring technique for possible problem.
■ Review personnel practices, gowning, sanitation, and sterilization.
■ If media fill environmental contaminant is the same as sterility test contaminant and if routine environmental contaminant is the same as sterility test contaminant:
■ Check environmental monitoring methods and techniques closely for problems.
■ Review personnel practices, gowning, sanitation, and sterilization.

If the failure repeated represents a potential for product concern, a corrective action system should be activated. This system should contain provision for the following:

SOP No. Val. 700.70

Effective date: mm/dd/yyyy
Approved by:

- Critical systems (HVAC, compressed air and gas, water, steam) should be reviewed for documented changes.
- Calibration records should be checked.
- All HEPA filters in the filling area should be inspected and rectified, if warranted.
- Training records for all individuals (production, maintenance, cleaning) involved in the fill should be reviewed to assure proper training was provided.

If the root cause is assignable, corrective action needs to be taken and documented. If three consecutive runs over action levels occur, a problem analysis corrective action report (PACAR) must be issued.

Frequency

Initial validation: three successful consecutive media fill runs required
Revalidation: twice per year, and additional tests should be performed in response to adverse trends or failures in the ongoing monitoring of the facilities or process such as:

- Continued critical area environmental monitoring results above the action levels
- Increased incidence of product sterility test failures
- Break of asepsis in the aseptic processing area, or to evaluate changes to procedures, practices or equipment configuration

Examples of such changes include:

- Major modification to the equipment or immediate product containers or closure
- Modification to equipment or facilities, which potentially affects the quality of air flow in the aseptic environment
- Major changes in the number of the production personnel or initiation of second (or third) shift production when the facility has been qualified only for single shift operations
- Major changes to the aseptic production process or procedures

SOP No. Val. 700.70   **Effective date: mm/dd/yyyy**
                                                   Approved by:

The process should be requalified when production line is not in operation for 3 months. In case of one run failure, three consecutive runs should be performed.

REASONS FOR REVISION

Effective date: mm/dd/yyyy

▪ First time issued for your company, affiliates, and contract manufacturers

YOUR COMPANY
VALIDATION STANDARD OPERATING PROCEDURE

SOP No. Val. 700.80

Effective date: mm/dd/yyyy
Approved by:

TITLE: Half-Automatic Inspection Machine

AUTHOR: _____
Name/Title/Department

_____
Signature/Date

CHECKED BY: _____
Name/Title/Department

_____
Signature/Date

APPROVED BY: _____
Name/Title/Department

_____
Signature/Date

REVISIONS:

No.	Section	Pages	Initials/Date

SOP No. Val. 700.80   **Effective date: mm/dd/yyyy**

Approved by:

SUBJECT: Half-Automatic Inspection Machine

PURPOSE

To describe the procedure for validation of the half-automatic inspection machine to ensure that it meets installation, operational, and performance qualification requirements

RESPONSIBILITY

It is the responsibility of the production manager and technical services manager to follow the procedure. The quality assurance manager is responsible for SOP compliance.

PROCEDURE

1. Installation Qualification

- Verify approved purchase order.
- Verify invoice.
- Check manufacturer and supplier.
- Verify model number and serial number.
- Check for any physical damage.
- Confirm location and installation requirements per recommendation of manufacturers.
- Verify that the utilities required are available.
- Installation shall be conducted per instructions provided in the manual.
- Ensure that all relevant documentation is received:
  - User manual.
  - Maintenance manual.
  - List of change parts.
  - Electrical drawings.
  - Mechanical drawings.

2. Operational Qualification

- Verify alarm control.
- Perform calibration requirements, identified in the manual or established by the validation team.
- Operate the equipment at low, medium, and high speed per operations manual to verify the operating control.
- Verify that all switches and push buttons are functioning properly.
- Establish procedures for operation, maintenance, and calibration.
- Establish training program for relevant staff.

3. Performance Qualification

The following parameters shall be evaluated for product to be inspected:

- Rotation velocity of ampules and vials prior entering the inspection cabin
- Rotation velocity of ampules and vials during visual inspection
- Setting of background illumination
- Setting of Tyndall lamps
- Coordination of brake moment, chain speed, and dropout position

After finishing these evaluations, the inspection efficiency of the machine should be compared with the visual inspection method.

REASONS FOR REVISION

Effective date: mm/dd/yyyy

- First time issued for your company, affiliates, and contract manufacturers

YOUR COMPANY
VALIDATION STANDARD OPERATING PROCEDURE

SOP No. Val. 700.90

Effective date: mm/dd/yyyy
Approved by:

TITLE: Ampule Crack Detection Machine

AUTHOR: _____
Name/Title/Department

_____
Signature/Date

CHECKED BY: _____
Name/Title/Department

_____
Signature/Date

APPROVED BY: _____
Name/Title/Department

_____
Signature/Date

REVISIONS:

No.	Section	Pages	Initials/Date

SUBJECT: Ampule Crack Detection Machine

PURPOSE

To describe the procedure for validation of the ampule crack detection machine to meet the specification; other machines available shall be validated accordingly.

RESPONSIBILITY

It is the responsibility of the production manager and technical services manager to follow the procedure. The quality assurance manager is responsible for SOP compliance.

PROCEDURE

1. Installation Qualification

- Verify approved purchase order.
- Verify invoice.
- Check manufacturer and supplier.
- Verify model number and serial number.
- Check for any physical damage.
- Confirm location and installation requirements per recommendation of manufacturers.
- Verify that the utilities required are available.
- Installation shall be conducted per the instructions provided in the manual.
- Ensure that all relevant documentation is received:
  - User manual
  - Maintenance manual
  - List of change parts
  - Electrical drawings
  - Mechanical drawings

SOP No. Val. 700.90   Effective date: mm/dd/yyyy

Approved by:

2. Operational Qualification

- Verify alarm control.
- Perform calibration requirements, identified in the manual or established by the validation team.
- Operate the equipment at low, medium, and high speed per operations manual to verify the operating control.
- Verify that all switches and push buttons are functioning properly.
- Establish procedures for operation, maintenance, and calibration.
- Establish training program for relevant staff.

3. Performance Qualification

Mix 50 ampules rejected by the dye vacuum with 50 accepted ampules to produce a calibration set. Each calibration set should contain a representative sample of ampules with different sizes of holes and cracks as shown by the intensity of the dye in the ampule. Number the bad ampules on the base depending on the intensity of the dye, from 1 to 50.

Pass the ampule sets through the machine at the manufacturer-recommended setting.

Requirements

All ampules rejected by the dye-vacuum method must also be rejected on the ampule crack detection machine.

Frequency

Initial validation: three test runs
Revalidation: after maintenance on the machine yearly (per company frequency)

4. Comparisons with the Dye-Vacuum Method on Actual Product Lots

Procedure

To compare the effectiveness of the machine with the dye-vacuum method, production batches are passed through the crack detection machine and then through the dye-vacuum chamber. The final check after that is carried out again on the crack detection machine.

Effective date: mm/dd/yyyy

Approved by:

The scheme of documentation and evaluation of the results is shown below.

Requirements

The crack detection machine shall recognize more bad ampules than the dye-vacuum method and shall reject an equivalent number of good ampules as the dye-vacuum method.

Frequency

Initial validation: three production batches of each product
Revalidation: one production batch of each product once every 3 years

5. Effect of Solution or Dried Product on the Outside of Ampules

To determine the effect of wet or dry product on the outside of the ampule, select 100 good ampules and pass through the machine. The ampules are then wetted (with water) and passed through the machine again. Rewet the ampules (with product solution); allow solution to dry on the outside of the ampules. Pass the ampules again through the crack detection machine.

Requirements

No requirements; for information only

Frequency

Initial validation: once per product
Revalidation: once every 2 years

6. Checking of Counting Device Accuracy

During the comparisons, three ampule trays shall be removed every hour and the count checked for accuracy over one working shift.

SOP No. Val. 700.90   Effective date: mm/dd/yyyy

                                       Approved by:

REASONS FOR REVISION

Effective date: mm/dd/yyyy

■ First time issued for your company, affiliates, and contract manufacturers

| YOUR COMPANY |
| VALIDATION STANDARD OPERATING PROCEDURE |

SOP No. Val. 700.100   Effective date: mm/dd/yyyy

Approved by:

TITLE: Laminar Flow Installations

AUTHOR: _____

Name/Title/Department

_____

Signature/Date

CHECKED BY: _____

Name/Title/Department

_____

Signature/Date

APPROVED BY: _____

Name/Title/Department

_____

Signature/Date

REVISIONS:

No.	Section	Pages	Initials/Date

SOP No. Val. 700.100   Effective date: mm/dd/yyyy

                                      Approved by:

SUBJECT: Laminar Flow Installations

PURPOSE

To ensure that laminar flow meets the specification

RESPONSIBILITY

The quality control manager is responsible for following the procedure. The quality assurance manager is responsible for SOP compliance.

PROCEDURE

1. Integrity Test of HEPA Filters

Procedure and requirements are according to SOP No. Val. 600.30.

Frequency

Initial validation: at the time of installation
Revalidation: once per year

2. Air Stream Profile and Air Velocity

Procedure

Use smoke cartridges to determine the air stream profile of the used laminar flow system. Measure the air velocity of at least 15 different locations below the LAF installation with the help of an anemometer.

Requirements

There should be undisturbed laminar air flow above the machinery and exhaust of the air. The average air velocities should be as follows: 0.4 to 0.5 m/s with a lower limit of 0.4 m/s at each measure point.

Frequency

Initial validation and revalidation once per year

SOP No. Val. 700.100   Effective date: mm/dd/yyyy

Approved by:

3. Surface Contamination

Procedure and requirements are according to SOP No. Val. 600.30.

Frequency

Initial validation and revalidation: once per year

REASONS FOR REVISION

Effective date: mm/dd/yyyy

▪ First time issued for your company, affiliates, and contract manufacturers

YOUR COMPANY
VALIDATION STANDARD OPERATING PROCEDURE

SOP No. Val. 700.110   Effective date: mm/dd/yyyy
                                        Approved by:

TITLE: Sterile Filtration Validation

AUTHOR: _____
                        Name/Title/Department

             _____
                        Signature/Date

CHECKED BY: _____
                        Name/Title/Department

             _____
                        Signature/Date

APPROVED BY: _____
                        Name/Title/Department

             _____
                        Signature/Date

REVISIONS:

No.	Section	Pages	Initials/Date

SOP No. Val. 700.110   Effective date: mm/dd/yyyy
Approved by:

SUBJECT: Sterile Filtration Validation

PURPOSE

To describe the procedure for validation of the sterile filtration to ensure filter integrity before and after the filtration

RESPONSIBILITY

The production manager is responsible for following the procedure. The quality assurance manager is responsible for SOP compliance.

PROCEDURE

1. Product Compatibility

Perform a test filtration with the selected product samples prior to and after filtration. Results should be examined with regard to chemical stability (i.e., active content, physical, color, etc.).

Requirements

Unfiltered and filtered product: no chemical and physical difference

2. Filter Integrity Test prior to and after Sterilization

Perform a pressure hold test with the nonsterilized filter. Expose the filter to the chosen sterilization conditions. Repeat the pressure hold test with the sterilized filter. For pressure hold test conditions, follow the recommendations of the filter supplier.

Requirements

Filter shall meet the pressure hold test prior to and after sterilization successfully per supplier specification.

Frequency

Repeat the procedure three times for each filter type.

SOP No. Val. 700.110   Effective date: mm/dd/yyyy

Approved by:

3. Filter Integrity Test after Product Filtration

Perform a pressure hold test with the sterilized filter. Perform the filtration of the product to be sterile filtered using normal production conditions. After the filtration step, the filter should be tested again with the bubble point test.

The filter shall meet the recommendation of filter supplier for pressure hold test.

Requirements

The filter must indicate integrity after undergoing a filtration at full production scale.

Frequency

For the largest batch size of each product: three times

4. Microbial Effectiveness (Bioburden)

The product bioburden must comply with the microbial retention capacity of the filter.

Frequency

For each product and each filter type: three times

REASONS FOR REVISION

Effective date: mm/dd/yyyy

- ▪ First time issued for your company, affiliates, and contract manufacturers

| YOUR COMPANY |
| VALIDATION STANDARD OPERATING PROCEDURE |

SOP No. Val. 700.120

Effective date: mm/dd/yyyy

Approved by:

TITLE: Cleaning Efficiency of Production Equipment for Parenterals

AUTHOR: _____

Name/Title/Department

_____

Signature/Date

CHECKED BY: _____

Name/Title/Department

_____

Signature/Date

APPROVED BY: _____

Name/Title/Department

_____

Signature/Date

REVISIONS:

No.	Section	Pages	Initials/Date

SOP No. Val. 700.120

Effective date: mm/dd/yyyy
Approved by:

SUBJECT: Cleaning Efficiency of Production Equipment for Parenterals

PURPOSE

To describe the procedure to ensure effective cleaning of parenteral manufacturing equipment to minimize the chances of cross-contamination and unacceptable viable count

RESPONSIBILITY

The validation manager and quality control manager are responsible for following the procedure. The quality assurance manager is responsible for SOP compliance.

PROCEDURE

1. Aqueous Products

1.1 Chemical cleanliness

Procedure

At the end of approval cleaning, the equipment should be rinsed with sterile water for injection. The water for final rinse shall be tested for its conductivity. As an alternative, run a placebo batch after a production batch and subsequent cleaning. Analyze the samples of placebo batch for the active ingredients of the previous run batch to ensure that there is no cross-contamination.

Requirements

Conductivity: <1.3 ms/cm at 25°C

Frequency

Three times for each piece of equipment and each different product

SOP No. Val. 700.120   Effective date: mm/dd/yyyy

                                                  Approved by:

1.2 Microbiological cleanliness

Procedure

After cleaning, the equipment should be rinsed with sterile water for injection. This water should be collected and tested for its viable count.

Requirements

Viable count: ≤10 cfu/100 ml

Frequency

Three times for each piece of equipment and product

2. Oily Products (Arachis Oil)

Procedure

Rinse the equipment with acetone after the cleaning. Analyze the rinsing using a UV spectrophotometer (sample size 20 ml). Wipe at each selected location a surface area of between 100 and 800 cm² or alternatively rinse the equipment with a suitable liquid and check.

Requirements

The residues found should not exceed the method's detection limit for the relevant active detergent.

Frequency

Initial validation: once at different equipment locations
Revalidation: once every 3 years

REASONS FOR REVISION

Effective date: mm/dd/yyyy

■ First time issued for your company, affiliates and contract manufacturers

SECTION

VAL 800.00

| YOUR COMPANY |
| VALIDATION STANDARD OPERATING PROCEDURE |

SOP No. Val. 800.10   Effective date: mm/dd/yyyy

                                        Approved by:

TITLE: Kneading Machine

AUTHOR: _____

                    Name/Title/Department

_____

                    Signature/Date

CHECKED BY: _____

                    Name/Title/Department

_____

                    Signature/Date

APPROVED BY: _____

                    Name/Title/Department

_____

                    Signature/Date

REVISIONS:

No.	Section	Pages	Initials/Date

SOP No. Val. 800.10   Effective date: mm/dd/yyyy
Approved by:

SUBJECT: Kneading Machine

PURPOSE

To describe the procedure for validation of the kneading machine to ensure product homogenization and that equipment meets installation, operational, and performance qualifications

RESPONSIBILITY

It is the responsibility of the validation manager, quality control manger and concerned departmental managers to follow the procedure. The quality assurance manager is responsible for SOP compliance.

PROCEDURE

1. Installation Qualification

- Verify approved purchase order.
- Verify invoice.
- Check manufacturer and supplier.
- Verify model number and serial number.
- Check for any physical damage.
- Confirm location and installation requirements per recommendation of manufacturers.
- Verify that the utilities required are available.
- Installation shall be conducted per the instructions provided in the manual.
- Ensure that all relevant documentation is received:
  - User manual
  - Maintenance manual
  - List of change parts
  - Electrical drawings
  - Mechanical drawings

1.1 Net capacity of the granulating chamber

Fill the granulating chamber with preweighed quantities of water.

SOP No. Val. 800.10

Effective date: mm/dd/yyyy
Approved by:

Requirements

The available net capacity should be equal to the manufacturer specification.

1.2 Velocity of the granulating device

Determine the velocity of the granulating device with an empty and a loaded granulator.

Requirements

Shall meet the manufacturer specification

1.3 Liquid dosing pump

1.3.1 Dead volume determination of the dosing system

Prepare a sample of the granulating liquid and pump it into a separate vessel. Determine the percentage of granulating liquid transferred. Determine the pump losses.

1.3.2 Dosing speed

Determine the effective liquid flow per minute by pumping a normal production granulating liquid at fixed dosing speeds.

2. Operational Qualification

- Verify alarm control.
- Perform calibration requirements, identified in the manual or established by the validation team.
- Operate the equipment at low, medium, and high speed per operations manual to verify the operating control.
- Verify that all switches and push buttons are functioning properly.
- Establish procedures for operation, maintenance, and calibration.
- Establish training program for relevant staff.

SOP No. Val. 800.10

Effective date: mm/dd/yyyy
Approved by:

2.1 End point determination of granulation

Determine procedure to establish end of granulation process for each product (for instance, by particle size distribution of granulate).

3. Performance Qualification

Manufacture one batch of the product, dry it and take samples for the analysis of the contents.

3.1 Compressing capabilities and tablet characteristics evaluation

The material obtained from the kneading process should be, after drying and blending, compressed to tablets. The compressing capabilities and the tablet characteristics (content uniformity, thickness, hardness, friability, weight variances, and disintegration time) should meet the finished product specification for the compressed tablets.

REASONS FOR REVISION

Effective date: mm/dd/yyyy

■ First time issued for your company, affiliates, and contract manufacturers

YOUR COMPANY
VALIDATION STANDARD OPERATING PROCEDURE

SOP No. Val. 800.20   Effective date: mm/dd/yyyy

                                        Approved by:

TITLE: Oscillating Granulator

AUTHOR: _____

                        Name/Title/Department

               _____

                        Signature/Date

CHECKED BY: _____

                        Name/Title/Department

               _____

                        Signature/Date

APPROVED BY: _____

                        Name/Title/Department

               _____

                        Signature/Date

REVISIONS:

No.	Section	Pages	Initials/Date

SOP No. Val. 800.20   Effective date: mm/dd/yyyy

Approved by:

SUBJECT: Oscillating Granulator

PURPOSE

To describe the procedure to ensure that the oscillating granulator meets installation, operational, and performance qualifications

RESPONSIBILITY

The validation manager and respective departmental managers are responsible for following the procedure. The quality assurance manager is responsible for SOP compliance.

PROCEDURE

1. Installation Qualification

- Verify approved purchase order.
- Verify invoice.
- Check manufacturer and supplier.
- Verify model number and serial number.
- Check for any physical damage.
- Confirm location and installation requirements per recommendation of manufacturer.
- Verify that the utilities required are available.
- Installation shall be conducted per the instructions provided in the manual.
- Ensure that all relevant documentation is received:
  - User manual
  - Maintenance manual
  - List of change parts
  - Electrical drawings

1.1 Calibration of recording equipment

Check the product flow vibrator and the oscillating rotor.

SOP No. Val. 800.20   Effective date: mm/dd/yyyy

                                       Approved by:

Procedure

Evaluate the effective frequency of the product flow vibrator and the oscillating rotor at the different possible adjustments; develop history and date.

2. Operational Qualification

- Verify alarm control.
- Perform calibration requirements, identified in the manual or established by the validation team.
- Operate the equipment at low, medium, and high speed per operations manual to verify the operating control.
- Verify that all switches and push buttons are functioning properly.
- Establish procedures for operation, maintenance, and calibration.
- Establish training program for relevant staff.

2.1 Evaluation of equipment capacity

Procedure

Verify the capacity of the equipment.

Requirements

Shall meet the manufacturer's specifications

3. Performance Qualification

Influence of machine variables on particle sizes and moisture content of the granulations:

- Speed of product flow through the granulator
- Flow vibrator frequency
- Product filling level in the granulating chamber
- Sieve mesh sizes
- Oscillating rotor frequency

SOP No. Val. 800.20   Effective date: mm/dd/yyyy

Approved by:

Procedure

Determine the moisture content of the nongranulated product with the granulated product.

Requirements

Product shall meet the product specification per SOP.

3.2 Compressing capabilities and tablet characteristics evaluation

After establishing the process variables for the granulation step, compressing capabilities and tablet characteristics (thickness, hardness, friability, weight variances, and disintegration time) should be reviewed.

Requirements

Product shall meet the specification per SOP for individual products.

REASONS FOR REVISION

Effective date: mm/dd/yyyy

- First time issued for your company, affiliates, and contract manufacturers

| YOUR COMPANY |
| VALIDATION STANDARD OPERATING PROCEDURE |

SOP No. Val. 800.30

Effective date: mm/dd/yyyy

Approved by:

TITLE: Milling Machine

AUTHOR: _____
Name/Title/Department

_____
Signature/Date

CHECKED BY: _____
Name/Title/Department

_____
Signature/Date

APPROVED BY: _____
Name/Title/Department

_____
Signature/Date

REVISIONS:

No.	Section	Pages	Initials/Date

SOP No. Val. 800.30   Effective date: mm/dd/yyyy

Approved by:

SUBJECT: Milling Machine

PURPOSE

To describe the procedure to ensure that the milling machine meets installation, operational, and performance qualifications

RESPONSIBILITIES

It is the responsibility of the validation manager, technical services manager and respective departmental managers to follow the procedure. The quality assurance manager is responsible for SOP compliance.

PROCEDURE

1. Installation Qualification

- Verify approved purchase order.
- Verify invoice.
- Check manufacturer and supplier.
- Verify model number and serial number.
- Check for any physical damage.
- Confirm location and installation requirements per recommendation of manufacturer.
- Verify that the required utilities are available.
- Installation shall be conducted per the instructions provided in the manual.
- Ensure that all relevant documentation is received:
  - User manual
  - Maintenance manual
  - List of change parts
  - Electrical drawings
  - Mechanical drawings

SOP No. Val. 800.30

Effective date: mm/dd/yyyy
Approved by:

2. Operational Qualification

- Verify alarm control.
- Perform calibration requirements, identified in the manual or established by the validation team.
- Operate the equipment at low, medium, and high speed per operations manual to verify the operating control.
- Verify that all switches and push buttons are functioning properly.
- Establish procedures for operation, maintenance, and calibration.
- Establish training program for relevant staff.

3. Performance Qualification

Influence of machine variables on granulation properties:

- Speed of product flow through the mill
- Rotation speed of the milling unit
- Influence of sieve mesh sizes (if sieves are used) on the granulation properties (e.g., particle size distribution, moisture content, poured density, tap density, and flow capacity)

3.1 Compressing capabilities and tablet characteristics evaluation

After fixing the process variables for the milling step, compressing capabilities and tablet characteristics (thickness, hardness, friability, weight variances, and disintegration time) should be investigated.

Requirements

The product shall meet the specification per SOP.

REASONS FOR REVISION

Effective date: mm/dd/yyyy

- First time issued for your company, affiliates, and contract manufacturers

YOUR COMPANY
VALIDATION STANDARD OPERATING PROCEDURE

SOP No. Val. 800.40

Effective date: mm/dd/yyyy

Approved by:

TITLE: Fluid Bed Drier

AUTHOR: _____

Name/Title/Department

_____

Signature/Date

CHECKED BY: _____

Name/Title/Department

_____

Signature/Date

APPROVED BY: _____

Name/Title/Department

_____

Signature/Date

REVISIONS:

No.	Section	Pages	Initials/Date

SOP No. Val. 800.40

Effective date: mm/dd/yyyy

Approved by:

SUBJECT: Fluid Bed Drier

PURPOSE

To describe the procedure for validation of the fluid bed drier to ensure that it meets installation, operation, and performance qualification requirements

RESPONSIBILITY

It is the responsibility of the production manager and technical services manager to follow the procedure. The quality assurance manager is responsible for SOP compliance.

PROCEDURE

1. Installation Qualification

- Verify approved purchase order.
- Verify invoice.
- Check manufacturer and supplier.
- Verify model number and serial number.
- Check for any physical damage.
- Confirm location and installation requirements per recommendation of manufacturer.
- Verify that the utilities required are available.
- Installation shall be conducted per instructions provided in the manual.
- Ensure that all relevant documentation is received:
  - User manual
  - Maintenance manual
  - List of change parts
  - Electrical drawings

Instruments for measuring temperature, humidity, time, air volume and pressure, as well as recording devices for these variables, should be calibrated.

SOP No. Val. 800.40

Effective date: mm/dd/yyyy

Approved by:

1.1 Air temperature distribution

Place several thermocouples at different locations in an empty fluid bed drier, e.g.:

- Inlet air channel below product container mesh bottom
- Product container
- Below filter bag
- Above filter bag
- Exhaust air channel

Measure the temperatures, letting in air of a constant temperature (e.g., 60°C).

1.3 Inlet air installation

1.3.1 Delay time for achieving constant air conditions

Determine, by use of a thermocouple and hygrometer, the necessary delay time required at an adjusted inlet air temperature (in relation to drying processes) for reaching constant air conditions. Determine these figures for the first use of the equipment at the working day, as well as for further use of the equipment at the same working day.

Also calculate from the obtained data the water content of the inlet air (g water per kg air) and compare with the previously fixed requirements.

1.3.2 Microbiological quality of the inlet air

Determine, by use of a biotest RCS centrifugal air sampler, the microbiological quality of the inlet air.

Sampling time 5 min = 8:1 air

Requirements

$\leq$200 cfu/m^3 inlet air

1.3.3 Influence of weather on inlet air conditions

1.3.4 Inlet air installation

SOP No. Val. 800.40   Effective date: mm/dd/yyyy

Approved by:

1.3.5 Delay time for achieving constant air conditions

Procedure

Determine, by use of a thermocouple and a hygrometer, the necessary delay time required at an adjusted inlet air temperature (in relation to granulating processes) for reaching constant air conditions. Determine these figures for first use of the equipment at the working day, as well as for further use of the equipment at the same working day.

Also calculate from the obtained data the water content of the inlet air (g water/kg air) and compare to the requirements.

2. Operational Qualification

- Verify alarm control.
- Perform calibration requirements, identified in the manual or established by the validation team.
- Operate the equipment at low, medium, and high speed per operations manual to verify the operating control.
- Verify that all switches and push buttons are functioning properly.
- Establish procedures for operation, maintenance, and calibration.
- Establish training program for relevant staff.

Procedure

Run three batches of each product and analyze for:

- Active ingredient homogeneity
- Moisture content
- Particle size distribution
- Percentage fines
- Tap density

Based on these data try to fix a drying end point of the process (e.g., correlation between moisture content of the product and the product bed temperature).

3. Performance Qualification

Run each product type.

Requirements

Each product shall meet the product characteristics per SOP.

REASONS FOR REVISION

Effective date: mm/dd/yy

- First time issued for your company, affiliates, and contract manufacturers

YOUR COMPANY
VALIDATION STANDARD OPERATING PROCEDURE

SOP No. Val. 800.50   Effective date: mm/dd/yyyy

Approved by:

TITLE: Blender

AUTHOR: _____

Name/Title/Department

_____

Signature/Date

CHECKED BY: _____

Name/Title/Department

_____

Signature/Date

APPROVED BY: _____

Name/Title/Department

_____

Signature/Date

REVISIONS:

No.	Section	Pages	Initials/Date

SOP No. Val. 800.50   Effective date: mm/dd/yyyy

Approved by:

SUBJECT: Blender

PURPOSE

To describe the procedure for validation of the blender to ensure that it meets installation, operational, and performance qualification requirements

RESPONSIBILITY

It is the responsibility of the production manager and technical services manager to follow the procedure. The quality assurance manager is responsible for SOP compliance.

PROCEDURE

1. Installation Qualification

- Verify approved purchase order.
- Verify invoice.
- Check manufacturer and supplier.
- Verify model number and serial number.
- Check for any physical damage.
- Confirm location and installation requirements per recommendation of manufacturer.
- Verify that the required utilities are available.
- Installation shall be conducted per the instructions provided in the manual.
- Ensure that all relevant documentation is received:
  - User manual
  - Maintenance manual
  - List of change parts
  - Electrical drawings
  - Mechanical drawings

SOP No. Val. 800.50
Effective date: mm/dd/yyyy
Approved by:

1.1 Calibration of the control and recording equipment

Instruments for measuring temperature, pressure, time, mixing chamber slope, and mixing velocity, as well as recording devices for these variables, should be calibrated.

2. Operational Qualification

- Verify alarm control.
- Perform calibration requirements, identified in the manual or established by the validation team.
- Operate the equipment at low, medium, and high speed per operations manual to verify the operating control.
- Verify that all switches and push buttons are functioning properly.
- Establish procedures for operation, maintenance, and calibration.
- Establish training program for relevant staff.

2.1 Net capacity of the mixing chamber

Procedure

Fill the mixing chamber with preweighed quantities of water.

Requirements

The available net capacity should be equal to the supplier specification.

2.2 Mixing or stirring velocity

Measure velocity three times at low, medium, and high speed and compare the average and deviation from the average of the single measurements with the supplier specification.

Requirements

Compliance with the supplier specification

SOP No. Val. 800.50

Effective date: mm/dd/yyyy

Approved by:

3. Performance Qualification

3.1 Product homogeneity

3.1.1 Mixing process

Procedure

Fix the mixing or stirring velocity, load the mixer with the product and switch the mixer on. After previously fixed intervals, the mixer should be switched off and samples should be taken from different locations of the product surface. The samples should be analyzed for their active content.

3.1.2 Unloading

Procedure

After determination of the suitable mixing time to achieve product homogeneity, the influence of the unloading process on the homogeneity should be evaluated. Samples should be taken and sent to QC for analysis.

Requirements

Homogeneity should remain consistent.

3.2 Water content of the product

Take samples of the product prior to mixing, after mixing, and after unloading (begin, mid, end). Determine the water content of all samples.

REASONS FOR REVISION

Effective date: mm/dd/yyyy

- First time issued for your company, affiliates, and contract manufacturers

---

YOUR COMPANY
VALIDATION STANDARD OPERATING PROCEDURE

SOP No. Val. 800.60

Effective date: mm/dd/yyyy
Approved by:

TITLE: Tablet Press

AUTHOR: _____
Name/Title/Department

_____
Signature/Date

CHECKED BY: _____
Name/Title/Department

_____
Signature/Date

APPROVED BY: _____
Name/Title/Department

_____
Signature/Date

REVISIONS:

No.	Section	Pages	Initials/Date

SOP No. Val. 800.60

Effective date: mm/dd/yyyy

Approved by:

SUBJECT: Tablet Press

PURPOSE

To describe the procedure for validation of the tablet press to ensure that it meets installation, operational, and performance qualification requirements

RESPONSIBILITY

It is the responsibility of the production manager, validation manager, and technical services manager to follow the procedure. The quality assurance manager is responsible for SOP compliance.

PROCEDURE

1. Installation Qualification

- Verify approved purchase order.
- Verify invoice.
- Check manufacturer and supplier.
- Verify model number and serial number.
- Check for any physical damage.
- Confirm location and installation requirements per recommenda-tion of manufacturers.
- Verify that the utilities required are available.
- Installation shall be conducted per the instructions provided in the manual.
- Ensure that all relevant documentation is received:
  - User manual
  - Maintenance manual
  - List of change parts
  - Electrical drawings
  - Mechanical drawings

SOP No. Val. 800.60   Effective date: mm/dd/yyyy
                                             Approved by:

1.1 Calibration

The following meters should be calibrated:

- Revolution speed of the tableting table
- Counterpressure at the precompression station
- Counterpressure at the main compression station

1.2 Adjustment of compression forces of the precompression and main compression stations

Use adequate pressure meter to adjust the compression forces.

1.3. Control of the product feeding unit

Determine the rotation speed of the product feeding unit at the variable adjustments per manufacturer specification.

1.4 Adjustment of the tablet outlet station

Prepare for each tablet diameter a test set of two upper punches (shorter and longer) as the standard punches (e.g., using a plaster layer). Build in the one punch and run the tablet machine with a placebo product. Then perform the same test with the other punch.

Requirements

All tablets compressed with manipulated punches should be discharged by the machine for 100%.

2. Operational Qualification

- Verify alarm control.
- Perform calibration requirements, identified in the manual or established by the validation team.
- Operate the equipment at low, medium, and high speed per operations manual to verify the operating control.

SOP No. Val. 800.60   **Effective date: mm/dd/yyyy**

                                           Approved by:

- Verify that all switches and push buttons are functioning properly.
- Establish procedures for operation, maintenance, and calibration.
- Establish training program for the relevant staff.

Run one pilot batch for each product on the tablet press and investigate the items detailed next.

2.1 Loading of granulation

Requirements

No sticking of the granulations in the feeding system

2.2 Segregation of granulation

Take samples from the granulation prior to tableting and during tableting (beginning, middle, and end of the pilot batch); analyze the particle size distribution of the samples taken.

Requirements

No significant deviations in particle size distribution should be found.

2.3 In-Process controls

In-charge the sample quantity of tablets to be taken at routine intervals to the available number of punches in the tableting table.

2.4 Control of the tablet outlet station

To evaluate the self-adjusting properties, collect all tablets discharged by the machine and evaluate them for their weight.

3. Performance Qualification

Evaluation of the compressing capabilities and tablet characteristics

SOP No. Val. 800.60   **Effective date: mm/dd/yyyy**
                                             Approved by:

Procedure

The compressing capabilities and tablet characteristics (i.e., content uniformity, thickness, hardness, friability, weight variation, and disintegration time) should be investigated.

REASONS FOR REVISION

Effective date: mm/dd/yyyy

■ First time issued for your company, affiliates, and contract manufacturers

---

YOUR COMPANY
VALIDATION STANDARD OPERATING PROCEDURE

SOP No. Val. 800.70

Effective date: mm/dd/yyyy

Approved by:

TITLE: Metal Check Device for Tablets

AUTHOR: _____

Name/Title/Department

_____

Signature/Date

CHECKED BY: _____

Name/Title/Department

_____

Signature/Date

APPROVED BY: _____

Name/Title/Department

_____

Signature/Date

REVISIONS:

No.	Section	Pages	Initials/Date

SOP No. Val. 800.70   **Effective date: mm/dd/yyyy**

                                                  Approved by:

SUBJECT: Metal Check Device for Tablets

PURPOSE

To describe the procedure to ensure that the metal check device meets installation, operational, and performance qualifications to detect metal parts accidentally added into the powder blend during and after sieving

RESPONSIBILITIES

It is the responsibility of the production manager, validation manager, and technical services manager to follow the procedure. The quality assurance manager is responsible for SOP compliance.

PROCEDURE

1. Installation Qualification

- Verify approved purchase order.
- Verify invoice.
- Check manufacturer and supplier.
- Verify model number and serial number.
- Check for any physical damage.
- Confirm location and installation requirements per recommendation of manufacturer.
- Verify that the utilities required are available.
- Installation shall be conducted per the instructions provided in the manual.
- Ensure that all relevant documentation is received:
  - User manual
  - Maintenance manual
  - List of change parts
  - Electrical drawings
  - Mechanical drawings

2. Operational Qualification

- Verify alarm control.
- Perform calibration requirements, identified in the manual or established by the validation team.
- Operate the equipment at low, medium, and high speed per operations manual to verify the operating control.
- Verify that all switches and push buttons are functioning properly.
- Establish procedures for operation, maintenance, and calibration.
- Establish training program for relevant staff.

3. Performance Qualification

3.1 Sensitivity determination

Prepare a test set of tablets with known contamination with stainless steel or iron wires or balls (e.g., used as sieve material for an oscillating granulator or a milling machine). Pass these tablets (falling free, directed through the center of the detector area, or directed along the side of the detector area) at the different adjustable sensitivity levels through the metal check device and document whether the device is approving or withdrawing the tablets. Determine the warm-up time of the device necessary to receive a constant device sensitivity.

Requirements

The sensor should detect a tablet with an iron ball of 0.5 μm diameter at a medium sensitivity adjustment.

3.2 Determination of the pass-through time of the different products

Determine the pass-through time necessary for the different products available and compare the calculated speeds with the requirements fixed by the supplier.

SOP No. Val. 800.70   Effective date: mm/dd/yyyy

Approved by:

3.3 Determination of the correct reaction time of the outlet mechanism for withdrawn tablets

Determine the minimum time necessary to ensure that test-set tablets will be sorted out in each case.

3.4 Determination of the pass-through capacity

Mix the tablet test set prepared for sensitivity determination with placebo tablets. Pass this mixture completely through the metal check sensor at a mass flow necessary to cover the output (kg tablets/min) of the connected tablet press.

Requirements

The number of manipulations of iron or stainless steel containing tablets withdrawn by the device should be determined and compared.

REASONS FOR REVISION

Effective date: mm/dd/yyyy

■ First time issued for your company, affiliates, and contract manufacturers

---

YOUR COMPANY
VALIDATION STANDARD OPERATING PROCEDURE

SOP No. Val. 800.80

Effective date: mm/dd/yyyy

Approved by:

TITLE: Tablet Coater

AUTHOR: _____
Name/Title/Department

_____
Signature/Date

CHECKED BY: _____
Name/Title/Department

_____
Signature/Date

APPROVED BY: _____
Name/Title/Department

_____
Signature/Date

REVISIONS:

No.	Section	Pages	Initials/Date

SOP No. Val. 800.80

Effective date: mm/dd/yyyy

Approved by:

SUBJECT: Tablet Coater

PURPOSE

To describe the procedure for validation of the tablet coater to ensure that it meets installation, operational, and performance qualification requirements

RESPONSIBILITY

It is the responsibility of the production manager and technical services manager to follow the procedure. The quality assurance manager is responsible for SOP compliance.

PROCEDURE

1. Installation Qualification

- Verify approved purchase order.
- Verify invoice.
- Check manufacturer and supplier.
- Verify model number and serial number.
- Check for any physical damage.
- Confirm location and installation requirements per recommendation of manufacturers.
- Verify that the utilities required are available.
- Installation shall be conducted per the instructions provided in the manual.
- Ensure that all relevant documentation is received:
  - User manual
  - Maintenance manual
  - List of change parts
  - Electrical drawings
  - Mechanical drawings

SOP No. Val. 800.80   Effective date: mm/dd/yyyy

                                                          Approved by:

1.1 Calibration

Instruments for measuring time, temperature, pressure, pressure differences, revolution speed, flow rate, air volume, and converters, as well as recording devices for these variables, should be calibrated.

1.2 Air volume

Determine the achievable air volume flow (m^3/h) for the empty and the loaded coater and compare the results with the previous set requirements and the supplier specification.

1.3 Delay time for achieving constant inlet air conditions

Determine, by use of a thermocouple and a hygrometer, the necessary delay time required at an adjusted inlet air temperature (with regard to the coating process) for reaching constant air conditions. Determine these figures for first use of the equipment at the working day, as well as for further use of the equipment at the same working day.

Also calculate from the obtained data the water content of the inlet air (g water per kg air) and compare with the requirements.

1.4 Microbiological quality of the inlet air

Determine, by use of biotest RCS centrifugal air sampler, the microbiological quality of the inlet air.

Sampling time 2 min = 80:1 air.

Requirements

Viable count ≤200 cfu/m^3 inlet air

1.5 Microbiological quality of the compressed air system

The outlet of the compressed air supply point is opened and purged for 5 min. Adjust to a flow volume of about 30 l/min. Then the compressed air is passed through an air sampler equipped with a 0.22 μm filter. Sampling time should be about 5 min. This procedure should be performed three times. One filter should later be incubated for anaerobic, the other for aerobic viable count. Viable count should be ≤ 200 CFU/m^3.

SOP No. Val. 800.80   Effective date: mm/dd/yyyy

Approved by:

1.6 Mixing properties of the coater

Load the coating pan for 90% of the determined usable capacity with product (e.g., white placebo tablets). Subsequently add additional 10% of a different product (e.g., colored placebo tablets). Measure the time necessary to achieve a homogenous mixture by visual examination.

2. Coating Solution Vessel

2.1 Capacity of the vessel

Fill the coating solution vessel with preweighed quantities of water.

Requirements

The available capacity should be equal to the manufacturer specification.

2.2 Mixing velocity of the stirring unit

Determine the effective stirring velocity at low, medium, and high speed stirring and compare it to the manufacturer specification.

3. Spraying Equipment

3.1 Spraying pattern

Establish the characteristics of the spraying pattern in relation to spraying pressure, spraying speed, viscosity of liquid to be sprayed, and nozzle sizes.

4. Operational Qualification

- Verify alarm control.
- Perform calibration requirements, identified in the manual or established by the validation team.
- Operate the equipment at low, medium, and high speed per operations manual to verify the operating control.
- Verify that all switches and push buttons are functioning properly.
- Establish procedures for operation, maintenance, and calibration.
- Establish training program for relevant staff.

SOP No. Val. 800.80 Effective date: mm/dd/yyyy

Approved by:

4.1 Optimization of process parameters

Optimization of process parameters should be based on use of placebo batches, environment equivalency factor, heat losses, or worst case simulation (too dry and too wet coating conditions).

5. Performance Qualification

Three production pilot batches shall be run and checked for:

- Physical attributes of inspection for coated tablets per SOP
- Thickness
- Diameter
- Weight
- Disintegration time
- Hardness
- Moisture content
- Actives content
- Dissolution time of actives

REASONS FOR REVISION

Effective date: mm/dd/yyyy

- First time issued for your company, affiliates, and contract manufacturers

SECTION

VAL 900.00

YOUR COMPANY
VALIDATION STANDARD OPERATING PROCEDURE

SOP No. Val. 900.10   Effective date: mm/dd/yyyy

                                        Approved by:

TITLE: Blistering Machine

AUTHOR: _____

                          Name/Title/Department

           _____

                          Signature/Date

CHECKED BY: _____

                          Name/Title/Department

               _____

                          Signature/Date

APPROVED BY: _____

                          Name/Title/Department

                _____

                          Signature/Date

REVISIONS:

No.	Section	Pages	Initials/Date

SOP No. Val. 900.10

Effective date: mm/dd/yyyy

Approved by:

SUBJECT: Blistering Machine

PURPOSE

To describe the procedure for validation of the blistering machine to ensure that it meets installation, operational, and performance qualification requirements

RESPONSIBILITY

It is the responsibility of the production manager and technical services manager to follow the procedure. The quality assurance manager is responsible for SOP compliance.

PROCEDURE

1. Installation Qualification

- Verify approved purchase order.
- Verify invoice.
- Check manufacturer and supplier.
- Verify model number and serial number.
- Check for any physical damage.
- Confirm location and installation requirements per recommendation of manufacturer.
- Verify that the required utilities are available.
- Installation shall be conducted per the instructions provided in the manual.
- Ensure that all relevant documentation is received:
  - User manual
  - Maintenance manual
  - List of change parts
  - Electrical drawings
  - Mechanical drawings
- Control panel
- Central EDP control unit
- Detector for presence of thermoforming foil and sticking blisters
- Preheating device thermoforming foil

SOP No. Val. 900.10   **Effective date: mm/dd/yyyy**

                                                          Approved by:

- Thermoforming station (pocket depth, pocket rupture)
- Forming head pressure and pressure distribution
- Cooling water supply for heated and cooled tools
- Vacuum exhaust system
- Empty hopper detector
- Fill control system
- Detector for presence of lidding foil and sticking blisters
- Light beam for steering of pulling off preprinted lidding foil
- Printer for printing blank lidding foil
- Sealing station (foil strain shift)
- Sealing station overload protection
- Cooling station
- Code embossing and printing station
- Perforating station (foil strain shift and perforation presence)
- Compensation loop before punching station
- Punching station (foil strain shift)
- Machine safety guards
- Blister ejector

2. Operational Qualification

- Verify alarm control.
- Perform calibration requirements, identified in the manual or established by the validation team.
- Operate the equipment at low, medium, and high speed per operations manual to verify the operating control.
- Verify that all switches and push buttons are functioning properly.
- Establish procedures for operation, maintenance, and calibration.
- Establish training program for relevant staff.

For a blistering process, the following critical points can generally be identified:

- Forming of blister
- Feeding of products into the blister pockets
- Fill control system
- Sealing of the blister pack
- Application of blister code

3. Performance Control

Select and run available products and confirm the following attributes:

- Identity check of used materials
- Visual inspection of blister appearance
- Correct function of fill control system
- Correct blister code embossed or printed
- Correct function of perforating station
- Blister seal quality (leak testing)
- Time interval: prior to daily production start; every hour during production

REASONS FOR REVISION

Effective date: mm/dd/yyyy

- First time issued for your company, affiliates, and contract manufacturers

---

YOUR COMPANY
VALIDATION STANDARD OPERATING PROCEDURE

SOP No. Val. 900.20

Effective date: mm/dd/yyyy

Approved by:

TITLE: Blister Filling Machine

AUTHOR: _____

                        Name/Title/Department

                _____

                        Signature/Date

CHECKED BY: _____

                        Name/Title/Department

                _____

                        Signature/Date

APPROVED BY: _____

                        Name/Title/Department

                _____

                        Signature/Date

REVISIONS:

No.	Section	Pages	Initials/Date

SOP No. Val. 900.20

Effective date: mm/dd/yyyy

Approved by:

SUBJECT: Blister Filling Machine

PURPOSE

To describe the procedure for validation of the blister filling machine to ensure that it meets installation, operational, and performance qualification requirements

RESPONSIBILITY

It is the responsibility of the production manager and technical services manager to follow the procedure. The quality assurance manager is responsible for SOP compliance.

PURPOSE

1. Installation Qualification

Depending on the type of the available fill control system, the following device items should be checked to ensure that they fulfill their stated function:

- Verify approved purchase order.
- Verify invoice.
- Check manufacturer and supplier.
- Verify model number and serial number.
- Check for any physical damage.
- Confirm location and installation requirements per recommendation of manufacturer.
- Verify that the utilities required are available.
- Installation shall be conducted per instructions provided in the manual.
- Ensure that all relevant documentation is received:
  - User manual
  - Maintenance manual
  - List of change parts
  - Electrical drawings
  - Mechanical drawings

SOP No. Val. 900.20   Effective date: mm/dd/yyyy

Approved by:

- Control panel
- Control camera
- Control software
- Blister ejector
- Blister ejector counter control
- Accuracy check while system is stationary

2. Operational Qualification

For a fill control system, the following critical points can generally be identified:

- Verify alarm control.
- Perform calibration requirements, identified in the manual or established by the validation team.
- Operate the equipment at low, medium, and high speed per operations manual to verify the operating control.
- Verify that all switches and push buttons are functioning properly.
- Establish procedures for operation, maintenance, and calibration.
- Establish training program for relevant staff.
- Check accuracy during production:
  - Effect of camera defect on fill control.
  - Effect of camera switch-off on fill control.
  - Effect of lamp defect on fill control.
  - Effect of power failure on fill control.
  - Effect of thermoforming foil shift on fill control.
  - Effect of thermoforming foil shift on fill control.
  - Effect of leading interruption between fill control and ejector.
  - Range tolerances for the different products.

3. Performance Qualification

- Working fill control
- Detection of an empty pocket
- Detection of a broken tablet (≥30% missing)
- Correct delay and ejection of the detected faulty blisters
- Correct function blister ejector counter control
- Time interval: prior to daily blistering machine start; every 2 h during production

SOP No. Val. 900.20   Effective date: mm/dd/yyyy

Approved by:

REASONS FOR REVISION

Effective date: mm/dd/yyyy

■ First time issued for your company, affiliates, and contract manufacturers

YOUR COMPANY
VALIDATION STANDARD OPERATING PROCEDURE

SOP No. Val. 900.30

Effective date: mm/dd/yyyy
Approved by:

TITLE: Code Reader

AUTHOR: _____

Name/Title/Department

_____

Signature/Date

CHECKED BY: _____

Name/Title/Department

_____

Signature/Date

APPROVED BY: _____

Name/Title/Department

_____

Signature/Date

REVISIONS:

No.	Section	Pages	Initials/Date

SOP No. Val. 900.30

Effective date: mm/dd/yyyy
Approved by:

SUBJECT: Code Reader

PURPOSE

To describe the procedure for validation of the code reader to ensure that it meets installation, operational, and performance qualification requirements

RESPONSIBILITY

It is the responsibility of the production manager and technical services manager to follow the procedure. The quality assurance manager is responsible for SOP compliance.

PROCEDURE

1. Installation Qualification

- Verify approved purchase order.
- Verify invoice.
- Check manufacturer and supplier.
- Verify model number and serial number.
- Check for any physical damage.
- Confirm location and installation requirements per recommendation of manufacturer.
- Verify that the utilities required are available.
- Installation shall be conducted per the instructions provided in the manual.
- Ensure that all relevant documentation is received:
  - User manual
  - Maintenance manual
  - List of change parts
  - Electrical drawings
  - Mechanical drawings
- Code reader should be checked as follows:
  - Light beam
  - Control panel display accuracy
  - Control panel

SOP No. Val. 900.30   Effective date: mm/dd/yyyy
                                             Approved by:

- Package ejector counter control
- Package ejector

2. Operational Qualification

- Verify alarm control.
- Perform calibration requirements, identified in the manual or established by the validation team.
- Operate the equipment at low, medium, and high speed per operations manual to verify the operating control.
- Verify that all switches and push buttons are functioning properly.
- Establish procedures for operation, maintenance, and calibration.
- Establish training program for relevant staff.

The following critical points shall be checked for operational performance effects:

- Power failure on code reader and package ejector function
- Switched-off code reader during production
- Defect light beam during production
- Leading interruption between light beam and control panel
- Leading interruption between code reader and package ejector
- Influence of incorrect adjusted code reader light beam

3. Performance Qualification

Evaluate the performance using actual product as follows

- Storage and control of correct code
- Detection of faulty code and ejection
- Correct function of ejector counter control
- Time interval: before daily production start-up; every 2 h during production

REASONS FOR REVISION

Effective date: mm/dd/yyyy

- First time issued for your company, affiliates, and contract manufacturers

---

| YOUR COMPANY |
| VALIDATION STANDARD OPERATING PROCEDURE |

SOP No. Val. 900.40

Effective date: mm/dd/yyyy

Approved by:

TITLE: Sachetting Machine

AUTHOR: _____

Name/Title/Department

_____

Signature/Date

CHECKED BY: _____

Name/Title/Department

_____

Signature/Date

APPROVED BY: _____

Name/Title/Department

_____

Signature/Date

REVISIONS:

No.	Section	Pages	Initials/Date

SOP No. Val. 900.40

Effective date: mm/dd/yyyy

Approved by:

SUBJECT: Sachetting Machine

PURPOSE

To describe the procedure for validation of the sachetting machine to ensure that it meets installation, operational, and performance qualification requirements

RESPONSIBILITY

It is the responsibility of the production manager and technical services manager to follow the procedure. The quality assurance manager is responsible for SOP compliance.

PURPOSE

1. Installation Qualification

- Verify approved purchase order.
- Verify invoice.
- Check manufacturer and supplier.
- Verify model number and serial number.
- Check for any physical damage.
- Confirm location and installation requirements per recommendation of manufacturer.
- Verify that the utilities required are available.
- Installation shall be conducted per the instructions provided in the manual.
- Ensure that all relevant documentation is received:
  - User manual
  - Maintenance manual
  - List of change parts
  - Electrical drawings
  - Mechanical drawings

SOP No. Val. 900.40   Effective date: mm/dd/yyyy
                                       Approved by:

The following machine characteristics should be checked to ensure proper installation:

- Control panel and push button
- Data processing control unit
- Sensor for sufficient blisters in the product feeder
- Sensor for presence of a blister on the conveyor
- Sensor for presence of correct stacking height
- Sealing unit
- Sealing unit for the latitudinal seal and cutting off the sachet foil
- Device for correct functioning of the heating cartridges in the sealing units
- Rejection bin for empty and faulty sachets
- Photocell-controlled pulling off of printed foil
- Code embossing unit
- Sensor for presence of print ribbon
- Inkjet code printing device

2. Operational Qualification

- Verify alarm control.
- Perform calibration requirements, identified in the manual or established by the validation team.
- Operate the equipment at low, medium, and high speed per operations manual to verify the operating control.
- Verify that all switches and push buttons are functioning properly.
- Establish procedures for operation, maintenance, and calibration.
- Establish training program for relevant staff.

The following critical points may be checked for their operational performance:

- Sachetting velocity and sealing unit adjustment
- Adjustment of sachet length
- Product feeding to the sealing units
- Pulling off preprinted foil
- Electrical and electronic check devices
- Code embossing and printing

SOP No. Val. 900.40

Effective date: mm/dd/yyyy
Approved by:

3. Performance Qualification

Run sleeted product of variable sachet size and evaluate for:

- Identity check of used materials
- Visual inspection of sachet appearance
- Correct code embossed or printed on the sachet
- Sachet leak testing
- Quality of sealing
- Time interval: before daily production start; every 2 h during production

REASONS FOR REVISION

Effective date: mm/dd/yyyy

- First time issued for your company, affiliates, and contract manufacturers

YOUR COMPANY
VALIDATION STANDARD OPERATING PROCEDURE

SOP No. Val. 900.50   Effective date: mm/dd/yyyy

                                       Approved by:

TITLE: Cartoning Machine

AUTHOR: _____

Name/Title/Department

_____

Signature/Date

CHECKED BY: _____

Name/Title/Department

_____

Signature/Date

APPROVED BY: _____

Name/Title/Department

_____

Signature/Date

REVISIONS:

No.	Section	Pages	Initials/Date

SOP No. Val. 900.50

Effective date: mm/dd/yyyy

Approved by:

SUBJECT: Cartoning Machine

PURPOSE

To describe the procedure for validation of the cartoning machine to ensure that it meets installation, operational, and performance qualification requirements

RESPONSIBILITY

It is the responsibility of the production manager and technical services manager to follow the procedure. The quality assurance manager is responsible for SOP compliance.

PROCEDURE

1. Installation Qualification

- Verify approved purchase order.
- Verify invoice.
- Check manufacturer and supplier.
- Verify model number and serial number.
- Check for any physical damage.
- Confirm location and installation requirements per recommendation of manufacturer.
- Verify that the required utilities are available.
- Installation shall be conducted per the instructions provided in the manual.
- Ensure that all relevant documentation is received:
  - User manual
  - Maintenance manual
  - List of change parts
  - Electrical drawings
  - Mechanical drawings

SOP No. Val. 900.50   **Effective date: mm/dd/yyyy**

                                              Approved by:

The following machine characteristics should be checked to ensure that they fulfill their stated function:

- Control panel
- Central EDP control unit
- Sensor for presence of product
- Sensor for presence of correct stacking height
- Empty feeder detector for cartons
- Detector for presence of cartons
- Code reader for cartons
- Empty feeder detector for leaflets
- Detector for presence of leaflets
- Code reader for leaflets (single side, double-sided)
- Overload protection for product feed-in
- Luminescence presence control for instance for leaflets
- Code embossing and printing station
- Detector for presence of coding
- Package ejector
- Machine safety guards

2. Operational Qualification

- Verify alarm control.
- Perform calibration requirements, identified in the manual or established by the validation team.
- Operate the equipment at low, medium, and high speed per operations manual to verify the operating control.
- Verify that all switches and push buttons are functioning properly.
- Establish procedures for operation, maintenance, and calibration.
- Establish training program for relevant staff.
- Verify:
  - Product feeding to the cartoner
  - Feeding of the product into the carton
  - Electronic check devices
  - Coding of the cartons with batch or packaging number, expiration or manufacturing date

SOP No. Val. 900.50

Effective date: mm/dd/yyyy

Approved by:

3. Performance Qualification

- Used materials identification check
- Visual inspection of package appearance
- Package contents complete and undamaged
- Correct functioning of code readers
- Correct code embossed or printed on the package
- Time interval: before daily production start; every 2 h during production

REASONS FOR REVISION

Effective date: mm/dd/yyyy

- First time issued for your company, affiliates, and contract manufacturers

| YOUR COMPANY |
| VALIDATION STANDARD OPERATING PROCEDURE |

SOP No. Val. 900.60

Effective date: mm/dd/yyyy

Approved by:

TITLE: Labeling Machine

AUTHOR: _____

Name/Title/Department

_____

Signature/Date

CHECKED BY: _____

Name/Title/Department

_____

Signature/Date

APPROVED BY: _____

Name/Title/Department

_____

Signature/Date

REVISIONS:

No.	Section	Pages	Initials/Date

SOP No. Val. 900.60   **Effective date: mm/dd/yyyy**

                                             Approved by:

SUBJECT: Labeling Machine

PURPOSE

To describe the procedure for validation of the labeling machine to ensure that it meets installation, operational, and performance qualification requirements

RESPONSIBILITY

It is the responsibility of the production manager and technical services manager to follow the procedure. The quality assurance manager is responsible for SOP compliance.

PROCEDURE

1. Installation Qualification

- Verify approved purchase order.
- Verify invoice.
- Check manufacturer and supplier.
- Verify model number and serial number.
- Check for any physical damage.
- Confirm location and installation requirements per recommendation of manufacturer.
- Verify that the utilities required are available.
- Installation shall be conducted per the instructions provided in the manual.
- Ensure that all relevant documentation is received:
  - User manual
  - Maintenance manual
  - List of change parts
  - Electrical drawings
  - Mechanical drawings

■ Depending on the type of the available labeling machine, the following machine items should be checked to ensure that they fulfill their stated function:
  ■ Control panel
  ■ Sensor for presence of sufficient label supply in the feeder
  ■ Sensor for presence of product
  ■ Code reader for identification of preprinted labels
  ■ Ring code reader for identification of ampules
  ■ Printing station for label text
  ■ Embossing station for codes
  ■ Heating station for heat seal labels
  ■ Glue applicator for wet glue labels
  ■ Coiling up station for heat seal, wet glue, and self-adhesive labels
  ■ Sensor for presence of printing or embossing on the label

2. Operational Qualification

■ Verify alarm control.
■ Perform calibration requirements identified in the manual or established by the validation team.
■ Operate the equipment at low, medium, and high speed per operations manual to verify the operating control.
■ Verify that all switches and push buttons are functioning properly.
■ Establish procedures for operation, maintenance, and calibration.
■ Establish training program for relevant staff.
■ The following critical points can be evaluated for their operational qualification:
  ■ Product feeding
  ■ Printing and embossing station
  ■ Heating station and glue applicator station

3. Performance Qualification

Run actual product and evaluate the following machine characteristics:

■ Identity check of used materials
■ Correct text or code printed or embossed
■ Quality of the printed or embossed texts or codes

SOP No. Val. 900.60   **Effective date: mm/dd/yyyy**
                                       Approved by:

- Correct functioning of code reader and luminescence presence control
- Time interval: before daily production start-up; every 2 h during production

REASONS FOR REVISION

Effective date: mm/dd/yyyy

- First time issued for your company, affiliates, and contract manufacturers

| YOUR COMPANY |
| VALIDATION STANDARD OPERATING PROCEDURE |

SOP No. Val. 900.70   Effective date: mm/dd/yyyy
                                       Approved by:

TITLE: Check Weigher

AUTHOR: _____
                        Name/Title/Department

              _____
                          Signature/Date

CHECKED BY: _____
                        Name/Title/Department

              _____
                          Signature/Date

APPROVED BY: _____
                        Name/Title/Department

              _____
                          Signature/Date

REVISIONS:

No.	Section	Pages	Initials/Date

SOP No. Val. 900.70

Effective date: mm/dd/yyyy

Approved by:

SUBJECT: Check Weigher

PURPOSE

To describe the procedure for validation of the check weigher to ensure that it meets installation, operational, and performance qualification requirements

RESPONSIBILITY

It is the responsibility of the production manager and technical services manager to follow the procedure. The quality assurance manager is responsible for SOP compliance.

PURPOSE

1. Installation Qualification

- Verify approved purchase order.
- Verify invoice.
- Check manufacturer and supplier.
- Verify model number and serial number.
- Check for any physical damage.
- Confirm location and installation requirements per recommendation of manufacturer.
- Verify that the utilities required are available.
- Installation shall be conducted per the instructions provided in the manual.
- Ensure that all relevant documentation is received:
  - User manual
  - Maintenance manual
  - List of change parts
- Electrical drawings
- Mechanical drawings
- Check control panel, weighing cell, and integrated printer to ensure their functionality.

SOP No. Val. 900.70 Effective date: mm/dd/yyyy

Approved by:

2. Operational Qualification

- Verify alarm control.
- Perform calibration requirements, identified in the manual or established by the validation team.
- Operate the equipment at low, medium, and high speed per operations manual to verify the operating control.
- Verify that all switches and push buttons are functioning properly.
- Establish procedures for operation, maintenance, and calibration.
- Establish training program for relevant staff.
- Adjustment of belt velocities
- Weighing accuracy during operational conditions
- Range of tolerance for the available product mix
- Adjustment of light barriers
- Package ejector

3. In-Process Controls during Production

Check correct function of weighing, package length check, sloping position check, and package ejector, as well as correct range of tolerance.

Time interval: before daily production start; every 2 h during production

REASONS FOR REVISION

Effective date: mm/dd/yyyy

- First time issued for your company, affiliates, and contract manufacturers

YOUR COMPANY
VALIDATION STANDARD OPERATING PROCEDURE

SOP No. Val. 900.80   Effective date: mm/dd/yyyy

Approved by:

TITLE: Shrink Wrapping and Bundling Machine

AUTHOR: _____

Name/Title/Department

_____

Signature/Date

CHECKED BY: _____

Name/Title/Department

_____

Signature/Date

APPROVED BY: _____

Name/Title/Department

_____

Signature/Date

REVISIONS:

No.	Section	Pages	Initials/Date

SOP No. Val. 900.80
Effective date: mm/dd/yyyy
Approved by:

SUBJECT: Shrink Wrapping and Bundling Machine

PURPOSE

To describe the procedure for validation of the shrink wrapping and bundling machine to ensure that it meets installation, operational, and performance qualification requirements

RESPONSIBILITY

It is the responsibility of the production manager and technical services manager to follow the procedure. The quality assurance manager is responsible for SOP compliance.

PROCEDURE

1. **Installation Qualification**

- Check approved purchase order.
- Verify invoice.
- Verify manufacturer and supplier.
- Check model and serial number.
- Verify that equipment is received undamaged.
- Confirm that installation requirements are available from utilities.
- Check list of change parts and spare parts.
- Verify that all applicable materials are received:
  - User manual
  - Operation manual
  - Mechanical drawings
- The following machine items should be checked to ensure that they fulfill their stated function:
  - Control panel
  - Detector for presence of product
  - Product feeding
  - Detector for presence of wrapping and bundling foils
  - Foil folding unit
  - Sealing unit
  - Thermal shrinking
  - Safety guards

SOP No. Val. 900.80   Effective date: mm/dd/yyyy
                                       Approved by:

2. Operational Qualification

For a shrink wrapping and bundling machine, the following critical points can generally be identified:

- Perform calibration required.
- Check equipment controls, push buttons, and switches per operational manual.
- Adjust motor speed.
- Check product feeding
- Check sealing unit
- Check electrical and electronic control devices

3. Performance Qualification

Identify a variable range of products to be shrinked wrapped (small, medium, and large) and shrink wrap. Check the following procedures:

- Operation
- Calibration
- Maintenance
- Cleaning
- Visual inspection of bundle appearance
- Correct adjustment of sealing temperature
- Time interval: before daily production start; every 2 h during production

REASONS FOR REVISION

Effective date: mm/dd/yyyy

- First time issued for your company, affiliates, and contract manufacturers

---

YOUR COMPANY
VALIDATION STANDARD OPERATING PROCEDURE

SOP No. Val. 900.90

Effective date: mm/dd/yyyy

Approved by:

TITLE: Tube Filling and Closing Machine

AUTHOR: _____

Name/Title/Department

_____

Signature/Date

CHECKED BY: _____

Name/Title/Department

_____

Signature/Date

APPROVED BY: _____

Name/Title/Department

_____

Signature/Date

REVISIONS:

No.	Section	Pages	Initials/Date

SOP No. Val. 900.90   **Effective date: mm/dd/yyyy**
                                              Approved by:

SUBJECT: Tube Filling and Closing Machine

PURPOSE

To describe the procedure for validation of the tube filling and closing machine to ensure that it meets installation, operational, and performance qualification requirements

RESPONSIBILITY

It is the responsibility of the production manager and technical services manager to follow the procedure. The quality assurance manager is responsible for SOP compliance.

PROCEDURE

1. Installation Qualification

- Verify approved purchase order.
- Verify invoice.
- Check manufacturer and supplier.
- Verify model number and serial number.
- Check for any physical damage.
- Confirm location and installation requirements per recommendation of manufacturer.
- Verify that the required utilities are available.
- Installation shall be conducted per the instructions provided in the manual.
- Ensure that all relevant documentation is received:
  - User manual
  - Maintenance manual
  - List of change parts
  - Electrical drawings
  - Mechanical drawings
- The following should be checked to ensure that they fulfill installation qualifications:

SOP No. Val. 900.90   Effective date: mm/dd/yyyy

Approved by:

- Control panel
- Cream pump control unit
- Cream stirrer control unit
- Outlet valve temperature control of cream feeding hopper
- Heating jacket temperature control of cream feeding hopper
- Detector for presence of empty tubes
- Detector for control of product presence in the hopper
- Feeding of empty tubes
- Code reader for checking of the position of the tube prior to filling station
- Code reader for tubes
- Closing station
- Tube closure screwing station
- Coding station
- Tube ejector

2. Operational Qualification

- Verify alarm control.
- Perform calibration requirements, identified in the manual or established by the validation team.
- Operate the equipment at low, medium, and high speed per operations manual to verify the operating control.
- Verify that all switches and push buttons are functioning properly.
- Establish procedures for operation, maintenance, and calibration.
- Establish training program for relevant staff.
- The following critical points should be evaluated:
  - Product feeding to the filler
  - Product filling
  - Electric and electronic check devices
  - Application of tube code
  - Tube closing station

SOP No. Val. 900.90   Effective date: mm/dd/yyyy

                                                 Approved by:

3. Performance Qualification

- Tube closure properly closed (torque force)
- Check of filled quantity
- Quality of tube closing
- Correct functioning of code readers
- Visual inspection of tube appearance
- Correct code embossed or printed on the tube
- Time interval: before daily production; every 2 h during packaging

REASONS FOR REVISION

Effective date: mm/dd/yyyy

- First time issued for your company, affiliates, and contract manufacturers

YOUR COMPANY
VALIDATION STANDARD OPERATING PROCEDURE

SOP No. Val. 900.100   Effective date: mm/dd/yyyy

Approved by:

TITLE: Liquid Filling and Closing Machine

AUTHOR: _____

Name/Title/Department

_____

Signature/Date

CHECKED BY: _____

Name/Title/Department

_____

Signature/Date

APPROVED BY: _____

Name/Title/Department

_____

Signature/Date

REVISIONS:

No.	Section	Pages	Initials/Date

SOP No. Val. 900.100

Effective date: mm/dd/yyyy

Approved by:

SUBJECT: Liquid Filling and Closing Machine

PURPOSE

To describe the procedure for validation of the liquid filling and closing machine to ensure that it meets installation, operational, and performance qualification requirements

RESPONSIBILITY

It is the responsibility of the production manager and technical services manager to follow the procedure. The quality assurance manager is responsible for SOP compliance.

PROCEDURE

1. **Installation Qualification**

- Verify approved purchase order.
- Verify invoice.
- Check manufacturer and supplier.
- Verify model number and serial number.
- Check for any physical damage.
- Confirm location and installation requirements per recommendation of manufacturers.
- Verify that the utilities required are available.
- Installation shall be conducted per the instructions provided in the manual.
- Ensure that all relevant documentation is received:
  - User manual
  - Maintenance manual
  - List of change parts
  - Electrical drawings
  - Mechanical drawings
- The following machine characteristics should be checked to ensure successful installation qualification:

- Control panel
- Feeding of empty bottles
- Detector for presence of bottles
- Detector for presence of liquid
- Terminal sterile filter unit
- Filling unit
- Detector for correct liquid level in the bottle
- Feeding of bottles to the closing machine
- Detector for presence of filled bottles
- Feeding of dropper and pouring ring (if applicable)
- Detector for presence of dropper or pouring ring prior to closing
- Feeding of closures
- Screwing unit

2. Operational Qualification

- Verify alarm control.
- Perform calibration requirements, identified in the manual or established by the validation team.
- Operate the equipment at low, medium, and high speed per operations manual to verify the operating control.
- Verify that all switches and push buttons are functioning properly.
- Establish procedures for operation, maintenance, and calibration.
- Establish training program for relevant staff.
- The following critical checks should be conducted:
  - Transfer of liquid to the filler
  - Filling unit
  - Closing unit
  - Electric and electronic control devices

SOP No. Val. 900.100   **Effective date: mm/dd/yyyy**

                                                      Approved by:

3. Performance Qualification

- Identity check of used materials
- Tamper evidence of closure fulfilled
- Bottle closure properly closed (torque force, vacuum test)
- Visual inspection of closed bottle appearance
- Check of bottle contents
- Thread quality of pilfer-proof closures
- Dropper or pouring ring not damaged
- Time interval: before daily production start; every 2 h during packaging

REASONS FOR REVISION

Effective date: mm/dd/yyyy

- First time issued for your company, affiliates, and contract manufacturers

---

YOUR COMPANY
VALIDATION STANDARD OPERATING PROCEDURE

SOP No. Val. 900.110

Effective date: mm/dd/yyyy

Approved by:

TITLE: Tablet Filling and Closing Machine

AUTHOR: _____

Name/Title/Department

_____

Signature/Date

CHECKED BY: _____

Name/Title/Department

_____

Signature/Date

APPROVED BY: _____

Name/Title/Department

_____

Signature/Date

REVISIONS:

No.	Section	Pages	Initials/Date

SOP No. Val. 900.110

Effective date: mm/dd/yyyy
Approved by:

SUBJECT: Tablet Filling and Closing Machine

PURPOSE

To describe the procedure for validation of the tablet filling and closing machine to ensure that it meets installation, operational, and performance qualification requirements

RESPONSIBILITY

It is the responsibility of the production manager and technical services manager to follow the procedure. The quality assurance manager is responsible for SOP compliance.

PURPOSE

1. Installation Qualification

- Verify approved purchase order.
- Verify invoice.
- Check manufacturer and supplier.
- Verify model number and serial number.
- Check for any physical damage.
- Confirm location and installation requirements per the recommendation of manufacturer.
- Verify that the utilities required are available.
- Depending on the type of the available tablet filling line, the following machine items should be checked to ensure that they fulfill their stated function:
  - Control panels
  - Unscrambler table for bottles
  - Detector for presence of empty bottles
  - Tablet filling and counting unit
  - Tablet feeder
  - Detector for presence of product
  - Electronic counting device
  - Cotton wool inserting unit
  - Detector for presence of filled bottles
  - Detector for presence of cotton wool
  - Capping unit

SOP No. Val. 900.110   Effective date: mm/dd/yyyy

Approved by:

- Ensure that all relevant documentation is received:
  - User manual
  - Maintenance manual
  - List of change parts
  - Electrical drawings
  - Mechanical drawings
- Installation shall be conducted per the instructions provided in the manual.

2. Operational Qualification

- Verify alarm control.
- Perform calibration requirements identified in the manual or established by the validation team.
- Operate the equipment at low, medium, and high speed per operations manual to verify the operational control.
- Verify that all switches and push buttons are functioning properly.
- Check that the tablet filling counter is calibrated.
- Establish procedures for operations, maintenance, and calibration.
- Provide training to identified staff.
- For a tablet filling line, the following critical points can generally be identified:
  - Tablet filling and counting unit
  - Capping unit
  - Electric and electronic control device

3. Performance Qualification

- Identify a variable range of products in numbers (filling units).
- Operate the line; verify the tablet filling machine for its accuracy in filling operation.
- Perform tablet filling machine capability study and demonstrate statistical control.
- Establish and validate cleaning procedure.
- Verify the training records.
- Identity check of used materials
- Visual inspection of bottle appearance
- Correct functioning of check devices

SOP No. Val. 900.110   **Effective date: mm/dd/yyyy**
                                              Approved by:

- Bottle closure properly fixed
- Time interval: prior to daily production start; every hour during production

REASONS FOR REVISION

Effective date: mm/dd/yyyy

- First time issued for your company, affiliates, and contract manufacturers

SECTION

VAL 1000.00

YOUR COMPANY
VALIDATION STANDARD OPERATING PROCEDURE

SOP No. Val. 1000.10   Effective date: mm/dd/yyyy
                                        Approved by:

TITLE: Installation Qualification of
          Computerized Equipment

AUTHOR: _____
                    Name/Title/Department

        _____
                    Signature/Date

CHECKED BY: _____
                    Name/Title/Department

        _____
                    Signature/Date

APPROVED BY: _____
                    Name/Title/Department

        _____
                    Signature/Date

REVISIONS:

No.	Section	Pages	Initials/Date

SUBJECT: Installation Qualification of Computerized Equipment

PURPOSE

To describe validation guideline for the computerized equipment to meet the installation qualification

RESPONSIBILITY

It is the responsibility of the production manager, technical service manager, and computer engineer to follow the procedure. The quality assurance manager is responsible for SOP compliance.

PROCEDURE

Installation qualification shall include the verification that user manuals, technical manuals, and instrument calibration reports of the computerized system are available, complete, appropriate, relevant, and up to date.

1. IQ Approach

The installation qualification of automatic control systems shall consist of a logical panoply of tests. The tests are carried out step-by-step on each component. The structure of the installation qualification shall be standardized for all installation qualifications of computerized pharmaceutical systems as follows.

2. Document Description

The documents description shall include a paragraph stating the objective of the document and a paragraph specifying the scope of the document, i.e., the exhaustive list of the concerned equipment, or categories of equipment.

SOP No. Val. 1000.10 **Effective date: mm/dd/yyyy**
Approved by:

3. IQ Summary

The goal here is to summarize the installation qualification document. This includes:

- An abbreviated description of the computerized system analyzed
- For protocols, the preapproval of the document
- For IQ reports, the certification of the document
- For IQ reports, a summary of the deviations observed during the installation qualification

4. System Description

The purpose here is to describe the as-built installation and to compare the findings with system specifications. Components of the system are compared with those specified. Any detected discrepancy is mentioned for investigation, correction, justification, and approval. When components found are neither specified nor approved, they will be mentioned as not specified.

4.2 General characteristics

The general system characteristics are reported, including:

- Manufacturer, main contractor, customer
- Intended purpose
- System identification (serial number, inventory number, etc.)
- System location
- Technical characteristics
- Main system limits
- Required utilities

4.3 Hardware

The hardware of the system is described, including:

- Summary block diagram
- List of main components, processors and modules, memories, storage, signal converters, networking, communications, and peripherals, including model and serial numbers when applicable

- List of all fuses (type, rated current, location, protected items)
- Operating controls and alarms
- Central commands and overrides
- Manual downgraded controls

4.4 Software

The software of the system is described, including:

- Software identification (name, version, etc.)
- Operating system (name, version, etc.)
- Source code availability
- Language and tools used

5. Documentation

The purpose here is to verify the documentation of the system. This documentation must be appropriate, up to date, relevant, and complete. The analyzed documentation may include:

- User requirements
- System specifications
- Purchase orders and related information
- As-built drawings (block diagrams, mechanical drawings, electrical schematics, wiring, and interconnection diagrams)
- Technical documentation (input–output list, alarms and safety list, automatic and manual control listings, regulation loop descriptions, calibration guide, maintenance guide, troubleshooting guide, etc.)
- Software technical documentation (programmer's guide, reference guide, software diagrams, etc.)
- User manuals (user's guide, guide to operations, etc.)
- Instrumentation calibration certificates (not expired)
- Source code availability and storage arrangements

6. Structural Testing

6.1 Introduction

This structural testing section of the installation qualification consists of verifying the internal integrity of the equipment software and hardware and should be done separately for each.

SOP No. Val. 1000.10

Effective date: mm/dd/yyyy

Approved by:

6.2 Hardware structural testing

This testing consists of a panoply of verifications pertaining to the hardware of the equipment. Tests include:

- Cabling and wiring
- Labeling of wires and components
- Grounding of the system
- Hardware testing performed by the supplier (normal and stress testing)
- Electromagnetic interference compatibility

6.3 Software structural testing

This testing consists of assessing the quality of the development process, the produced code, and the testing process.

Verify quality of the development process by auditing the supplier's internal development process. The aspects analyzed cover:

- Specification methodology
- Diagramming techniques used
- Design verification
- Existence and adherence to production SOPs
- Enforced naming conventions

The method used to verify the quality of the produced source code consists of group reviews, called *structured walk-through* by experts in the field. In summary, these reviews consist of a presentation, made by the designers of the software, of source code sections selected by the reviewers according to their influence on the process or safety of the system. The acceptance criteria shall be based on existing internationally accepted engineering standards that the supplier must meet (delivery timing, budgets, personnel, methods used).

The aspects analyzed cover:

- Adherence of produced functions to system specifications
- Existence and quality of comments into the source code
- Code organization and structure
- Proper indentation

SOP No. Val. 1000.10   Effective date: mm/dd/yyyy

                                               Approved by:

- Out-of-range or out-of-context input detection, including sensor fail detection
- Data edits
- Timeouts
- Abnormal conditions recovery
- Absence of dead code

The quality of the supplier testing process is verified by auditing the supplier's internal testing process. The aspects analyzed cover:

- Testing methodology, including stress testing
- Existence of and adherence to testing SOPs
- Testing reports

REASONS FOR REVISION

Effective date: mm/dd/yyyy

- First time issued for your company, affiliates, and contract manufacturers

YOUR COMPANY
VALIDATION STANDARD OPERATING PROCEDURE

SOP No. Val. 1000.20

Effective date: mm/dd/yyyy

Approved by:

TITLE: Operational Qualification of Computerized Equipment

AUTHOR: _____

Name/Title/Department

_____

Signature/Date

CHECKED BY: _____

Name/Title/Department

_____

Signature/Date

APPROVED BY: _____

Name/Title/Department

_____

Signature/Date

REVISIONS:

No.	Section	Pages	Initials/Date

SOP No. Val. 1000.20

Effective date: mm/dd/yyyy

Approved by:

SUBJECT: Operational Qualification of Computerized Equipment

PURPOSE

To provide the guideline for the operational qualification of computerized equipment to ensure that it meets the equipment operational requirement

RESPONSIBILITY

It is the responsibility of the production manager, technical service manager, and computer manager to follow the procedure. The quality assurance manager is responsible for SOP compliance.

PROCEDURE

1. Introduction

A specific operational qualification protocol shall be prepared for each piece of equipment undergoing operational qualification.

2. OQ Approach

The operational qualification of computerized systems consists of a group of tests, pooled in functional checks. The tests are carried out step by step on each component. The general approach used is of the gray-box type. More precisely, study the input and output data transmission at intermediate points. The structure of the operational qualification may be standardized for all operational qualification of computerized pharmaceutical equipment.

3. Document Description

The objective of the document and the scope of the document, i.e., the exhaustive list of the concerned equipment or categories of equipment, are the operational qualification documents to be included.

SOP No. Val. 1000.20   **Effective date: mm/dd/yyyy**
                                                   Approved by:

4. OQ Summary

This summary includes:

- An abbreviated description of the computerized system analyzed
- For protocols, the preapproval of the document
- For OQ reports, the certification of the document
- For OQ reports, a summary of the deviations observed during the operational qualification

5. Reporting the Initial Setpoint

Each user-accessible setpoint of the installation, equipment, or control system will be reported. Results shall be written in a report.

Most of these setpoints are fixed by the supplier, detailed in the purchase order, or are part of the installation qualification file. If the setpoints are fixed by the supplier, there are no acceptance criteria unless the supplier documents the setpoint in written form. When available from supplier's documents or when specified in the bill of order or the installation qualification report, the observed setpoints shall be compared with the documented ones. Any detected discrepancy is mentioned for investigation, justification, approval, or correction. When reported parameters are neither specified nor approved, they shall be mentioned as not specified.

Reporting the initial setpoints enables definition of the characteristics of the system as it is during qualification and will evidence any subsequent modifications of the setpoints.

6. Checking Digital Transmissions: Input

Qualification tests shall be carried out in order to check the transmission of one-bit binary information sent by sensors and safeties to the computerized control system or programmable logic controller.

Each sensor or safety tested is activated or deactivated a few times to visualize clearly the transmission. Signal visualization at the controller level is done either directly on the controller using an appropriate submenu or by using adequate validation measuring equipment.

7. Checking Digital Transmissions: Output

The qualifications tests consists of checking the transmission of the one-bit binary information sent by the controller to the contactors, relays, electrovalves, pilot lights, and other digital outputs of the system.

Each tested output is activated or deactivated a few times to visualize clearly the transmission. The visualization of the transmitted signals is made either on the activated system itself (mechanical indicators of electrovalves, start of motors or pumps) or by using an adequate validation measuring equipment.

8. Checking Analog Transmissions: Input

Tests shall be carried out to check the transmission of analog information sent by the sensors of the equipment or utilities to the system controller.

These tests check the integrity of measurement chains between the sensor and the equipment as well as along the measurement chain. For example, a temperature signal carried over by a current loop is checked against the exact temperature level and the exact conversion levels on the current loop. However, if the temperature sensor contains its own local temperature indicator, this indication will be compared to the temperature available on the control system.

Unless technically impossible, the accuracy and linearity of each measurement chain shall be checked at least at two different points of the measurement range. Visualization of values transmitted from the sensors to the system controller is made most often directly on the controller itself (screen or printer), using an appropriate submenu of the controller. The accuracy of the values read on the equipment or utilities instrumentation is checked by means of appropriate calibrated reference instrumentation traceable to national official standards.

9. Checking Analog Transmission: Output

Tests shall be performed to check the transmission of analog information sent by the system controller to analog peripheral systems (proportional valves, recorders, etc.). These tests check the integrity of the analog transmission chain equally between the control system and the peripheral system

SOP No. Val. 1000.20   **Effective date: mm/dd/yyyy**
                                          Approved by:

as well as along the transmission chain. For example, the command of a proportional valve carried over by a current loop is checked both at the exact opening level and at the exact level of conversion on the current loop.

Unless technically impossible, accuracy and linearity of each output chain are checked. Visualization of values sent to the equipment is made either directly on the checked peripheral system or using appropriate equipment. Accuracy of values transmitted from the analog outputs is checked by using an appropriate calibrated reference instrumentation traceable to national official standards.

10. Data Entry and Boundary Testing

Tests shall be performed to check the data entry functions and the proper rejection of out-of-boundaries values. Where applicable, the mouse, graphic digitizer, or pen interface is checked for correct reaction to the user's commands. These tests include cursor movement checks, button verifications (simple- and double-click, left, right, and center, or special functions when applicable), and dragging operations.

The tests are conducted on critical data entry fields only. Each tested data entry field is challenged, including special key actions, control keys, invalid data type, out-of-range data, incorrect syntax or semantic, etc.

11. Access Control Testing

Tests shall be performed to check the computer system access control functions, including access level differentiation. The tests are conducted on critical functions only. Each tested function is verified against each access level.

For example, three different passwords are created to access the software system controlling lyophilizer. Each password is given a different access level. Using these passwords and an existing one, selected system functions requiring different authorization levels are initiated. The systems must refuse to start a tested function unless a proper password having correct access rights is supplied.

SOP No. Val. 1000.20   Effective date: mm/dd/yyyy

Approved by:

REASONS FOR REVISION

Effective date: mm/dd/yyyy

- First time issued for your company, affiliates, and contract manufacturers

YOUR COMPANY
VALIDATION STANDARD OPERATING PROCEDURE

SOP No. Val. 1000.30

Effective date: mm/dd/yyyy

Approved by:

TITLE: Performance Qualification of Computerized Equipment

AUTHOR: _____

Name/Title/Department

_____

Signature/Date

CHECKED BY: _____

Name/Title/Department

_____

Signature/Date

APPROVED BY: _____

Name/Title/Department

_____

Signature/Date

REVISIONS:

No.	Section	Pages	Initials/Date

SOP No. Val. 1000.30

Effective date: mm/dd/yyyy

Approved by:

SUBJECT: Performance Qualification of Computerized Equipment

PURPOSE

To describe the validation guideline for performance qualification of computerized equipment to ensure that it meets the performance requirement

RESPONSIBILITY

It is the responsibility of the production manager, technical service manager, and computer manager to follow the procedure. The quality assurance manager is responsible for SOP compliance.

PROCEDURE

1. Introduction

A specific performance qualification protocol shall be prepared for each piece of equipment undergoing performance qualification. The performance qualification of the computerized system consists of accumulating enough evidence that the computerized system is in compliance with its intended specifications, when functioning for the concerned process at the production premises.

2. Performance Qualification (PQ) Approach

The performance qualification of computerized control systems consists of a group of tests pooled in functional checks. The tests shall be carried out on the integrated system. The structure of the performance qualification shall be standardized for all performance qualification of computerized pharmaceutical equipment.

3. Document Description

Performance qualification documents shall include a paragraph stating the objective of the document and a paragraph specifying the scope of the document, i.e., the exhaustive list of the concerned equipment, or categories of equipment.

SOP No. Val. 1000.30   **Effective date: mm/dd/yyyy**

                                                       Approved by:

4. PQ Summary

The goal here is to summarize the performance qualification document. This includes:

- An abbreviated description of the computerized system analyzed
- For protocols, the preapproval of the document
- For PQ reports, the certification of the document
- For PQ reports, a summary of the deviations observed during the operational qualification

5. Standard Operating Procedure (SOP)

The purpose here is to verify the standard operating procedure at the user's premises. The procedures analyzed include:

- Installation operation SOPs
- Installation preventive maintenance SOPs, including virus checks
- Installation corrective maintenance SOPs
- Change control SOPs (planned and unplanned)
- SOPs for audits and on-going evaluations

These procedures must exist and be appropriate, up-to-date, relevant, and complete. They must be easily available to the users concerned.

6. Training

The purpose of the training section is to check that all people involved with the computerized equipment have received adequate training on operation of the computerized system.

The key aspects of this evaluation shall include checking:

- Existence of training programs for users, both technical and staff personnel
- Contents of training (basic, advanced, in-depth)
- Employee training history

7. System Recovery Procedures

The purpose of these procedures is to verify the system recovery procedures at the user's premises. The procedures analyzed include:

- Periodic backup and archival SOPs
- Data restoring SOPs
- Program restoring SOPs
- Disaster recovery SOPs

These procedure must exist and be appropriate, up-to-date, relevant, and complete. They must be easily available to the users concerned.

8. Computerized System Environment

The purpose here is to qualify the computerized system environment at the user's premises. The key aspects analyzed include:

- Quality of the electric power supplied to the equipment
- Efficiency of uninterruptible power supply (UPS) or safety power group (reaction delay, activity time)
- General environment: temperature, humidity, dust, etc.

9. Checking Process Control and Regulation Loops

The purpose here is to check the efficiency of process control and regulation loops critical to the process.

10. Checking Alarms and Safeties

Tests aim to check the alarm conditions and safeties available on the computerized equipment, even those with low activation probability. Each alarm or safety checked is activated between one and three times, possibly substituting some sensor with an appropriate calibrator simulator.

REASONS FOR REVISION

Effective date: mm/dd/yyyy

- First time issued for your company, affiliates, and contract manufacturers

SECTION

VAL 1100.00

YOUR COMPANY
VALIDATION STANDARD OPERATING PROCEDURE

SOP No. Val. 1100.10   Effective date: mm/dd/yyyy
                                              Approved by:

TITLE: Validation of Microbiological Methods

AUTHOR: _____
                          Name/Title/Department

                _____
                          Signature/Date

CHECKED BY: _____
                          Name/Title/Department

                _____
                          Signature/Date

APPROVED BY: _____
                          Name/Title/Department

                _____
                          Signature/Date

REVISIONS:

No.	Section	Pages	Initials/Date

SOP No. Val. 1100.10   Effective date: mm/dd/yyyy

Approved by:

SUBJECT: Validation of Microbiological Methods

PURPOSE

To provide the guideline for validation of the microbiological methods to ensure analytical accuracy and precision and that the methods are suitable for the intended use

RESPONSIBILITY

It is the responsibility of the quality control manager to follow the procedure. The quality assurance manager is responsible for SOP compliance.

PROCEDURE

The main objective of validation of an analytical procedure is to demonstrate that the procedure is suitable for its intended purpose. The procedures presented in this SOP provide basic guidelines for the validation of methods for microbiological assay, estimation of the number of microorganisms, detection of indicators of objectionable microorganisms, validation of preservative efficacy testing, and validation of the sterility testing and endotoxin test (LAL test).

Microbiological Assay of Antibiotics

It is an essential condition of biological assay methods that the tests on the standard preparation and on the sample whose potency is being determined should be carried out at the same time and, in all other respects, under strictly comparable conditions. The validation of microbiological assay method includes performance criteria (analytical parameters) such as linearity, range, accuracy, precision, specificity, etc.

Specificity is usually difficult to assess with microbial assay methods, because the tests measure the total activity and this will represent the synergetic action of all active components in the mixture under test.

SOP No. Val. 1100.10   Effective date: mm/dd/yyyy

                                        Approved by:

1. Linearity

The correlation coefficient, y-intercept, and slope of the regression line should be included. A plot of the data should be included. For the establishment of linearity, a minimum of five concentrations are recommended. It may be demonstrated directly on the active substance (by dilution of a standard stock solution).

Acceptance criteria

A correlation coefficient ≥0.95 is acceptable.

2. Range

The specified range is derived normally from linearity studies. It is established by confirming that the analytical procedure provides an acceptable degree of linearity, accuracy, and precision when applied to samples containing amounts of analyzed material within or at the extremes of the specified range of the procedure.

Acceptance criteria

The minimum specified range considered for assay of an active substance or a finished product is normally from 80 to 120% of the test concentration.

3. Accuracy

Accuracy may be determined by application of the analytical procedure to synthetic mixtures of the product components to which known quantities of the substance to be analyzed have been added. The accuracy should be assessed by using a minimum of nine determinations over a minimum of three concentration levels covering the specified range (i.e., three concentrations and three replicates of each concentration).

Accuracy should be reported as percent recovery by the assay of known added amount of active ingredient in the sample.

Acceptance criteria

Accuracy within ±10% of the true value is accepted.

4. Precision

The precision of an analytical method is usually expressed as the standard deviation or relative standard deviation (coefficient of variation) of a series of measurements. Precision represents repeatability or reproducibility of the analytical method under normal operating conditions. Precision determinations permit an estimate of the reliability of single determinations and are commonly in the range of ±0.3 to 3% for dosage form assays.

Repeatability may be assessed using a minimum of nine determinations covering the specified range for the procedure (e.g., three concentrations and three replicates of each concentration) or a minimum of six determinations at 100% of the test concentration. (This can be determined by performing six replicate assays on six aliquots of the same homogeneous sample.)

Acceptance criteria

Relative standard deviation of ≤5% is acceptable.

Consider an assay as preliminary if its computed potency is less than 80% or more than 125% of assumed potency; adjust the assumed potency accordingly and repeat the assay. Overage should be taken into account when determining target assumed potency.

In routine use, the combined result of a series of independent assays spread over a number of days is a more reliable estimate of potency than that from a single assay. Minimum requirement for routine microbiological testing is duplicate assay.

Potency assays require comprehensive statistical packages and for a standard large plate assay this could include all the relevant statistical parameters. The *EP* (European Pharmacopoeia) or *BP* (British Pharmacopoeia) is recommended as a source of reference. For a high-precision large plate bioassay, the following parameters should be included: analysis of variance, tests of validity, estimation of potency and fiducial limits, missing values, and combination of potency estimates. Potency assays should not be performed using low-precision designs.

Bioassay may also be of low-precision design (multiple samples on large plates, i.e., >3 manual turbidimetric assays, small-plate assays). These types of assay are useful for trace analyses (cleaning validation), and are often used for the analysis of samples in body fluids as they are capable of dealing with the large numbers of samples that may be generated in these studies.

SOP No. Val. 1100.10   Effective date: mm/dd/yyyy
                                        Approved by:

The amount of data generated by low-precision analyses is often insufficient for sophisticated statistical analysis. Even so, it is important to minimize manual data handling as this allows subjective interpretation to enter the interpretive stages.

All potency assays, from the simplest designs to the most complex Latin square design, necessitate potency estimation by computer. Low-precision assays employing plotting of zone sizes (response) against concentration of standards must be dealt with using computerized regression analysis, with the potency (standard equivalent) estimation calculated from the computed equation of the line. In this way, all opportunity for operator subjectivity is minimized.

For low-precision design the statistical package should include statistical rejection of outlying or aberrant observation. (EP makes no provision for this; USP has a test — USP 24, standard deviation, regression analysis, and potency estimation.)

The most important aspects of data handling for potency assays and low-precision assays are that the data is handled by validated computer programs and that the acceptance and rejection criteria incorporated are clear and based upon statistical or proven (at validation) limits.

All programs must be validated by comparison vs. manual calculation.

5. Validation of Microbial Recovery from Compendial Article

The antimicrobial properties of a product may be due to the presence of preservatives or its formulation. This antimicrobial property must be neutralized to recover viable microorganisms present. The neutralization of antimicrobial property of a pharmaceutical article can be achieved by:

■ Using specific neutralizer (chemical inhibition)
   Table 1 shows known neutralizers for a variety of chemical antimicrobial agents and the reported toxicities of some chemical neutralizers to specific microorganisms. Antibiotics may not be susceptible to neutralization by chemical means, but rather by enzymatic treatment (e.g., penicillinase). These enzymes may be used where required (for β-lactum antibiotics).

SOP No. Val. 1100.10

Effective date: mm/dd/yyyy
Approved by:

Table 1 Some Common Neutralizers for Chemical Biocides

Neutralizer	Biocide Class	Potential Action of Biocides
Bisulfate	Glutaraldehyde, Mercurials	Non-sporing bacteria
Dilution	Phenolics, Alcohol, Aldehydes, Sorbate	—
Glycine	Aldehydes	Growing cells
Lecithin	Quaternary Ammonium Compounds (QACs), Parabens, *Bis*-biguanides	Bacteria
Mg^{+2} or Ca^{+2} ions	EDTA	—
Polysorbate	QACs, Iodine, Parabens	—
Thioglycollate	Mercurials	Staphylococci and spores
Thiosulfate	Mercurials, Halogens, Aldehydes	Staphylococci

■ Dilution
The relationship between concentration and antimicrobial effect differs among bactericidal agents but is constant for a particular antimicrobial agent. The relationship between concentration and antimicrobial effect is exponential in nature, with the general formula

$$C^{\eta}t = k$$

where
C = concentration
t = time required to kill a standard inoculum
k = constant
η = the slope of the plot of log t vs. log C.
Antimicrobial agents with high η values are rapidly neutralized by dilution, while those with low η values are not good candidates for neutralization by dilution.
■ Filtration and washing
This approach is used especially in sterility testing.
■ Combination of washing and dilution
■ Any combination of these methods

SOP No. Val. 1100.10   **Effective date: mm/dd/yyyy**

                                              Approved by:

6. Validation of Neutralization Methods

The validation method for neutralizing the antimicrobial properties of a product must meet two criteria — neutralizer efficacy and neutralizer toxicity. The validation study documents that the neutralization method applied is effective in inhibiting the antimicrobial properties of the product (neutralizer efficacy) without impairing the recovery of viable micro-organisms (neutralizer toxicity). Validation protocol may meet these two criteria by comparing recovery results for three treatment groups.

The following challenge organisms may be used as appropriate (*USP 24, BP* Vol. II, 1999, or *EP* 3rd edition, 1997):

- *Aspergillus niger* (ATCC 16404)
- *Candida albicans* (ATCC 10231, NCPF 3179)
- *Bacillus subtilis* (ATCC 6633, NCIMB 8054)
- *Escherichia coli* (ATCC 8739, NCIMB 8545)
- *Staphylococcus aureus* (ATCC 6538, NCTC 10788)
- *Pseudomonas aeruginosa* (ATCC 9027, NCIMB 8626)
- *Salmonella typhimurium* (or nonpathogenic strain, such as *Salmonella agona* NCTC 6017).

In a **test group** the product is subjected to the neutralization method, and then a low level of challenge microorganism (less than 100 cfu) is inoculated for recovery. In a **peptone control group** the neutralization method is used with peptone or diluting fluid A (Sterility test 71) as the test solution. In a **viability group** the actual inoculum is used without exposure to the neutralization method.

Similar recovery between the test group and peptone group demonstrates adequate neutralizer efficacy; similar recovery between the peptone group and the viability group demonstrates adequate neutralizer toxicity. In principle, the protocol must show that recovery of a low inoculum (less than 100 cfu) is not inhibited by the test sample and the neutralization method. Validation protocols may meet these two criteria by comparing recovery among three distinct test groups: neutralized product with inoculum, challenge inoculum control in buffered solution, and inoculum in the absence of product or neutralizer.

This can be established by directly comparing the result in the treated solution to the inoculum above. If the growth on the treated solution is not comparable to the growth on the inoculum group, it should be determined whether the neutralization method is toxic to the microorganisms.

7. Recovery on Agar Medium

At least three independent replicates of the experiment should be performed, and each should demonstrate that the average number of cfu recovered from the challenge product is not less than 70% of that recovered from the inoculum control.

In the event that a greater number of replicates is required in the validation study, the comparisons may be evaluated by transforming the numbers of cfu to their logarithmic values and analyzing the data statistically.

Acceptance criteria

■ Similar recovery between the first and second group demonstrates adequate neutralizer efficacy
■ Similar recovery between the second and third group demonstrates adequate neutralizer toxicity.
■ At least three independent replicates of the experiment should be performed, and each should demonstrate that the average number of cfu recovered from the challenge product is not less than 70% of that recovered from the inoculum control.

8. Absence of Specified Organisms: (*S. aureus, P. aeruginosa, E. coli* and *Salmonella* spp.)

A similar approach to that employed in aerobic microbial count validation is employed but quantification is not possible. A low level (≤100 cells) of specified organism is added to various product and broth mixtures and recovery viewed on the resultant selective plates. For the method to be considered valid, growth on plates must be comparable to that derived from parallel control cultures containing no product. Parallel controls not only must be run at validation stage but also as a matter of routine to indicate acceptable preparation and performance of media.

pH checks on broth and product mixtures are important owing to the protracted period of contact. Failure to recover organisms from dilutions

SOP No. Val. 1100.10   Effective date: mm/dd/yyyy
                                        Approved by:

of product in broth in excess of 1:1000 is indicative of inability of that organism to contaminate the product; testing on a routine basis would not be recommended.

9. Recovery by Membrane Filtration

This validation follows the procedure described for validation tests for bacteriostasis and fungistasis under sterility tests in USP XXIV with the exception of plating on solid medium to quantitative recovery.

Three 100-ml rinses are assumed, but the volume and number of rinses are subject to validation. Each validation run should be performed independently at least three times.

1. In the test solution group, the product is filtered through the membrane filter, followed by two 100-ml portions of diluting-neutralizing fluid.
2. After the second rinse has been filtered, a final 100-ml portion containing less than 100 cfu of the specific challenge microorganism is passed through the filter. This filter is then transferred to the appropriate recovery agar medium and incubated for recovery.
3. The inoculum is directly plated onto the solid medium (to check viability).
4. Diluting fluid A is used as the dilution medium without exposing the filter to the product.
5. After addition of the low level of inoculum to the final rinse, the filter is plated as above.
6. Technique-specific loss of microorganisms can be estimated by comparing the recovery in the diluting fluid A group to the inoculum count.
7. If it is necessary to solubilize the test sample (in case of ointments, suspensions, etc.), the effects of the solubilization method on viable microorganisms must be determined.

Acceptance criteria

The method can be considered validated if the recovery rate in the three independent replicates is similar for the test solution and the diluting fluid A control.

SOP No. Val. 1100.10   Effective date: mm/dd/yyyy

Approved by:

10. Recovery in Liquid Medium

It is assumed in the direct transfer method under sterility tests that the recovery medium will allow for growth of all surviving microorganisms. The liquid medium in that test must serve to neutralize any antimicrobial properties of the test solution and to support the growth of the microorganisms. The treatment groups described above (antimicrobial neutralization for recovery on agar medium) can be used for validation of the recovery method, with the proportions of product and recovery medium varied to achieve adequate neutralization.

Acceptance criteria

The method can be considered validated if all groups show copious growth within 7 days for all microorganisms.

11. Estimating the Number of Colony-Forming Units

The accepted range for countable colonies on a standard agar plate is between 25 and 250 cfu for most bacteria and *Candida albicans*. It is not optimal for all environmental isolates. The recommended counting range for *Aspergillus niger* is between 8 and 80 cfu per plate. The use of membrane filtration to recover challenge microorganisms or the use of environmental isolates as challenge microorganisms in antimicrobial effectiveness testing requires validation of the countable range. This validation may be performed by statistical comparison of estimated cfu from successive pairs in a dilution series. Prepare a suspension so that plating will provide approximately 1000 cfu per plate, and then dilute two-fold to a theoretical concentration of approximately 1 cfu per plate. Plate all the dilutions in the series in duplicate, and incubate for recovery under the conditions of the antimicrobial effectiveness testing. Compare the estimates of cfu per ml from paired tubes in the dilution series by the formula:

$$\frac{\left[2L_{cfu} - H_{cfu}\right]}{\sqrt{2L_{cfu} + H_{cfu}}} \leq 1.96$$

where L_{cfu} = the number of colonies on the plate with the lower count (greater dilution) and H_{cfu} = the number of colonies on the plate with the

SOP No. Val. 1100.10   Effective date: mm/dd/yyyy

Approved by:

higher count (lesser dilution). The estimates of the cfu per ml provided by L_{cfu} and H_{cfu} should agree within the limits of the formula with a critical value of 1.96. The upper limit of the plate counts is then defined as the number (H_{cfu}) that reproducibly passes this test.

This study should be independently repeated a sufficient number of times to establish an upper limit of cfu for the particular plating conditions. There is a lower limit at which the ability of the antimicrobial effectiveness test to recover microorganisms becomes unreasonable. If the first plating is performed with 1 ml of 10^{-1} dilution, cfu in the range of 1 to 10 per ml would not be seen. On this dilution plating, only the lower number of cfu may be reduced to three, allowing as few as 30 cfu/ml survivors to be reported.

Lower counting thresholds for the greatest dilution plating in series must be justified. Numbers of colonies on a plate follow the Poisson distribution, so the variance of the mean value equals the mean value of counts. Therefore, as the mean number of cfu per plate becomes lower, the percentage error of the estimate increases (Table 2). Three cfu per plate at the 10^{-1} dilution provide an estimate of 30 cfu per ml, with an error of 58% of the estimate.

12. Bacterial Endotoxin (LAL) Test

Validation is accomplished by performing the inhibition or enhancement test. Appropriate negative control should be included. Validation must be repeated if the LAL reagent source, method of manufacture, or formulation of the product is changed. Confirm the labeled sensitivity of each new lot of LAL reagent prior to use in the test.

Comprehensive tests described in USP 24, *BP* 1999 and *EP* 3rd edition, 1997 cover validation requirements.

See also U.S. Department of Health's "Guideline on Validation of the LAL Test as an End-Product Endotoxin Test for Human and Animal Parenteral Drugs, Biological Products and Medical Devices."

Specific guidance on the initial quality control procedure for a testing laboratory is available as an FDA addendum. The addendum and other practical guidelines are available from Associates of Cape Cod, Inc.

Effective date: mm/dd/yyyy

Approved by:

Table 2 Error as Percentage of Mean for Plate Counts

cfu per Plate	Standard Error	Error as % of Mean
30	5.48	18.3%
29	5.39	18.6%
28	5.29	18.9%
27	5.20	19.2%
26	5.10	19.6%
25	5.00	20.0%
24	4.90	20.4%
23	4.80	20.9%
22	4.69	21.3%
21	4.58	21.8%
20	4.47	22.4%
19	4.36	22.9%
18	4.24	23.6%
17	4.12	24.3%
16	4.00	25.0%
15	3.87	25.8%
14	3.74	26.7%
13	3.61	27.7%
12	3.46	28.9%
11	3.32	30.2%
10	3.16	31.6%
9	3.00	33.3%
8	2.83	35.4%
7	2.65	37.8%
6	2.45	40.8%
5	2.24	44.7%
4	2.00	50.0%
3	1.73	57.7%
2	1.41	70.7%
1	1.00	100.0%

SOP No. Val. 1100.10   **Effective date: mm/dd/yyyy**

                                            Approved by:

REASONS FOR REVISION

Effective date: mm/dd/yyyy

- First time issued for your company, affiliates, and contract manufacturers

YOUR COMPANY
VALIDATION STANDARD OPERATING PROCEDURE

SOP No. Val. 1100.20

Effective date: mm/dd/yyyy

Approved by:

TITLE: Validation of Analytical Methods

AUTHOR: _____
Name/Title/Department

_____
Signature/Date

CHECKED BY: _____
Name/Title/Department

_____
Signature/Date

APPROVED BY: _____
Name/Title/Department

_____
Signature/Date

REVISIONS:

No.	Section	Pages	Initials/Date

SOP No. Val. 1100.20   Effective date: mm/dd/yyyy

Approved by:

SUBJECT: Validation of Analytical Methods

PURPOSE

To provide validation of analytical methods to ensure accuracy and reliability

RESPONSIBILITY

It is the responsibility of the quality control analyst and quality control manager to follow the procedure. The quality assurance manager is responsible for SOP compliance.

PROCEDURE

For detailed description, refer USP 24 (1225): Validation of Compendial Methods.

1. Validation

1.1 Definition of validation parameters

1.1.1 Precision

The precision of an analytical method can be defined as the pattern of variation of single assays on a uniform sample. The precision serves to identify random errors and is described by the repeatability (variability within a laboratory) and reproducibility (variation between different laboratories).

1.1.2 Accuracy

A procedure is accurate if — on the average — the method provides the true answer.

1.1.3 Selectivity

In providing evidence of selectivity, it must be shown that an analytical method exclusively determines the desired compound.

1.1.4 Linearity

Providing evidence of the linearity of an analytical method is necessary for quantitative determinations.

The direct proportional relation between the measured signal and the concentration of the sample has to be proven.

1.1.5 Limit of detection

The lowest value of a compound with which a defined statistical probability can be clearly identified for an analytical method is the limt of detection.

1.1.6 Limit of quantification

The limit of quantification is the concentration of a compound with which a defined precision and accuracy is only quantifiable.

1.1.7 Ruggedness

An analytical method is rugged when it shows, under different circumstances (different laboratory, different laboratory assistants, different times, etc.), the same results.

1.1.8 Robustness

The robustness of an analytical procedure is a measure of its capacity to remain unaffected by small but deliberate variations in method parameters; it provides an indication of its reliability during normal usage.

2. Validation Parameters to Be Realized

Overview

Compendial assay procedures vary from highly exacting analytical determinations to subjective evaluation of attributes. Considering this variety of assays, it is only logical that different test methods require different validation schemes. Only the most common categories of assays for which validation data should be required are covered. These categories are as follows (see also Table 1):

SOP No. Val. 1100.20   **Effective date: mm/dd/yyyy**
Approved by:

Table 1   Data Elements Required for Assay Validation

Analytical Performance Parameter	Assay Category I	Assay Category II		Assay Category III
		Quantitative	Limit Tests	
Accuracy	Yes	Yes	*	
Precision	Yes	Yes	No	Yes
Specificity	Yes	Yes	Yes	*
Limit of Detection	No	No	Yes	*
Limit of Quantitation	No	Yes	No	*
Linearity	Yes	Yes	No	*
Range	*	*	*	*
Ruggedness	Yes	Yes	Yes	Yes

* May be required, depending on the nature of the specific test.

Category I — Analytical methods for quantitation of major components of bulk drug substances or active ingredients (including preservatives) in finished pharmaceutical products

Category II — Analytical methods for determination of impurities in bulk drug substances or degradation compounds in finished pharmaceutical products. These methods include quantitative assays and limit tests.

Category III — Analytical methods for determination of performance characteristics (e.g., dissolution, drug release)

Category IV — Identification tests

3. Precision

3.1 Precision of the system

Procedure

Perform the analysis based on one standard solution (with a concentration equal to the expected sample concentration) six consecutive times. Calculate the relative standard deviation of the obtained values.

Criteria

The relative standard deviation must be less than 1.5%; it is inevitably more than 1.5%, the reason has to be explained.

- Documentation
- Obtained values
- Calculation of the relative standard deviation (RSD)
- Justification for a higher RSD

3.2 Precision of the Method

3.2.1 Repeatability

The repeatability is the precision determined under equal conditions with one homogeneous sample. This sample should be analyzed in six-fold increments. Calculate the relative standard deviation of the obtained values.

Criteria

The RSD must be less than 2%; if it is more than 2%, the reason must be explained. The repeatability (r) can be calculated from the relative standard deviation, using the equation:

$$r = 2.83 \times RSD$$

The value thus found for r represents the difference between the results that is not exceeded more than once in every 20 cases.

3.2.2 Reproducibility

Reproducibility is the precision determined under different conditions (laboratory, reagents, analysts, days, equipment) with one homogeneous sample. Reproducibility may also be established retrospectively, using data obtained from earlier release procedures.

Criteria

The relative standard deviation should not exceed the 4% level.

SOP No. Val. 1100.20   **Effective date: mm/dd/yyyy**

                                                   Approved by:

Documentation

- Obtained values
- Calculation of the RSD
- Justification of exceedingly high RSD levels

3.2.3 Accuracy (addition and recovery)

Accuracy is a measure for the difference between the average value found in the analyses and the theoretical value. Accuracy studies should be performed at a level of active ingredient equal to 100% of the established label concentration of the product tested.

Prepare a series of six placebo mixtures to which the components are added in concentrations equal to the values expected for the sample. After the analysis of these mixtures, perform a statistical evaluation.

Criteria

1. The theoretical value should be within the 95% confidence limits of the average found.
2. The average found should be between 99 and 101% of the theoretical value.
3. The relative standard deviation should be less than 2%; if it is more than 2%, the reason must be explained.
4. The accuracy is acceptable if either criteria 1 and 3 or criteria 2 and 3 are fulfilled. If not, the reason must be given.

Documentation

- Method of the preparation of the placebo with the added components
- Obtained values
- Statistical evaluation
- Justification of a less accurate method

3.2.4 Selectivity

It must be ensured that the analytical method exclusively determines the desired compound.

The selectivity of an analytical method is determined by comparing test results from the analysis of samples containing impurities (related compound), degradation products (originated from samples submitted to stress conditions), or placebo ingredients with those obtained from the analysis of samples without impurities, degradation product, or placebo ingredients.

When impurities or degradation products are unidentified or unavailable, selectivity may be demonstrated by analysis by the method in question of samples containing impurities or degradation products and comparing the results to those from additional purity assays. The degree of agreement of test results is a measure of the selectivity.

3.2.5 Linearity

Procedure

Prepare by dilution from one solution five standard solutions with concentrations between 0 and 200% (e.g., 20%, 60%, 100%, 140%, 180%) of the expected sample concentration. Each solution must be tested at least twice.

Plot the average response against the quantity (or concentration) of the component and use linear regression analysis to calculate the calibration line. If an internal standard is used, the linearity of its curve has to be determined similarly.

Criteria

No lack of fit shall occur fitting a first order polynome through the measured points. The 95% confidence interval shall include zero. If the 95% confidence interval does not include zero, the 80% and 120% points on the calculated line shall not deviate more than 1% from a straight line through 0 and the calculated 100% point.

Calculation of the correlation coefficient

If the calibration curve is linear and the origin of the coordinate system lies within the 95% confidence limits of the curve, the use of one standard concentration (equal to the expected sample concentration) is allowed for the analysis.

SOP No. Val. 1100.20   **Effective date: mm/dd/yyyy**

Approved by:

If the deviation of the curve from the line 0.0 to 100% (at 80 and 120%) is less than 1% of the y value, the use of one standard concentration (equal to the expected sample concentration) is allowed.

If the curve is linear but does not comply with the criteria mentioned above, three standard concentrations around the expected sample value have to be used.

In all other cases, a second degree calibration curve has to be modified to five standard concentrations.

Documentation.

- All the data plots
- Statistical evaluation of the data
- Same procedure should be applied to an internal standard, if used

3.2.6 Limit of detection

If the objective of the analytical method is to detect trace components, the limit of detection (LOD) must be determined.

Procedure

Perform the standard analysis with a blank sample and calculate the standard deviation of the measured response at the position where the substance to be determined is expected. This has to be performed over a distance of 20 times the peak width at the middle of the peak height.

The detection limit is equal to two to three times the noise of the system.

Documentation

- Analysis results of the blank determination
- Calculation of the standard deviation

3.2.7 Limit of quantification

If the objective of the analytical method is to determine trace components, the limit of quantification (LOQ) must be determined.

SOP No. Val. 1100.20   Effective date: mm/dd/yyyy

                                            Approved by:

Procedure

Perform the standard analysis with a blank sample and calculate the standard deviation of the measured response at the position the substance to be determined is expected. This has to be performed over a distance of 20 times the peak width at the middle of the peak height.

The quantification limit is equal to five to ten times the noise.

Criteria

The LOQ must be lower than the release requirement.

Documentation

- Analysis results of the blank determination
- Calculation of the standard deviation

3.2.8 Ruggedness

The ruggedness of an analytical method is determined by analysis of aliquots from homogeneous samples in different laboratories, by different analysts, using operational and environmental conditions that may differ but are still within the specified parameters of the assay. The degree of reproducibility of test results is then determined as a function of the assay variables. This reproducibility may be compared to the precision of the assay under normal conditions to obtain a measure of the ruggedness of the analytical method.

3.2.9 Stability indicating

Known degradation products — When the route of degradation is not known, or if samples of known or postulated degradation products are not available, the sample should be degraded artificially by heat, light, oxidation or reduction, acid or base, etc. Suitable conditions should be employed such that measurable, but not complete, degradation is induced, in order that the residual main component and likely levels of degradation are detected. It is important that the complete formulation is degraded, in case components of the product matrix contribute to the degradation,

SOP No. Val. 1100.20   Effective date: mm/dd/yyyy
                                                     Approved by:

or themselves degrade. One should also degrade a placebo (or mixture or excipients) as a control.

Example: stability indicating assay for product which is expected to be susceptible to hydrolysis and oxidation. Only one degradate (A) is readily available.

1. Demonstrate separation of A from major component.
2. To show that separation from other degradates is achieved, degrade the formulation and subject it to assay. Samples of the formulation should be stressed as follows:
    Heat — 105° for 24 h
    Light — high intensity UV for 24 h
    Hydrolysis — reflux an aqueous solution for 2 h
    Oxidation — reflux and aqueous solution containing hydrogen peroxide for 1 h
    Acid — reflux in 1 M sodium hydroxide for 1 h
    Alkali — reflux in 1 M sodium hydroxide for 1 h

Before testing, the solutions should be neutralized if necessary. If the method is suitable for stability studies, then the peak due to the major components will be reduced and other peaks will probably be represented. These degradation products should be resolved from the parent peak.

REASONS FOR REVISION

Effective date: mm/dd/yyyy

■ First time issued for your company, affiliates, and contract manufacturers

SECTION

VAL 1200.00

YOUR COMPANY
VALIDATION STANDARD OPERATING PROCEDURE

SOP No. Val. 1200.10

Effective date: mm/dd/yyyy

Approved by:

TITLE: Vendor Certification

AUTHOR: _____
Name/Title/Department

_____
Signature/Date

CHECKED BY: _____
Name/Title/Department

_____
Signature/Date

APPROVED BY: _____
Name/Title/Department

_____
Signature/Date

REVISIONS:

No.	Section	Pages	Initials/Date

SOP No. Val. 1200.10   Effective date: mm/dd/yyyy

Approved by:

SUBJECT: Vendor Certification

PURPOSE

To describe the procedure for evaluation of suppliers to ensure that the materials purchased are of consistent quality

RESPONSIBILITY

It is the responsibility of the quality assurance manager to develop the vendor approval system and maintain SOP compliance.

PROCEDURE

After successful vendor auditing, it can be determined whether purchased ingredients and materials can be accepted on the basis of suppliers' certificates, with minimized inspections of incoming goods to a certain level.

Vendor certification leads to reduction of costs and release times.

VENDOR CERTIFICATION

The vendor certification procedure may include a list of selected vendors, historical review of test results of previous suppliers, and formal inspection on site and decision making.

1. Selection of Vendors to be Certified

The selection of vendors to be certified should be jointly made by the heads of purchasing and production and the quality assurance manager.

2. Review of Historical Data and Test Results

Summarize the quality data of batches delivered during the last 3 years and prepare trend analysis. Report deviations with regard to normal failure levels, out-of-specification situations, and corrective actions. The quality control and quality assurance managers shall review the trend.

SOP No. Val. 1200.10   Effective date: mm/dd/yyyy

                                                Approved by:

3. Site Audit

The quality assurance manager or the system in charge may perform an on-site audit. The audit should specifically:

- Determine the accuracy, precision, and reliability of test and inspection data of the vendor.
- Review the process reproducibility and the batch records for process variations.
- Perform general GMP compliance inspection. Review the potential for contamination and mix-ups thoroughly.
- Ensure that vendors' in-process controls include the use of statistical process control critical product parameters that are significant and may affect the final product quality
- Ensure the absence of significant online problems.

4. Recommendations

It is not essential to perform on-site inspections. As an alternative, evaluation questionnaires can be used. Vendors can also be certified based on an extensive review of historical analytical inspection data and their performance over the last 3 years. Alternatively, third-party audits may be conducted for a predefined period.

5. Decision on Certification

The data obtained as a result of these reviews and audits shall be reviewed by the QA manager and sent for approval to quality control, production, and purchasing. Final release must be authorized by quality control.

6. Steps after Certification

- After vendor approval, quality control or quality assurance will reduce the number of tests and inspections of incoming goods as agreed in the certification report, e.g., one out of ten batches.
- For packaging materials certification, it is sufficient to review the results of three suppliers.
- If during this process of verification no discrepancies appear, the verification may be discontinued.

- For incomplete certification, a provisional classification report shall be published by the QA manager for components.
- Materials to be used in production without complete testing must be supported with acceptable certificates of analysis by manufacturers.
- Active ingredients should be checked for their identity.
- All deviations regarding purchased materials encountered by production must be reported to the quality assurance manager for referral to the manufacturer or supplier.

7. Recertification

Recertification of an active and excepient manufacturer may be performed on request. Recertification can be requested by quality control and production. The certification is valid for a period determined by the QA manager. Certification of packaging materials is valid for 5 years.

REASONS FOR REVISION

Effective date: mm/dd/yyyy

- First time issued for your company, affiliates, and contract manufacturers

SECTION

VAL 1300.00

YOUR COMPANY
VALIDATION STANDARD OPERATING PROCEDURE

SOP No. Val. 1300.10   Effective date: mm/dd/yyyy

Approved by:

TITLE: Facility Qualification

AUTHOR: _____

Name/Title/Department

_____

Signature/Date

CHECKED BY: _____

Name/Title/Department

_____

Signature/Date

APPROVED BY: _____

Name/Title/Department

_____

Signature/Date

REVISIONS:

No.	Section	Pages	Initials/Date

SOP No. Val. 1300.10

Effective date: mm/dd/yyyy

Approved by:

SUBJECT: Facility Qualification

PURPOSE

To provide guidelines for checking facility construction and finishing to ensure that they meet the facility design qualification requirement

RESPONSIBILITY

It is the responsibility of all concerned managers and contractors to follow the procedure. The quality assurance manager is responsible for SOP compliance.

PROCEDURE

The conceptual design of the facility (approved civil layout) should be available and be compared with the actual construction. Details should be summarized in a tabular form describing:

- Room number
- Activity
- Room classification (e.g., class 100, 10,000, etc.)
- Utilities required and their quality (e.g., WFI, distilled water, compressed air, etc.)
- Room finishes (floor, walls, ceilings, partitions)
- Illumination
- Safety features
- Pass-throughs
- Communicators
- Prefabricated partitions

1. Materials

The quality of materials used shall be confirmed from the purchasing specification and the materials receiving reports. Generally, the use of wooden materials is not recommended.

SOP No. Val. 1300.10   **Effective date: mm/dd/yyyy**

                                               Approved by:

1.1 Materials for floor construction

Epoxy resin-welded PVC sheets and epoxy terrazzo are generally recommended.

1.2 Materials for walls

Generally acceptable materials are as follows:

- Epoxy coating
- Polyester coating
- Seamless PVC coating
- Welded sheets of PVC
- Prefabricated wall panels

1.3 Materials for ceilings

Commonly used materials are gypsum board with epoxy, polyester coating, and seamless PVC coating.

1.4 Materials for doors

The following materials are generally used:

- Standard painted timber
- Timber with plastic lamination
- Stainless steel
- Glass

2. Measurement Checks of Construction and Finishes

The area of each room shall be checked to ensure that it meets specifications (length, width, and height).

- Rooms should be designed and constructed so that air leakage through openings or penetration is kept to a minimum.
- All surfaces must be completely cleanable and resistant to germicidal solutions.

SOP No. Val. 1300.10   Effective date: mm/dd/yyyy

                                        Approved by:

■ Horizontal or other surfaces on which dust can accumulate shall be kept to an absolute minimum, and they shall be completely avoided above areas where the product or washed product containers are exposed.

3. Quality Checks of Floors

The material used for the construction of floor shall be free from seams, cracks, or holes, and shall be durable, washable, and cleanable.

■ Floor–wall joints, building columns, equipment pads, and other obstructions should be completely sealed.
■ Floor drains should be provided only where large volumes of liquids are anticipated.
■ The floor should be pitched toward the drain (if available) to prevent accumulation of liquids.
■ The floor drain should not be located in the aseptic clean room because of concerns about microbiological growth in the trap.
■ Trapped floor drains should be provided.

4. Quality Checks of Walls

■ Construction materials shall be free from seams, cracks, or holes, and also washable, cleanable, and free from rust or corrosion.
■ Walls should be smooth, rigid, and resistant to impact and abrasion.
■ Walls should be protected by physical barriers such as stainless steel subrails (if necessary).
■ Wall surfaces should be easy to clean and resistant to repeated exposure to disinfectants.
■ Floor–wall, wall–wall, and ceiling–wall intersections should be properly sealed to prevent dirt accumulation and pest entry.
■ Floor–wall and wall–wall joints should be coved.
■ Ledges should be sloped to reduce dust accumulation.

5. Quality Checks of Ceilings

■ Construction materials shall be durable, cleanable, and washable.

SOP No. Val. 1300.10   **Effective date: mm/dd/yyyy**

                                        Approved by:

- Ceilings should be smooth, nonperforated, and properly sealed to the framing.
- All penetration through ceilings must be designed to establish and maintain the integrity of a 100% sealed system.
- The suspension system should be heavy-duty aluminum with anodized or leaked-on, even-finish stainless steel.

6. Quality Checks of Entryway, Doors, and Air Locks

- Construction materials, door openings and door gaskets should be satisfactory.
- Glass fixing should be satisfactory (if used).
- Entry of all personnel, materials, and equipment should be through a suitable air lock.
- The design and arrangement should minimize migration of particulate and other contaminants into the controlled process area.
- Door should have automatic closing systems.
- Locks and latches should be avoided.

7. Quality Checks of Windows

- Materials shall be durable, washable, and cleanable.
- Windows between class 100 areas and other areas should have sloping sills on both sides of glass to facilitate cleanliness.

8. Quality Checks of Curtains and Partitions

- To maintain the cleanliness classification and protect the critical areas of equipment from air turbulence, the following equipment should be surrounded by a curtain:
  - Equipment
  - Curtain Material
  - Curtain Height

9. Quality Checks of Conveyor Passages and Partition Holes

The passage of conveyors through partition is adjusted using polycarbonate sheets cut as needed.

10. Equipment Maintenance Provision

Access to the critical parts of the equipment for maintenance and cleaning should be available.

11. Critical Checks of Illumination

Check the units of lux and compare with the standard for each room.

12. Utility Lines Check

Check and ensure that utilities are installed per the room requirement list:

- Industrial steam
- Pure steam
- Chilled water
- Tap water (warm)
- Tap water (cold)
- Distilled water
- Water for injection
- Compressed air
- Nitrogen gas
- Exposed piping and conduit should be avoided as much as possible.
- Exposed piping should be completely cleanable by being set out approximately 1 to 1.5 in. from the wall or other surfaces.
- Check that all piping is properly labeled with flow directions.

13. Safety Features Checks

Check and ensure that safety equipment is installed per room requirement list:

- Location, condition, and number of water showers available
- Location and number of fire extinguishers available
- Location, condition, and number of smoke detectors available
- Location and number of emergency doors available

SOP No. Val. 1300.10   Effective date: mm/dd/yyyy

Approved by:

14. Drainage System Checks

- Drainage should be designed and constructed per approved layout.
- Verify the drainage locations using building specification.
- The floor should be pitched toward drain holes to prevent accumulation of liquids.
- All surfaces must be cleanable.
- All drains should have deep seal traps to avoid back-flow (nonreturn valves).
- Open drains should be avoided in manufacturing and packaging areas.
- Process and sanitary drains should be separate to avoid the possibility of introducing sanitary waste into process systems.
- Drains must be covered with secured lids.
- Check that drain flushing procedure is established and identified.
- Check that drainage lines are sanitized according to procedure and that records are available.
- Check that drain pipe slope is toward outside.

REASONS FOR REVISION

Effective date: mm/dd/yyyy

- First time issued for your company, affiliates, and contract manufacturers

SECTION

VAL 1400.00

YOUR COMPANY
VALIDATION STANDARD OPERATING PROCEDURE

SOP No.: Val. 1400.10

Effective date: mm/dd/yyyy

Approved by:

TITLE: Sterilization Assurance Information and Data

AUTHOR: _____

Name/Title/Department

_____

Signature/Date

CHECKED BY: _____

Name/Title/Department

_____

Signature/Date

APPROVED BY: _____

Name/Title/Department

_____

Signature/Date

REVISIONS:

No.	Section	Pages	Initials/Date

SOP No.: Val. 1400.10   Effective date: mm/dd/yyyy

Approved by:

SUBJECT: Sterilization Assurance Information and Data

PURPOSE

This procedure provides the information required to support the sterility assurance of the drug product (product name), USP, manufactured by ABC Pharmaceutical Industries. It references the FDA Guidance titled "Guidance for Industry for the Submission of Documentation for Sterilization Process Validation in Applications for Human and Veterinary Drug Products" prepared by the Sterility Technical Committee of the Chemistry Manufacturing Controls Coordinating Committee of the Center for Drug Evaluation and Research (CDER) and the Center for Veterinary Medicine (CVM) in November of 1994.

The required information as outlined in sections IV and V of the FDA Guidance is in boldface type, followed by ABC Pharmaceutical Industries' information. It is submitted in the ANDA file for the manufacturing of (product name) USP. Other sections of the Guidance (sections I to III) pertain to the Guidance's introduction and terminal sterilization processes.

This document contains information related to the liquid aseptic fill operation used in the manufacture of (product name), USP, at ABC Pharmaceutical Industries located at (provide postal address). Additional information to support the liquid aseptic filling validation includes but is not limited to environmental monitoring and controls, as well as product-specific testing such as bioburden and sterility testing. The main subsections are:

1. **Information for aseptic fill manufacturing processes** (reference FDA Guidance Section IV, A to J)
2. **Maintenance of microbiological control and quality** (reference FDA Guidance Section V, A to C)

1. Sterility Assurance

1.1 Information for aseptic fill manufacturing processes

1.1.1 Buildings and facilities

ABC Pharmaceutical Industries is located (provide postal address). ABC Pharmaceutical Industries manufactures sterile pharmaceutical dosage forms in strict compliance to the cGMP, United States FDA guidelines. ABC Pharmaceutical Industries is supplying its products to its customers around the world.

SOP No.: Val. 1400.10   Effective date: mm/dd/yyyy

Approved by:

Manufacturing Facilities. The plant is located in (provide city name and country name). The facility is dedicated to the manufacturing of sterile products.

Total built area: _____ m²
Production capacity: ____ million **ampoules**
                      ____ million **liquid vials**
                      ____ million **lyophilized products**
                      ____ million **prefilled syringes**

Construction. The sterile manufacturing facility is built-up concrete walls with final oil antifungal paint on the exterior boundary walls. All interior walls and ceilings are built up by special clean room panels, which have a final finish of smooth white polyester lacquer. All 90° wall/wall and wall/ceiling interfaces are constructed with enough coving to aid in cleaning and decontamination. Windows are placed in a standard dimension of 900 × 1200 mm with 6 mm thickness at both sides flush with the panel facing.

Floors. All manufacturing area floors have been furnished with epoxy flooring with 2- to 4-in. integral cove base.

Lighting. Lighting fixtures in the manufacturing areas are dust- and air-proof surfaces, flush-mounted, clean room types. They are sealed to the ceiling with pharmaceutical-grade silicone, serviced from above, outside the manufacturing area.

Address
ABC Pharmaceutical Industries (postal address)

1.1.1.1 Floor plan

The floor plans of the ABC Pharmaceutical Industries production facility located in (city and country name) are included under (provide reference to attachment no.). The production areas are color coded as to their environmental classification. The site plan highlights the production areas, rooms, and area classifications. The following drawings are included under (provide reference to attachment no.):

■ Drawing titled "Floor Plan Showing Room Classes and Location of Critical Equipment"
■ Drawing highlighting the flow of material titled "Floor Plan Showing Material Flow"

SOP No.: Val. 1400.10   Effective date: mm/dd/yyyy

                                          Approved by:

- Drawing highlighting the flow of people titled "Floor Plan Showing People Flow"
- Drawing highlighting the flow of the product titled "Floor Plan Showing Product Flow"
- Drawing highlighting the "Machine Layout"
- Drawing highlighting "Water Piping and Distribution"
- A list of key equipment used in the manufacturing and processing of (product name), USP

For sterile product manufacturing, five areas that have a direct impact on the quality of the manufacturing are defined as:

- Critical areas: class 100 (level I)
- Critical areas: class 100 (level II)
- Controlled access areas: class 10,000 (level III)
- Controlled areas: class 100,000 (level IV).
- Noncontrolled areas: level V

The plant facility is divided into these five levels of environmentally controlled areas. The levels are defined as follows.

Level I. Areas classified as class 100, critical areas, involve operations in which the sterilized product containers and closures are exposed to the environment. The most stringent quality standards are imposed upon areas directly over the filling/closing operation (in the filling suite). Curtains or shroud borders are attached to the perimeter of the HEPA filter housing to maintain the air quality of the critical area. This type of area includes the aseptic filling suites and the sterility test LAFH. The areas classified as level I are as follows:

- **Area A:** HEPA unidirectional air flow hood with shroud border (vial-filling machine), room no. _____
- **Area B:** HEPA unidirectional air flow hood with shroud border (loading/unloading freeze dryer), room no. _____
- **Area C:** HEPA unidirectional air flow hood with shroud border (conveyors), room no. _____
- **Area D:** HEPA unidirectional air flow hood with shroud border (unloading autoclave), room no. _____
- **Area E:** HEPA unidirectional air flow hood with shroud border (sterilization and depyrogenation tunnel), no. _____

SOP No.: Val. 1400.10 Effective date: mm/dd/yyyy
Approved by:

- **Area F:** HEPA unidirectional air flow hood with shroud border (pass-through to clean rooms), room no. _____
- **Area G:** HEPA unidirectional air flow hood with shroud border (Q.C, microbiological lab)

Shroud construction is of Plexiglas or vinyl material to enhance the unidirectional airflow to the working surface. The ceiling of the area is constructed of the HEPA filter face with lighting projected below the HEPA filter face. The average illumination at work surfaces is approximately 500 lux. The mechanical drawings that provide specific details are on file.

HEPA filter integrity testing is performed at 6-month intervals on level I and level II HEPA filters. HEPA filter testing includes "Measuring (Machines, Filters, Diffusers) Inlet and Outlet Air Velocity or Volume" and "Integrity Test of HEPA Filters Using Aerosol and Photometer." Refer to (provide reference to attachment no.).

Level II. Level II areas are classified as level I class 100 controlled-access areas. They are work zones that may surround class 100 areas or be adjacent to class 10,000 areas. The examples of this type of area include freeze drying loading/unloading, syringe filling, autoclave loading, auto-claved unloading, and vial-filling rooms. However, the areas surrounding the LAFU are considered class 1000 at dynamic conditions.

The heating, ventilation, and air conditioning (HVAC) system is designed to create airflow, which has a cascading effect. It provides a pressure differential between the aseptic filling suites (critical/controlled access areas, class 100 level I and level II and the sterile storage unit of at least 15 Pa. In addition, a pressure differential is maintained between the sterile storage area (controlled access area, class 100, level II) and the controlled areas (class 10,000 and class 100,000) of at least 20 Pa at all times. The areas classified as level II are as follows:

- **Room no.** _____: autoclave unloading area and sterile storage suit (surrounding)
- **Room no.** _____: vial-filling room (surrounding)
- **Room no.** _____: loading/unloading of the freeze dryer (sur-rounding)

Level III. Level III areas are classified as class 10,000 controlled access areas. These work zones include areas for operations involving bulk solution preparation, filtration, preparation of containers, closures and

SOP No.: Val. 1400.10

Effective date: mm/dd/yyyy

Approved by:

primary contact filling equipment, etc. Actual washing of containers and primary contact filling equipment is performed in class 100 areas. The areas classified as level III are as follows:

- **Room no.** _____ : Clean room personnel gowning room
- **Room no.** _____ : Airlock (formulation materials entrance and exit)
- **Room no.** _____ : Airlock (formulation personnel entrance and exit)
- **Room no.** _____ : Solution preparation room
- **Room no.** _____ : Capping room
- **Room no.** _____ : Glassware (ampoule/vial) washing and preparation; stopper preparation, filter preparation, and filling equipment washing and preparation
- **Room no.** _____ : Solution holding room
- **Room no.** _____ : Freeze drying unloading process

Level IV. Level IV areas are classified as class 100,000 controlled access areas. They are work zones that may surround class 100 areas or be adjacent to class 10,000 areas. The level IV areas are specified as follows:

- **Room no.** _____ : Production and packaging corridor
- **Room no.** _____ : Air lock (clean room personnel entrance and exit)
- **Room no.** _____ : Air lock (materials/personnel entrance and exit to room _____)
- **Room no.** _____ : Packaging area
- **Room no.** _____ : Gown washing and preparation
- **Room no.** _____ : Visual inspection area

Level V. Level V areas are noncontrolled. They include research and development laboratories, offices, quality control laboratories, microbiological laboratories (with the exception of the sterility test area), lunchroom, toilet facilities, and other nonmanufacturing areas of the facility.

Table 1 summarizes the environmental control characteristics for the aseptic processing complex. The areas' classifications are based on activity

SOP No.: Val. 1400.10   Effective date: mm/dd/yyyy

Approved by:

and the limits are based on review of collective data over a period of 3 months.

Utilities for the facility include: (The text used to define the utilities is generic. Each company should define its particular system.)

■ *Steam system.* Clean steam is used for all equipment that comes into contact with containers, solution, or closures prior to product assembly. Pure steam is produced by a generator fed by deionized water. Steam traps are installed to collect condensate when necessary. The quality of pure steam condensate is the same as water for injection. The quality of pure steam is monitored through a quality analyzer system that measures the conductivity of condensed pure steam.

■ *Compressed air.* Oil-free compressed air is produced in rotary screw compressors. It is stored in a stainless steel receiver, then passes through a 1-μm filter for particle removal and two air dryers to assure complete removal of moisture traces. It is delivered to the plant via a stainless steel loop that supplies all use points and has a filter and regulator.

Five use points where the compressed air quality is considered of medical grade, USP, are defined to be critical:

1. (Define use point)
2. (Define use point)
3. (Define use point)
4. (Define use point)
5. (Define use point)

■ *Nitrogen system.* Nitrogen is produced by ABC Pharmaceutical Industries at the manufacture site (minimum purity of 99%), where compressed air is fed into the "Permea" unit and passed through a three-stage filtration. Clean air is heated to 45°C, then flows into separators where oxygen is removed and dry nitrogen is generated. Continuous on-stream analysis of nitrogen purity is monitored within the unit, and whenever purity is below specification, nitrogen gas is automatically diverted to vent line. The nitrogen is stored in a stainless steel tank and distributed to the plant through a stainless steel loop to all use points.

SOP No.: Val. 1400.10

Effective date: mm/dd/yyyy

Approved by:

Table 1   Summary of Environmental Control Characteristics

Room Description	*Work Areas Encompassed in Description*		*Clean Room Classification*
Critical Areas			
Level I Sterilized product and sterilized product contact parts may be exposed to environment	1	**Area A** Aseptic fill area Vial filling room LFH (room no. ___)	Class 100
	2	**Area B** Loading freeze dryer LAF (room no. ___)	Class 100
	3	**Area C** Conveyer from filling to freeze dryer room (room no. ___)	Class 100
	4	**Area D** Unloading autoclave LFH (room no. ___)	Class 100
	5	**Area E** Vial sterilization and depyrogenation tunnel (room no.___)	Class 100
	6	**Area F** Solution room (pass box) LFH	Class 100
	7	**Area G** Sterility test LFH (quality control microbiological lab)	Class 100
Critical Areas			
Level II Areas restricted to personnel gowned for aseptic processing area immediately surrounding class 100 laminar flow hood and laminar air flow maintained through plastic curtains	1	Unloading autoclave and storage area (room no. ___)	Class 100 at rest Class 1,000 at work
	2	Vial filling LFH surrounding filling LFH (room no. ___)	Class 100 at rest Class 1,000 at work
	3	Loading freeze dryer LAF surrounding LAF (room no. ___)	Class 100 at rest Class 1,000 at work

SOP No.: Val. 1400.10

Effective date: mm/dd/yyyy

Approved by:

Airborne Nonviable 0.5-μ Particles per Cubic ft	Airborne Viable Microbiological Level (cfu/m³)	Minimum No. Air Changes Required per Hour	Minimum Differential Air Pressure between Work Areas
NMT 100 particles/ft³	1 cfu/m³	NLT 20 air changes/h	Work zones maintained with differential air pressure to ensure positive air cascade to class 10,000 work areas
NMT 100 particles/ft³	1 cfu/m³		
NMT 100 particles/ft³	1 cfu/m³		
NMT 100 particles/ft³	1 cfu/m³		
NMT 100 particles/ft³	1 cfu/m³		
NMT 100 particles/ft³	1 cfu/m³		
NMT 100 particles/ft³	1 cfu/m³		
NMT 100 particles/ft³ NMT 1000 particles/ft³	1 cfu/m³ 3 cfu/m³	NLT 20 air changes/h	Work zones maintained with differential air pressure to ensure positive air cascade to class 10,000 work areas
NMT 100 particles/ft³ NMT 1000 particles/ft³	1 cfu/m³ 3 cfu/m³		
NMT 100 particles/ft³ NMT 1000 particles/ft³	1 cfu/m³ 3 cfu/m³		

SOP No.: Val. 1400.10   Effective date: mm/dd/yyyy

                                         Approved by:

Table 1 (continued)   Summary of Environmental Control Characteristics

Room Description	Work Areas Encompassed in Description		Clean Room Classification
Controlled Access Areas			
Level III	1	Personnel gowning (room no. ____)	Class 10,000
a) Areas used to process product materials before sterilization or depyrogenation	2	Air lock (material transfer to solution preparation) (room no. ____)	Class 10,000
b) Rooms surrounding class 100 levels I and II	3	Air lock (personnel entrance and exit to sol. preparation) (room no. ____)	Class 10,000
	4	Bulk solution preparation room (room no. ____)	Class 10,000
	3	Vial capping (room no. ____)	Class 10,000
	6	Commodity washing, filling machine parts washing, and autoclave loading (glassware, filling equipment preparation) (room no. ____)	Class 10,000
	7	Filtration of bulk solution (close system) room (room no. ____)	Class 10,000
	8	Freeze dryer during unloading process (room no. ____)	Class 10,000

SOP No.: Val. 1400.10   **Effective date: mm/dd/yyyy**

                                             Approved by:

Airborne Nonviable 0.5-μ Particles per Cubic ft	Airborne Viable Microbiological Level (cfu/m³)	Minimum No. Air Changes Required per Hour	Minimum Differential Air Pressure between Work Areas
NMT 10,000 particles/ft³	3 cfu/m³	NLT 20 air changes/h	Work zones maintained with differential air pressure to ensure positive air cascade to class 100,000 work areas
NMT 10,000 particles/ft³	10 cfu/m³		
NMT 10,000 particles/ft³	10 cfu/m³		
NMT 10,000 particles/ft³	10 cfu/m³		
NMT 10,000 particles/ft³	10 cfu/m³		
NMT 10,000 particles/ft³	10 cfu/m³		
NMT 10,000 particles/ft³	10 cfu/m³		
NMT 10,000 particles/ft³	10 cfu/m³		

　　　　　　　　　Effective date: mm/dd/yyyy

Approved by:

Table 1 (continued)　Summary of Environmental Control Characteristics

Room Description	Work Areas Encompassed in Description		Clean Room Classification
Controlled Areas			
Level IV Limited access areas with particular gowning. Blue lint-free shirt and paint as undergarment and lint-free frock over it with dedicated clogs and head cover.	1	Production and packaging area access corridor (room no. ___)	Class 100,000
	2	Air lock (clean room personnel entrance and exits) (room no. ___)	Class 100,000
	3	Air lock (material and personnel access to commodity preparation) (room no. ___)	Class 100,000
	4	Vial packaging (room no. ___)	Class 100,000
	5	Washing area for gown (prior to sterilization) (room no. ___)	Class 100,000
	6	Vial visual inspection (room no. ___)	Class 100,000

- *Water system.* ABC Pharmaceutical Industries manufactures two levels of water quality: purified water and water for injection. City water is supplied from a municipal source. After passing through a backflow preventer, it is diverted to general plant use or to the purified water pretreatment system, which includes the RO (reverse osmosis) system. This system supplies purified (DI) water to the pure steam generator, the ampoule/vial washer, cleaning use points, and the distillation unit used to produce water for injection.
- *Water for injection (WFI) system.* The condensate of the heated vapor (free distillate) is collected into the condenser, where the vapor is cooled and condensed by the incoming cooling water. The WFI is collected in the main storage tank (3000-l capacity) from where two loops start:
  - One for the CIP of the freeze dryer
  - One for the distribution inside the plant
    Temperature indicators are used to monitor the temperature continuously. The temperature requirement for the WFI tank is >85°C and for the water distribution system >80°C. Conductivity

SOP No.: Val. 1400.10   **Effective date: mm/dd/yyyy**
                                                 Approved by:

Airborne Nonviable 0.5-µ Particles per Cubic ft	Airborne Viable Microbiological Level (cfu/m³)	Minimum No. Air Changes Required per Hour	Minimum Differential Air Pressure between Work Areas
NMT 100,000 particles/ft³	50 cfu/m³	NLT 20 air changes/h	Work zones maintained with differential air pressure to ensure positive air cascade to unclassified areas
NMT 100,000 particles/ft³	35 cfu/m³		
NMT 100,000 particles/ft³	50 cfu/m³		
NMT 100,000 particles/ft³	50 cfu/m³		
NMT 100,000 particles/ft³	50 cfu/m³		
NMT 100,000 particles/ft³	50 cfu/m³		

and temperature at the return of the loop are monitored and registered on the control panels. Sampling points are available near each main point of use.

■ Weekly chemical and physical monitoring of WFI from room no. _____ (commodity washing) and room no. _____ (solution preparation).

■ Daily microbiological monitoring of WFI from room no. _____ (commodity washing), room no. _____ (solution preparation) and parenteral area WFI inlet and outlet.

■ Heat ventilation and air conditioning. Reference:
   a. HVAC principal scheme drawing no. _____
   b. HVAC ducting layout ground floor drawing no._____
   c. HVAC ducting layout first floor drawing no._____
   (Provide reference to allocation.)

The plant areas are supplied with adequate ventilation sufficiently controlled for pressure, temperature, relative humidity, dust, and microorganisms and to sustain personnel comfort. The design of this system has filters, which aid in the control of microorganisms.

SOP No.: Val. 1400.10   Effective date: mm/dd/yyyy

Approved by:

The air systems consist of various heating, ventilation, and cooling units to produce the desired temperatures and pressure differentials and other air quality attributes in the various areas of operation. High efficiency particulate air (HEPA) filters are installed in accordance with manufacturing site design documents to assure that the air quality will comply with federal and industry standards. The HEPA filters are qualified after installation and are requalified semiannually, generally during scheduled plant shutdowns. The qualification and requalification include integrity testing using an aerosol challenge and air velocity testing.

1.1.1.2 Location of equipment

A floor plan of the facility that includes locations (or identification of locations) of equipment, including laminar flow hoods, sterilizers, lyophilizers, filling heads and equipment segregated by barrier or isolation systems, is included in (reference attachment no.). The filling equipment is located in the level I unidirectional laminar flow hood areas.

1.1.2 Overall MANUFACTURING OPERATION

FDA Guidance: The overall manufacturing operation, including, for example, material flow, filling, capping, and aseptic assembly, should be described. The normal flow (movement) of product and components from formulation to finished dosage form should be identified and indicated on the floor plan described above. The following information should be considered when describing the overall manufacturing operation:

ABC Pharmaceutical Industries information. Weighing of a lot follows detailed step-by-step instructions as outlined in the batch production record. Weighing of all ingredient materials is accomplished in a dedicated room within the class 100 air quality. The preweighed ingredients are transported to the mixing/compounding area.

Covered stainless steel tanks are used for compounding. Each ingredient is checked and verified per manufacturing site standard operating procedures by qualified personnel and documented in the batch record. Final q.s. of the tank is performed and the contents mixed using pre-established specifications. The bulk product is analyzed by the Quality Control as required by the in-process specifications. After verifications to meet specifications, the bulk product is released for batch filling. It is prefiltered into a sterile mobile vessel in the compounding area. This vessel is transported to the solution holding room. Final sterile filtration is done during product filling.

SOP No.: Val. 1400.10	Effective date: mm/dd/yyyy
	Approved by:

Components are received and are quarantined in ABC Pharmaceutical Industries stores until all testing and certificate of analysis requirements are reviewed and have met the acceptance criteria set forth in manufacturing site standard operating procedures. When all acceptance criteria have been met, the components are released by Quality Control and are ready to be issued for production using the procedures specified in manufacturing site standard operating procedures. **Glass vials are received from the warehouse and enter the washing area.** The vials are washed in a validated vial washer, using deionized water and water for injection with sterile compressed air. The vials are depyrogenated in a validated laminar flow dry heat tunnel and unloaded into the sterile filling room under laminar flow.

Stoppers are received from the warehouse and enter the commodities preparation area. The stoppers used for (product name) are purchased as "ready to sterilize" (i.e., washed, siliconized, and depyrogenated by the vendor). The stoppers are placed in a double-door component sterilizer and then subjected to a validated sterilization cycle. All stoppers are unloaded into the sterile storage area and stored until needed for production.

Primary contact (product contact) filling equipment used in the filling process is washed (a detergent may be utilized), rinsed with WFI, and wrapped or packaged for sterilization. The cleaning process for all product contact, nondisposable equipment has been validated. Equipment is sterilized in a cycle validated to deliver a $F_0 \geq 20$ minutes using "overkill model" sterilization process design. Sterilized equipment is then unloaded into the sterile storage area (class 100 level II) until use.

The lyophilizers are cleaned using a validated clean-in-place (CIP) cycle using hot WFI. After the cleaning process, a validated sterilization cycle is run. The qualification of the steam sterilization cycle was performed during the operation qualification (OQ) of the lyophilizer.

Any personnel entering classified areas must have successfully completed clean room training and gowning qualification. Periodically, filling suite operator behavior may be observed for appropriateness (i.e., whether it conforms to the training previously given) as a part of the on-going education program for the operator, in addition to routine re-education training program as described in plant site standard operating procedures.

Full gowning is required for all filling suite personnel and support personnel. Standard operating procedures identify the method and type of gowning for each classified room.

SOP No.: Val. 1400.10 **Effective date: mm/dd/yyyy**
Approved by:

The vials, stoppers, and primary contact filling equipment are transferred from the sterile storage area into a filling room for the filling process. The filtration set-up and filling equipment are aseptically assembled under class 100 in the filling room. Sterilized product tubing is transferred up to the solution holding area for connection to the sterile holding tank. Filling starts after a successful filter integrity test, fill volume check, and other quality checks (i.e., verification of filling room, freeze dryer loading room, and completion of documentation — prefilling checklists).

Filled vials are partially stoppered and place in the auto load/unload system (ALUS) to load the lyophilizer. After lyophilization is complete, the vials are stoppered while still in the lyophilizer using an internal stoppering mechanism. They are then unloaded onto conveyer belts to the capping area outside the filling room where aluminum seals are applied to the stoppered vials and crimped.

Upon completion of filling, integrity of product final filter and vent filters is checked. Empty vial and stopper containers, left-over unused containers, and primary product contact equipment are returned from the filling room via interlocking pass-through into the preparation area. Any remaining stoppers are not reprocessed. Vials may be reprocessed by repeating the complete vial preparation operation (i.e., vial washing followed by a validated depyrogenation cycle in a dry heat tunnel).

All filled and stoppered vials are 100% visually inspected, labeled, and packaged. The product is placed in quarantine until all process and product release testing is completed. At this time, the final disposition of the product is determined.

Material Flow. Floor plans showing the material and personnel flows of the production process are provided under reference attachment no. ____. Positive pressure is maintained in each of the clean rooms.

Step 1. The preweighed raw materials are received in the solution preparation room no. ____.

In-process checks:

- Solution preparation area clearance before compounding including checks for the processing materials for availability, completeness, release, and per the approved material requisition
- Area temperature, humidity, and pressure differential verification
- Nonviable particulate monitoring of the solution preparation room
- Other GMP compliance checks, per manufacturing site

SOP No.: Val. 1400.10 Effective date: mm/dd/yyyy
Approved by:

Step 2. Primary commodities (vials and stoppers) are received from the store and then transferred to preparation room no. ____. The vials are washed in the washing machine using purified water and finally rinsed with WFI. Then the vials pass through the tunnel for sterilization and depyrogenation before they reach filling room no. ____.

The stoppers are received ready to be sterilized. They are loaded into the sterilizing autoclave. After successful sterilization cycle, the stoppers are unloaded into the sterile storage area room no. ____.

In-process checks:

- Area clearance (also includes verification for QC release of primary commodities)
- Nonviable particulate monitoring of the area prior to commencing the process
- Nonviable particulate monitoring of sterilization tunnel
- GMP compliance checks

Step 3. The compounded bulk solution is prefiltered through a 0.2-μm sterile filter to a sterile mobile vessel, which is then transferred to the solution filtration room no. ____.

In-process checks:

- Verification of SIP process for sterile mobile vessel.

Step 4. Final sterile filtration with sterile 0.2-μm filter is done during the product filling in room no. ____.

In-process checks:

- Area clearance
- Continuous nonviable particles monitoring during filling process
- Hourly fill checks (by weight)
- Hourly physical attributes checks of filled vials
- GMP compliance checks

Step 5. Filled vials are then conveyed to the freeze drying chamber in room no. ____ by the ALUS (automatic loading and unloading system).

In-process checks:

- Area clearance
- Nonviable particulate monitoring (continuous)
- GMP compliance checks

Step 6. Lyophilized products are then visually inspected according to the related SOPs in room no. ____.
In-process checks:

- Area clearance prior to visual inspection
- Rejection verification
- Reconstitution time test for lyophilized vials
- Liquid-borne particulate test of the reconstituted vials
- Random sample inspection from the accepted vials (good) per the sampling plan and disposition (OK for packaging/reinspection)

Step 7. The accepted vials are labeled using the dedicated labeling machine model no. ____. The product label includes all the necessary information along with the batch number and expiry date.
In-process checks:

- Area clearance of packing area (also include commodity checks for QC release, appropriateness as per the master packaging instructions)
- Hourly checks of the printing, labeling, and packaging process

Step 8. Finished products are then finally packed according to the related SOPs and the master packaging instructions and transferred to the stores.
In-process checks:

- Area clearance of packing area (also include commodity checks for QC release, appropriateness per the master packaging instructions)
- Hourly checks of the printing, labeling, and packaging process

Quality control in-process checks.

- Bioburden bulk solution (preceding filtration)
- Microbiological air monitoring of the environment
- Surface monitoring of class 100 (all the machine parts are made to perform after filling)
- Personnel hygiene: gown and finger tips (glove print)
- Weekly chemical and physical monitoring of DI water and WFI from room no. ____ (commodity washing) and room no. ____ (solution preparation)
- Daily microbiological monitoring of WFI from room no. ____ (commodity washing), room no. _____ (solution preparation), and parenteral area WFI inlet and ____

SOP No.: Val. 1400.10

Effective date: mm/dd/yyyy

Approved by:

Currently, the following rooms are utilized as indicated:

- **Solution preparation room no._____:** product solution compounding
- **Washing room no. _____:** washing of vials; sterilization/depyrogenation of vials
- **Filling room no._____:** filling of vials
- **Freeze dryer room no._____:** loading and unloading of the freeze dryer
- **Capping room no._____:** capping of lyophilized vials and liquid vials
- **Inspection room no._____:** inspection of filled containers (vials and ampoules)
- **Packaging room no._____:** packaging of sterile products
- **Solution holding room no._____:** sterile filtration to the filling machine
- **Sterile gowning room no._____:** gowning to filling rooms

1.1.2.1 Drug product solution filtration

FDA Guidance: The specific bulk product solution filtration processes, including tandem filter units, prefilters, and bacterial retentive filters, should be described. A summary should be provided containing information and data concerning the validation of the retention of microbes and compatibility of the filter used for the specific product. Any effects of the filter on the product formulation should be described (e.g., adsorption of preservatives or active drug substance, or extractable).

ABC Pharmaceutical Industries information. A copy of the filter retention validation report for (product name) USP is included in (reference attachment no. ____). (Please note that this report is supplied by the filter manufacturer, who should be identified by name.)

1.1.2.2 Specifications concerning holding periods

FDA Guidance: Section 211.111 of the Code of Federal Regulations requires, in part, when appropriate, the establishment of time limits for completing each phase of production to ensure the quality of the drug product. Therefore, specifications concerning any holding periods between the compounding of the bulk drug product and its filling into final containers should be provided. These specifications should include, for example, holding tanks, times, temperatures, and conditions of storage.

SOP No.: Val. 1400.10 Effective date: mm/dd/yyyy

Approved by:

Procedures used to protect microbiological quality of the bulk drug during these holding periods should be indicated. Maintenance of the microbiological quality during holding periods may need verification.

ABC Pharmaceutical Industries information. When appropriate, time limits are established for completing each phase of production to ensure the quality of the drug product. Holding times and conditions such as area classification and temperature are specified for each aspect of manufacturing and are listed in manufacturing site SOPs. Holding times are supported by validation studies, media fills, or retrospective data review, which ensure the microbiological quality of the product during this hold time.

Holding times for (product name) USP are as follows:

Phase of Production	Parameters for Hold Time	Hold Time (Not to Exceed)
Bulk solution	Ambient temperature in a covered tank in a controlled access area	_____ hours from bulk solution q.s. to end of filtration
Product filling	Ambient temperature in a sterile mobile vessel in a critical fill zone in aseptic fill suite	_____ hours from start of filtration to the last vial filled

Requirements for hold time limits and protection of product microbiological quality are described in manufacturing site SOPs. The hold times for (product name) USP will be validated during manufacturing of process validation lots.

The time limit for completing the compounding and filling operation for (product name) USP is 24 hours (from bulk solution q.s. to end of filtration).

1.1.2.3 Critical operations

(Product or product contact surfaces exposed to the environment)
FDA Guidance: The critical operations that expose product or product contact surfaces to the environment (such as transfer of sterilized containers or closures to the aseptic filling areas) should be described. Any barrier or isolation systems should be described.

SOP No.: Val. 1400.10   Effective date: mm/dd/yyyy

Approved by:

ABC Pharmaceutical Industries information. The following operations are considered critical operations in the overall manufacturing operation as described in Section 1.1.2:

- The compounded product is prefiltered into a sterile mobile vessel in the compounding area and transported to the solution holding room for final filtration.
- Vials are washed with deionized water; final rinse is with water for injection. They are depyrogenated in a validated laminar flow dry heat tunnel and unloaded into the sterile filling room (level I, class 100) under laminar flow prior to production.
- The stoppers are sterilized in a validated component sterilizer and unloaded into the sterile storage area (level II, controlled access class 100) until needed for production. Supplier (provide name) ships rubber stoppers to ABC Pharmaceutical Industries in autoclavable bags.
- (Ready to sterilize) Filling occurs in controlled access rooms (level I, class 100) equipped with vertical flow HEPA filtration at the process steps. HEPA filters are located in the classified area over and around the container in-feed and filling/stoppering equipment. Shrouds are attached to the perimeter of the HEPA filter housing and extend below the equipment work surface to maintain air quality of the critical area (level I, class 100) for the filling lines.
- The stoppers and primary contact filling equipment are transferred from the sterile storage area into a filling room. For the filling process, the filtration set-up and filling equipment are aseptically assembled under class 100 conditions in the filling room.
- Partially stoppered vials are transported to the lyophilizer using the auto load/unload system (ALUS) and loaded into the chamber of the lyophilizer.
- All product contact surfaces (tubing, filling pumps, and accessories) are cleaned, wrapped, and sterilized prior to use. The sterilized product contact parts are briefly exposed to the level I environment during equipment set-up in which aseptic connections are made with the tubing to transport the sterile drug product to the filling equipment.

(Product name) USP is manufactured using equipment in the following manufacturing process:

SOP No.: Val. 1400.10   Effective date: mm/dd/yyyy
                                         Approved by:

a. The filling line
    a.1 Vial washing machine
    a.2 Laminar flow dry heat tunnel
    a.3 Filling and closing machine
    a.4 Capping machine
b. The freeze dryer
    b.1 Freeze dryer
        b.1.1   Drying chamber with product shelf assembly
        b.1.2   Hydraulic stoppering device
        b.1.3   Sterilizable piston rod of hydraulic stoppering device
        b.1.4   Ice condenser
        b.1.5   Heat transfer circuit
        b.1.6   Refrigeration system
        b.1.7   Vacuum pump set
        b.1.8   Venting system
        b.1.9   Chamber clean-in-place system (CIP)
        b.1.10 Sterilization-in-place system (SIP)
        b.1.11 Control system
    b.2 Automatic loading/unloading system (ALUS)
    b.3 Lyophilization process description: loading, freezing, primary
        drying, secondary drying, unloading, stoppering, and unloading

The filling line (manufactured by [vendor name] in [country of origin])
consists of:

a.1   Vial washing machine (model no.: _____)

Located in level III area room no. ____. Vials are loaded to the machine
through the in-feed conveyor. Then they go through the following cleaning
cycle:

Step 1: the vials are rinsed inside the ultrasonic tank.
Step 2: the vials are cleaned with deionized water from outside and
        inside.
Step 3: the vials are flushed with clean compressed air.
Step 4: the vials are finally rinsed with water for injection.
Step 5: the vials are dried with clean compressed air.

SOP No.: Val. 1400.10	Effective date: mm/dd/yyyy
	Approved by:

a.2   Laminar flow dry heat tunnel (model no.: _____)

The dry heat tunnel is connected directly after the washing machine. Starting at this point, the vials will be processed under class 100 laminar flow areas. First, the washed vials are loaded directly to the preheating zone of the tunnel, which is covered by HEPA filtered laminar flow. The vials are heated and dried properly before passing to the second stage, sterilization and depyrogenation zones. Finally, the vials will be cooled down to room temperature at the cooling zone.

The tunnel speed and temperature are controlled for sizes of vials according to the validation studies.

a.3   Filling and closing machine (model no.: _____)

Located in level II area, room no. ____ and equipped with HEPA filtered laminar flow units above the processing area to provide level I (critical zone class 100)

Vials are coming from the tunnel out-feed, which is connected to rotary table at the in-feed of the filling machine.

The first station is the filling station. This station has four nozzles. Each pump can be adjusted independently. The filling operation uses four rotary piston stainless steel pumps. The dosing system, consisting of buffer tank, filling nozzles, and dosing pumps, is easily disassembled for cleaning and is steam sterilized before use.

The second station is the stoppering station. The filled vials are partially stoppered, keeping enough opening required for the lyophilization process. For liquid products, the vials are fully stoppered. The vibratory bowels and stoppering assembly are easily disassembled for cleaning and sterilization before use.

a.4   Capping machine (model no.: _____), located in level III area, room no. _____

The vials are fed by the machine upstream via a conveyor belt, which brings the vials to the ABC 007. As the vials enter the machine on the in-feed belt, they are picked up by the continuously rotating in-feed star wheel. The flip-off caps are transferred loosely to the containers as they pass under the cap transfer track. From the in-feed star wheel, the containers pass to one of the nine rotary support plates of the central transport system.

SOP No.: Val. 1400.10   Effective date: mm/dd/yyyy
                                         Approved by:

First, the spring-loaded pressure pin comes into contact with the cap and retains it. During the following transport, the plunger continues to descend and presses the cap fully onto the container.

The container is then rotated and pressed against the crimping rail as the plunger continues to descend. The lower edge of the cap is turned inward and gradually firmly crimped. The plunger then moves upwards and leaves the container.

The closed container is picked up by the discharge star wheel and conveyed onto the discharge belt to the next turntable.

The vibratory bowl and the tracks can be easily removed from the unit for cleaning and sterilization. The bulk caps are loaded into the vibratory bowl and conveyed to the capping-up stations. The capping operation is controlled by the "no stopper/no cap" sensor. If a stopper is not sensed at the station, a cap will not be placed on the vial.

a.   The freeze dryer is manufactured by (vendor name) in (country of origin)

b.1   Freeze dryer type (model no.: _____) includes the following:

b.1.1 Drying chamber with product shelf assembly

The drying chamber is rectangular in shape and is equipped with a door. It is vacuum- and pressure-resistant and designed for the freeze-drying process and sterilization with pure steam at an overpressure of (provide value) bar corresponding to (provide temperature in °C).

Material: stainless steel AISI316L with inner surface finish of $R_z \leq$ (provide value: ____) µm.

The chamber floor is slanted toward the condensate discharge. The floor is further reinforced with a pressure plate.

The chamber is equipped with one full door on the sterile side. The door is opened/closed manually. The door is used for loading and unloading the unit. Elevation is coordinated with the automatic loading/unloading system (ALUS).

The door is equipped with a hydraulic bolt locking system.

Product shelf assembly. There is one shelf assembly installed in the chamber. The shelf assembly consists of six product shelves with a total effective charging area of (provide value) m^2.

Material of shelves is stainless steel AISI316 L with surface roughness $R_z \leq$ (provide value) µm.

Temperature difference between incoming and outgoing heat transfer medium under static operating condition per shelf is ± (provide value) °C. Per-shelf package is ± (provide value) °C.

The shelves are provided with protective rails on three sides. The rails are installed in such a way that cleaning water and condensate can run freely.

The bottom shelves can be latched and raised to facilitate cleaning the bottom of the chamber.

b.1.2   Hydraulic stoppering device

The hydraulic stoppering device is equipped with one hydraulic cylinder. Stoppering is effected centrally from the top of the chamber toward the bottom. The piston rod seal in the chamber has a double seal:

■ Atmosphere against vacuum and excess pressure (sterilization)
■ Atmosphere against the oil pump (piston rod of the cylinder)

This combination of seals makes sure that if there should ever be a leak in the hydraulic cylinder, it will not be possible for oil to penetrate into the inner area of the drying chamber.

b.1.3   Sterilizable piston rod of hydraulic stoppering device

This device makes it possible to sterilize the piston rod independently of the sterilization of the freeze-drying plant. The part of the piston rod that enters the chamber during the stoppering process is placed in a sterilization container, the so-called dome. This dome is equipped with a steam inlet line, condensate drain, temperature measuring point, pressure transmitter, etc. Sterilization takes place automatically in one of two possible cycles:

■ Sterilization together with the entire freeze-drying plant. This is usually done if the hydraulic system is used for loading the product shelves.
■ Sterilization alone and independent from the freeze-drying plant. This is usually done after loading has been completed. This makes sure that the piston rod is sterile for the next stoppering process.

b.1.4   Ice condenser

The ice condenser is a horizontal condenser. It has a storage capacity of approximately (provide value) kg. The condenser cooling capacity has

SOP No.: Val. 1400.10   Effective date: mm/dd/yyyy

Approved by:

been dimensioned in such a way that the cooling time from ambient temperature to (provide value) °C takes place in (provide value) minutes (under vacuum).

The evaporator insert consists of a smooth, seamless, stainless steel pipe (AISI316L). The evaporator is divided into four separate refrigerant injection parts with the required refrigerating capacity.

The condenser housing is round and horizontal with vacuum- and pressure-resistant design for the freeze-drying and steam sterilization processes. The condenser is slanted toward the drain ports.

The ice condenser valve is a hydraulically operated valve (provide DN requirement standard) with stable positioning and vacuum tight on both sides against atmosphere. The piston rod seal in the condenser has a double seal:

■ Atmosphere against vacuum and excess pressure (sterilization)
■ Atmosphere against the oil sump (piston rod of the cylinder)

This combination of seals makes sure that if there should ever be a leak in the hydraulic cylinder, it will not be possible for oil to penetrate into the inner area of the drying chamber. In addition, it prevents microbial contamination in the condenser.

b.1.5   Heat transfer circuit

This system is used for heating and cooling the product shelves in the drying chamber to effect the following functions:

■ Remove the solidification heat from the product
■ Supply the required sublimation heat to the product during the drying phase
■ Recool the product shelves down to charging temperature after a cycle

Operating temperature range: (provide value) °C to (provide value) °C
Regulating range during drying: (provide value) °C to (provide value) °C
Cooling rate from +20 to −40°C: approximately (provide value) minutes
Heating rate, average temperature gradient: approximately (provide value) °C/minute

The temperature difference between the inlet and outlet heater transfer medium under static conditions is ± (provide value) °C. With the shelves fully loaded and under static conditions, the temperature difference between the shelf inlet and outlet is ± (provide value) °C.

b.1.6   Refrigerating system

The refrigeration system is required to cool the ice condenser and the heat exchanger for the heat transfer medium. The system consists of four separate two-stage compressors with water-cooled liquefiers.

(Provide information) refrigerant is used. Each unit has separate circuits divided as follows:

Refrigerators (compressors):

■ Cooling of the heat transfer circuit during freezing
■ Cooling of the condenser during primary drying and secondary drying

All the compressors are made by _____ (provide manufacturer name) model no. _____.

Refrigerating equipment:

The required refrigerating equipment is divided into individual circuits as described above between the cooling circuits and the operating groups. It includes the connecting piping and valves such as liquid magnetic valves and thermostatic expansion valves (provide manufacturer information).

b.1.7   Vacuum pump set

This is a three-stage pump set with the corresponding accessories.

Backing pumps: (provide information)
Ultimate pressure is approximately: _____ mbar
Roots pump: (provide information)
Nominal pumping speed is approximately 1 (provide value) m^3/h
Operating data: lowering of pressure in the drying chamber from (provide value) to 10^{-1} mbar: approximately (provide value) minutes;

ultimate total pressure with gas ballast (measured in the empty chamber and cold condenser): approximately (provide value) mbar

b.1.8 Venting system

The system consists mainly of the following: gas inlet piping with the necessary valves and accessories; pall sterile filter. The venting system is designed in such away that it is also completely sterilized (SIP) at the same time that the freeze-drying plant is sterilized.

b.1.9 Chamber clean-in-place system (CIP)

This system is designed to clean the freeze-drying chamber and condenser with water for injection. Spraying with WFI at an inlet temperature of (provide value) °C effects cleaning of the chamber. There are jet rods installed in the drying chamber and the pathway to the condenser. The position of these rods as well as the distribution of the spray jets provides good coverage all over the interior surfaces of the chamber with WFI. The jet rods are equipped with full cone and flat jet nozzles.

After the cleaning process has been completed, the jet rods are emptied automatically. The water ring pump is used for drying when the cycle has completed.

b.1.10 Steam sterilization (SIP)

The freeze dryer is designed for steam sterilization of the chamber and condenser, as well as defrosting of the condenser with steam. The steam sterilization is carried out at an overpressure of 1 (provide value) bar.

All parts of the plant, such as venting filter, pressure-measuring sensors, temperature-measuring sensors, and their lead-throughs, within the sterilization area are also sterilized.

The sterilization process is completely controlled and monitored by the automatic control unit and registered by the (provide numbers) channels (provide manufacturer name) recorder.

b.1.11 Control system

The freeze-drying plant is controlled taking into account the menu data specific to the product and entered by the user.

SOP No.: Val. 1400.10   Effective date: mm/dd/yyyy

                                         Approved by:

Valves and components are switched on and off automatically in accordance with the requirements of the process. These requirements are defined in a batch recipe for each product. All functions and operations are continuously monitored. The heating capacity is adapted automatically to the required sublimation capacity.

Continuous control of the measured values and of the sensors, as well as initiation of special measures in case of failure, guarantees the safe and reliable course of the process.

Rise in pressure measurements determines the progress and the end of the drying process.

b.2 Automatic Loading/Unloading System (ALUS)

The movable transfer table includes a buffer belt conveyor leading to the filling machine. The transfer table is docked to the chamber; the vials coming from the filling machine are collected on the transfer table and then pushed into the chamber.

The loading/unloading is effected at a variable height. While the shelf is loaded, the vials coming from the filling machine are collected on the buffer collecting table.

The ALUS is located in the level I area (class 100 under laminar flow). The whole loading/unloading sequence is fully automatic. The vials are transferred from the filling line and charged into the freeze dryer chamber automatically without operator interference.

b.3 Lyophilization process description

The vials arrive at the automatic loading/unloading system (ALUS) of the freeze dryer located in room no. ____. They arrive filled with (product name) USP solution and partially stoppered. The vials are loaded into the freeze dryer chamber automatically by the ALUS, which is under laminar air flow (LAF) class 100 area. Upon completion of the chamber loading, the freeze dryer door is closed and the cycle is started from the control system located in control room no. ____. At this step and forward, the process is fully computerized. The freeze-drying program will start the lyophilization cycle following the recipe developed for (product name) as described in the batch record.

SOP No.: Val. 1400.10   Effective date: mm/dd/yyyy
                                          Approved by:

1.1.3 Sterilization and depyrogenation of containers, closures, equipment, and components

FDA Guidance: The sterilization and depyrogenation processes used for containers, closures, equipment, components, and barrier systems should be described. A description of the validation of these processes should be provided including, where applicable, heat distribution and penetration summaries, biological challenge studies (microbiological indicators and endotoxin), and routine monitoring procedures. Validation information for sterilization processes other than moist heat should also be included. Methods and data (including controls) demonstrating distribution and penetration of the sterilant and microbiological efficacy of each process should be submitted. The section of this guidance concerning terminal sterilization contains information that may be of further assistance.

ABC Pharmaceutical Industries information. The following information describes the sterilization and depyrogenation processes used for containers, closures, equipment, and component validation.

Closures — washing and depyrogenation validation. Note: This section is not applicable for the "ready-to-sterilize" stoppers used for (product name) USP. They are washed and depyrogenated in accordance with the manufacturer's DMF.

Component sterilizers: steam sterilization process. A steam autoclave is used at ABC Pharmaceutical Industries to sterilize equipment and components used in aseptic processing. The following system description applies to the (model no. _____) autoclave used for processing equipment and components used for (product name) USP. The autoclave is located as depicted on reference attachment no. _____. It dispenses sterile components into level II area, room no. _____.

Autoclave. (ABC 009) serial no. (provide number) autoclave is utilized for sterilization of solution buffer vessels, filters, filter housings, closures, filling machine dosing system, and support equipment. The unit has a chamber size of (width × length × height) mm and (provide volume) liter capacity with dual interlocking doors.

Clean steam is supplied to the sterilizer jacket and the chamber via the clean steam generator. Compressed air is supplied to the chamber through a (provide filter pore size) μm hydrophobic filter by an oil-free (manufacturer name) compressor.

The sterilization process is controlled by the microprocessor-based control system. Temperatures and pressures are measured from the chamber

SOP No.: Val. 1400.10   Effective date: mm/dd/yyyy

                               Approved by:

via pt100 temperature sensors and I/P transducers. The correct sterilization temperature is achieved by controlling the pressure of saturated steam. There is a correlation between the temperature and the pressure. The steam pressure is controlled by a control valve and PID regulator. There can be several PID loops in the application program. A diagnostic is built into the autoclave. Several alarms in the control system inform the operator or interrupt the cycle when the conditions in the chamber are not within certain limits or if there is some other risk for unacceptable sterilizing result. This is to guarantee that the sterilized products really are sterile. The control system also updates the operator interface, informing the operator about the cycle going on and advising, if required, any cycle disruptions. The printer connected to the control system prints a document of each sterilization cycle (temperatures, pressures, cycle phases, etc.); also alarms that occurred during a cycle will be printed. Therefore, each sterilizing cycle can be found afterwards from printer report. The following is a list of the parameters for different cycles:

Filling machine parts and filtration assemblies. Cycle description:

- Cycle number = (provide number)
- Number of RTDS = (provide validated parameters)
- Cycle label = M/C PART + FLT
- Cycle type = (provide validated parameters)
- Ramps (y/n) = (provide validated parameters)
- No. of pvac pulses = (provide validated parameters)
  1. Prevac level = (provide validated parameters)
  1. Steam level = (provide validated parameters)
  2. Prevac level = (provide validated parameters)
  2. Steam level = (provide validated parameters)
  3. Prevac level = (provide validated parameters)
- Exposure temp = (provide validated parameters)
- Exposure type = (provide validated parameters)
- Exposure time = (provide validated parameters)
- Temp. dev. limit = (provide validated parameters)
- Drying type = (provide validated parameters)
- Drying level = (provide validated parameters)
- Drying time = (provide validated parameters)
- Dry JK heat (y/n) = (provide validated parameters)
- Drying temp. = (provide validated parameters)
- Loading side = (provide validated parameters)

- Unloading side = (provide validated parameters)
- Print interval = (provide validated parameters)

Load configuration. Vial filling machine parts:

- Four dosing pumps, disengaged
- Manifold
- Buffer tank (with accessories)
- Four prepump silicone hoses
- Four postpump silicone hoses, with needles attached
- Two vibratory sorters for stoppers, inverted
- Stopper feed track
- Final insertion station for stoppers

Filtration assembly:

- Filter housings fitted with (provide filter pore size) μ filter cartridges (product specific)
- Upstream hose, sterile filtration
- Upstream hose, prefiltration
- Downstream hose, sterile filtration
- Downstream hose, sterile filtration
- Triclamps, gaskets, triclover ends with stainless steel collar
- Stainless steel container to collect wetting fluid
- Two silicone hoses for collecting wetting fluid

Gowns. Cycle description:

- Cycle number = (provide number)
- Number of RTDS = (provide validated parameters)
- Cycle label = gowns
- Cycle type = (provide validated parameters)
- Ramps (y/n) = (provide validated parameters)
- No. of pvac pulses = (provide validated parameters)
  1. Prevac level = (provide validated parameters)
  1. Steam level = (provide validated parameters)
  2. Prevac level = (provide validated parameters)
  2. Steam level = (provide validated parameters)
  3. Prevac level = (provide validated parameters)

SOP No.: Val. 1400.10

Effective date: mm/dd/yyyy

Approved by:

- Exposure temp = (provide validated parameters)
- Exposure type = (provide validated parameters)
- Exposure time = (provide validated parameters)
- Temp. dev. limit = (provide validated parameters)
- Drying type = (provide validated parameters)
- No. drying pulses = (provide validated parameters)
- Drying vac level = (provide validated parameters)
- Drying vac time(s) = (provide validated parameters)
- Drying air level = (provide validated parameters)
- Drying air time(s) = (provide validated parameters)
- Dry JK heat (y/n) = (provide validated parameters)
- Drying temp. = (provide validated parameters)
- Loading side = (provide validated parameters)
- Unloading side = (provide validated parameters)
- Print interval = (provide validated parameters)

Load configuration:

- (Define numbers) gowns. Each gown is distributed over two auto-clave bags (one with a pair of boots and one with a suit and a hood). The (define numbers) bags are then distributed evenly over the autoclave trolley (_____ no. of bags on each tray).
- Sterilization date is written on each bag.
- Validity of sterilization is 1 week from the sterilization date, pro-vided that the pack is intact. If the gown is not used within 1 week, it should be resterilized.

Rubber stoppers. Cycle description:

- Cycle number
- Number of RTDS = (provide validated parameters)
- Cycle label = stoppers
- Cycle type = (provide validated parameters)
- Ramps (y/n) = (provide validated parameters)
- No. of pvac pulses = (provide validated parameters)
  1. Prevac level = (provide validated parameters)
  2. Steam level = (provide validated parameters)
  3. Prevac level = (provide validated parameters)
  4. Steam level = (provide validated parameters)
  5. Prevac level = (provide validated parameters)

SOP No.: Val. 1400.10　　　　　　　　　　Effective date: mm/dd/yyyy

Approved by:

- Exposure temp = (provide validated parameters)
- Exposure type = (provide validated parameters)
- Exposure time = (provide validated parameters)
- Temp. dev. limit = (provide validated parameters)
- Drying type = (provide validated parameters)
- No. drying pulses = (provide validated parameters)
- Drying vac level = (provide validated parameters)
- Drying vac time(s) = (provide validated parameters)
- Drying air level = (provide validated parameters)
- Drying air time(s) = (provide validated parameters)
- Dry JK heat (y/n) = (provide validated parameters)
- Drying temp. = (provide validated parameters)
- Loading side = (provide validated parameters)
- Unloading side = (provide validated parameters)
- Print interval = (provide validated parameters)

Load configuration: (define numbers) Tyvek bags, each containing (define numbers) stoppers

Steam sterilization validation. All autoclave operations conform to the Master Validation Plan employed at ABC Pharmaceutical Industries. The autoclave has undergone installation qualification, operational qualification, and performance qualification. The autoclave is revalidated on an annual basis. The equipment steam sterilization cycle revalidation data for the autoclave are provided in validated archives.

The following load configurations have been validated on the autoclave:

- Vial-filling machine parts and filtration assembly: steam sterilization temperature = ___ °C, exposure time = _____ minutes
- Gowns: steam sterilization temperature = _____°C, exposure time = _____ minutes and pulse drying
- Stoppers: steam sterilization temperature = _____°C, exposure time = _____ minutes and pulse drying

Sterilization validation reports are provided in validated archives.

The heating regimen is developed using a KAYE validator or equivalent and copper-constantan thermocouples. The determination of the load cold spot is achieved by actual experiments. A microbiological challenge test was performed to verify the sterilization conditions. Biological indicators

SOP No.: Val. 1400.10   Effective date: mm/dd/yyyy

Approved by:

with 10^6 populations of *B. sterothermophillus* were used to guarantee the required 6-log reduction.

The sterilization specifications developed are transferred to production personnel by SOP as time and temperature specifications.

Depyrogenation process. A dry heat laminar flow tunnel is used at ABC Pharmaceutical Industries for the sterilization and depyrogenation of vials used in aseptic processing. The following system description applies to (model no. ___) dry heat tunnel used for processing the (provide vial size) ml vials used for (product name) USP. The dry heat tunnel is located as depicted on attachment no. ___. It dispenses sterile/depyrogenated vials into the level I area in room no. ___.

Dry heat tunnel. A dry heat tunnel to (model no.), serial no. (provide number), is utilized for sterilization and depyrogenation of glass vials. The dry heat-sterilizing tunnel operates according to the short-time sterilizing process, using the laminar flow principle. This unit is built into a process line and operates between a washing machine (model no.) and an integrated vial filling and closing machine (model no.). The washed and wet vials are transported on a conveyor belt through three different zones in the tunnel:

Preheating zone. In this zone, the vial containers coming from the upstream washing machine are flooded with sterile air from a laminar flow unit. This air is circulated in a circulating duct and flows through the HEPA filter before reaching the containers. During production, the circulating air is heated by heat radiated from the following sterilizing zone so that the containers are already preheated at the in-feed.

Drying and sterilizing zone. The preheated containers are fully dried and sterilized in the sterilizing zone. The sterilization is effected with a low-turbulence air stream, heated up to 350°C by heating elements. The heated air is then circulated in another circulating channel and flows through the HEPA filter before reaching the containers. The hot air leaving the HEPA filter flows vertically along the vial neck, thus ensuring the uniform heating of all containers.

Cooling zone. In the cooling zone, an additional HEPA-filtered laminar flow unit cools down the sterilized containers. When leaving the tunnel, the container temperature is about 10 to 20°C above room temperature. The laminar flow unit and exhaust fan of the cooling zone are adjusted so that the tunnel remains pressurized, thus preventing any unsterilized air from entering. All critical process parameters are fully controlled by a dedicated PLC unit. The following parameters are continuously monitored throughout the process:

- Preheating zone temperature
- Heating zone temperature
- Cooling zone temperature
- Belt speed

Dry heat sterilization validation. The tunnel operations conform to the master validation plan employed at ABC Pharmaceutical Industries. The sterilizing tunnel has undergone installation qualification, operational qualification, and performance qualification. The sterilizing tunnel is revalidated on an annual basis. The sterilization/depyrogenation cycle revalidation data for the tunnel are provided in validation archives. Sterilization validation reports are provided in validation archives.

The heating regimen is developed using a Kaye validator and copper-constantan thermocouples. The determination of the load cold spot is achieved by actual experiments. Microbiological challenge test was performed to verify the sterilization and depyrogenation conditions. Endotoxin *E. coli* of concentration >10,000 EUs/ml was used for the challenge test performed to guarantee the required minimum 4-log reduction.

The sterilization/depyrogenation specifications developed are transferred to production personnel by appropriate SOP procedures as time and temperature specifications.

1.1.3.1 Bulk drug solution components that are sterilized separately

FDA guidance: If the bulk drug solution is aseptically formulated from components that are sterilized separately, information and data concerning the validation of each of these separate sterilization processes should be provided.

ABC Pharmaceutical Industries information. The bulk solution is not aseptically formulated; therefore, it does not require any supporting validation data.

1.1.3.2 Sterilization information in the batch records

FDA guidance: The completed batch record supplied with the chemistry, manufacturing, and controls section of the application should identify the validated processes to be used for sterilization and depyrogenation of any container-closure components. This information may be included in the batch record by reference to the validation protocol or SOP.

SOP No.: Val. 1400.10

Effective date: mm/dd/yyyy
Approved by:

ABC Pharmaceutical Industries information. The equipment preparation pages of the master batch record specify the validated sterilization processes to be employed in the preparation of the equipment for (product name) USP. Cycle sterilization parameters are defined along with attributes such as loading patterns and the mechanics of operating the sterilizing equipment. The following lists the sterilization cycles utilized for the equipment required in the processing of (provide product name) USP:

- Vial machine parts and filtration assembly: the autoclave cycle is specified as _____°C for _____ minutes.
- Rubber stoppers: the autoclave cycle is specified as _____°C for _____ minutes.
- Mobile bulk vessel: the sterilization in place (SIP) cycle is specified as _____ °C for _____ minutes.
- Freeze dryer: the sterilization in place (SIP) cycle is specified as _____ °C for _____ minutes.

The equipment sterilization charts are included in the batch production record. The equipment sterilization charts for stability batch are produced in support of this submission. These sterilization charts shall be reviewed by Quality Assurance for adherence to the sterilization cycles specified in the batch records.

1.1.4 Procedures and specifications for media fills

FDA guidance: The validation of an aseptic processing operation should include the use of a microbiological growth nutrient medium in place of product. This has been termed a "media fill" or "process simulation." The nutrient medium is exposed to product contact surfaces of equipment, container systems, critical environments, and process manipulation to closely simulate the same exposure that the product itself will undergo. The sealed containers filled with the media are then incubated to detect microbial contamination. The results are interpreted to determine the potential for any given unit of drug product to become contaminated during actual operations (e.g., start-up, sterile ingredient additions, aseptic connections, filling, closing). Environmental monitoring data are integral to the validation of an aseptic processing operation.

ABC Pharmaceutical Industries information. A series of media fills (three consecutive runs) were performed to initially validate the aseptic processing of the filled vials in filling room no. ___ and loading/unloading

SOP No.: Val. 1400.10   Effective date: mm/dd/yyyy

Approved by:

the freeze dryer in room no. _____. The validation was designed to include the largest and smallest unit sizes of containers manufactured at ABC Pharmaceutical Industries. The validation included the execution of three media fills of (provide vial size). Media fill data summaries for the initial qualification are presented in the plant validation archives.

Our normal frequency for media fills is defined in the manufacturing site SOP "Process Simulation (Media Fill) Test." At a minimum, media fills are performed twice a year for the filling line and each shift.

1.1.4.1 The filling room

(Product name) USP is filled in room no. _____ under the HEPA unidirectional airflow hood, which is classified as a level I (class 100) area. The location of room no. _____ is identified on the floor plan of the facility (ABC-011) (provide reference attachment number).

1.1.4.2 Container closure type and size

The (provide vial size) vial used for the filling of (product name) USP and the media fill run study are the same in dimension and in conformity to glass container requirements as defined by USP <661> and EP3.2.1. The stoppers are made up of bromobutyle-based robber formulation, conform to EP 3.2.9 (type I closure), ISO 8871, Current USP, JPXII, and are sterilizable Tyvek bags. The aluminum flip-cap seals used meet the FDA regulation 21 CFR 177.1520 (c) (1.16).

1.1.4.3 Volume of medium used in each container

The volume for each media fill is listed in the data summary. The practice at (ABC Pharmaceutical Industries) has been to simulate production by using the 50% volume of the container (defined as the practical fill capacity of the container by the manufacturer) for the media fill. This method simulates the filling process utilized in normal production.

1.1.4.4 Type of medium used

All media fills have been performed with tryptic soy broth.

1.1.4.5 Number of units filled

The minimum number of units required by (ABC Pharmaceutical Industries) is _____ units and for (product name) _____ units have been used, to simulate two shift operations.

1.1.4.6 Number of units incubated

The entire media fill is incubated. Damaged containers were not considered in the evaluation (acceptance) of the aseptic processing.

1.1.4.7 Number of units positive

The three media fill runs have achieved the target "zero positives." The target is zero positive units. The current alert level is one positive per **5000 to 10,000** units and the current action level is two positives in **5000 to 10,000** units (provide reference attachment number).

1.1.4.8 Incubation parameters

After filling, the units are stored in a controlled temperature environment for 7 days at 20 to 25°C and 14 days at 30 to 35°C. The units are inspected for growth positives per SOPs for "Process Simulation (Media Fill) Test and Media Fill Run Investigation."

1.1.4.9 Date of each media fill

The date of each media fill is included in each data summary in (provide reference attachment number).

1.1.4.10 Simulations

Media fills are conducted under worst-case conditions encompassing the following conditions:

- Slowest allowable line operating speed for a particular fill machine to ensure maximum exposure of unsealed vial or fastest allowable line operating speed to ensure maximum manipulations of vials. Line speeds depend on the machine and vial size. Typical line speeds range from (provide number) vials per minute to (provide number) vials per minute.
- Largest size vial with largest corkage regularly run on a particular fill machine, to ensure maximum exposure of unsealed vial and longest fill time or smallest size vial run on a particular fill machine, to ensure maximum manipulations of unsealed vial. The ranges of sizes typically run are (provide volume) ml to (provide volume) ml. The size used is specified in the media fill summary (provide reference attachment number).

SOP No.: Val. 1400.10   **Effective date: mm/dd/yyyy**

                                             Approved by:

- Maximum occupancy of the fill room to ensure highest degree of airflow disruption and exposure to bacterial and particulate sources. Typical occupancy is three operators.
- An incursion into the fill room by a quality control technician to perform routine environmental monitoring is conducted to simulate actual production monitoring.
- During each media fill, simulations of all major and minor interventions are logged to document personnel involved, time of occurrence.
- Semiannual requalifications are filled over 24 hours (encompassing all shifts).
- Lyophilization simulation: simulation of all filling and handling procedures required for processing lyophilized product. Media fills for fill room no. _____. Process simulation tests are of 24-hour filling duration to allow enough containers to be filled to determine the contamination rate (for process simulation) properly. Process simulation tests include a representative number of atypical interventions that might occur during an actual production filling operation, i.e., aseptic manipulations. The following activities were involved for simulating the lyophilization process:
  - Media fill protocol and manufacturing direction were established and approved per SOPs to simulate process for lyophilized product.
  - All equipment was prepared and sterilized per approved manufacturing direction and SOPs.
  - Media solution was prepared per approved manufacturing direction.
  - Rubber stoppers were sterilized per manufacturing direction and approved cycle.
  - Vials were washed, sterilized, and depyrogenated per manufacturing direction and validated cycle.
  - Filtration was performed per approved manufacturing direction and SOPs.
  - Filling over _____ hours per approved manufacturing direction, SOPs, and validated filling speed.
  - Interventions were categorized to major and minor and simulated per approved manufacturing direction and SOPs:

Power shutdown	Major
Fill needle replacement	Major

SOP No.: Val. 1400.10

Effective date: mm/dd/yyyy

Approved by:

Upright vials	Minor
Load stoppers	Minor
Clear vials from line	Minor
Fill needle adjustment	Minor
Jammed stopper head	Minor
Fill pump manifold adjustment	Minor
Bleeding of sterilizing filter	Minor
Movement of stalled vials	Minor
Employee break	Minor

■ Lyophilizer was sterilized and tested for integrity prior to use.
■ The partially stoppered vials were loaded into the lyophilizer transfer cart of the ALUS transferred to the lyophilizer and loaded.
■ The first shift of media-filled vials is left in the lyophilizer for next day filling and the door is left open during this period to simulate loading of the lyophilizer.
■ After this time, a partial vacuum is generated to simulate actual processing of product.
■ The vacuum is held for a period of _____ hours and then released.
■ The shelves are lowered and vials are stoppered.
■ The media fill units are then unloaded and crimped.
■ Visual inspection is performed and good vials are sent to the Microbiological Section for incubation
■ The following in-process tests were performed during the media fill process:

Fill Volume Check

Viable total particulate count, surfaces and personnel hygiene
Airborne viable counts
Nonviable particulate counts
WFI and DI water testing from point of use (microbiological and chemical)

Additional media fill information for injection products manufactured in room no. _____ is described in (provide attachment reference number).

1.1.4.11 Microbiological monitoring

The microbiological monitoring data are included in the media fill data summaries. The type of data collected during media fill operations is also collected during normal production runs. The environmental data include RODAC plates, RCS sampling, settling plates, particulate counting, and personnel bioburden (fingerprint and gowning).

1.1.4.12 Process parameters

Routine media fills are conducted in each filling suite at least twice annually. The test schedule includes all operational shifts, operating parameters, and fill room personnel. The following table identifies the difference between productions and media fill settings.

Parameter	Production Settings	Media Fill Settings
Line speed	Line speeds are validated for the fill line	Slowest allowable speed for large vials or fastest allowable speed for small vials
Fill volume	Fill volumes are validated for the fill line	Fill volume of approximately half vial capacity
Vial size/ corkage	Fill line is validated for vial size	Maximum vial size with largest corkage or minimum vial size
No. units filled	Containers filled dependent on batch size	>5000 units
Duration of fill	Depends on batch size; maximum time based on media fill validation	_____ hours

The parameters used for media fills are meant to simulate actual production runs. The line speed was the same as the actual production run. Recent media fills have been conducted at fill volume as that of product. The number of containers filled for a media fill is approximately _____. The personnel involved in the media fills take breaks or go to lunch as they would in a normal production run. Other normal filling operations are done, such as checking the volume in the containers periodically during the media fill operation as is done during normal production.

SOP No.: Val. 1400.10   **Effective date: mm/dd/yyyy**
                                              Approved by:

1.1.5 Actions concerning product when media fills fail

The alert and action levels are defined below.

No. Units in Media Fill	Alert Level (No. Positive Units)	Action Level — Media Fill Failure (No. Positive Units)
≤5000	N/L	1
5000	1	2
>10,000+	1	2

Media fill units are treated in the same manner as product and require batch record documentation. All aseptically assembled products manufactured subsequent to a media fill failure are placed in quarantine until acceptable media fill results are obtained.

A media fill failure will require an investigation and a determination, if possible, of assignable cause if the alert or the action level is reached.

In the case of an unacceptable media fill, product remains in quarantine pending investigation of results and establishment of the system as being in a state of control. The status of products produced prior to the media fill failure is also evaluated. Products manufactured for a period of at least 60 days preceding the media fill failure will be evaluated. Products manufactured prior to this 60-day period may also be evaluated based on the results of the media fill failure investigation.

Media fill failures are investigated in accordance with manufacturing site SOPs. The investigation may include, but is not limited to, a review of the aseptic fill data, review of the environmental monitoring data, review of the component sterilization results, review of all intervention activities, and identification of the contaminants found. Also, additional media fills may be performed to demonstrate that the area and processes are under aseptic control. If the system cannot be shown to be under control, product produced subsequent to the media fill failure on that fill line will be placed under management review for final disposition.

The disposition of product made before or after a failed media fill would depend upon the results of an investigation to determine the cause of the media fill failure. The investigation would be conducted in accordance with manufacturing site SOP for "Media Fill Failure Investigations."

If the investigation concluded that other products might have been affected, appropriate action would be taken, including batch rejection.

1.1.6 Microbiological monitoring of the environment

ABC Pharmaceutical Industries routinely monitors the microbial content of the air, inanimate surfaces, personnel, water systems, and product component bioburden. Microbiological monitoring of these areas generally reflects on the efficiency of cleaning and sanitization procedures and employee practices. Continuous environmental monitoring provides the assurance that product is produced by a controlled process that will maximize the sterility and quality of the manufactured sterile product.

Monitoring requirements are specified for each manufacturing area. Routine monitoring for these specifications flags potential problems early and initiates investigation and corrective actions in advance of a real problem occurring in the manufactured product. These specifications provide data for analysis that will assure that the manufacturing process remains in control and that product will be manufactured under conditions that maximize product quality. The frequency of monitoring, type of monitoring, sites monitored, alert and action level specifications, and precise descriptions of the actions taken when specifications are exceeded are included in each section. Data are given for all areas regardless of which particular filling room is used for this product to give a more comprehensive picture of environmental control monitoring

A documented environmental control program is followed by company management. The program defines responsibility, authority, alert and action limits for viable and nonviable particulates, temperature, humidity, and pressure differential specifications for the aseptic core of the facility.

No differential pressure is required between rooms of the same level. All differential pressures are to be measured with all openings closed. The alert and action levels for pressure differentials are defined based on individual air handling system performance.

Sampling plan. To determine the number of samples, the areas are reviewed for criticality. Samples are taken from representative location at frequencies specified in the respective procedures. The sampling frequencies are based on the environmental classifications per current USP. The sampling plans shall be dynamic with monitoring frequencies and sample plan locations adjusted based on trending performance. It is appropriate to increase or decrease sampling based on this performance.

SOP No.: Val. 1400.10

Effective date: mm/dd/yyyy
Approved by:

Viable and Total Particulate Counts Surfaces and Personnel Hygiene

Sampling Area	Frequency of Sampling	Personnel Gowning and Fingerprinting
Level I	Each	—
Aseptic fill area vial filling room LFH room no.: _____	operating shift	
Loading freeze dryer LFH room no.: _____		
Conveyors between room no.: _____ and room no.: _____ LFH		
Unloading freeze dryer LAF, room no.: _____		
Unloading autoclave LFH room no.: _____		
Solution room pass box LFH room no.: _____		
Level II supporting areas immediately adjacent level I	Each operating shift	All staff that pass into the clean room
Clean room personnel gowning room, room no.: _____		
Vial filling room, room no.: _____		
Loading/unloading of the freeze dryer, room no.: _____		
Potential product/container contact areas	Each fill	Twice/week
Room no.: _____ autoclave unloading area and personnel entrance loopy to sterile suit		
Room no.: _____ solution holding room		
Room no.: _____ solution preparation room		
Room no.: _____ glassware (ampoule/vial) washing and preparation		
Stopper preparation, filter preparation, and filling equipment preparation		
Other support areas (class 100,000)	Twice/week	Twice/week
Room no.: _____ changing room (clean room personnel entrance and exit)		
Room no.: _____ vial capping room		
Other support areas to aseptic processing areas but nonproduct contact (class 100,000 or lower)	Once/week	Once/week
Room no.: _____ production corridor		
Room no.: _____ air lock (materials/personnel entrance to room no.: _____		
Room no.: _____ packaging area		

SOP No.: Val. 1400.10

Effective date: mm/dd/yyyy

Approved by:

Viable and Total Particulate Counts Surfaces and Personnel Hygiene (continued)

Sampling Area	Frequency of Sampling	Personnel Gowning and Fingerprinting
Room no.: _____ packaging area and visual inspection area		
Room no.: _____ gown washing and preparation		
Room no.: _____ airlock (formulation material entrance and exit)		
Room no.: _____ airlock (formulation personnel entrance and exit)		

Testing for obligatory anaerobes is performed every 3 months routinely. However, more frequent testing is indicated, especially when the identification of these organisms occurs in sterility testing.

Response to alert and action levels. The following is a summary of the actions to be taken when alert and action levels are exceeded. More specific details regarding the actions to be taken in the different controlled level environments are included in the manufacturing site SOP for "Environmental Control Programs."

Alert levels. Every time an alert level is reached, the Manufacturing and Quality Assurance managers are notified. Verification of the satisfactory operation of all components of the environmental control system, including but not limited to: HVAC systems, equipment operation procedures, and material handling procedures is performed. The results of the investigation and any actions taken are documented.

Action levels. Every time the action level is reached, an investigation shall be initiated to determine the cause(s) for excursion. All the responses as defined under the alert level are executed. Segregation and quarantine of all exposed materials are performed until determination of the cause(s) for the excursion. Final decision is executed by the quality assurance management regarding product release for commercial distribution. Results of the investigation and any actions taken are documented.

The area can only be released for use upon approval of the Quality Assurance Department.

SOP No.: Val. 1400.10

Effective date: mm/dd/yyyy

Approved by:

Airborne Viable — Action Levels

Work Areas Encompassed in Description	Airborne Viable Microbiological Level (CFUs/m³)
Level I	
1 Aseptic fill area, vial filling room LFH room no.: ____	1
2 Loading freeze dryer LAF room no.: ____	1
3 Conveyer from room no.: ____ to room no.: ____	1
4 Unloading autoclave LFH room no.: ____	1
5 Vial sterilization and depyrogenation tunnel room no.: ____	1
6 Solution room (pass box) LFH room no.: ____	1
7 Sterility test, LFH (quality control microbiological lab)	1
Level II	3
1 Unloading autoclave LFH, surrounding room no.: ____	3
2 Vial filling LFH, surrounding filling LFH room no.: ____	3
3 Loading freeze dryer LAF, surrounding LAF room no.: ____	
Level III	3
1 Personnel gowning, room no.: ____	10
2 Air lock (material transfer to solution preparation), room no.: ____	10
3 Air lock (personnel entrance and exit to sol. preparation), room no.: ____	10
4 Bulk solution preparation, room no.: ____	10
5 Vial capping room no.: ____.	10
6 Commodity washing, vial filling machine parts washing and autoclave loading (glassware, filling equipment preparation), room no: ____	10
7 Filtration of bulk solution (close system), room no.: ____	10
Freeze dryer during unloading process, room no.: ____	
Level IV	50
1 Production and packaging area access corridor room no.: ____	
2 Air lock (clean room personnel entrance and exits) room no.: ____	35

SOP No.: Val. 1400.10

Effective date: mm/dd/yyyy

Approved by:

Airborne Viable — Action Levels (continued)

Work Areas Encompassed in Description	Airborne Viable Microbiological Level (CFUs/m³)
3 Air lock (material and personnel access to commodity preparation) room no.: _____	50
4 Vial packaging room no.: _____	50
5 Washing area for gown (prior to sterilization), room no.: _____	50
6 Vial visual inspection room no.: _____	50

Particle Matter — Action Levels

	Maximum Allowable No. Particles (per Cubic Foot of Air)			
	Action Limit		Alert Limit	
Class	0.5 µm	5.0 µm	0.5 µm	5.0 µm
Class 100	100	0	50	0
Class 1000	1,000	7	500	3.5
Class 10,000	10,000	70	5,000	35
Class 100,000	100,000	700	50,000	350

Additional procedures to ensure environmental control. In addition to those procedures directly associated with the Environmental Control Program, other procedures ensure that the equipment used to collect the test data is functioning properly per manufacturing site SOP for "Calibration of Biotest RCS Air Sampler" and "Calibration of Model FKA-P1 Differential Pressure Sensors," which provides assurance that the data collected are accurate and reliable.

■ Summary reports for year (specify) for viable, nonviable, and inanimate surface is provided as (provide reference attachment number) for summary report of airborne microorganism for the year (specify).
■ (Provide reference attachment number) for summary of microorganism on in animate surfaces for the year (specify) is provided.
■ (Provide reference attachment number) for nonviable environmental monitoring summary report for year (specify) is provided.

SOP No.: Val. 1400.10

Effective date: mm/dd/yyyy

Approved by:

1.1.6.1 Microbiological methods

Airborne microorganisms. The airborne microorganisms are collected via RCS sampling and settling plates. Data are collected for each batch and the results are included in the batch records. Sampling is performed per manufacturing site SOP. Samples are taken in the level I and level II areas during filling operation. All samples are incubated at 32 ± 2°C for 2 days and at 22 ± 2°C for more than 3 days.

All colony-forming units (CFUs) are enumerated if present. Isolates recovered in the level I and level II areas are identified as to genus and species. The summary of airborne microorganisms is monitored during the manufacturing of (product name) stability batch.

Results of all sampling are recorded and achieved in the Quality Control Department.

Microorganisms on inanimate surfaces. Microorganisms on inanimate surfaces are collected per manufacturing site SOP. Samples are taken in the level I and level II areas during the beginning and concluding segments of each filling operation. All samples are incubated at 32°C for 2 days and at 22°C for more than 3 days. All CFUs are enumerated if present. Isolates recovered in the level I and level II areas are identified as to genus and species where possible.

Results of all sampling are recorded and achieved in the Quality Control Department.

Inanimate Surface Sampling Action Levels

Class	Surface, Action Level
Level I	1 CFU/plate or swab
Supporting areas immediately adjacent level I	3 CFUs/plate or swab
Potential product/container contact areas	All sites 10 CFUs/plate
	Floor 20 CFUs/plate
Other support areas (class 100,000)	All sites 25 CFUs/plate
	Floor 30 CFUs/plate
Other support areas to aseptic processing areas but no product contact (class 100,000 or lower)	All sites 50 CFUs/plate
	Floor 100 CFUs/plate

SOP No.: Val. 1400.10   Effective date: mm/dd/yyyy

Approved by:

Overaction levels (OALs) results will have an investigation initiated to determine the cause and source of the contamination in order to apply efficient sanitization procedures or appropriate manufacturing controls or systems. Affected product is placed on release hold while this investigation is in progress. The investigation may include, but is not limited to, a review of other environmental data, identification of the isolate(s), and a review of the aseptic fill data and sanitization practices. This may also require a change in the frequency of sanitization and/or monitoring.

Personnel hygiene monitoring. All personnel participating in a production run shall be tested daily for gloves and gowns bioburden upon exit from level II areas. Other personnel involved in the aseptic process and other supporting areas are monitored for gloves and gowns bioburden per manufacturing site SOP (provide reference attachment number). Summary of personnel bioburden recovered prior to exiting the clean room areas for the year (specify years).

The results of personnel gloves and gowns bioburden testing are maintained in a logbook. The alert and action levels for the personnel monitoring bioburden are defined in manufacturing site SOP. If any personnel exceed the action limit and were involved in the aseptic filling of the drug, it is required that this person be retrained and recertified before re-entry into the clean rooms.

Action Limit for Personnel Hygiene Monitoring

Class	Fingerprint	Gowning
Level I	1 CFU/10 fingers	1 CFU/plate
Supporting areas immediately adjacent level I	3 CFUs/10 fingers	5 CFUs/plate
Potential product/container contact areas	10 CFUs/10 fingers	20 CFUs/plate
Other support areas (class 100,000)	20 CFUs/10 fingers	50 CFUs/plate
Other support areas to aseptic processing areas but no product contact (class 100,000 or lower)	50 CFUs/10 fingers	50 CFUs/plate

SOP No.: Val. 1400.10   **Effective date: mm/dd/yyyy**

                                                 Approved by:

Water systems. The water systems at ABC Pharmaceutical Industries are monitored per manufacturing site SOP for "Microbiological Monitoring of Water." The levels of water tested are city water, purified water (deionized water), and water for injection (WFI). Manufacturing site SOP describes the procedures for obtaining samples, sampling frequencies, test procedures, and acceptance criteria (alert and action limits).

Frequency of Water Monitoring

Room	Sampling Point	Activity	Frequency[a]
Water distribution loop	WFI inlet WFI outlet	Station	Each manufacturing day
Room no.: _____	WFI	Washing	Twice weekly
Room no.: _____	WFI	Solution preparation	Each manufacturing day
Room no.: _____	WFI	Washing machine	Each manufacturing day
		Vessel machine	Each manufacturing day
Plant	Steam	Station	Weekly
Room no.: _____	DI water	Hand washing	Twice weekly
Room no.: _____	DI water	Steam sterilizer	As required
Room no.: _____	DI water	Washing	Once weekly
Room no.: _____	DI water	Washing	Once weekly
Room no.: _____	DI water	Solution preparation	Twice weekly
Room no.: _____	DI water	Washing machine	Twice weekly
Room no.: _____	Feed water	Station	Once monthly
Room no.: _____	R.O.	Station	Once monthly
Room no.: _____	Deionizer	Station	Once monthly
Room no.: _____	After filtration	Station	Once monthly
Room no.: _____	Chilled	Station	Once/3 months

[a] It is appropriate to increase or decrease sampling based on the trend performance. The time and date will be determined according to the activity. The (specify year) microbiological water testing report, "Water Systems Bioburden and Bacterial Endotoxin Summary Report for Year (specify)," is also included in (provide reference attachment number).

Effective date: mm/dd/yyyy
Approved by:

Type of Water	Alert Limit	Action Limit
Pure steam	≤1 CFU/100 ml Endotoxin: LT 0.06 USP EU/ml	≤5 CFUs/100 ml Endotoxin: LT 0.125 USP EU/ml
Water for injection	≤2 CFUs/100 ml Endotoxin: LT 0.125 USP EU/ml	≤5 CFUs/100 ml Endotoxin: LT 0.25 USP EU/ml
Purified water used for highly bacteria-critical products	≤10 CFUs/100 ml	≤20 CFUs/100 ml Pseudomonas: absence in 100 ml Enterobacteriaceae: absence in 100 ml
Purified water used for rinsing	≤50 CFUs/100 ml	≤100 CFUs/100 ml
Purified water used for washing	≤25 CFUs/ml	≤100 CFUs/ml Pseudomonas: absence in 1 ml
R.O. water Feed water Drinking water	NA	Total viable count: ≤500 CFUs/ml Coliform: absence in 100 ml Pseudomonas: absence in 1 ml Fecal streptococci: absence in 1 ml

The data recovered from the sampling are reviewed for conformance to specifications. Maintenance and production management are notified immediately when action levels are exceeded. OALs results cause initiation of an investigation to identify the source of the contaminant. Bacteria recovered are identified as to genus whenever possible. Sampling may be performed more frequently during the investigation or until counts fall back within limits. System sanitization will also be performed. The specifications for monitoring of water for injection are available.

Product component bioburden. Endotoxin analysis involves testing twenty (20) pieces. Twenty (20) vials or stoppers are filled with or immersed in WFI, mixed and sonicated. Duplicate 0.1 mL aliquots are removed and tested for endotoxin per the USP <85> Bacterial Endotoxins Test.

SOP No.: Val. 1400.10   **Effective date: mm/dd/yyyy**
                                             Approved by:

Acceptance Criteria

Test	Ready-to-Sterilize Stoppers	Unprocessed Vials
Bioburden	NMT[a] 25 CFUs/stopper	NMT 100 CFUs/vial
Spore bioburden	Spore bioburden data are collected to screen for heat-resistant organisms. Organisms surviving heat resistance testing are to be submitted to the terminal sterilization laboratory for D-value analysis. D-values are then compared to established models for the component sterilization process.	
Endotoxin	NMT 2.5 EUs/stopper	NMT 10 EUs/vial

[a] NMT = not more than.

Bioburden testing is performed on components *prior to steam sterilization* as verification that the sterilization parameters are sufficient for sterilization of components. The method for determining the bioburden level of components is included in the manufacturing site standard test method. For product component bioburden summary, refer to (provide reference attachment number).

1.1.6.2 Yeasts, molds, and anaerobic microorganisms

Testing for anaerobes is performed every 3 months per manufacturing site standard operating procedure "Testing in Level I Environment." However, more frequent testing is indicated, especially when these organisms are identified in sterility testing. No anaerobic organisms have been recovered in year (specify period at least 3 consecutive years). The environmental monitoring program employs media suitable to detect yeasts and molds. For the detection and quantification of yeast and molds, modified Sabouraud's dextrose agar or inhibitory agar is used. Other media used for promoting growth fungi are TSA for RCS + testing agar strip YM (BIOTEST).

1.1.6.3 Exceeded limits

The actions taken in response to alert and action limits are defined in the environmental control program per manufacturing site SOP. Each time an action level is reached, an investigation is performed to determine the cause. If appropriate, product is segregated and quarantined until the cause of the failure has been determined and a decision is made by Quality Assurance regarding release.

1.1.7 Container closure and package integrity

Container-closure integrity of (product name) USP was performed on the stability batches produced in support of this submission per standard test method no. (specify number), "Container/Closure Integrity Testing with Analysis via UV Spectrophotometry," included as (provide reference attachment number). The testing of the (product name) USP vials was performed under static conditions. Vials were immersed in a dye bath. The product in the vials was then tested for the presence of dye. The container/closure integrity testing yielded acceptable results. The final report for the container/closure integrity test for (product name) is included in (provide reference attachment number).

Additional container/closure integrity testing is to be performed on the first commercial batch at the end of the expiration date as indicated in the stability protocol.

The integrity of the product container/closure is assessed by physical tests or microbiological challenge tests and long-term product sterility tests. These tests are specific for container size, fill volume, and closure type. All integrity tests are performed after sterilization and are defined in manufacturing site SOPs.

Initial validation requires the performance of the dye ingress challenge or the microbial ingress challenge. Matrixing is permissible with the following conditions:

- Testing must be performed on glass and plastic vials separately.
- Molded vials and tubing vials must be tested separately. Color of vial (e.g., amber vs. clear) has no impact on container closure integrity.
- If multiple manufacturers are used for the same vial, each manufacturer does not need to be tested.
- Within a type of vial, the minimum and maximum vial size must be tested for each neck finish size. Within a size of vial, the maximum fill for that vial size is considered worst case, due to its having the smallest headspace.
- Each stopper formulation must be tested.
- Each stopper configuration must be tested for each finish size, i.e., additive, serum.
- Methods of sterilization (aseptic filling vs. terminal sterilization) must be tested separately.

SOP No.: Val. 1400.10   **Effective date: mm/dd/yyyy**
                                                     Approved by:

Dye ingress challenge test. Green dye: a solution of green dye with a validated concentration is placed in a leak test chamber to a height at which the vials are covered with solution. The chamber is covered and a vacuum is pulled for a period of time. The vials (test vials = 10) are removed from the vacuum chamber and tested for the presence of green dye using a UV/VS spectrophotometer. Positive controls (= 10 vials that have been compromised by placing a hole in the stopper) are also placed in the vacuum chamber for each test. The measured absorbance for the positive controls must exceed the absorbance of the test vials. Negative controls (= 10 vials) are not placed in the chamber for each test.

Each product is tested for interaction absorbance at 630 nm (the same wavelength used for green dye determination). If product interferes with the detection of green dye at 630 nm, the limit for green dye absorption may be adjusted to compensate for this interference. Acceptance criteria for green dye ingress are that the absorbance of the sample is less than or equal to zero or NMT 0.002 in absorbance difference between the negative control sample and the test sample. The acceptance specification was set at two times the sensitivity of the method per USP <1225> "Validation of Compendial Methods."

Integrity over the product shelf life. The following testing is performed to assess the microbiological characteristics over the stability of the product.

Container closure integrity: test methodology is per (specify standard test method number). All units for container closure integrity testing are processed using routine production parameters for preparation. Worst-case parameters are used for sterilization.

The dye ingress challenge test is performed at the end of expiry to show container-closure integrity over the production shelf life. Dye ingress testing is performed on vials having rubber stoppers exposed to the maximum exposure time and temperature during sterilization cycle.

1.1.8 Sterility testing methods and release criteria

Samples of drug product are sampled for sterility testing per manufacturing site SOP, "Collection of Finished Product Samples," included as (provide reference attachment number).

The sterility testing area is located in an area of the microbiology laboratory. The area contains a gowning room and sterility testing room. Each testing room contains a unidirectional flow HEPA hood.

SOP No.: Val. 1400.10   Effective date: mm/dd/yyyy

Approved by:

The method for sterility testing of (product name) USP is manufacturing site SOP and the microbiological sterility method validation summary report is provided in (provide reference attachment number). The USP bacteriostasis/fungistasis test was performed to validate the sterility test method

All finished products are tested for sterility in a controlled-access clean room environment in a laminar flow hood, which provides no greater level of a microbial challenge than that encountered in the fill environment. The testing environment consists of a gowning room and sterility test suite that meets the same air quality standards as the filling suite.

Sterility test validations are performed per current USP Section <71>. The validation is performed on a single lot of the largest volume, highest product strength to simulate the maximum concentration of active drug substance on the membrane filter.

Finished product samples are selected from the beginning, middle, and end of each fill of commercial product. The number of samples tested is as directed in the current USP. The samples are disinfected by soaking in a validated antimicrobial agent before introduction into the sterility suite. All sterility testing manipulations are performed in a laminar flow hood using membrane filtration (Millipore's Steritest System). At the completion of the test, the test units are removed from the sterility suite and incubated for 14 days for aseptically filled product. The units are candled at the completion of the incubation period and observed for turbidity. Test units that do not exhibit any growth meet USP acceptance criteria. If microbial growth is found, the lot is placed on hold and investigated per current USP. Additional actions required are identified in manufacturing site SOP.

1.1.9 Bacterial endotoxin test and method

Bacterial endotoxin analysis is performed on finished product where specified in the individual monographs of the current USP. Additionally, endotoxin analysis is performed on water for injection, raw materials, and incoming components to detect contamination even in the absence of recovery of viable bacteria. Bacterial endotoxin analysis is performed per current USP, using gel clot analysis or other suitable endotoxin analysis methods (provide reference attachment number).

The inhibition and enhancement testing and determination of noninhibitory concentration and maximum valid dilution is performed per standard test method.

SOP No.: Val. 1400.10 Effective date: mm/dd/yyyy
Approved by:

Qualification of the laboratory. Qualification of the laboratory includes qualifying analysts initially by demonstrating their proficiency at preparing the necessary reagents, standards, and controls. Analyst qualification is necessary for each type of bacterial endotoxin analysis test prior to performance of any testing on actual samples.

Inhibition enhancement validation testing is performed on each raw material and finished product as part of the validation for bacterial endotoxin analysis in accordance with the current USP and the FDA Guidelines. Limits are determined from USP monographs or the FDA Guidelines and are used in calculating the maximum valid dilution (MVD). Serial dilutions are performed of spiked and unspiked product in order to determine the dilution required to overcome inhibition or enhancement (DROIE). If the DROIE does not exceed the MVD, the test is determined to be valid. Confirmation of the DROIE is performed on three finished product or raw material lots in order to complete the product validation. Products are tested at a dilution greater than the DROIE but less than the MVD.

Limulus amebocyte lysate is obtained from a licensed manufacturer. Each lot of reagent is tested per USP for release. Endotoxin used in all bacterial endotoxin analysis testing is obtained from a licensed manufacturer and is standardized against the USP reference standard.

Finished product samples for bacterial endotoxin analysis are selected from the beginning, middle, and end of each fill of commercial product requiring a bacterial endotoxin test. Samples are pooled or tested individually per the current USP and the FDA Guidelines using the appropriate standards and controls. Products found to be within specification per the current USP and FDA Guidelines are acceptable for release. Products found to exceed the limit for endotoxins are placed on hold and subject to an investigation. The investigation may include, but is not limited to, a review of the analyst and the reagents used for test and a review of the raw materials used, including the water used for formulation. Additional actions required are identified in manufacturing site SOPs.

1.1.10 Evidence of formal written procedures

The following list identifies the sections of this document and standard procedures written to support the document.

Aseptic Filling Sterilization Program Section	SOP No.
I.A	
CIP cycle of the giusti preparation/mobile tanks	
Cleaning procedure for filtration assembly and filling machine parts (specify model)	
Operating procedure for automatic ultrasonic cleaning machine (specify model)	
Operating procedure for hot air sterilization tunnel (specify model)	
Operating procedure for fully automatic filling and closing machine for vials (specify model)	
Operating procedure for fully automatic continuous motion closing machine (specify model)	
Operating procedure for CIP of the freeze dryer	
Operating procedure for freeze dryer plant test	
Operating procedure SIP cycle of freeze dryer	
Operating procedure for freeze dryer	
Operating procedure for Truflow integrity tester	
Operating procedure for SIP cycle of the preparation/mobile tank	
Operating procedure for GMP sterilizer	
Process filtration	
Operating of the freeze dryer loading/unloading system (ALUS)	
Visual inspection of small volume parenterals	
Packaging line clearance and release	
In-process checks for finishing	
Good manufacturing practices in the premises	
Labeling for incoming material, in-process products, and complete processed products	
Particulate monitoring (nonviable) of injectable area	
Fill volume/weights and other checks for parenteral products during filling	

SOP No.: Val. 1400.10   Effective date: mm/dd/yyyy

Approved by:

Aseptic Filling Sterilization Program Section	SOP No.
Area clearance procedure for parenterals QA inspector responsibilities in injectable area Particulate matter in injectables, USP criteria Visual inspection of lyophilized products parenterals	
I.B Sampling reporting format for density/weight calculation for lyophilized products Microbiological monitoring of waters Monitoring of personal hygiene Microbiological monitoring of injectables Chemical, physical monitoring of DI water, WFI, and RO waters Inspection, sampling, and checking of flint tubing vials Inspection, sampling, and checking of seal (aluminum flip cap seal) Inspection, sampling, and checking of stoppers	
I.C Process and equipment validation certification guideline Hot air sterilization tunnel certification/validation guideline Steam sterilization certification/validation guideline Validation and certification of hot air sterilization tunnel Validation and certification of SIP preparation/mobile vessel Validation/revalidation of sterilization cycles in sterilizer by Kaye validator Validation/revalidation of SIP cycle freeze dryer Validation/revalidation for uniformity of heat and chilled temperature distribution in freeze dryer Qualification/requalification of vial washer (specify model) Validation of LAL test	
I.D Process simulation (media fill) test Media fill run investigation	
I.E Environmental control program Microbiological monitoring of injectable area Particulate monitoring of injectable area	

SOP No.: Val. 1400.10   Effective date: mm/dd/yyyy

                                             Approved by:

Aseptic Filling Sterilization Program Section	SOP No.
Media fill failure investigation Calibration of biotest plus RCS air sampler Calibration procedure for pressure sensor (specify model) Determination of components' bioburden before sterilization Sterility testing of bacitracin for injection solution, USP Sterility test results failure investigation Endotoxin test method Endotoxin test method validation Validation of membrane filtration and direct inoculation sterility test methods Monitoring of bioburden, spore bioburden, and endotoxin present in rubber stoppers and unprocessed vials	
I.F Training procedure of employees	

2. Maintenance of Microbiological Control and Quality

The following testing is performed to assess the microbiological characteristics over the stability of the product.

- **Container closure integrity.** Test methodology is per ABC Pharmaceutical Industries. The dye ingress challenge test is performed as described in the manufacturing site stability protocol.
- **Preservative effectiveness.** No preservative is used (optional).
- **Endotoxin testing.** Endotoxin analysis, whenever applicable, is conducted as described in the stability protocol in (provide section number specified in ANDA file).

BIBLIOGRAPHY

Guidance for Industry Sterile Drug Products, produced by Aseptic Processing Current Good Manufacturing Practices, U.S. Department of Health and Human Services, Food and Drug Administration, Center for Drug Evaluation and Research, Center for Biologics Evaluation and Research, Office of Regulatory Affairs, September 2004 Pharmaceutical cGMPs.

SOP No.: Val. 1400.10   Effective date: mm/dd/yyyy

Approved by:

Guideline on General Principles of Process Validation. May 1987. Center for Drugs and Biologics, Food and Drug Administration, 5600 Fishers Lane, Rockville, MD, 20857.

Guideline on PDA (Parenterals Drug Association Inc.) of Validation of Pharmaceutical and Biopharmaceuticals, February 1993.

International Standard of ISO 13408-1 for aseptic processing of health care products.

ISO 14644-1. Clean rooms and Associated Controlled Environments, Classification of Air Cleanliness.

Parenterals Drug Association, Inc., 2002.

Part 211 — Current Good Manufacturing Practice for Fished Pharmaceuticals.

Technical Report No. 26, Sterilization Filtration of Liquids, Parenterals Drug Association, Inc. 1998.

Technical Report No. 36, Current Practices in the Validation of Aseptic Processing.

United States Pharmacopoeia.

List of Attachments

Attachment No.		Description	Drawing No.
1.	1.1	Civil layout — building (ground floor)	
	1.2	Civil layout — building (first floor)	
	1.3	Drawing titled "Floor Plan Showing Room Classes and Location of Critical Equipment"	
	1.4	Drawing highlighting the flow of material titled "Floor Plan Showing Material Flow"	
	1.5	Drawing highlighting the flow of the people titled "Floor Plan Showing People Flow"	
	1.6	Drawing highlighting the flow of the product titled "Floor Plan Showing Product Flow"	
	1.7	Drawing highlighting the machine layout	
	1.8	P & I drawing for deionized water systems	
	1.9	P & I drawing for deionized water piping	
	1.10	Pure steam network piping production and distribution	
	1.11	P & I for water for injection production and distribution	
	1.12	Key equipment list used in manufacturing and processing	
2.		HEPA filter and machine layout in building	
3.	3.1	HVAC principal schema	
	3.2	HVAC ducting layout ground floor	
	3.3	HVAC ducting layout first floor	

4.   Filter retention validation report
5.   Qualification and requalification summary reports
      (A) Example of vial washer performance qualification
      (B) Example of vial washer requalification
      (C) Example of depyrogenation tunnel qualification
      (D) Example of depyrogenation tunnel requalification
      (E) Example of solution holding tanks (sip cycle) requalification
      (F) Example of rubber stopper qualification
      (G) Example of rubber stopper requalification
      (H) Example of equipments sterilization qualification
      (I) Example of equipments sterilization requalification
      (J) Example of freeze dryer qualification
      (K) Example of freeze dryer requalification
      (l) Example of container/closure combination qualification for bacitracin
6.   Equipment preparation pages of the manufacturing direction (MD) no.
7.   Equipment sterilization charts for the stability batch no.
8.   Process simulation (media fill) test report
9.   Additional media information for injection products manufactured in (filling room reference)
10.   Summary report of airborne microorganisms for the year (specify)
11.   Summary of microorganisms on inanimate surfaces for the year (specify)
12.   Nonviable environmental monitoring summary report (specify period)
13.   Summary of personnel bioburden recovered prior to exiting the clean room areas for the year (specify)
14.   Water systems bioburden and bacterial endotoxin summary report for year (specify)
15.   Product component bioburden summary reports
16.   Container closure integrity testing for (product name) USP, batch no.
17.   Sterility testing of (specify name) USP microbiological sterility method validation summary report
18.   Bacterial endotoxin validation report

SOP No.: Val. 1400.10   Effective date: mm/dd/yyyy

                                                  Approved by:

REASONS FOR REVISION

Effective date: mm/dd/yyyy

- First time issued for your company, affiliates, and contract manu-facturers.

SECTION

VAL 1500.00

YOUR COMPANY
VALIDATION STANDARD OPERATING PROCEDURE

SOP No.: Val. 1500.10

Effective date: mm/dd/yyyy

Approved by:

TITLE: Qualification and Requalification Matrix

AUTHOR: _____

Name/Title/Department

_____

Signature/Date

CHECKED BY: _____

Name/Title/Department

_____

Signature/Date

APPROVED BY: _____

Name/Title/Department

_____

Signature/Date

REVISIONS:

No.	Section	Pages	Initials/Date

SOP No.: Val. 1500.10

Effective date: mm/dd/yyyy

Approved by:

SUBJECT: Qualification and Requalification Matrix

PURPOSE

The purpose is to establish the criteria and frequency for requalification (validated process, equipment, facility, systems, utilities, and area) for any significant change or on a periodic basis to maintain the validation status and to ensure the adequacy of performance.

RESPONSIBILITY

It is the responsibility of validation team members to follow the procedures. The quality assurance (QA) manager is responsible for SOP compliance.

PROCEDURE

■ Revalidation is the repetition of the validation process or a specific portion of it.

■ Revalidation may be because of a change that might affect the validated status or on a periodic basis even when no change is believed to have occurred, to demonstrate state of control over the equipment, process, or the system.

■ Change control is a formal monitoring system by which concerned representatives of appropriate departments review proposed or actual changes that might affect the validated status and cause corrective action to be taken that will assure the system retains its validated state of control.

■ Attachment No. 1500.10(A) details the frequency of revalidation and the portion of qualification to be performed on a periodic basis, with responsibilities when there is no significant change.

■ Attachment No. 1500.10(B) describes the examples of significant change and the portion of qualification to be carried out with responsibilities, wherever there is significant change.

■ For change control required and process deviation control, refer to SOPs (provide number) and (provide number), respectively. This may be postponed subject to the management decision as appropriate and justified.

SOP No.: Val. 1500.10   **Effective date: mm/dd/yyyy**
                                             Approved by:

Note: The revalidation guideline is not the final rule and can be altered subject to the priorities and may be postponed subject to the management decision and appropriate justification.

REASONS FOR REVISION

Effective date: mm/dd/yyyy

- First time issued for your company, affiliates, and contract manufacturers.

SOP No.: Val. 1500.10 Effective date: mm/dd/yyyy

Approved by:

Attachment No. 1500.10(A)
REVALIDATION MATRIX

Criteria: <u>Periodic Revalidation:</u> When there is no significant change to affect the validation status as governed by change control

Scope	Revalidation Frequency	Department/Person Responsible	Responsible
Equipment			
Tunnel			
■ Heat Penetration Study for 1-ml ampoules	Once in a year		
■ Heat Penetration Study for 10-ml ampoules	Once in a year		
■ Heat Penetration Study for 15-ml vials	Once in a year		
■ Heat Penetration Study for 30-ml vials (20-ml vial bracketed in-between 10- to 30-ml vials)	Once in a year Once in a year		
Autoclave (Steam Sterilizer)			
■ Heat Penetration Study of ampoule filling machine parts, filtration assembly, and accessories	Once in a year		
■ Hear Penetration Study of 1-ml ampoule filling machine parts, filtration assembly, and accessories	Once in a year		
■ Heat Penetration Study of vial filling machine parts, filtration assembly, and accessories	Once in a year		
■ Heat Penetration Study of uniforms containing 24 bags	Once in a year		
■ Heat Penetration Study of rubber stoppers (lyophilized vials)	Once in a year		

SOP No.: Val. 1500.10   **Effective date: mm/dd/yyyy**

                                             Approved by:

Scope	Comments	Re-Validation	Responsibility
Equipment	a) Extensive maintenance	OQ, PQ	QA + Concerned dept. managers
	b) Shifted (transfer to another area)	IQ, OQ, PQ	QA + Concerned dept. managers
Utilities Qualification	a) Extension of water treatment capacity	IQ, OQ, PQ	QA + Concerned dept. managers
	b) Add an additional loop		QA + Concerned dept. Managers
	c) Change of supplier for filter, resin, activated carbon, RO membrane, etc.		QA + Concerned dept. managers
	d) Sanitization procedure or frequency change		QA + Concerned dept. managers
Cleaning	a) Cleaning agent change	Cleaning Val	QA + Concerned dept. managers
	b) Excipient change in the product		
	c) Procedure change		

SOP No.: Val. 1500.10   Effective date: mm/dd/yyyy

Approved by:

Attachment No. 1500.10(B)
REVALIDATION

Criteria: <u>Significant change affecting the validation status</u>

Frequency: Promptly to preserve validation status. Monitoring through change
control procedure

Scope	Comments	Revalidation	Responsibility
Processes:	a) Formula change (addition, deletion, excess) b) Equipment change c) Procedure/direction change	Process Qualification	QA + Concerned dept. managers
Facility Qualification	a) Any alteration in area specification b) Sanitization procedure or frequency change	Area Qualification	QA + Concerned dept. managers
HVAC Qualification	a) Extension of HVAC capacity. b) Add an additional AHU, chiller, etc. c) Change of supplier for filter, room class, etc. d) Change in air pressures, flow, air changes, etc.	OQ, PQ	QA + Concerned dept. managers

SOP No.: Val. 1500.10

Effective date: mm/dd/yyyy

Approved by:

Scope	Revalidation Frequency	Department/Person Responsible	Responsible
Freeze Dryer			
■ Heat and chilled temperature distribution on shelves	Once in a year		
■ SIP (heat distribution and microbiological challenge)	Once in a year		
Manufacturing Vessels			
■ Heat Penetration Study for 500-liter solution prepration vessel	Once in a year		
■ Heat Penetration Study for 500-liter Mobile Tank-1	Once in a year		
■ Heat Penetration Study for 500 liter Mobile Tank-2	Once in a year		
■ Heat Penetration Study for Pfaudler mobile vessel	Once in a year		
Washing Machine			
■ Vials and ampoules	Once in a year		
Media Fill Simulation			
■ Ampoules filling line (1 ml and 3 ml)	Twice in a year		
■ Vials filling line (including freeze dryer) (10 ml and 30 ml)	Twice in a year		

YOUR COMPANY
VALIDATION STANDARD OPERATING PROCEDURE

SOP No.: Val. 1500.20

Effective date: mm/dd/yyyy

Approved by:

TITLE: Vial/Ampoule Washer Performance Qualification Protocol

AUTHOR: _____

Name/Title/Department

_____

Signature/Date

CHECKED BY: _____

Name/Title/Department

_____

Signature/Date

APPROVED BY: _____

Name/Title/Department

_____

Signature/Date

REVISIONS:

No.	Section	Pages	Initials/Date

SOP No.: Val. 1500.20   Effective date: mm/dd/yyyy

Approved by:

SUBJECT: Vial/Ampoule Washer Performance Qualification Protocol

PURPOSE

The purpose of this protocol is to provide the methodology and acceptance criteria to qualify and requalify the cleaning machine for vials or ampoules.

RESPONSIBILITY

It is the responsibility of validation team members to follow the procedures. The quality assurance (QA) manager is responsible for SOP compliance.

1. Background

The performance qualification of vial/ampoule washer is to be performed on all sizes of vials/ampoules. The tests are to be performed on X ml vial/ampoules size for pyroburden analysis on washed vials/ampoules, particulate reduction challenge test on washed vials/ampoules, and soil test to ensure vial/ampoule washer efficacy.

2. Studies

2.1 Pyroburden of washed components

This test is to assess the efficacy of the vial/ampoule washer in producing washed vials/ampoules that have low pyroburden. The X ml vials/ampoules are to be washed according to the operational parameters described for each vial/ampoule size. The results of endotoxin for each vial/ampoule size are to be entered in Table 1.

SOP No.: Val. 1500.20   **Effective date: mm/dd/yyyy**

Approved by:

Table 1

Pyrogen Test	No. of Vials	Acceptance Criteria	X ml Vials/Ampoules Results
Unwashed vials	1	LT 10.00 EU/ml of rinse water	
Washed vials	1	LT 0.25 EU/ml of rinse water	
	2	LT 0.25 EU/ml of rinse water	
	3	LT 0.25 EU/ml of rinse water	
	4	LT 0.25 EU/ml of rinse water	
	5	LT 0.25 EU/ml of rinse water	
	6	LT 0.25 EU/ml of rinse water	
	7	LT 0.25 EU/ml of rinse water	
	8	LT 0.25 EU/ml of rinse water	

2.1.1 Vial washer operational parameters for vials

Parameter	X ml Vials	X ml Ampoules
DI water pressure	Bar min	Bar min
WFI pressure	Bar min	Bar min
Compressed air pressure	Bar	Bar
Compressed air pressure after regulator	Bar	Bar
Machine speed	/Minute	/Minute
DI water temperature	°C	°C
WFI temperature	°C	°C
DI water rinsing time	Seconds	Seconds
DI water volume	Milliliters per vial	Milliliters per vial
WFI rinsing time	Seconds	Seconds
WFI volume	Milliliters per vial	Milliliters per vial
Study no.		
Applicable SOP		

SOP No.: Val. 1500.20   Effective date: mm/dd/yyyy

Approved by:

2.1.2 Conclusion

The studies are to be performed on vial/ampoule washer using X ml vials/X ml ampoules. The vial/ampoule washer should meet the acceptance criterion for pyrogen, i.e., LT 0.25 EU/ml of rinse water in X ml vials/ampoules after washing.

2.2 Particulate level reduction of spiked vials

This test is to assess the efficacy of the vial/ampoule washer in reducing the particulate level in contaminated vials. Particulate matter to be determined in eight vials/ampoules, each spiked with approximately 500 particles of 40-μm glass beads. A sample is to be taken from each individual needle at the end of vial/ampoule washing at set operational parameter of vial/ampoule washer and analyzed to determine the reduction in particulate level. The X ml vials/ampoules are to be washed according to the operational parameters described for each vial/ampoule size. The results are to be compared with negative and positive controls. The test results for each size of vial/ampoule are to be entered in Table 2.

Table 2

Vial/ Amp. Size	Vial/Amp. No. (Corresponding to Machine Needles)	"S"[a] Spiked Vials/Amp. Results after Washing 40 μm no. of Particles	"U"[b] Unspiked, Unwashed Particulate Level 40 μm no. of Particles	% Reduction	Acceptance Criterion
	1				Particulate level must be reduced by not less than 90%
	2				
	3				
	4				
	5				
	6				
	7				

[a] S = spiked vials
[b] U = unspiked vials

	Vial/Ampoule 1	Vial/Ampoule 2	Average
Negative control: 40 μm			
Positive control: 40 μm			
Disposition: accept/no test/reject			
Performed and analyzed by			

2.2.1 Vial/ampoule washer operational parameters for vials/ampoules

Parameter	X ml	X ml	X ml
DI water pressure	Bar min	Bar min	Bar min
WFI pressure	Bar min	Bar min	Bar min
Compressed air pressure	Bar	Bar	Bar
Compressed air pressure after regulator	Bar	Bar	Bar
Machine speed	/Minute	/Minute	/Minute
DI water temperature	°C	°C	°C
WFI temperature	°C	°C	°C
DI water rinsing time	Seconds	Seconds	Seconds
DI water volume	Milliliters per vial	Milliliters per vial	Milliliters per vial
WFI rinsing time	Seconds	Seconds	Seconds
WFI	Milliliters per vial	Milliliters per vial	Milliliters per vial
Study no.	1	2	3
Applicable SOP			

2.2.2 Conclusion

In the three studies performed using X ml vials/ampoules, the vial/ampoule washer should meet the acceptance criterion of NLT 90% reduction in particulate matter in each spiked vial/ampoule after the completion of washing.

SOP No.: Val. 1500.20

Effective date: mm/dd/yyyy

Approved by:

2.3 Cleaning efficiency of soiled vials

This test is to assess the efficacy of the vial/ampoule washer in cleaning chemically contaminated vials/ampoules. The cleaning efficiency of vial/ampoule washer is to be determined qualitatively by spiking eight individual vials/ampoules with 10% sodium chloride. After drying, the vials/ampoules are to be washed at the set machine operational parameters. The X ml vials/ampoules are to be washed according to the operational parameters described for each vial/ampoule size. The results are to be compared with negative and positive controls. The test results for each size of vial/ampoule are to be entered in Table 3.

Table 3

Vial/Amp. Size	Challenge Vial (Corresponding to Machine Needles)	Results	Acceptance Criterion
	1		All vials spiked with the test soil must be negative
	2		
	3		
	4		
	5		
	6		
	7		
	8		

	Vial/Ampoule 1	Vial/Ampoule 2
Negative control test result		
Positive control test result		
Disposition: accept/no test/reject		
Performed and analyzed by		

2.3.1 Vial/ampoule washer operational parameters for vials/ampoules

Table 4

Parameter	X ml	X ml	X ml
DI water pressure	Bar min	Bar min	Bar min
WFI pressure	Bar min	Bar min	Bar min
Compressed air pressure	Bar	Bar	Bar
Compressed air pressure after regulator	Bar	Bar	Bar
Machine speed	/Minute	/Minute	/Minute
DI water temperature	°C	°C	°C
WFI temperature	°C	°C	°C
DI water rinsing time	Seconds	Seconds	Seconds
DI water volume	Milliliters per vial	Milliliters per vial	Milliliters per vial
WFI rinsing time	Seconds	Seconds	Seconds
WFI	Milliliters per vial	Milliliters per vial	Milliliters per vial
Study no.	1	2	3
Applicable SOP			

2.3.2 Conclusion

In the three studies performed on vial/ampoule washer using X ml size, the vial/ampoule washer should meet soil test acceptance criterion, i.e., negative (free from precipitates).

3. Acceptance Criteria

3.1 Pyroburden analysis

Washed components must contain no more than 0.25 EU/ml of rinse water.

3.2 Particulate removal challenge test

Particulate level must be reduced by not less than 90%.

SOP No.: Val. 1500.20   **Effective date: mm/dd/yyyy**
                                         Approved by:

3.3 Soil test challenge test

All vials spiked with the test soil must test negative.

4. Qualification Status

Qualification completed on: Date_____

5. Requalification Status

Requalification due on: Date_____

6. Deviation

Found   ☐   Not found   ☐
If found, reason: _____
_____

7. Corrective Action

Brief description of corrective action taken: _____
_____
_____

8. Documentation

Qualification/requalification summary report is to be prepared.

REASONS FOR REVISION

Effective date: mm/dd/yyyy

- ■ First time issued for your company, affiliates, and contract manufacturers.

| YOUR COMPANY |
| VALIDATION STANDARD OPERATING PROCEDURE |

SOP No.: Val. 1500.30

Effective date: mm/dd/yyyy

Approved by:

TITLE: Vial/Ampoule Washer Performance
Requalification Protocol

AUTHOR: _____

Name/Title/Department

_____

Signature/Date

CHECKED BY: _____

Name/Title/Department

_____

Signature/Date

APPROVED BY: _____

Name/Title/Department

_____

Signature/Date

REVISIONS:

No.	Section	Pages	Initials/Date

SOP No.: Val. 1500.30

Effective date: mm/dd/yyyy

Approved by:

SUBJECT: Vial/Ampoule Washer Performance Requalification Protocol

PURPOSE

The purpose of this protocol is to provide the methodology and acceptance criteria to requalify the cleaning machine for vials or ampoules.

RESPONSIBILITY

It is the responsibility of validation team members to follow the procedures. The quality assurance (QA) manager is responsible for SOP compliance.

1. Background

The performance requalification of vial washer is to be performed on all sizes of vials/ampoules. The tests are to be performed on a single run using X ml vial or ampoule size for pyroburden analysis on washed vials, particulate reduction challenge test on washed vials, and soil test to ensure vial/ampoule washer efficacy.

Note: The vial washer can also be used for ampoules washing. Therefore, either term can be selected as applicable to your company's operations. The document mostly refers to the term "vials."

2. Studies

2.1 Pyroburden of washed components

This test is to assess the efficacy of the vial/ampoule washer in producing washed vials that have low pyroburden. The X ml vials/ampoules are to be washed according to the operational parameters described for each vial size. The results of endotoxin for each vial size are to be entered in Table 1.

SOP No.: Val. 1500.30 **Effective date: mm/dd/yyyy**

Approved by:

Table 1

Pyrogen Test	No. of Vials	Acceptance Criteria	X ml Vials/Ampoules Results
Unwashed vials	1	LT 10.00 EU/ml of rinse water	
Washed vials	1	LT 0.25 EU/ml of rinse water	
	2	LT 0.25 EU/ml of rinse water	
	3	LT 0.25 EU/ml of rinse water	
	4	LT 0.25 EU/ml of rinse water	
	5	LT 0.25 EU/ml of rinse water	
	6	LT 0.25 EU/ml of rinse water	
	7	LT 0.25 EU/ml of rinse water	
	8	LT 0.25 EU/ml of rinse water	

2.1.1 Vial washer operational parameters for vials

Parameter	X ml Vials	X ml Ampoules
DI water pressure	Bar min	Bar min
WFI pressure	Bar min	Bar min
Compressed air pressure	Bar	Bar
Compressed air pressure after regulator	Bar	Bar
Machine speed	/Minute	/Minute
DI water temperature	°C	°C
WFI temperature	°C	°C
DI water rinsing time	Seconds	Seconds
DI water volume	Milliliters per vial	Milliliters per vial
WFI rinsing time	Seconds	Seconds
WFI	Milliliters per vial	Milliliters per vial
Study no.		
Applicable SOP		

SOP No.: Val. 1500.30

Effective date: mm/dd/yyyy

Approved by:

2.1.2 Conclusion

The single study is to be performed on vial washer using X ml vials/ampoules. The vial washer should meet the acceptance criterion for pyrogen, i.e., LT 0.25 EU/ml of rinse water in X ml vials after washing.

2.2 Particulate level reduction of spiked vials

This test is to assess the efficacy of the vial washer in reducing the particulate level in contaminated vials. Particulate matter to be determined in eight vials, each spiked with approximately 500 particles of 40-μm glass beads. A sample is to be taken from each individual needle at the end of vial washing at set operational parameter of vial washer and analyzed to determine the reduction in particulate level. The X ml vials are to be washed according to the operational parameters described for each vial size. The results are to be compared with negative and positive controls. The test results for each size of vial are to be entered in Table 2.

Table 2

Vial Size	Vial No. (Corresponding to Machine Needles)	"S"[a] Spiked Vial Results after Washing 40 μm no. of Particles	"U"[b] Unspiked, Unwashed Particulate Level 40 μm no. of Particles	% Reduction	Acceptance Criterion
	1				Particulate level must be reduced by not less than 90%
	2				
	3				
	4				
	5				
	6				
	7				
	8				

[a] S = spiked vials
[b] U = unspiked vials

SOP No.: Val. 1500.30   **Effective date: mm/dd/yyyy**

Approved by:

	Vial 1	Vial 2	Average
Negative control: 40 μm			
Positive control: 40 μm			
Disposition: accept/no test/reject			
Performed and analyzed by			

2.2.1 Vial washer operational parameters for vials

Parameter	X ml
DI water pressure	Bar min
WFI pressure	Bar min
Compressed air pressure	Bar
Compressed air pressure after regulator	Bar
Machine speed	/Min
DI water temperature	°C
WFI temperature	°C
DI water rinsing time	Seconds
DI water volume	Milliliters per vial
WFI rinsing time	Seconds
WFI	Milliliters per vial
Study no.	1
Applicable SOP	

2.2.2 Conclusion

In the single study performed using X ml vials, the vial washer should meet the acceptance criterion of NLT 90% reduction in particulate matter in each spiked vial after the completion of washing.

SOP No.: Val. 1500.30   Effective date: mm/dd/yyyy

Approved by:

2.3 Cleaning efficiency of soiled vials

This test is to assess the efficacy of the vial washer in cleaning chemically contaminated vials. The cleaning efficiency of vial washer is to be determined qualitatively by spiking eight individual vials with 10% sodium chloride. After drying, the vials are to be washed at the set machine operational parameters. The X ml vials are to be washed according to the operational parameters described for each vial size. The results are to be compared with negative and positive controls. The test results for each size of vial are to be entered in Table 3.

Table 3

Vial Size	Challenge Vial (Corresponding to Machine Needles)	Results	Acceptance Criterion Soil Challenge Test
	1		All vials spiked with the test soil must be negative
	2		
	3		
	4		
	5		
	6		
	7		
	8		

	Vial 1	Vial 2
Negative control test result		
Positive control test result		
Disposition: accept/no test/reject		
Performed and analyzed by		

2.3.2 Vial/ampoule washer operational parameter for vials

Parameter	X ml
DI water pressure	Bar min
WFI pressure	Bar min
Compressed air pressure	Bar
Compressed air pressure after regulator	Bar
Machine speed	/Minute
DI water temperature	°C
WFI temperature	°C
DI water rinsing time	Seconds
DI water volume	Milliliters per vial
WFI rinsing time	Seconds
WFI	Milliliters per vial
Study no.	1
Applicable SOP	

2.3.3 Conclusion

In the single study performed on vial washer using X ml size, the vial washer should meet soil test acceptance criterion, i.e., negative (free from precipitates).

3. Acceptance Criteria

3.1. Pyroburden Analysis

Washed components must contain no more than 0.25 EU/ml of rinse water.

3.2 Particulate removal challenge test

Particulate level must be reduced by not less than 90%.

3.3 Soil test challenge test

All vials spiked with the test soil must test negative.

SOP No.: Val. 1500.30

Effective date: mm/dd/yyyy

Approved by:

4. Qualification Status

Qualification completed on: Date_____

5. Requalification Status

Requalification due on: Date_____

6. Deviation

Found ☐   Not found ☐

If found, reason: _____

_____

7. Corrective Action

Brief description of corrective action taken: _____

_____

_____

8. Documentation

Qualification/requalification summary report to be prepared.

REASONS FOR REVISION

Effective date: mm/dd/yyyy

- First time issued for your company, affiliates, and contract manufacturers.

YOUR COMPANY
VALIDATION STANDARD OPERATING PROCEDURE

SOP No.: Val. 1500.40

Effective date: mm/dd/yyyy

Approved by:

TITLE: Depyrogenation Tunnel Performance Qualification Protocol

AUTHOR:

_____

Name/Title/Department

_____

Signature/Date

CHECKED BY:

_____

Name/Title/Department

_____

Signature/Date

APPROVED BY:

_____

Name/Title/Department

_____

Signature/Date

REVISIONS:

No.	Section	Pages	Initials/Date

SOP No.: Val. 1500.40   **Effective date: mm/dd/yyyy**

                                         Approved by:

SUBJECT: Depyrogenation Tunnel Performance Qualification Protocol

PURPOSE

The purpose is to describe the format and contents of the depyrogenation tunnel performance qualification protocol.

RESPONSIBILITY

It is the responsibility of validation team members to follow the procedures. The quality assurance (QA) manager is responsible for SOP compliance.

PROCEDURE

1. Document Description

1.1 Responsibility

(Name): _____

1.2 Objective

To establish the confidence that fully automatic hot air sterilizing and depyrogenation tunnel and auxiliary systems perform in accordance with the manufacturer specifications and with GMP principles, and are capable of operating within established limits and tolerances.

1.3. Scope

The performance qualification is applicable to:
Equipment name: fully automatic hot air sterilizing tunnel
All concerned equipment (list):

- Revolving vane anemometer for indicating the air speed
- Connections for particle counter
- Operating hours counter
- Night reduction: the (LF) fans continued operation at a reduced speed capacity when the tunnel is switched off

Effective date: mm/dd/yyyy

Approved by:

- Automatic timer for switching on the heating so that the tunnel heating turns on at a required time
- Cross flow control for automatic compensation of pressure fluctuations between the sterile and washing areas
- Sliding plate adjustment to raise the separating "baffle plates" between the individual zones to a required height
- Filter monitoring system to indicate when a filter has reached its maximum degree of contamination

Others:

_____

_____

Written by

(Validation)

Signature: _____ Date: _____

Verified by

(Validation engineer)

Signature: _____ Date: _____

2. Standard Operating Procedure

2.1 Responsibility

(Name) validation: _____

Verify that following approved standard operating procedures are available at the user's premises:

Description	Result
Equipment Operation SOP's	_____
Equipment cleaning SOPs/equipment sanitization SOPs	_____
Preventive maintenance SOPs	_____
Calibration SOPs	_____
Training SOPs	_____

Checks performed by: _____ Date: _____

SOP No.: Val. 1500.40   **Effective date: mm/dd/yyyy**

       Approved by:

2.2 Conclusion

Acceptance criteria met   ☐
Acceptance criteria not met   ☐

2.3 Deviation

None ☐
Found: _____

2.4 Comments

None ☐
Found: _____

2.5 Verified by

(Validation officer)

Verified for the completeness of the contents.

Signature: _____ Date: _____

3. Training

3.1 Responsibility (name)

Validation engineer: _____

Production manager: _____

Calibration officer: _____

3.2 Objective

Verify that employees involved for equipment operation, maintenance, calibration, and cleaning have received adequate training.

Description   *Result*

For equipment operation:

Name: _____   _____

Job title: _____

SOP No.: Val. 1500.40   **Effective date: mm/dd/yyyy**
                                   Approved by:

Name: _____   _____

Job title: _____

For equipment cleaning/equipment sanitization:

Name: _____   _____

Job title: _____

For calibration:

Name: _____   _____

Job title: _____

For equipment maintenance:

Name: _____   _____

Job title: _____

Checks performed by: _____ Date: _____

3.3 Conclusion

Acceptance criteria met   ☐
Acceptance criteria not met   ☐

3.4 Deviation

None   ☐
Found: _____

3.5 Comments

None   ☐
If any, reason: _____

3.6 Verified by

(Validation officer)

Verified for the completeness of the contents.

Signature: _____ Date: _____

SOP No.: Val. 1500.40

Effective date: mm/dd/yyyy
Approved by:

4. Process Performance Qualification Checks

4.1 Responsibility (name)

Validation engineer: _____

Validation officer: _____

Operations manager: _____

4.2 Performance checks

Verify the performance of the hot sterilization tunnel during sterilization process.

Note 1: Separate study should be performed for different loads/sizes.
Note 2: Heat penetration and pyrogen test should be performed at beginning, middle, and end of the operations.

4.2.1 Description

Empty tunnel heat distribution

4.2.2 Test objective

The objective of the empty tunnel heat distribution run(s) will be

- To evaluate the heating characteristics of the sterilizer, product carrier system, and the sterilization medium employed
- To determine the temperature profile in the heating zone and to check the temperature uniformity

SOP no.: _____

Equipment used: portable Kaye validator

Equipment SOP no.: _____

Calibration valid up to: _____

4.2.3 Acceptance criteria

The distribution run(s) must meet the time and temperature requirements of the corresponding specifications. All function initiations required during the operating modes must have occurred as specified.

SOP No.: Val. 1500.40

Effective date: mm/dd/yyyy

Approved by:

- $\Delta T < 10°C$
- $\Delta T = T_{MAX} - T_{MIN}$

Results

Set Parameters		Actual Results		
Tests	Requirements	Run 1	Run 2	Run 3
Set temperature				
Base temperature				
Machine speed				

$\Delta T < 10°C$ from the set temperature.

Actual Results		
Run 1	Run 2	Run 3

Conclusion.

Checks performed by: _____ Date: _____

4.3 Heat penetration

4.3.1 Test objective

- To evaluate the heating characteristics of items within the tunnel when subjected to the sterilization medium
- To determine coolest points for each load configuration/size
- To evaluate the relative heating characteristics of items and reference thermocouples where applicable
- To establish production work order sterilization parameters

SOP no.:   _____

Equipment used: <u>portable Kaye validator or equivalent</u>

Equipment SOP no.:   _____

Calibration valid up to: <u>mm/dd/yyyy</u>

SOP No.: Val. 1500.40   **Effective date: mm/dd/yyyy**

                                             Approved by:

4.3.2 Acceptance criteria

- All heat penetration data collected during each run must meet the requirement for the corresponding specification.
- The production operating ranges/windows established from the heat penetration runs must assure all products in the test runs will meet the calculated requirement for the corresponding specification.
- If a satisfactory operating range is not established using minimum and maximum loading parameters, intermediate-loading conditions must be tested.
- Where tunnel peak dwell temperature and time are to be used for routine production cycle control, or as back-up control, correlation of sterilizer peak dwell time and temperature with the hottest and coldest profile container must be shown for each run, where applicable.
- FD > 1000 minutes at all locations and runs (Z = ____°C and ref. T ____°C). FD values of each load must show that the average of the lowest FD values is more than 1000 minutes at a confidence level of 95%.

Results

Load	Set Parameters		Actual Results			Test Performed on
	Tests	Requirements	Run 1	Run 2	Run 3	
A ml vial	Set temperature					
	Base temperature					
	Machine speed					
	Min FD value	NLT 6 min				
B ml vial	Set temperature					
	Base temperature					
	Machine speed					
	Min FD value	NLT 6 min				

Conclusion.

Checks performed by: _____ Date: _____

4.4 Componentry microbial or pyrogen challenge

4.4.1 Test objective

To confirm the biological relationship between parametrically determined process lethalities by demonstrating the ability of the sterilizer to effectively reduce the challenge material to an acceptable level

4.4.2 Acceptance criteria

- A minimum pyrogen challenge must be equal to or greater than four log reductions for each run.
- At least 15% of the required functional container time and temperature values must show subminimal process conditions per run.
- Cold zone temperature correction must be used where applicable.

Results

Load	Actual Results			Test Performed on
	Run 1	Run 2	Run 3	
A ml vial				
B ml vial				

4.5 Conclusion

Acceptance criteria met ☐
Acceptance criteria not met ☐

4.6 Deviation

None ☐
Found: _____

4.7 Comments

None ☐
If any, reason: _____

SOP No.: Val. 1500.40   **Effective date: mm/dd/yyyy**
                                         Approved by:

4.8 Verified by:

(Validation officer)

Verified for the completeness of the contents.

Signature: _____ Date: _____

REASONS FOR REVISION

Effective date: mm/dd/yyyy

- First time issued for your company, affiliates, and contract manufacturers.

| YOUR COMPANY |
| VALIDATION STANDARD OPERATING PROCEDURE |

SOP No.: Val. 1500.50   Effective date: mm/dd/yyyy

Approved by:

TITLE: **Depyrogenation Tunnel Performance Requalification Protocol**

AUTHOR: _____

Name/Title/Department

_____

Signature/Date

CHECKED BY: _____

Name/Title/Department

_____

Signature/Date

APPROVED BY: _____

Name/Title/Department

_____

Signature/Date

REVISIONS:

No.	Section	Pages	Initials/Date

SOP No.: Val. 1500.50   Effective date: mm/dd/yyyy

                                        Approved by:

SUBJECT: Depyrogenation Tunnel Performance Requalification Protocol

PURPOSE

The purpose is to describe the format and contents of the depyrogenation tunnel performance requalification protocol.

RESPONSIBILITY

It is the responsibility of validation team members to follow the procedures. The quality assurance (QA) manager is responsible for SOP compliance.

PROCEDURE

1. Purpose

The purpose is to requalify the sterilization and depyrogenation cycle of the A-ml and B-ml molded clear glass vials in hot sterilization tunnel by one heat penetration study and endotoxin challenge test.

2. Background

To ensure sterility during the aseptic processing, it was decided to revalidate the A-ml and B-ml vials' sterilization and depyrogenation cycle by one heat penetration study and endotoxin challenge test once in a year.

3. Location

Injectable building, room no.: ＿＿＿＿＿＿＿＿＿＿

4. Test Instruments Required

4.1   Anemometer

Serial no.: ＿＿＿＿＿＿＿＿＿＿
Calibrated on: ＿＿＿＿＿＿＿＿＿＿
Calibration due on: ＿＿＿＿＿＿＿＿＿＿

4.2 Laser particulate counter

Serial no.: _____
Calibrated on: _____
Calibration due on: _____

4.3 HTR-400

Serial no.: _____
Calibrated on: _____
Calibration due on: _____

4.4 Portable Kaye validator

Serial no.: _____
Calibrated on: _____
Calibration due on: _____

5. Procedure

5.1 Air velocity of the tunnel HEPA filters

The purpose is to demonstrate that the air speed is homogeneous in each section of the tunnel (preheating, heating, and cooling zones). The air speed values and homogeneity are important for uniform heating (sterilization) and uniform cooling of glass containers. Refer to plant SOP for checking and limit.

5.1.1 Acceptance criteria

- Air velocity should be 0.50 ± 0.1 m/s (preheating and cooling zones)
- Air velocity should be 0.90 ± 0.1 m/s (sterilization zone)

5.2 Tunnel particle counts

The purpose is to demonstrate that the air quality is in compliance with a class 100 air on all the surfaces of the conveyor (preheating, heating, and cooling zones). Use plant SOP for monitoring.

SOP No.: Val. 1500.50

Effective date: mm/dd/yyyy

Approved by:

5.2.1 Acceptance criteria

Particulate count — all locations with particle counts:

- ≤100 particles/ft^3 for particle size of ≥0.5 μm
- 0 for size 5.0 μm

5.3 Heat penetration

Heat penetration is the most critical component of the entire validation process.

The heat penetration study will be performed per plant SOP.

In each study, 15 thermocouples and 15 endotoxin vials will be distributed in the tunnel at different positions identified in the following sketch.

Endotoxin Vials (by Number) and Thermocouples (by CH) Location Identification

	11 (CH-11)	6 (CH-6)	1 (CH-1)	
Washing room	12 (CH-12)	7 (CH-7)	2 (CH-2)	Sterile room
	13 (CH-13)	8 (CH-8)	3 (CH-3)	
	14 (CH-14)	9 (CH-9)	4 (CH-4)	
	15 (CH-15)	10 (CH-10)	5 (CH-5)	

Tunnel direction ⟩

The F_D value will be calculated based on the temperature recorded by the thermocouples inside the vials.

Precalibration and postcalibration should be performed for all thermocouples before and after the study, respectively. The calibration range should cover the operating temperature of the tunnel.

5.3.1 Acceptance criteria

- F_D value NLT 6 minutes
- Minimum four-log reduction of endotoxin

5.4 Tunnel speed

Tunnel speed should be recorded at the following conditions:

- Beginning of the production: when tunnel speed will be controlled by the washing machine in terms of % age (theoretical speed) and vials per minute
- End of the production: when washing will disconnect from the line. The maximum tunnel speed should be noted in terms of mm/min.

The validated tunnel speed should be stipulated in the manufacturing protocol for 20-ml vial depyrogenation.

6. Operating Condition of Tunnel

- Set temperature: 330°C
- Base temperature: 250°C
- Z-value: 54°C

7. Documentation

- Thermocouples calibration report before test
- Qualification report by Kaye validator
- Thermocouples calibration check report after test
- Microbiological test report
- Tunnel chart (speed and temperature)
- Air velocity check report
- Particulate count report
- Validation summary report

REASONS FOR REVISION

Effective date: mm/dd/yyyy

- First time issued for your company, affiliates, and contract manufacturers.

| YOUR COMPANY |
| VALIDATION STANDARD OPERATING PROCEDURE |

SOP No.: Val. 1500.60

Effective date: mm/dd/yyyy

Approved by:

TITLE: SIP Cycle for Holding Vessel Requalification Protocol

AUTHOR:
_____

Name/Title/Department

_____

Signature/Date

CHECKED BY:
_____

Name/Title/Department

_____

Signature/Date

APPROVED BY:
_____

Name/Title/Department

_____

Signature/Date

REVISIONS:

No.	Section	Pages	Initials/Date

SOP No.: Val. 1500.60   **Effective date: mm/dd/yyyy**

Approved by:

SUBJECT: SIP Cycle for Holding Vessel Requalification Protocol

PURPOSE

The purpose is to describe the format and contents of the SIP cycle for holding vessel requalification protocol.

RESPONSIBILITY

It is the responsibility of validation team members to follow the procedures. The quality assurance (QA) manager is responsible for SOP compliance.

PROCEDURE

1. Purpose

The purpose is to requalify the mobile vessel (serial no. _____) by one heat penetration study and microbial challenge test using *Bacillus stearo-thermophilus* strip.

2. Background

The sterile solutions are filled in different sizes of ampoules/vials under aseptic conditions. For this purpose, all operations and conditions require validation according to the approved protocols and subsequent requalification on an annual basis by one heat penetration study and microbial challenge test using a *B. stearothermophilus* strip.

3. Location

Injectable building, solution room

4. Test Instruments

- Portable Kaye validator
  - Serial no.: _____
  - Calibrated on: _____
  - Calibration due on: _____

SOP No.: Val. 1500.60

Effective date: mm/dd/yyyy
Approved by:

- Validator parameters used for calculation: _____
- Base temperature: _____°C
- Z-value: _____°C
- LTR-140
  - Serial no.: _____
  - Calibrated on: _____
  - Calibration due on: _____
- Microbiological challenge
  - *B. stearothermophilus* strip (10^6 population)

5. Process Parameters

- Set steam pressure: _____ bar
- Exposure temperature: not less than _____°C
- Exposure time: not less than _____ minutes

6. Procedure/Acceptance Criteria

6.1 Heat penetration

- Heat penetration study will be performed per plant SOP.
- Sterilization of the vessel (including dip tube) will be performed per plant SOP.
- Seven thermocouples will be positioned in different locations of the vessel along with one *B. stearothermophilus* strip (population 10^6) for microbiological challenge test (for thermocouples' location, refer to attached layout).
- The F_0 value will be calculated based on the base temperature (_____°C) and recorded by the thermocouples inside the vessel.
- Thermocouples should be calibrated before performing the heat penetration study and the calibration should be checked again after completion of the study.

6.2 Acceptance criteria

- Steam pressure should be _____ bar during the sterilization.
- Exposure temperature not be less than _____°C.
- Exposure time should not be less than _____ minutes.

▪ Minimum (F_0) lethality should be NLT 20.
▪ Minimum spore log reduction should be six.

7. Documentation

▪ Thermocouple calibration report before test
▪ Qualification report by Kaye validator
▪ Thermocouple calibration check report after test
▪ Microbiological test report
▪ *B. stearothermophilus* supplier certificate
▪ Sterilization charts
▪ Validation summary report

REASONS FOR REVISION

Effective date: mm/dd/yyyy

▪ First time issued for your company, affiliates, and contract manufacturers.

SOP No.: Val. 1500.60

Effective date: mm/dd/yyyy

Approved by:

Mobile Vessel
Thermocouples Location

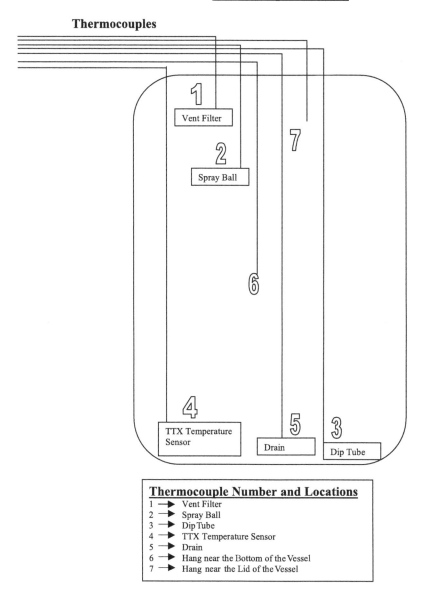

Figure 1   Mobile Vessel Thermocouple Location

YOUR COMPANY
VALIDATION STANDARD OPERATING PROCEDURE

SOP No.: Val. 1500.70

Effective date: mm/dd/yyyy

Approved by:

TITLE: Stopper Sterilization Performance
               Qualification Protocol

AUTHOR: _____

Name/Title/Department

_____

Signature/Date

CHECKED BY: _____

Name/Title/Department

_____

Signature/Date

APPROVED BY: _____

Name/Title/Department

_____

Signature/Date

REVISIONS:

No.	Section	Pages	Initials/Date

SOP No.: Val. 1500.70

Effective date: mm/dd/yyyy

Approved by:

SUBJECT: Stopper Sterilization Performance Qualification Protocol

PURPOSE

The purpose is to describe the format and contents of the stopper sterilization performance qualification protocol.

RESPONSIBILITY

It is the responsibility of validation team members to follow the procedures. The quality assurance (QA) manager is responsible for SOP compliance.

PROCEDURE

1. Purpose

The purpose is to qualify the sterilization cycle of _____-mm rubber stoppers in steam sterilizer (model _____, serial no. _____) by three heat penetration studies and microbial challenge test using *B. stearothermophilus* strip.

2. Background

To ensure sterility during the aseptic processing, it was decided to validate A-mm stopper sterilization cycle by three heat penetration studies and a *B. stearothermophilus* strip considering maximum load size.

3. Procedure

3.1 Location

Injectable building, sterilization room

3.2 Test instruments

- Portable Kaye validator
  - Serial no.: _____
  - Calibrated on: _____
  - Calibration due on: _____

SOP No.: Val. 1500.70   **Effective date: mm/dd/yyyy**
                                             Approved by:

- LTR-140
  - Serial no.: _____
  - Calibrated on: _____
  - Calibration due on: _____

3.3 Heat penetration

Heat penetration is the most critical component of the entire validation process. Heat penetration studies will be performed per plant SOP. Using four shelves of the trolley, distribute the load per following sequence:

- Shelf 1   Eight bags of stoppers
- Shelf 2   Eight bags of stoppers
- Shelf 3   Eight bags of stoppers
- Shelf 4   Eight bags of stoppers

A *B. stearothermophilus* strip and 16 thermocouples will be inserted in bags that are placed in different positions in the chamber. The F_0 value will be calculated based on the base temperature and recorded by the thermocouples inside the commodity. Thermocouples should be calibrated before performing the heat penetration study and the calibration should be checked again after completion of the study.

3.3.1 Stopper load

- Load quantity = _____ stoppers (Tyvek/PE bags)
- No. of bags = _____ bags
- Each bag contains = _____ stoppers (bromobutyle, gray color)

3.4 Acceptance criteria

- Minimum lethality should be NLT _____.
- Cycle parameters should be followed according to plant SOP.
- Minimum spore log reduction should be six.

3.4.1 Operating condition of sterilizer:

- Set temperature: _____°C
- Base temperature: _____°C
- Z-value: _____
- Exposure time: _____ minutes

SOP No.: Val. 1500.70   Effective date: mm/dd/yyyy
                                         Approved by:

4. Documentation

- Thermocouple calibration report before test
- Qualification report by Kaye validator
- Thermocouple calibration check report after test
- Microbiological test report
- Sterilization charts
- Validation summary report

REASONS FOR REVISION

Effective date: mm/dd/yyyy

- First time issued for your company, affiliates, and contract manufacturers.

**YOUR COMPANY
VALIDATION STANDARD OPERATING PROCEDURE**

SOP No.: Val. 1500.80

Effective date: mm/dd/yyyy

Approved by:

TITLE: **Stopper Sterilization Requalification Protocol**

AUTHOR: _____
                         Name/Title/Department

   _____
                         Signature/Date

CHECKED BY: _____
                         Name/Title/Department

   _____
                         Signature/Date

APPROVED BY: _____
                         Name/Title/Department

   _____
                         Signature/Date

REVISIONS:

No.	Section	Pages	Initials/Date

SOP No.: Val. 1500.80

Effective date: mm/dd/yyyy
Approved by:

SUBJECT: Stopper Sterilization Requalification Protocol

PURPOSE

The purpose is to describe the format and contents of the stopper sterilization requalification protocol.

RESPONSIBILITY

It is the responsibility of validation team members to follow the procedures. The quality assurance (QA) manager is responsible for SOP compliance.

PROCEDURE

1. Purpose

The purpose is to requalify the sterilization cycle of _____-mm rubber stoppers in steam sterilizer (model _____, serial no. _____) by two heat penetration studies and microbial challenge test using *B. stearothermophilus* strip. If cycle study fails once, proceed as per initial performance qualification protocols.

2. Background

To ensure sterility during the aseptic processing, it was decided to revalidate A-mm stopper sterilization cycle once in a year by one heat penetration study and a *B. stearothermophilus* strip considering maximum load size.

3. Procedure

3.1 Location

Injectable building, sterilization room

3.2 Test instruments

- Portable Kaye validator
  - Serial no.: _____
  - Calibrated on: _____
  - Calibration due on: _____

SOP No.: Val. 1500.80

Effective date: mm/dd/yyyy
Approved by:

- LTR-140
  - Serial no.: _____
  - Calibrated on: _____
  - Calibration due on: _____

3.3 Heat penetration

Heat penetration is the most critical component of the entire validation process. A heat penetration study will be performed per plant SOP. Using four shelves of the trolley, distribute the load per following sequence:

- Shelf 1   Eight bags of stoppers
- Shelf 2   Eight bags of stoppers
- Shelf 3   Eight bags of stoppers
- Shelf 4   Eight bags of stoppers

A *B. stearothermophilus* strip and 16 thermocouples will be inserted in bags that are placed in different positions in the chamber. The F_0 value will be calculated based on the base temperature and recorded by the thermocouples inside the commodity. Thermocouples should be calibrated before performing the heat penetration study and the calibration should be checked again after completion of the study.

3.3.1 Stopper load

- Load quantity = _____ stoppers (Tyvek/PE bags)
- No. of bags = _____ bags
- Each bag contains = _____ stoppers (bromobutyle, gray color)

3.4 Acceptance criteria

- Minimum lethality should be NLT _____.
- Cycle parameters should be followed according to plant SOP.
- Minimum spore log reduction should be six.

3.4.1 Operating condition of sterilizer:

- Set temperature: _____°C
- Base temperature: _____°C
- Z-value: _____
- Exposure time: _____ minutes

4. Documentation

- Thermocouple calibration report before test
- Qualification report by Kaye validator
- Thermocouple calibration check report after test
- Microbiological test report
- Sterilization charts
- Validation summary report

REASONS FOR REVISION

Effective date: mm/dd/yyyy

- First time issued for your company, affiliates, and contract manufacturers.

YOUR COMPANY
VALIDATION STANDARD OPERATING PROCEDURE

SOP No.: Val. 1500.90

Effective date: mm/dd/yyyy

Approved by:

TITLE: **Equipment Sterilization Performance Qualification Protocol**

AUTHOR: _____

Name/Title/Department

_____

Signature/Date

CHECKED BY: _____

Name/Title/Department

_____

Signature/Date

APPROVED BY: _____

Name/Title/Department

_____

Signature/Date

REVISIONS:

No.	Section	Pages	Initials/Date

SOP No.: Val. 1500.90   Effective date: mm/dd/yyyy
                                         Approved by:

SUBJECT: Equipment Sterilization Performance Qualification Protocol

PURPOSE

The purpose is to describe the format and contents of the equipment sterilization performance qualification protocol.

RESPONSIBILITY

It is the responsibility of validation team members to follow the procedures. The quality assurance (QA) manager is responsible for SOP compliance.

PROCEDURE

1. Purpose

The purpose is to qualify the sterilization cycle of the A-ml vial filling machine parts, filtration assembly, and hose pipes in steam sterilizer (model _____, serial no. _____) by three heat penetration studies and a microbial challenge test using *B. stererothermophilus* strip.

2. Background

The sterile solutions are filled in different sizes of ampoules/vials under aseptic conditions. For this purpose, all operations and conditions should be validated according to the approved protocols to ensure the sterility during the aseptic processing to validate the sterilization cycle for A-ml vial filling machine parts and accessories by three heat penetration studies and a microbial challenge test using *B. stererothermophilus* strip.

3. Location

Injectable building, sterilization room no.: _____

4. Test Instruments Used

- ■ Portable Kaye validator
  - ■ Serial no.: _____
  - ■ Calibrated on: _____
  - ■ Calibration due on: _____

SOP No.: Val. 1500.90

Effective date: mm/dd/yyyy

Approved by:

- LTR-140
  - Serial no.: _____
  - Calibrated on: _____
  - Calibration due on: _____

5. Process Parameters

- Sterilization temperature: _____°C
- Exposure time: _____ minutes

6. Procedure/Acceptance Criteria/Results/Conclusion

6.1 Procedure for heat penetration study

Heat penetration is the most critical component of the entire validation process. Heat penetration studies will be performed per plant SOP. Using three shelves of the trolley, distribute the load per following sequence:

- Shelf 1
  Buffer tank (amber Duran bottle) with buffer tank manifold
  Vent filter
  _____ liter Sartorius pressure vessel
  _____ liter Sartorius pressure vessel
  Kleen pack filter (_____ no.)
  Stainless steel container (_____ no.)
- Shelf 2
  Filling needles with silicone hoses
  Spare needles
  Dosing pumps (_____ nos.) disengaged
- Shelf 3
  Before filling gassing assembly:
    Gassing needles
    Silicone hoses
    Manifolds
    Upstream silicon hoses
    Downstream silicon hoses
    Disc filter
  After filling gassing assembly:
    Gassing needles
    Silicone hoses

SOP No.: Val. 1500.90   Effective date: mm/dd/yyyy

Approved by:

Manifolds
Upstream silicon hoses
Downstream silicon hoses
Disc filter
Pump hoses
Triclamps, gaskets, clamps, and forceps

A *B. stearothermophilus* strip and 16 thermocouples will be inserted in bags/items placed in different positions in the chamber. The F_0 value will be calculated based on the base temperature and recorded by the thermocouples inside the commodity. Thermocouples should be calibrated before performing the heat penetration study and the calibration should be checked again after completion of the study.

6.2 Acceptance criteria

- Minimum lethality should be NLT 20 minutes.
- Cycle parameters should be followed according to plant SOP.
- Minimum spore log reduction should be six.

7. Documentation

- Thermocouple calibration report before test
- Qualification report by Kaye validator
- Thermocouple calibration check report after test
- Microbiological test report
- Sterilization charts

REASONS FOR REVISION

Effective date: mm/dd/yyyy

- First time issued for your company, affiliates, and contract manufacturers.

| YOUR COMPANY |
| VALIDATION STANDARD OPERATING PROCEDURE |

SOP No.: Val. 1500.100

Effective date: mm/dd/yyyy
Approved by:

TITLE: Lyophilizer Performance Qualification Protocol

AUTHOR: _____
Name/Title/Department

_____
Signature/Date

CHECKED BY: _____
Name/Title/Department

_____
Signature/Date

APPROVED BY: _____
Name/Title/Department

_____
Signature/Date

REVISIONS:

No.	Section	Pages	Initials/Date

SOP No.: Val. 1500.100

Effective date: mm/dd/yyyy
Approved by:

SUBJECT: Lyophilizer Performance Qualification Protocol

PURPOSE

The purpose is to describe the format and contents of the lyophilizer performance qualification protocol.

RESPONSIBILITY

It is the responsibility of validation team members to follow the procedures. The quality assurance (QA) manager is responsible for SOP compliance.

PROCEDURE

1. Document Description

1.1 Responsibility

(Name) validation officer: _____

1.2 Objective

The objective is to ensure that the freeze drying unit will operate and produce product quality results under load conditions. This will give a representation of long-term performance, verification of vendor guarantees for availability and failure rates, and a better debugged and more reliable system.

1.3 Scope

The performance qualification protocols are applicable to:

- Equipment name: freeze drying unit — model _____
- All concerned equipment (list):
  - Drying chamber with product shelf assembly
  - Ice condenser
  - Heat transfer circuit
  - Refrigeration system
  - Vacuum pump set
  - Venting system

SOP No.: Val. 1500.100   Effective date: mm/dd/yyyy
                                              Approved by:

- Chamber clean-in-place (CIP) system
- Steam sterilization
- Electrical equipment with Falco control system
- Operating and safety devices

Others:

_____
_____
_____
_____

1.4 Written by

(Validation officer)

Signature: _____ Date: _____

1.5 Verified by

(Validation engineer)

Signature: _____ Date: _____

2. Standard Operating Procedures

2.1 Responsibility

(Name) validation officer: _____

2.2 Objective

Verify that the following approved standard operating procedures are available at the user's premises.

Description   *Result*

Equipment operation SOPs

_____   _____

Equipment cleaning SOPs/
Equipment sanitization SOPs

_____   _____

SOP No.: Val. 1500.100

Effective date: mm/dd/yyyy

Approved by:

Preventive maintenance SOPs

_____   _____

Calibration SOPs

_____   _____

Training SOPs

_____   _____

Other SOPs

_____   _____

Checks performed by: _____ Date: _____

2.3 Conclusion

Acceptance criteria met ☐
Acceptance criteria not met ☐

2.4 Deviation

None ☐
Found: _____

2.5 Comments

None ☐
If any: _____

2.6 Verified by

(Validation)

Verified for the completeness of the contents.

Signature: _____ Date: _____

3. Training

3.1 Responsibility

(Name) validation officer: _____

3.2 Objective

Verify that all people involved with equipment have received adequate training on the operation of the system.

Description   *Result*

For equipment operation

Name: _____

Job title: _____

For equipment cleaning/equipment sanitization

Name: _____

Job title: _____

For calibration

Name: _____

Job title: _____

For equipment maintenance

Name: _____

Job title: _____

Name: _____

Job title: _____

Checks performed by: _____ Date: _____

3.3 Conclusion

Acceptance criteria met   ☐
Acceptance criteria not met ☐

3.4 Deviation

None   ☐
Found: _____

3.5 Comments

None   ☐
If any: _____

3.6 Verified by

(Validation officer)

Verified for the completeness of the contents.

Signature: _____ Date: _____

4. System Recovery Procedures

4.1 Responsibility

Name (validation engineer): _____

4.2 Objective

Verify the following system recovery procedures:

Description	Status
Periodic backups and archival SOPs	_____
Data restoring SOPs	_____
Programs restoring SOPs	_____
Disaster recovery SOPs	_____

Checks performed by: _____ Date: _____

4.3 Conclusion

Acceptance criteria met ☐
Acceptance criteria not met ☐

4.4 Deviation

None ☐
Found: _____

4.5 Comments

None ☐
If any: _____

4.6 Verified by

(Validation)

Verified for the completeness of the contents.

Signature: _____ Date: _____

5. Performance Checks

5.1 Responsibility

(Name) validation engineer: _____

5.2 Sterilization of chamber and condenser

5.2.1 Heat distribution

Test objective.

Perform three heat distribution studies.
   The objective of the empty distribution run(s) shall be to evaluate:

- The heating characteristics of the chamber and condenser sterilization
- The ability of the chamber and condenser to hold the required sterilization pressures without leaks
- The ability of the sterilization cycle control mechanisms to operate as intended

Acceptance criteria.

(a) The heat distribution run(s) must meet the time, temperature, and pressure requirements. T_{MIN} = not less than 121.1°C.
(b) $T_{MAX} - T_{MIN} < 1°C$ during stabilized dwell period.

Result.

- SOP no.: _____
- Equipment used: ___ portable Kaye validator _____
- Equipment SOP no.: _____
- Calibration valid up to: _____

SOP No.: Val. 1500.100

Effective date: mm/dd/yyyy

Approved by:

(a)

Test	Requirement	Actual Results		
		Run 1	*Run 2*	*Run 3*
Time	____ minutes			
Temperature	____°C			
Minimum lethality	LT ____			

(b) Results for study 1: _____
Results for study 2: _____
Results for study 3: _____

Conclusion.

Checks performed by: _____ Date: _____

5.2.2 Microbiological challenge test

Test objective.

Perform microbiological challenge studies to determine the degree of process lethality provided by the sterilization cycle. The microorganisms most frequently utilized to challenge steam sterilizer cycles are *Bacillus stearothermophilus* and ATCC 7953. The Kaye validator equipped with 12 (minimum) thermocouples and biological indicators (10^6) shall be positioned in the detected cool points of the chamber and condenser. After the sterilization cycle is complete, the B.I's are recovered and subjected to microbiological test procedures.

Acceptance criteria.

A minimum spore log reduction (SLR) of equal to or greater than six must be shown for each run.

Result.

Actual Results		
Run 1	*Run 2*	*Run 3*

Conclusion.

Checks performed by: _____ Date: _____

5.3 Shelf temperature distribution test

Test.

Five thermocouples are distributed over shelf no. 6. Two thermocouples are placed on the other shelves. The measurement is affected at two temperatures, −40 and +40°C, while the system is in static condition, i.e., approximately 2 hours after the temperature has been reached.

Acceptance criteria.

- Temperature deviation from average value per shelf ± 1°C
- Temperature deviation from average value within the shelf assembly ±2°C

TC-location on the shelves

Left Rear	LR	RR	Right Rear
		C (Center)	
Left Front	LF	RF	Right Front

Results.

Heat distribution studies were conducted according to approved protocols; actual results found during heat distribution studies are:

SOP No.: Val. 1500.100

Effective date: mm/dd/yyyy

Approved by:

Cycle No.	−40°C Set Point Temperature		+40°C Set Point Temperature		Performed on
	Maximum Temperature Deviation from Average Value of Each Shelf	Maximum Temperature Deviation from Average Value within All Shelves	Maximum Temperature Deviation from Average Value of Each Shelf	Maximum Temperature Deviation from Average Value within All Shelves	
Run 1					
Run 2					
Run 3					

Detail reports are attached.

Conclusion.

Checks performed by: _____ Date: _____

5.4 Conclusion

Acceptance criteria met   □
Acceptance criteria not met   □

5.5 Deviation

None   □
Found: _____

5.6 Comments

None   □
If any: _____

5.7 Verified by

(Validation officer)

Verified for the completeness of the contents.

Signature: _____ Date: _____

SOP No.: Val. 1500.100

Effective date: mm/dd/yyyy

Approved by:

REASONS FOR REVISION

Effective date: mm/dd/yyyy

- First time issued for your company, affiliates, and contract manufacturers.

YOUR COMPANY
VALIDATION STANDARD OPERATING PROCEDURE

SOP No.: Val. 1500.110   **Effective date: mm/dd/yyyy**

                                              Approved by:

TITLE: **Lyophilizer Requalification Protocol**

AUTHOR: _____

                              Name/Title/Department

                    _____

                              Signature/Date

CHECKED BY: _____

                              Name/Title/Department

                    _____

                              Signature/Date

APPROVED BY: _____

                              Name/Title/Department

                    _____

                              Signature/Date

REVISIONS:

No.	Section	Pages	Initials/Date

SOP No.: Val. 1500.110   Effective date: mm/dd/yyyy

Approved by:

SUBJECT: Lyophilizer Requalification Protocol

PURPOSE

The purpose is to describe the format and contents of the lyophilizer requalification protocol.

RESPONSIBILITY

It is the responsibility of validation team members to follow the procedures. The quality assurance (QA) manager is responsible for SOP compliance.

PROCEDURE

1. Purpose

The purpose is to requalify the freeze drying unit (model _____, serial no. _____) by:

- Sterilization cycle one heat penetration study and microbial challenge test using *B. stearothermophilus* strip
- Heat/chilled temperature distribution on the shelves by one heat distribution

2. Background

The sterile solutions are filled in different sizes of vials and lyophilized in freeze dryer under aseptic conditions. For this purpose, all operations and conditions were validated according to the approved protocols. To ensure sterility during aseptic processing, the frequency was established to revalidate all sterilization cycles by one-heat penetration/distribution study and microbial challenge test using *B. stearothermophilus* strip.

3. Procedure

3.1 Location

Injectable building, lyophilization room no. _____

SOP No.: Val. 1500.110

Effective date: mm/dd/yyyy

Approved by:

3.2 Test instruments

- Portable Kaye validator
- LTR-140/HTR-400/CTR-80

3.3 Freeze drying unit revalidation is divided into two parts

- Revalidation of SIP cycle
- Heat and chilled temperature distribution on the shelves

4. Revalidation of SIP Cycle

4.1 Instruments required

- Portable Kaye validator
- LTR-140

4.2 Heat penetration

- Heat penetration is the most critical component of the entire validation process.
- Heat penetration study will be performed per plant SOP.
- *B. stearothermophilus* strip and 10 to 20 thermocouples will be distributed in the chamber.
- The F_0 value will be calculated based on the base temperature and recorded by the thermocouples inside the chamber.
- Thermocouples should be calibrated before performing the heat penetration study and the calibration should be checked again after completion of the study.

4.3 Acceptance criteria

- All over the chamber and condenser temperature should be minimum 122°C for 30 minutes.
- Minimum lethality should be NLT 20.
- Minimum spore log reduction should be six.

4.4 Operating conditions for sterilization

- Set temperature: 122°C
- Sterilization time: 30 minutes
- Base temperature: 121°C
- Z-value: 10°C

5. Heat and Chilled Temperature Distribution on the Shelves

5.1 Instruments required

- Portable Kaye validator
- HTR-400
- CTR-80

5.2 Heat penetration

- Heat distribution in the freeze drying unit is the most critical component of the entire validation process.
- Heat distribution study will be performed per plant SOP.
- Minimum of 27 thermocouples will be distributed in the chamber per attached plan.
- The measurement is effected at two temperatures: −40 and +40°C, while the system is in static condition, i.e., approximately 2 hours after the temperature has been reached.
- The F_0 value will be calculated based on the base temperature and recorded by the thermocouples inside the chamber.
- Thermocouples should be calibrated before performing the heat penetration study and the calibration should be checked again after completion of the study.

5.3 Acceptance criteria

- Maximum temperature deviation from average value per shelf: mean ±1°C
- Maximum temperature deviation from average value within all shelf: mean ±2°C

5.4 Operating conditions

- Set temperature: −40 and +40°C

SOP No.: Val. 1500.110   **Effective date: mm/dd/yyyy**
                                             Approved by:

6. Leak Rate Test and Vacuum Pump Down Test

Perform the leak test using the following parameters:

- Condenser temperature: –50°C
- Target vacuum: 17%
- Leak rate phase I: 17 to 23% in 120 minutes
- Leak rate phase II: 23 to 27% in 120 minutes

7. Documentation

- Thermocouple calibration report before test (SIP and temperature distribution study)
- Qualification report by Kaye validator (SIP and temperature distribution study)
- Thermocouple calibration check report after test (SIP and temperature distribution study)
- Microbiological test report for SIP
- Sterilization and drying charts
- Leak rate test and vacuum pump down test report
- Validation summary report

REASONS FOR REVISION

Effective date: mm/dd/yyyy

- First time issued for your company, affiliates, and contract manufacturers.

TC-location on the shelves

Effective date: mm/dd/yyyy

Approved by:

–40°C Set Point Temperature

Shelf No.	Location	TC No.	Reading	Max. Deviation Each Shelf Mean		Max. Deviation from All Shelves Mean	
				Mean	Difference	Mean	Difference
1	RR C						
2	LR C						
3	RR LR						
4	LR RF						
5	RR LR						
6	RR LR C RF LF						
7	C RF						
8	RR C						
9	LR LF						
10	C RR						
11	LR C						
12	C LT						

Note: Thermocouples should be inserted in the chamber per previously mentioned locations. Thermocouple readings after temperature stabilization will be recorded in related columns and the mean and differences calculated.

SOP No.: Val. 1500.110

Effective date: mm/dd/yyyy

Approved by:

+40°C Set Point Temperature

Shelf No.	Location	TC No.	Reading	Max. Deviation Each Shelf Mean		Max. Deviation from All Shelves Mean	
				Mean	Difference	Mean	Difference
1	RR C						
2	LR C						
3	RR LR						
4	LR RF						
5	RR LR						
6	RR LR C RF LF						
7	C RF						
8	RR C						
9	LR LF						
10	C RR						
11	LR C						
12	C LT						

SECTION

VAL 1600.00

YOUR COMPANY
VALIDATION STANDARD OPERATING PROCEDURE

SOP No.: Val. 1600.10

Effective date: mm/dd/yyyy

Approved by:

TITLE: Qualification and Requalification Summary Report

AUTHOR: _____

Name/Title/Department

_____

Signature/Date

CHECKED BY: _____

Name/Title/Department

_____

Signature/Date

APPROVED BY: _____

Name/Title/Department

_____

Signature/Date

REVISIONS:

No.	Section	Pages	Initials/Date

SOP No.: Val. 1600.10
Effective date: mm/dd/yyyy
Approved by:

SUBJECT: Qualification and Requalification Summary Report

PURPOSE

The purpose is to establish the inclusion list of equipment and processes requiring formal written qualification and requalification summary reports.

RESPONSIBILITY

It is the responsibility of validation team members to follow the procedures. The quality assurance (QA) manager is responsible for SOP compliance.

PROCEDURE

The following table provides the list of critical equipment, utilities, and processes to be supported with a qualification summary report.

S. No.	Equipment/Process/Utilities

REASONS FOR REVISION

Effective date: mm/dd/yyyy

- First time issued for your company, affiliates, and contract manufacturers.

| YOUR COMPANY |
| VALIDATION STANDARD OPERATING PROCEDURE |

SOP No.: Val. 1600.20   Effective date: mm/dd/yyyy

                                         Approved by:

TITLE: Vial/Ampoule Washer Performance
            Qualification Summary Report

AUTHOR: _____

                    Name/Title/Department

            _____

                         Signature/Date

CHECKED BY: _____

                    Name/Title/Department

            _____

                         Signature/Date

APPROVED BY: _____

                    Name/Title/Department

            _____

                         Signature/Date

REVISIONS:

No.	Section	Pages	Initials/Date

SOP No.: Val. 1600.20

Effective date: mm/dd/yyyy

Approved by:

SUBJECT: Vial/Ampoule Washer Performance Qualification Summary Report

PURPOSE

The purpose is to describe the format and contents of the vials/ampoule performance qualification.

RESPONSIBILITY

It is the responsibility of validation team members to follow the procedures. The quality assurance (QA) manager is responsible for SOP compliance.

PROCEDURE

1. Purpose

The purpose of this report is to provide the methodology and acceptance criteria to qualify the cleaning machine for vials/ampoules.

2. Table of Contents

2.1 Vial/ampoule washer number and make

Name: Fully automatic ultrasonic cleaning machine for vials/ampoules
Model: ABC   Serial no: 222   Manufactured by: XYZ

2.2 Pyrogen data on different vial/ampoule sizes, acceptance criteria, and results

The performance qualification of vial/ampoule washers is to be performed on X-ml vials/ampoules. The tests performed on X-ml vial/ampoule size are for pyroburden analysis on washed vials/ampoules, particulate reduction challenge test on washed vials/ampoules, and soil test to ensure vial/ampoule washer efficacy.

- Pyroburden in washed vials
- Particulate reduction in washed vials
- Soil test to ensure vial washer efficacy

SOP No.: Val. 1600.20   **Effective date: mm/dd/yyyy**

                                    Approved by:

The X-ml vials/ampoules were washed according to the operational parameters described for each vial size in Section 2.2.

2.2.1 Pyroburden in washed vials/ampoules

The level of endotoxin in X-ml vials/ampoules after washing was found
to be LT _____ EU/ml of rinse water. For results, refer to Table 1.

Table 1

Pyrogen Test	No. Vials/ Ampoules	Acceptance Criteria	X-ml Vials/Ampoules[a]		
			Results 1	Results 2	Results 3
Unwashed vials/ ampoules	1	LT 10.00 EU/ml of rinse water	LT __ EU/ml	LT __ EU/ml	LT __ EU/ml
Washed vials/ ampoules	1	LT 0.25 EU/ml of rinse water	LT __ EU/ml	LT __ EU/ml	LT __ EU/ml
	2	LT 0.25 EU/ml of rinse water	LT __ EU/ml	LT __ EU/ml	LT __ EU/ml
	3	LT 0.25 EU/ml of rinse water	LT __ EU/ml	LT __ EU/ml	LT __ EU/ml
	4	LT 0.25 EU/ml of rinse water	LT __ EU/ml	LT __ EU/ml	LT __ EU/ml
	5	LT 0.25 EU/ml of rinse water	LT __ EU/ml	LT __ EU/ml	LT __ EU/ml
	6	LT 0.25 EU/ml of rinse water	LT __ EU/ml	LT __ EU/ml	LT __ EU/ml
	7	LT 0.25 EU/ml of rinse water	LT __ EU/ml	LT __ EU/ml	LT __ EU/ml
	8	LT 0.25 EU/ml of rinse water	LT __ EU/ml	LT __ EU/ml	LT __ EU/ml

[a] Specify the results obtained.

Effective date: mm/dd/yyyy

Approved by:

2.2.2 Vial/ampoule washer operational parameters for X ml vials/ampoules

Machine Operational Parameters

DI water pressure	(___ bar min)	___ bar
WFI pressure	(___ bar min)	___ bar
Compressed air pressure	(___ bar)	___ bar
Compressed air pressure after regulator	(___ bar)	___ bar
Machine speed		___ % (___ /min)
DI water temperature		___ °C
WFI temperature		___ °C
DI water rinsing time/volume		___ Seconds/___ milliliters per vial/ampoule
WFI rinsing time/volume		___ Seconds/___ milliliters per vial/ampoule

2.2.3 Conclusion

The vial/ampoule washer met the acceptance criterion for pyrogen, i.e., LT _____ EU/ml of rinse water in X ml vials/ampoules after washing.

2.3 Particulate reduction in X-ml vials/ampoules

2.3.1 Particulate level reduction of spiked vials/ampoules

Particulate matter was determined in ten vials/ampoules, each spiked with approximately 500 particles of 40-μm glass beads. Samples were taken from each individual needle at the end of vial washing at set operational parameters of vial/ampoule washer and analyzed to determine the reduction in particulate level. The results were compared with negative and positive controls. The summary of test results is provided in Table 2.

SOP No.: Val. 1600.20

Effective date: mm/dd/yyyy

Approved by:

Table 2

Vial/ Amp. Size	Vial No. (Corresponding to Machine Needles)	"S" Spiked Vials Results after Washing 40 μm no. of Particles	"U" Unspiked, Unwashed Particulate Level 40 μm no. of Particles[a]	% Reduction	Acceptance Criterion
	1	—	—	—	Particulate level must be reduced by not less than 90%
	2	—	—	—	
	3	—	—	—	
	4	—	—	—	
	5	—	—	—	
	6	—	—	—	
	7	—	—	—	
	8	—	—	—	
	9	—	—	—	
	10	—	—	—	

Note: S = spiked vials; U = unspiked vials.
[a] Data reported only for 40-μm particulate size. Other particulate size burdens were monitored for information only. Data are available in original printouts.

	Vial 1	Vial 2	Average
Negative control: 40 μm:			
Positive control: 40 μm:			
Disposition: accept/~~no test/reject~~:	_Accept_		
Performed and analyzed by:	_AAAAAAAAA_		

Machine Operational Parameters

DI water pressure	(___ bar min) ___ bar
WFI pressure	(___ bar min) ___ bar
Compressed air pressure	(___ bar) ___ bar
Compressed air pressure after regulator	(___ bar) ___ bar
Machine speed	___ % (___ /min)
DI water temperature	___ °C
WFI temperature	___ °C
DI water rinsing time/volume	___ Seconds/___ milliliters per vial/ampoule
WFI rinsing time/volume	___ Seconds/___ milliliters per vial/ampoule

2.3.2 Conclusion

The vial/ampoule washer met the acceptance criterion of NLT 90% reduction in particulate matter in each spiked vial after the completion of washing. The vial/ampoule washer was approved for X ml vial/ampoule washing.

2.4 Soil test data of X-ml vials/ampoules

2.4.1 Cleaning efficacy of soiled vials/ampoules

The cleaning efficacy of vial/ampoule washer ABC was determined qualitatively by spiking ten individual vials/ampoules with 10% sodium chloride. After drying, the vials were washed at the set machine operational parameters for X ml vials/ampoules. The results were compared with negative and positive controls. The summary of test results is provided in Table 3.

2.4.2 Conclusion

The vial/ampoule washer met the soil test acceptance criterion. All ten vials/ampoules were found negative (free from precipitates). The vial/ampoule washer was approved for X ml vial/ampoule washing.

SOP No.: Val. 1600.20   Effective date: mm/dd/yyyy

Approved by:

Table 3

Vial/Ampoule Size	Challenge Vials (Corresponding to Machine Needles)	Results			Acceptance Criterion
		1	2	3	
X ml	1				All vials spiked with the test soil must be negative
	2				
	3				
	4				
	5				
	6				
	7				
	8				
	9				
	10				

	Vial 1	Vial 2
Negative control test result:		
Positive control test result:		
Sensitivity test result:		
Disposition: accept/~~no test/reject~~:	Accept	
Performed and analyzed by:	AAAAAAA	

Machine Operational Parameters

DI water pressure	(___ bar min)	___ bar
WFI pressure	(___ bar min)	___ bar
Compressed air pressure	(___ bar)	___ bar
Compressed air pressure after regulator	(___ bar)	___ bar
Machine speed	___ % (___ /min)	
DI water temperature	___ °C	
WFI temperature	___ °C	
DI water rinsing time/volume	___ Seconds/___ milliliters per vial/ampoule	
WFI rinsing time/volume	___ Seconds/___ milliliters per vial/ampoule	

2.5 Disposition: study disposition, approval signatures, washer status (release to manufacturing)

Study disposition: the vial/ampoule washer meets the following acceptance criteria and is approved for X ml vials/ampoules:

- Pyrogen: LT 0.25 EU/ml of rinse water
- Percent particulate reduction: NLT 90%
- Soil test: negative

Status: X-ml vials/ampoules qualified on mm/dd/yyyy
Requalification due on: mm/dd/yyyy

REASONS FOR REVISION

Effective date: mm/dd/yyyy

- First time issued for your company, affiliates, and contract manufacturers.

YOUR COMPANY
VALIDATION STANDARD OPERATING PROCEDURE

SOP No.: Val. 1600.30

Effective date: mm/dd/yyyy

Approved by:

TITLE: Vial/Ampoule Washer Requalification Summary Report

AUTHOR:

_____

Name/Title/Department

_____

Signature/Date

CHECKED BY:

_____

Name/Title/Department

_____

Signature/Date

APPROVED BY:

_____

Name/Title/Department

_____

Signature/Date

REVISIONS:

No.	Section	Pages	Initials/Date

SOP No.: Val. 1600.30   Effective date: mm/dd/yyyy

Approved by:

SUBJECT: Vial/Ampoule Washer Requalification Summary Report

PURPOSE

The purpose is to describe the format and contents of the vial/washer requalification report. The same machine can wash vials and ampoules. Therefore, one report format is applicable to either.

RESPONSIBILITY

It is the responsibility of validation team members to follow the procedures. The quality assurance (QA) manager is responsible for SOP compliance.

PROCEDURE

1. Purpose

The purpose of this report is to provide the methodology and acceptance criterion to requalify the cleaning machine for vials/ampoules. The requalification has to be performed on a single run.

2. Table of Contents

2.1 Vial/ampoule washer number and make

Name: Fully automatic ultrasonic cleaning machine for vials/ampoules
Model: ABC   Serial no: 222   Manufactured by: XYZ

2.2 Pyrogen data on different vial sizes, acceptance criterion, and results

The performance qualification of vial washers is to be performed on X-ml vials. The tests performed on X-ml vial size are for pyroburden analysis on washed vials, particulate reduction challenge test on washed vials, and soil test to ensure vial washer efficacy.

- Pyroburden in washed vials
- Particulate reduction in washed vials
- Soil test to ensure vial washer efficacy

SOP No.: Val. 1600.30 **Effective date: mm/dd/yyyy**
Approved by:

The X-ml vials were washed according to the operational parameters described for each vial size in Section 2.2.2.

2.2.1 Pyroburden in washed vials

The level of endotoxin in X-ml vials after washing was found to be LT _____ EU/ml of rinse water. For results, refer to Table 1.

Table 1

Pyrogen Test	No. Vials	Acceptance Criterion	X-ml Vials Results
Unwashed vials	1	LT 10.00 EU/ml of rinse water	LT __ EU/ml
Washed vials	1	LT 0.25 EU/ml of rinse water	LT __ EU/ml
	2	LT 0.25 EU/ml of rinse water	LT __ EU/ml
	3	LT 0.25 EU/ml of rinse water	LT __ EU/ml
	4	LT 0.25 EU/ml of rinse water	LT __ EU/ml
	5	LT 0.25 EU/ml of rinse water	LT __ EU/ml
	6	LT 0.25 EU/ml of rinse water	LT __ EU/ml
	7	LT 0.25 EU/ml of rinse water	LT __ EU/ml
	8	LT 0.25 EU/ml of rinse water	LT __ EU/ml

2.2.2 Vial washer operational parameters for X-ml vials

Machine Operational Parameters

Parameters	X-ml Vial	X-ml Vial
DI water pressure	__ bar min	__ bar
WFI pressure	__ bar min	__ bar
Compressed air pressure	__ bar	__ bar
Compressed air pressure after regulator	__ bar	__ bar
Machine speed	__ % (__ /min)	
DI water temperature	__ °C	
WFI temperature	__ °C	
DI water rinsing time/volume	__ Seconds/__ milliliters per vial/ampoule	
WFI rinsing time/volume	__ Seconds/__ milliliters per vial/ampoule	

SOP No.: Val. 1600.30

Effective date: mm/dd/yyyy

Approved by:

2.2.3 Conclusion

The vial washer met the acceptance criterion for pyrogen, i.e., LT _____ EU/ml of rinse water in X-ml vials after washing on a single run.

2.3 Particulate reduction in X-ml vials

2.3.1 Particulate level reduction of spiked vials

Particulate matter was determined in ten vials, each spiked with approximately 500 particles of 40-μm glass beads. Samples were taken from each individual needle at the end of vial washing at set operational parameters of vial washer and analyzed to determine the reduction in particulate level. The results were compared with negative and positive controls. The summary of test results is provided in Table 2.

Table 2

Vial Size	Vial No. (Corresponding to Machine Needles)	"S" Spiked Vials Results after Washing 40 μm no. of Particles	"U" Unspiked, Unwashed Particulate Level 40 μm no. of Particles[a]	% Reduction	Acceptance Criterion
	1	–	–	–	Particulate level must be reduced by not less than 90%
	2	–	–	–	
	3	–	–	–	
	4	–	–	–	
	5	–	–	–	
	6	–	–	–	
	7	–	–	–	
	8	–	–	–	
	9	–	–	–	
	10	–	–	–	

Note: S = spiked vials; U = unspiked vials.
[a] Data reported only for 40-μm particulate size. Other particulate size burdens were monitored for information only. Data are available in original printouts.

SOP No.: Val. 1600.30

Effective date: mm/dd/yyyy

Approved by:

	Vial 1	Vial 2	Average
Negative control: 40 μm:			
Positive control: 40 μm:			
Disposition: accept/~~no test/reject~~:	Accept		
Performed and analyzed by:	AAAAAAAAA		

Machine Operational Parameters

Parameters	X-ml Vial	X-ml Vial
DI water pressure	___ bar min	___ bar
WFI pressure	___ bar min	___ bar
Compressed air pressure	___ bar	___ bar
Compressed air pressure after regulator	___ bar	___ bar
Machine speed	___ % (___ /min)	
DI water temperature	___ °C	
WFI temperature	___ °C	
DI water rinsing time/volume	___ Seconds/___ milliliters per vial/ampoule	
WFI rinsing time/volume	___ Seconds/___ milliliters per vial/ampoule	

2.3.2 Conclusion

The vial washer met the acceptance criterion of NLT 90% reduction in particulate matter in each spiked vial after the completion of washing. The vial washer was approved for X-ml vial washing.

2.4 Soil test data of X-ml vials

2.4.1 Cleaning efficacy of soiled vials

The cleaning efficacy of vial washer ABC was determined qualitatively by spiking ten individual vials with 10% sodium chloride. After drying, the vials were washed at the set machine operational parameters for X-ml vials. The results were compared with negative and positive controls. The summary of test results is provided in Table 3.

Table 3

Vial Size	Challenge Vials (Corresponding to Machine Needles)	Results	Acceptance Criterion
X ml	1		All vials spiked with the test soil must be negative
	2		
	3		
	4		
	5		
	6		
	7		
	8		
	9		
	10		

	Vial 1	Vial 2
Negative control test result:		
Positive control test result:		
Sensitivity test result:		
Disposition: accept/no test/reject:	Accept	
Performed and analyzed by:	AAAAAAAAA	

Machine Operational Parameters

Parameters	X-ml Vial	X-ml Vial
DI water pressure	___ bar min	___ bar
WFI pressure	___ bar min	___ bar
Compressed air pressure	___ bar	___ bar
Compressed air pressure after regulator	___ bar	___ bar
Machine speed	___ % (___ /min)	
DI water temperature	___ °C	
WFI temperature	___ °C	
DI water rinsing time/volume	___ Seconds/___ milliliters per vial/ampoule	
WFI rinsing time/volume	___ Seconds/___ milliliters per vial/ampoule	

2.4.2 Conclusion

The vial washer met the soil test acceptance criterion. All ten vials were found negative (free from precipitates). The vial washer was approved for X-ml vial washing, based on a single run.

2.5 Disposition: study disposition, approval signatures, washer status (release to manufacturing)

Study disposition: the vial washer meets the following acceptance criteria and is approved for X-ml vials:

- Pyrogen: LT 0.25 EU/ml of rinse water
- Percent particulate reduction: NLT 90%
- Soil test: negative
- Status: X-ml vials qualified on mm/dd/yyyy
- Requalification due on: mm/dd/yyyy

REASONS FOR REVISION

Effective date: mm/dd/yyyy

- First time issued for your company, affiliates, and contract manufacturers.

YOUR COMPANY
VALIDATION STANDARD OPERATING PROCEDURE

SOP No.: Val. 1600.40

Effective date: mm/dd/yyyy

Approved by:

TITLE: Depyrogenation Tunnel Performance
Qualification Summary Report

AUTHOR: _____

Name/Title/Department

_____

Signature/Date

CHECKED BY: _____

Name/Title/Department

_____

Signature/Date

APPROVED BY: _____

Name/Title/Department

_____

Signature/Date

REVISIONS:

No.	Section	Pages	Initials/Date

SOP No.: Val. 1600.40

Effective date: mm/dd/yyyy

Approved by:

SUBJECT: Depyrogenation Tunnel Performance Qualification Summary Report

PURPOSE

The purpose is to describe the content of the depyrogenation tunnel performance qualification summary report.

RESPONSIBILITY

It is the responsibility of validation team members to follow the procedures. The quality assurance (QA) manager is responsible for SOP compliance.

PROCEDURE

1. Performance Qualification Summary Reports

1.1 Heat distribution and penetration studies for depyrogenation of vials in tunnel

1.1.1 Table of contents

1.1.1.1 Tunnel number and make

Name: Fully automatic hot sterilizing tunnel
Model: ABC
Serial no.: 1234
Year of manufacture: YYYY
Manufactured by: EFG

1.1.2 Study date, component type, study status, signatures, and dates defined in Table 1

Table 1

Study Dates	Component Type	Study Status	Signatures and Dates
YYYY	A-ml vials	Found satisfactory	
YYYY	B-ml vials	Found satisfactory	

SOP No.: Val. 1600.40

Effective date: mm/dd/yyyy

Approved by:

1.1.3 Purpose and scope

1.1.3.1 Three heat distribution studies per protocol

The purpose of the empty tunnel heat distribution run(s) will be:

- ■ To evaluate the heating characteristics of the tunnel and the sterilization medium employed
- ■ To determine the temperature profile in the heating zone and to check the temperature uniformity

1.1.3.2 Three penetration studies per protocol

- ■ To evaluate the heating characteristics of glass vials within the tunnel when subjected to the sterilization medium
- ■ To determine the cold spot inside the sterilization zone, if any
- ■ To assure depyrogenation and the required log reduction

1.1.3.3 Four maximum-load penetration studies per protocol

This is not applicable. Because the depyrogenation tunnel is a continuous line activity, minimum and maximum load cannot be defined. The validation study simulated all possible production scenarios and load configurations. Thermocouples were distributed at the beginning of production, middle of production, and end of production.

1.1.3.4 Four minimum-load penetration studies per protocol

This is not applicable. Because the depyrogenation tunnel is a continuous line activity, minimum and maximum load cannot be defined. The validation study simulated all possible production scenarios and load configurations. Thermocouples were distributed at the beginning of production, middle of production, and end of production.

1.1.4 General documentation summary

For personnel identification list, see Table 2.

1.1.4.1 Test equipment and materials

Equipment used: portable Kaye validator/HTR-400
Material used: endotoxin (pyroquant)

Table 2

Study No.	Personnel Name	ID No.
1	AAAAAAAA	11111
2	BBBBBBBB	22222
3	CCCCCCC	33333

1.1.4.2 Deviations

No deviation was observed during A- and B-ml vial validation. The desired F_D value was achieved at the belt speed _____mm/min with the tunnel temperature of _____°C.

1.1.5 Test summary

The heat distribution studies were conducted in YYYYY. Heat distribution study results are described in Table 3.

Table 3

Set Parameters		Actual Results		
Tests	Acceptance Criterion	Run 1	Run 2	Run 3
Set temperature	____°C			
Temperature deviation from the average	NMT ____°C			
Maximum temperature deviation individual TC	NMT ____°C			
Base temperature	_____°C			
Belt speed	_____ mm/min			

SOP No.: Val. 1600.40   **Effective date: mm/dd/yyyy**

                                        Approved by:

For thermocouple (by CH) location identification, see Table 4.

Table 4

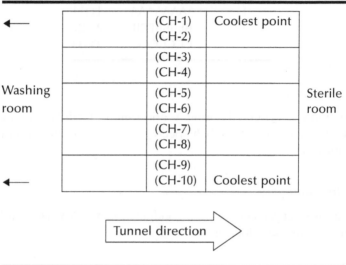

Conclusion: All parameters during heat distribution are meeting the acceptance criterion and determine the coolest point of the tunnel. For supporting documents, e.g., Kaye validator prints, calibration of thermocouples (before and after the study).

The heat penetration studies were conducted in YYYY and January YYYY for A-ml and B-ml vials, respectively. The validation operating parameters are shown in Table 5.

Table 5

Activities/Parameters	A-ml Vials	B-ml Vials
Set temperature	_____°C	_____°C
Base temperature	_____°C	_____°C
Z-value	____°C	____°C
F_D value required	Not less than ___ minutes	Not less than ___ minutes

SOP No.: Val. 1600.40   **Effective date: mm/dd/yyyy**
                                          Approved by:

1.1.5.1 Actual results

During heat penetration studies, the following speeds were used.

■ Belt speed for A-ml vials:
  Beginning and during production: the tunnel speed is controlled by the washing machine output, _____ vials per minute.
  End of the production: when the washing machine was disconnecting from the tunnel, the maximum tunnel speed was _____ mm/min.
■ Belt speed for B-ml vials:
  Beginning and during production: the tunnel speed is controlled by the washing machine output, _____ vials per minute.
  End of the production: when the washing machine was disconnecting from the tunnel, the maximum tunnel speed was _____ mm/min.

1.1.5.2 Heat penetration

The heat penetration study was performed per plant SOP. As identified in Table 6, 15 thermocouples and 15 endotoxin vials were distributed in the tunnel at different positions in each study.

Table 6 Endotoxin Vials (by Number) and Thermocouples (by CH) Location Identification

	11 (CH-11)	6 (CH-6)	1 (CH-1)	
	12 (CH-12)	7 (CH-7)	2 (CH-2)	
Washing room	13 (CH-13)	8 (CH-8)	3 (CH-3)	Sterile room
	14 (CH-14)	9 (CH-9)	4 (CH-4)	
	15 (CH-15)	10 (CH-10)	5 (CH-4)	

Tunnel direction ⟩

SOP No.: Val. 1600.40   Effective date: mm/dd/yyyy
                                         Approved by:

The F_D value was calculated by the Kaye validator automatically based on the temperature recorded by the thermocouples inside the vials. Precalibration and postcalibration for all thermocouples were performed before and after the study, respectively, and found to be within limits.

1.1.5.3 Test results

Validation study results are described in Table 7.

Table 7

Vial Size	Study No./Date	Minimum F_D Value (Min)	Maximum F_D Value (Min)	Endotoxin Reduction
A ml	01/mm/dd/yyyy			4 log
	02/mm/dd/yyyy			4 log
	03/mm/dd/yyyy			4 log
B ml	01/mm/dd/yyyy			4 log
	02/mm/dd/yyyy			4 log
	03/mm/dd/yyyy			4 log

Note: F_D calculated with base temperature of 250°C and Z-value of 54°C.

1.1.5.4 Acceptance criteria

■ F_D value NLT 6 minutes
■ Minimum 4 log reduction of endotoxin

Conclusion: The cycle was run per previously mentioned parameters for sterilization and depyrogenation of A-ml and B-ml vials. Minimum F_D value was obtained C and D minutes for A-ml and B-ml vials, respectively, during three heat penetration studies. All results are found in compliance with the acceptance criteria.

1.1.5.5 Other tests

The following other tests were conducted during the heat penetration study.

■ Air velocity of tunnel HEPA filters
   This test was conducted to demonstrate that the air speed is homogeneous in each section of the tunnel (preheating, heating, and cooling zones). The air speed values and homogeneity are

SOP No.: Val. 1600.40   Effective date: mm/dd/yyyy

Approved by:

important for uniform heating (sterilization) and uniform cooling of glass containers.

■ Test results: the test was conducted during A-ml and B-ml vial qualification, respectively. For results, see Table 8.

Table 8

Zone	Acceptance Criterion	Results A-ml Vials	B-ml Vials
Preheating zone	_____ m/s		
Cooling zone	_____ m/s		
Sterilization zone	_____ m/s		

■ Conclusion: air velocity in all three areas (preheating, cooling, and sterilization zones) of the tunnel was checked and found in compliance and meeting the acceptance criteria recommended by the manufacturer.

■ Tunnel particle counts
This test was conducted to demonstrate that the air quality is in compliance with class 100 air on all the surfaces of the conveyor (preheating, heating, and cooling zones).

■ Acceptance criteria — particulate count:
All locations with particle counts: ≤ 100 particles/ft^3 for particle size ≥ 0.5 μm; 0 for size 5.0 μm

■ Results: nonviable particulate counts were performed on all areas of the tunnel conveyor (preheating, heating, and cooling zones) per SOP. Results are reported in Table 9.

Table 9

Vial Size	Particle Size	Preheating Zone	Cooling Zone	Sterilization Zone
A ml	0.5 μm			
	5.0 μm			
B ml	0.5 μm			
	5.0 μm			

■ Conclusion: Nonviable particles for all three areas (preheating, cooling, and sterilization zones) of the tunnel were checked and found in compliance and meeting the acceptance criteria.

1.1.6 Discussion and conclusion

All acceptance criteria for set temperature, base temperature, machine speed, minimum FD value, and endotoxin challenge test were met during the heat penetration studies for A-ml and B-ml vials. For supporting documents, e.g., Kaye validator prints, calibration of thermocouples (before and after the study).

1.1.7 Follow-up actions

Revalidation will be performed in YYYY (A-ml vials) and YYYY (B-ml vials), respectively.

1.1.8 Overall conclusion

All parameters during heat distribution are meeting the acceptance criteria. For supporting documents, e.g., Kaye validator prints, calibration of thermocouples (before and after the study).

1.1.9 Documents

- Air velocity check report
- Particulate count report
- Thermocouple calibration report before test
- Qualification report by Kaye validator with tunnel chart (speed and temperature)
- Thermocouple calibration check report after test
- Microbiological test report
- Validation summary report
- Final summary report including approval signatures

REASONS FOR REVISION

Effective date: mm/dd/yyyy

- First time issued for your company, affiliates, and contract manufacturers.

YOUR COMPANY
VALIDATION STANDARD OPERATING PROCEDURE

SOP No.: Val. 1600.50   Effective date: mm/dd/yyyy

Approved by:

TITLE: Depyrogenation Tunnel Performance
Requalification Summary Report

AUTHOR: _____

Name/Title/Department

_____

Signature/Date

CHECKED BY: _____

Name/Title/Department

_____

Signature/Date

APPROVED BY: _____

Name/Title/Department

_____

Signature/Date

REVISIONS:

No.	Section	Pages	Initials/Date

SUBJECT: Depyrogenation Tunnel Performance Requalification Summary Report

PURPOSE

The purpose is to describe the contents of the depyrogenation tunnel performance requalification summary report.

RESPONSIBILITY

It is the responsibility of validation team members to follow the procedures. The quality assurance (QA) manager is responsible for SOP compliance.

PROCEDURE

1. Purpose

The purpose is to requalify the sterilization and depyrogenation cycle of the A-ml tubular clear glass vials in hot sterilization tunnel (model AAAA, serial no. BBBB) by one heat penetration study and endotoxin challenge test.

2. Background

The sterile solutions are filled in different sizes of vials and ampoules. One dosage form is lyophilized injection and filled in A-ml vials. To ensure the sterility during the aseptic processing, it was decided to revalidate the A-ml vials' sterilization/depyrogenation cycle once in a year by one heat penetration study and endotoxin challenge test.

3. Location

Injectable building, sterilization room no.: _____

4. Test Instruments Used

4.1 HTR-400

Serial no.: CCCCCCC
Calibrated on: mm/dd/yyyy
Calibration due on: mm/dd/yyyy

SOP No.: Val. 1600.50   **Effective date: mm/dd/yyyy**

                                               Approved by:

4.2 Portable Kaye validator

Serial no.: DDDDDDD
Calibrated on: mm/dd/yyyy
Calibration due on: mm/dd/yyyy

5. Operating Condition of Tunnel

- Set temperature: _____°C
- Base temperature: _____°C
- Z-value: _____°C
- Machine speed:
  Beginning of the production: when tunnel speed was controlled by
    the washing machine, washing machine speed was _____%,
    _____ vials per minute.
  End of the production: when washing machine was disconnected
    from the line, maximum tunnel speed was 100 mm/minute.

6. Procedure/Result/Acceptance Criteria/Conclusion

6.1 Heat penetration

- Heat penetration study was performed per plant SOP.
- In each study, 15 thermocouples and 15 endotoxin vials were
  distributed in the tunnel at different positions as identified in the
  following table.

Endotoxin Vials (by number) and Thermocouples (by CH) Location Identification

	11 (CH-11)	6 (CH-6)	1 (CH-1)	
Washing Room	12 (CH-12)	7 (CH-7)	2 (CH-2)	Sterile Filling Room
	13 (CH-13)	8 (CH-8)	3 (CH-3)	
	14 (CH-14)	9 (CH-9)	4 (CH-4)	
	15 (CH-15)	10 (CH-10)	5 (CH-5)	

- The F_D value was calculated by the Kaye validator automatically based on the temperature recorded by the thermocouples inside the vials.
- Precalibration and postcalibration for all thermocouples are performed before and after the study, respectively.

6.1.1 Acceptance criteria

- F_D value NLT 6 minutes
- Minimum four-log reduction of endotoxin

6.1.2 Test results

The following results were obtained after validation.
Test performed on _____

SOP No.: Val. 1600.50   **Effective date: mm/dd/yyyy**

Approved by:

Testing Stage	Minimum F_D Value (Min)	Maximum F_D Value (Min)	Endotoxin Reduction
Beginning of production			
Middle of production			
End of production			

6.1.3 Conclusion

The cycle was run per the previously mentioned parameters for sterilization and depyrogenation of A-ml vials. Minimum F_D value was found _____ and maximum value was found _____ during heat penetration study. All results were found in compliance with the acceptable criteria and meeting the FDA requirements.

6.2 Tunnel speed

Tunnel speed was checked and recorded during heat penetration studies and results are:

- Beginning of the production: when tunnel speed was controlled by the washing machine, machine speed was _____ (___%) vials per minute.
- End of the production: when washing machine was disconnecting from the washing line, the maximum tunnel speed was _____ mm/min.

7. Documentation

- Thermocouple calibration report before test
- Qualification report by Kaye validator with tunnel chart (speed and temperature)
- Thermocouple calibration check report after test
- Microbiological test report

8. Tunnel Particulate Count

In addition to sterilization and depyrogenation cycle revalidation of the vials, tunnel particulate counts on all the surfaces of the conveyor (pre-heating, heating, and cooling zones) was also checked after heat penetration study. Following are test details:

SOP No.: Val. 1600.50   Effective date: mm/dd/yyyy

Approved by:

9. Test Instruments Used

CI-500-003A laser particulate counter
Serial no.: SSSSSSS
Calibrated on: mm/dd/yyyy
Calibration due on: mm/dd/yyyy

10. Procedure/Test Result/Acceptance Criteria/Conclusion

10.1 Tunnel particle counts

Demonstrate that the air quality is in compliance with a class 100 air on all the surface of the conveyor (preheating, heating, and cooling zones). Use plant SOP for monitoring.

10.2 Acceptance criteria

All locations with particle counts: $\leq$100 particles/ft^3 for particle size $\geq$0.5 μm; 0 for particle size 5.0 μm.

10.3 Results

Nonviable particulate testing was performed on all areas of tunnel conveyor (preheating, heating, and cooling zones) by CI-500 per plant SOP. Results are found in the following table.

Particle Size	Preheating Zone	Cooling Zone	Sterilization Zone
0.5 μm	0.0	0.0	0.0
5.0 μm	0.0	0.0	0.0

10.4 Conclusion

The nonviable counts in preheating, cooling, and sterilization zones were checked and found in compliance with the acceptable criteria of class 100 per plant SOP.

REASONS FOR REVISION

Effective date: mm/dd/yyyy

■ First time issued for your company, affiliates, and contract manufacturers.

YOUR COMPANY
VALIDATION STANDARD OPERATING PROCEDURE

SOP No.: Val. 1600.60   Effective date: mm/dd/yyyy

Approved by:

TITLE: SIP Cycle for Holding Vessels Requalification
Summary Report

AUTHOR: _____
Name/Title/Department

_____
Signature/Date

CHECKED BY: _____
Name/Title/Department

_____
Signature/Date

APPROVED BY: _____
Name/Title/Department

_____
Signature/Date

REVISIONS:

No.	Section	Pages	Initials/Date

SOP No.: Val. 1600.60   Effective date: mm/dd/yyyy
Approved by:

SUBJECT: SIP Cycle for Holding Vessels Requalification Summary Report

PURPOSE

The purpose it to describe the contents of the SIP cycle for holding vessel requalification summary report.

RESPONSIBILITY

It is the responsibility of validation team members to follow the procedures. The quality assurance (QA) manager is responsible for SOP compliance.

PROCEDURE

1. Requalification of SIP Cycle for Holding Tank

1.1 Table of contents

SIP no. and make:
Name: A Mobile Vessel
Model: ABC
Manufactured by: DEF

2. Study Date, Component Type, Study Status, Signatures, and Dates

Study date: mm/dd/yyyy
Component type: sterilization of mobile tank
Status: validated
Protocol no.: RQP-000

3. Purpose and Scope

■ Performed one heat penetration study to assure that the required temperature is reaching all areas of the mobile tank by distributing the thermocouples. A Kaye validator equipped with seven thermocouples (minimum) was used to provide an equal representation among layers in the vessel.

SOP No.: Val. 1600.60

Effective date: mm/dd/yyyy

Approved by:

■ Performed microbiological challenge studies to determine the degree of process lethality provided by the sterilization cycle. *Bacillus stearothermophilus* strips were used for microbiological challenge. A Kaye validator equipped with seven (minimum) thermocouples is distributed in the vessel along with biological indicator. After the sterilization cycle is complete, the *B. stearothermophilus* strips were recovered and subjected to microbiological test procedures.

4. General Documentation Summary

4.1 Personnel identification list

Table 1

Study No.	Personnel Name	ID No.
1		
2		
3		
4		

4.2 Test equipment and materials

Material: *B. stearothermophilus* strips
Equipment: portable Kaye validator; TR-140

4.3 Deviations

No deviation was observed.

4.4 Adjustable parameters

No adjustment was required.

4.5 Test summary

4.5.1 Sterilization cycle for mobile tanks

The sterilization cycle exposure time, exposure temperature, minimum lethality, and spore log reduction with acceptance criteria and results are provided in Table 2.

Table 2

Test	Acceptance Criteria	Results Vessel A	Vessel B
Exposure time	Not less than ____ minutes		
Exposure temperature	Not less than ____°C		
Minimum lethality	Not less than ____minutes		
Microbial challenge test	Minimum 6 spore log reduction		

4.5.2 Deviations and results

All parameters during revalidation were found in compliance and results were found satisfactory.

4.5.3 Discussion and conclusion

All parameters during SIP cycle, heat distribution, and microbiological challenge test are meeting the acceptance criteria.

6. Follow-Up Items

Follow-up for revalidation will be due next year, mm/dd/yyyy.

7. Overall Conclusion

All parameters and test results for SIP cycle were found in compliance. The SIP cycle of the mobile vessel is considered requalified and approved for regular sterilization.

REASONS FOR REVISION

Effective date: mm/dd/yyyy

■ First time issued for your company, affiliates, and contract manufacturers.

YOUR COMPANY
VALIDATION STANDARD OPERATING PROCEDURE

SOP No.: Val. 1600.70   Effective date: mm/dd/yyyy

Approved by:

TITLE: Stopper Sterilization Performance
Qualification Summary Report

AUTHOR: _____

Name/Title/Department

_____

Signature/Date

CHECKED BY: _____

Name/Title/Department

_____

Signature/Date

APPROVED BY: _____

Name/Title/Department

_____

Signature/Date

REVISIONS:

No.	Section	Pages	Initials/Date

SOP No.: Val. 1600.70   Effective date: mm/dd/yyyy
                                         Approved by:

SUBJECT: Stopper Sterilization Performance Qualification
           Summary Report

PURPOSE

The purpose is to describe the contents of the stopper sterilization performance qualification summary report.

RESPONSIBILITY

It is the responsibility of validation team members to follow the procedures. The quality assurance (QA) manager is responsible for SOP compliance.

PROCEDURE

1. Stopper Sterilization Qualification in Autoclave

2. Table of Contents

2.1 Component sterilizer no. and make

Name: steam sterilizer (autoclave)
Model: AAAA
Serial No.: BBBB
Manufacturer: Finn Aqua

2.2 Study date, component type, study status

Study date: mm/dd/yyyy
Component type: _____ mm rubber stoppers, _____color

- Load quantity = _____ stoppers
- No. of bags = _____ bags
- Each bag contains = _____ stoppers
- Bag type = Tyvek/PE

Study status: stopper sterilization cycle qualified all acceptance criteria.

2.3 List of operation parameters *(cycle events/times, TC distribution probes, biological indicator test, installation qualification review, sterilizer SOP review) acceptable limits, and study results*

2.3.1 Cycle events/times (Table 1)

Table 1

Event	Run 1	Run 2	Run 3

No. Vacuum Pulses
    Prevacuum Level
    Steam Level
    Prevacuum Level
    Steam Level
    Prevacuum Level			

Exposure temperature			
Exposure type			
Exposure time			
Temp. deviation limit			
Drying type			
No. drying pulses			
Drying vacuum level			
Drying vacuum time(s)			
Drying air level			
Dry air time(s)			
Dry jacket heat (y/n)			
Drying temperature			
Loading side			
Unloading side			

SOP No.: Val. 1600.70

Effective date: mm/dd/yyyy
Approved by:

2.3.2 TC distribution probes

The distribution of the probes in shelves for stoppers is described in Table 2.

Table 2

Shelf 1: Eight Bags of Stoppers (LF, CR, RR)	Bag 3	Bag 5 TC 3	Bag 8 TC 2
	Bag 2	Bag 4	Bag 7
	Bag 1 TC 1		Bag 6
Shelf 2: Eight Bags of Stoppers (LR, CF, RF, CC)	Bag 11 TC 4	Bag 13 TC 6	Bag 16
	Bag 10	Bag 12 TC 5	Bag 15
	Bag 9		Bag 14 TC 7
Shelf 3: Eight Bags of Stoppers (LC, LF, CR, RR)	Bag 19	TC 10 Bag 21	Bag 24 TC 11
	TC 9 Bag 18	Bag 20	Bag 23
	Bag 17 TC 8		Bag 22
Shelf 4: Eight Bags of Stoppers (LR, CC, LC)	Bag 27 TC 12	Bag 29 TC 13	Bag 32
	Bag 26	Bag 28	Bag 31
	Bag 25		Bag 30 TC14

Notes: TC 15 — hang in the center; TC 16 — drain. Total thermocouples = 16.
Abbreviations: LR = left rear; RR = right rear; LF = left front; RF = right front; CR = center rear; CC = center center; CF = center front.

SOP No.: Val. 1600.70   Effective date: mm/dd/yyyy
Approved by:

2.3.3 Biological indicator test

During sterilization, 16 strips of *B. stearothermophilus* (10^6 population) were used with each thermocouple and sent to microbiological section for testing.

2.3.4 Installation qualification review

The following installation documents were reviewed and found in compliance:

- Facility checks
- Environmental checks
- Utilities checks
- Utility line identification checks
- Equipment installation records
- Manuals
- List of change parts
- List of spare parts
- Calibration

2.3.5 Sterilizer SOP review

The following sterilizer SOPs were reviewed during revalidation and found in compliance:

- Equipment operation SOPs
- Equipment cleaning/sanitization SOPs
- Preventive maintenance SOPs
- Calibration SOPs

2.3.6 Study results

For rubber stopper requalification, three satisfactory heat penetration studies were performed including a microbiological challenge test. The results of three runs met the acceptance criteria of exposure time, exposure temperature, minimum lethality, and microbiological test. See Table 3.

Table 3

		Actual		
Test	Acceptance Criteria	Run 1	Run 2	Run 3
Exposure time	Minimum _____ minutes			
Exposure temperature	_____°C			
Minimum lethality	Not less than ___ minutes			
Microbial challenge test	Minimum 6 spore log reduction			

Note: F_0 calculated with base temperature _____°C and Z-value _____°C.

2.3.6 Conclusion

The qualification studies conducted indicated that the GMP steam sterilizer (autoclave) performance is satisfactory and in accordance with the operational parameters defined previously.

2.4 Documents

- Printout for thermocouple calibration before study
- Printout for qualification study including temperature, time for exposure, and F_0
- Printout for thermocouple calibration checks after study
- Autoclave cycle printouts
- Microbial challenge test report
- Final summary report including approval signatures

REASONS FOR REVISION

Effective date: mm/dd/yyyy

- First time issued for your company, affiliates, and contract manufacturers.

| YOUR COMPANY |
| VALIDATION STANDARD OPERATING PROCEDURE |

SOP No.: Val. 1600.80   Effective date: mm/dd/yyyy

Approved by:

TITLE: Stopper Sterilization Requalification Summary Report

AUTHOR: _____

Name/Title/Department

_____

Signature/Date

CHECKED BY: _____

Name/Title/Department

_____

Signature/Date

APPROVED BY: _____

Name/Title/Department

_____

Signature/Date

REVISIONS:

No.	Section	Pages	Initials/Date

SOP No.: Val. 1600.80   Effective date: mm/dd/yyyy

Approved by:

SUBJECT: Stopper Sterilization Requalification Summary Report

PURPOSE

The purpose is to describe the contents of the stopper sterilization requalification summary report.

RESPONSIBILITY

It is the responsibility of validation team members to follow the procedures. The quality assurance (QA) manager is responsible for SOP compliance.

PROCEDURE

1. Purpose

The purpose is to requalify the sterilization cycle of ___ mm rubber stoppers in steam sterilizer (model AAAA, serial no. BBBB) by one heat penetration study and microbial challenge test using *B. stearothermophilus* strip.

2. Background

To ensure the sterility during the aseptic processing _____ mm stopper sterilization cycle was validated by three heat penetration studies and *B. stearothermophilus* strip considering maximum load size. It was also decided to revalidate the rubber stopper cycle once in a year by one heat penetration study and *B. stearothermophilus* strip.

3. Location

Injectable building, sterilization room no.: ____

4. Test Instruments Used

4.1 Portable Kaye validator

Serial no.: SSSSSS
Calibrated on: mm/dd/yyyy
Calibration due on: mm/dd/yyyy

SOP No.: Val. 1600.80   Effective date: mm/dd/yyyy

Approved by:

4.2 LTR-140

Serial no.: SSSSSSS
Calibrated on: mm/dd/yyyy
Calibration due on: mm/dd/yyyy

5. Process Parameters

- Exposure temperature: ____°C ± 1°C
- Base temperature: ____°C
- Exposure time: not less than ____ minutes
- Z-value: ___°C

6. Study Conducted on

mm/dd/yyyy

7. Procedure/Acceptance Criteria/Results/Conclusion

7.1 Procedure for heat penetration

7.1.1 Stopper load

- Load quantity = _____ stoppers
- Number of bags = _____ bags (Tyvek/HDPE)
- Each bag contains = _____ stoppers

7.1.2 Heat penetration

Heat penetration study was performed per plant SOP. Using four shelves of the trolley, distribute the load and thermocouples were distributed per the following sequence. *Note:* All bags were to be positioned with Tyvek side down and HDPE side up.

Shelf 1: Eight Bags of Stoppers (LF, CR, RR)	Bag 3	Bag 5 TC 3	Bag 8 TC 2
	Bag 2		Bag 7
		Bag 4	
	Bag 1 TC 1		Bag 6

Shelf 2: Eight Bags of Stoppers (LR, CF, RF, CC)	Bag 11 TC 4	Bag 13 TC 6	Bag 16
	Bag 10		Bag 15
		Bag 12 TC 5	
	Bag 9		Bag 14 TC 7

Shelf 3: Eight Bags of Stoppers (LC, LF, CR, RR)	Bag 19	TC 10 Bag 21	Bag 24 TC 11
	TC 9 Bag 18		Bag 23
		Bag 20	
	Bag 17 TC 8		Bag 22

Shelf 4: Eight Bags of Stoppers (LR, CC, LC)	Bag 27 TC 12	Bag 29 TC 13	Bag 32
	Bag 26		Bag 31
		Bag 28	
	Bag 25		Bag 30 TC14

Notes: TC 15 — hang in the center; TC 16 — drain. Total thermocouples = 16.
Abbreviations: LR = left rear; RR = right rear; LF = left front; RF = right front; CR = center rear; CC = center center; CF = center front.

SOP No.: Val. 1600.80   **Effective date: mm/dd/yyyy**

                                                           Approved by:

Per the previously mentioned layout, 16 thermocouples and *B. stearothermophilus* strips were inserted in the bags. The F_0 value was calculated based on the base temperature recorded by the thermocouples inside the bags.

7.2 Acceptance criteria

- Minimum lethality should be NLT ____.
- Cycle parameters should be followed according to plant SOP.
- Minimum spore log reduction should be six.

7.3 Test results

The following results were achieved after revalidation of rubber stopper sterilization cycle:

Test	Requirement	Actual
Exposure time	Not less than ____ minutes	
Exposure temperature	____°C ± 1°C	
Minimum lethality	Not less than ____ minutes	
Microbial challenge test	Minimum six spore log reduction	

7.4 Conclusion

The cycle was run per the previously mentioned parameters and plant SOP for sterilization of rubber stoppers. Minimum F_0 value is obtained AAAA during three heat penetration studies and a six-log reduction was achieved. All results are found in compliance and within the acceptance criteria and meeting the requirements. On the basis of satisfactory revalidation results/data, the cycle is considered revalidated.

8. Review of Documents

During validation, the following documents were reviewed and found in compliance.

8.1 Installation qualification review

The following installation documents were reviewed and found in compliance:

- Facility checks
- Environmental checks
- Utilities checks
- Utility line identification checks
- Equipment installation records
- Manuals
- Calibration

8.2 Sterilizer SOP review

The following sterilizer SOPs were reviewed during revalidation and found in compliance:

- Equipment operation SOPs
- Equipment cleaning/sanitization SOPs
- Preventive maintenance SOPs
- Calibration SOPs

9. Documentation

- Thermocouple calibration report before test
- Qualification report by Kaye validator
- Thermocouple calibration check report after test
- Microbiological test report
- Sterilization charts

REASONS FOR REVISION

Effective date: mm/dd/yyyy

- First time issued for your company, affiliates, and contract manufacturers.

YOUR COMPANY
VALIDATION STANDARD OPERATING PROCEDURE

SOP No.: Val. 1600.90   Effective date: mm/dd/yyyy

Approved by:

TITLE: Equipment Sterilization Performance
Qualification Summary Report

AUTHOR: _____

Name/Title/Department

_____

Signature/Date

CHECKED BY: _____

Name/Title/Department

_____

Signature/Date

APPROVED BY: _____

Name/Title/Department

_____

Signature/Date

REVISIONS:

No.	Section	Pages	Initials/Date

SUBJECT: Equipment Sterilization Performance Qualification Summary Report

PURPOSE

The purpose is to describe the contents of the equipment sterilization performance qualification summary report.

RESPONSIBILITY

It is the responsibility of validation team members to follow the procedures. The quality assurance (QA) manager is responsible for SOP compliance.

PROCEDURE

1. Example of Equipment Sterilizer Requalification

2. Table of Contents

2.1 Component sterilizer no. and make

Name: steam sterilizer (autoclave)
Model: AAAAA
Serial no.: BBBBB
Manufacturer: Finn Aqua

2.2 Study date, component type, study status

Study date: mm/dd/yyyy
Component type:

2.2.1 Vial-filling machine parts, filtration assemblies, and accessories

Study status: vial-filling machine parts and accessories cycle validated.

SOP No.: Val. 1600.90   Effective date: mm/dd/yyyy

Approved by:

2.3 List of operation parameters *(cycle events/times, thermocouple (TC) distribution probes, biological indicator test, installation qualification review, sterilizer SOP review), acceptable limits, and study results*

2.3.1 Cycle events/times (Table 1)

Table 1

Event	Run 1	Run 2	Run 3
Cycle label			
Cycle type			
RAMPS (Y/N)			

No. vacuum pulses
  Prevacuum level
  Steam level
  Prevacuum level
  Steam level
  Prevacuum level			

Exposure temperature			
Exposure type			
Exposure time			
Temperature deviation limit			
Drying type			
Drying level			
Drying time			
Dry jacket heat (y/n)			
Drying temperature			
Loading side			
Unloading side			
Print interval			

2.3.2 TC distribution probes

The distribution of the probes in shelf for vial filling machine parts and accessories is described in Table 2.

Table 2

Shelf No.	Item No.	Component	Location	TC
Shelf 1	1	Buffer tank with accessories	RR	TC 1
	2	Vent filter	RC; RF	TC 2
	3	Stainless hose	CF	TC 3
	4	Pipe to manifold etc.	CC	TC 4
	5	Filter housings (prefilter, sterile filter, pressure vessels)	CR; LR	TC 5; TC 7
	6	Upstream and downstream hoses for filtration assembly	LC	TC 6
	7	Stopper vibrating bowls (two nos.)	LF	TC 8
Shelf 2	8	Manifold	RF	TC 9
	9	Filter cartridge and spares	RR	TC 10
	10	Dosing pumps (four nos.) disengaged	CC; LR	TC 11
	11	Stopper feed tracks and insertion station	LF	TC 12
Shelf 3	12	Gassing assembly (disc filters and silicon hoses)	RF	TC 13
	13	Silicon tubes and needles (prepumps four nos. and postpumps four nos.)	CC LR	TC 14 TC 15

Notes: TC 15 — hang in the center; TC 16 — drain. Total thermocouples = 16.
Abbreviations: LR = left rear; RR = right rear; LF = left front; RF = right front; CR = center rear; CC = center center; CF = center front.

SOP No.: Val. 1600.90

Effective date: mm/dd/yyyy

Approved by:

PLACEMENT OF COMPONENTS & LOCATION OF TCs & BIs

Shelf-1 (Upper Rack)

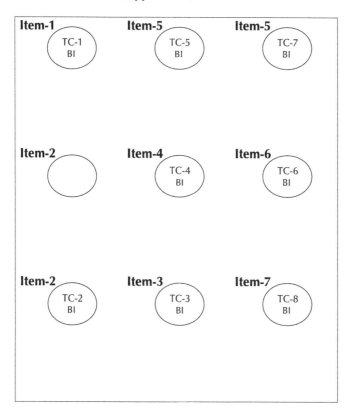

2.3.3 Biological indicator test

During heat penetration studies, 16 strips of *B. stearothermophilus* (10^6 population) were used with each thermocouple and after completion of sterilization, BI strips were sent to microbiological section for testing.

2.3.4 Installation qualification review

The following installation documents were reviewed and found in compliance:

Shelf-2 (Middle Rack)

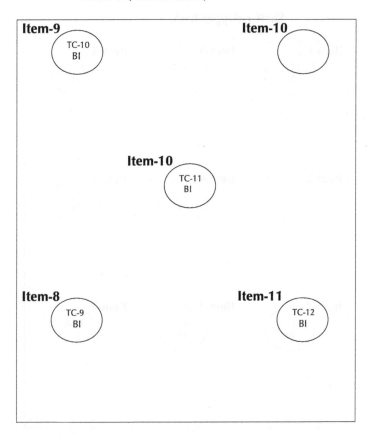

- Facility checks
- Utility line identification checks
- Equipment installation records
- Manual
- List of change parts
- List of spare parts
- Calibration

SOP No.: Val. 1600.90   **Effective date: mm/dd/yyyy**
                                             Approved by:

Shelf-3 (Lower Rack)

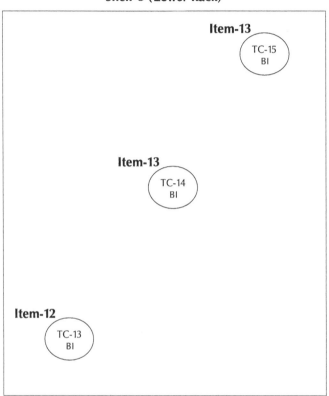

2.3.5 Sterilizer SOP review

The following sterilizer SOPs were reviewed during revalidation and found in compliance:

- Equipment operation SOPs
- Equipment cleaning/sanitization SOPs
- Preventive maintenance SOPs
- Calibration SOPs

2.3.6 Study results

For vial filling machine parts, three heat penetration studies were performed and found satisfactory including microbiological challenge test. The results of exposure time temperature, minimum lethality, and microbiological challenge test met the acceptance criteria. See Table 3.

Table 3

Test	Acceptance Criteria	Results		
		Run 1	Run 2l	Run 3
Exposure time	Minimum 30 minutes			
Exposure temperature	___°C ± ___°C			
Minimum lethality	Not less than ___			
Microbial challenge test	Minimum six spore log reduction			

Note: F_0 calculated with base temperature ____°C and Z-value ____°C.

2.3.7 Conclusion

The requalification studies conducted indicated that the GMP steam sterilizer (autoclave) performance is satisfactory and in accordance with the operational parameters defined previously.

4. Documents

- Printout for thermocouple calibration before study
- Printout for qualification study including temperature, time for exposure, and F_0
- Printout for thermocouple calibration checks after study
- Autoclave cycle printouts
- Microbial challenge test report
- Final summary report including approval signatures

REASONS FOR REVISION

Effective date: mm/dd/yyyy

- First time issued for your company, affiliates, and contract manufacturers.

YOUR COMPANY
VALIDATION STANDARD OPERATING PROCEDURE

SOP No.: Val. 1600.100

Effective date: mm/dd/yyyy

Approved by:

TITLE: Lyophilizer Performance Qualification Summary Report

AUTHOR: _____

Name/Title/Department

_____

Signature/Date

CHECKED BY: _____

Name/Title/Department

_____

Signature/Date

APPROVED BY: _____

Name/Title/Department

_____

Signature/Date

REVISIONS:

No.	Section	Pages	Initials/Date

SOP No.: Val. 1600.100   Effective date: mm/dd/yyyy

                                      Approved by:

SUBJECT: Lyophilizer Performance Qualification Summary Report

PURPOSE

The purpose is to describe the template procedure for an example of lyophilizer performance qualification.

RESPONSIBILITY

It is the responsibility of validation team members to follow the procedures. The quality assurance (QA) manager is responsible for SOP compliance.

PROCEDURE

1. **Example of Lyophilizer Performance Qualification**

2. **Table of Contents**

3. **Lyophilizer No. and Make**

Name: freeze drying unit
Model: AAAAA
Serial no.: BBBBB
Manufactured by: Finn Aqua GmbH

4. **Study Date, Component Type, Study Status, Protocol No.**

Study date: mm/dd/yyyy
Component type:

- Sterilization cycle for freeze-drying unit
- Heating and freezing temperature uniformity checks at −40 and +40°C
- Status: validated and meeting the acceptance criteria
- Protocol no.:

5. **Purpose**

The purpose of the empty chamber heat distribution run is to evaluate:

SOP No.: Val. 1600.100   Effective date: mm/dd/yyyy

Approved by:

- The heating characteristics of the chamber and condenser sterilization
- The ability of the chamber and condenser to hold the required sterilization pressures without leaks
- The ability of the sterilization cycle control mechanisms to operate as intended
- Verify that the sterilization cycle is capable of obtaining F_0 value and log reduction
- Verify heating and freezing temperature uniformity on the shelves, i.e., –40 and +40°C

6. Conclusion and Recommendations

All parameters during temperature uniformity study and microbiological challenge test are meeting the acceptance criteria.

7. Materials and Methods

Material: *B. stearothermophilus* strips
Equipment: portable Kaye validator/LTR-140 /CTR-80
Methods include:

- Calibration of instruments, analytical/microbiological test methods, and qualification runs
- Calibration procedure for thermometer
- Calibration procedure for time controllers and timers
- Calibration procedure for pressure gauges
- Operating procedure for portable Kaye validator
- Equipment performance qualification guideline
- Validation/revalidation of SIP cycle of freeze dryer
- Validation/revalidation of uniformity of heat and chilled temperature distribution in freeze dryer
- Validation and certification of sterilization cycle

8. Results and Data

8.1 Sterilization cycle of freeze-drying unit (chamber and condenser)

Following is the summary of test results achieved on three individual sterilization cycles of the freeze-drying unit. The results of the three runs met the acceptance criteria of exposure time, exposure temperature,

SOP No.: Val. 1600.100   Effective date: mm/dd/yyyy

Approved by:

minimum lethality, and microbiological challenge test. The summary of test results is provided in Table 1.

Table 1

		Actual		
Test	Acceptance Criteria	Run 1	Run 2	Run 3
Exposure time	Not less than ____ minutes			
Exposure temperature	Not less than ____°C			
Minimum lethality	Not less than _____ minutes			
Microbial challenge test	Minimum 6 spore log reduction			

8.2 Heating and freezing temperature uniformity check at –40 and +40°C

Heating and freezing temperature uniformity studies were conducted on three individual runs using 27 thermocouples meeting the acceptance criteria for maximum temperature deviation from average value of each shelf and maximum temperature deviation from average value within all shelves at –40 and +40°C. A summary of results is provided in Table 2.

Table 2

	–40°C Set Point Temperature		+40°C Set Point Temperature		
Cycle No.	Maximum Temperature Deviation from Average Value of Each Shelf	Maximum Temperature Deviation from Average Value within All Shelves	Maximum Temperature Deviation from Average Value of Each Shelf	Maximum Temperature Deviation from Average Value within All Shelves	Performed on
Run 1					
Run 2					
Run 3					
Acceptance criteria	NMT ___°C	NMT ___°C	NMT ___°C	NMT ___°C	

9. Discussion

All three cycle runs for heat/chilled temperature uniformity on shelves met the acceptance criteria at –40 and +40°C temperature.

9.1 Leak rate test and vacuum pump down test

Acceptance criteria: 17 to 22% in 120 minutes; 22 to 50% in 600 minutes
Actual = _____ vacuum pump down test
Acceptable = ____% to ____% in ____ minutes
Actual = ____% to ____% in ____ minutes

10. Documentation

10.1 SIP

- Printout for thermocouple calibration before study
- Printout for qualification study including temperature, time for exposure, and F_0
- Printout for thermocouple calibration checks after study
- SIP cycle printouts
- Microbial challenge test report
- Final summary report including approval signatures

10.2 Temperature uniformity on shelves

- Printout for thermocouple calibration before study
- Printout for qualification study including temperature and time
- Printout for thermocouple calibration checks after study
- Freeze dryer cycle printouts
- Final summary report including approval signatures

REASONS FOR REVISION

Effective date: mm/dd/yyyy

- First time issued for your company, affiliates, and contract manufacturers.

YOUR COMPANY
VALIDATION STANDARD OPERATING PROCEDURE

SOP No.: Val. 1600.110

Effective date: mm/dd/yyyy

Approved by:

TITLE: Lyophilizer Requalification Summary Report

AUTHOR: _____

Name/Title/Department

_____

Signature/Date

CHECKED BY: _____

Name/Title/Department

_____

Signature/Date

APPROVED BY: _____

Name/Title/Department

_____

Signature/Date

REVISIONS:

No.	Section	Pages	Initials/Date

SOP No.: Val. 1600.110

Effective date: mm/dd/yyyy

Approved by:

SUBJECT: Lyophilizer Requalification Summary Report

PURPOSE

The purpose is to describe the contents of the lyophilizer requalification summary report.

RESPONSIBILITY

It is the responsibility of validation team members to follow the procedures. The quality assurance (QA) manager is responsible for SOP compliance.

PROCEDURE

1. **Example of Lyophilizer Requalification**

2. **Table of Contents**

3. **Lyophilizer No. and Make**

Name: freeze-drying unit
Model: AAAAA
Serial no.: BBBBB
Manufactured by: Finn Aqua GmbH

4. **Study Date, Component Type, Study Status, Protocol No.**

Study date: mm/dd/yyyy
Component type:

- Sterilization cycle for freeze-drying unit
- Heating and freezing temperature uniformity checks at −40 and +40°C

Status: freeze dryer revalidated. All results found in compliance.
Protocol no.: RQP

5. **List of parameters** (Shelf-to-Shelf Temperature Uniformity, Steam Sterilization Study — Shelf Monitoring, Steam Sterilization Study — Chamber Monitoring — Last 15 minutes of Exposure, Steam Sterilization Study — F_0 Minutes, Steam Sterilization — Biological Indicator Test Results, Leak Rate Test, Vacuum Pump Down Test) Acceptable Limits, and Study Results

5.1 Shelf-to-shelf temperature uniformity (–40 and +40°C)

Heating and freezing temperature uniformity checks were conducted on two individual runs using 30 thermocouples. The maximum temperature deviation from average value of each shelf and maximum temperature deviation from average value within all shelves at –40 and +40°C was found within acceptance criteria. A summary of results is provided in Table 1.

Table 1

Cycle No.	–40°C Set Point Temperature		+40°C Set Point Temperature		
	Maximum Temperature Deviation from Average Value of Each Shelf	Maximum Temperature Deviation from Average Value within All Shelves	Maximum Temperature Deviation from Average Value of Each Shelf	Maximum Temperature Deviation from Average Value within All Shelves	Performed on
Run 1 results					
Run 2 results					
Acceptance criteria	NMT ___°C	NMT ___°C	NMT ___°C	NMT ___°C	

5.1.1 Conclusion

The cycle was run at –40 and +40°C temperature; temperature was held for 2 hours of freeze-drying unit. All temperatures during the heat distribution were found in compliance and within the acceptance criteria.

SOP No.: Val. 1600.110

Effective date: mm/dd/yyyy
Approved by:

5.2 Sterilization cycle for freeze-drying unit

The lyophilizer requalification was performed at set exposure time, exposure temperature, defined lethality. and spore log reduction. The results met the acceptance criteria. For details, see Table 2.

Table 2

Freeze Dryer Monitoring Parameters

Event	Parameters
Sterilization time	____ minutes
Sterilization temperature	_____°C
Drying time	_____ hours
Chamber wall cool down to	_____°C
Shelves cool down to	_____°C

Requalification Results (Chamber and Condenser)

Test	Acceptance Criteria	Actual
Exposure time	_____minutes	
Exposure temperature	Not less than ____°C	
Minimum lethality	Not less than _____	
Microbial challenge test	Minimum six spore log reduction	

5.2.1 Conclusion

The cycle was run per previously mentioned parameters and related SOPs for sterilization of the freeze-drying unit. Minimum F_0 value AAA and six-log reduction were achieved during the heat penetration study of the freeze-drying unit. All results during sterilization were found in compliance and within the acceptance criteria.

5.3 Leak rate test and vacuum pump down test

5.3.1 Leak rate test

Acceptable test rate = ____% to ____% in ____ minutes (____ mbar L S^{-1})
Actual calculation rate = _____ mbar L S^{-1}, i.e., acceptable

5.3.2 Vacuum pump down test

Acceptable = _____% to _____% in 30 minutes
Actual = _____% to _____% in 20.5 minutes

5.3.3 Conclusion

The vacuum pump down test result was found in compliance and within the acceptance criteria. The leak rate test result meets the acceptance criteria.

6. Documents

6.1 SIP

- Printout for thermocouple calibration before study
- Printout for requalification study including temperature, time for exposure, and F_0
- Printout for thermocouple calibration checks after study
- SIP cycle printouts
- Microbial challenge test report
- Final summary report including approval signatures

6.2 Temperature uniformity check

- Printout for thermocouple calibration before study
- Printout for requalification study including temperature and time
- Printout for thermocouple calibration checks after study
- Freeze dryer cycle printouts
- Final summary report including approval signatures

REASONS FOR REVISION

Effective date: mm/dd/yyyy

- First time issued for your company, affiliates, and contract manufacturers.

SECTION

VAL 1700.00

YOUR COMPANY
VALIDATION STANDARD OPERATING PROCEDURE

SOP No.: Val. 1700.10   Effective date: mm/dd/yyyy

                                               Approved by:

TITLE: Evidence of Formal Written Procedures

AUTHOR: _____
                Name/Title/Department

                _____
                Signature/Date

CHECKED BY: _____
                Name/Title/Department

                _____
                Signature/Date

APPROVED BY: _____
                Name/Title/Department

                _____
                Signature/Date

REVISIONS:

No.	Section	Pages	Initials/Date

SUBJECT: Evidence of Formal Written Procedures

PURPOSE

The purpose is to provide the index of critical SOPs submitted to FDA as a part of ANDA and NDA submissions.

RESPONSIBILITY

It is the responsibility of all concerned officers and the manager to follow the procedure. The quality assurance (QA) manager is responsible for SOP compliance.

PROCEDURE

The evidence of formal written procedures is one of the requirements of cGMP as well as the SOP stated in the Abbreviated New Drug Application (ANDA) and (NDA) New Drug Application to ensure compliance to cGMP requirements.

The following index identifies the company's key standard operating procedure (SOP) particularly applicable to the ANDA file submitted. Individual companies shall identify their own critical procedures and list them accordingly.

Critical Procedures Index

SOP No.	Subject	Issue Date	Revision No.

SOP No.: Val. 1700.10 **Effective date: mm/dd/yyyy**

Approved by:

Critical Procedures Index (continued)

SOP No.	Subject	Issue Date	Revision No.

REASONS FOR REVISION

Effective date: mm/dd/yyyy

■ First time issued for your company, affiliates, and contract manufacturers.

YOUR COMPANY
VALIDATION STANDARD OPERATING PROCEDURE

SOP No.: Val. 1700.20

Effective date: mm/dd/yyyy

Approved by:

TITLE: **Particulate Monitoring (Nonviable) of Injectable Area**

AUTHOR:

_____

Name/Title/Department

_____

Signature/Date

CHECKED BY:

_____

Name/Title/Department

_____

Signature/Date

APPROVED BY:

_____

Name/Title/Department

_____

Signature/Date

REVISIONS:

No.	Section	Pages	Initials/Date

SOP No.: Val. 1700.20

Effective date: mm/dd/yyyy

Approved by:

SUBJECT: Particulate Monitoring (Nonviable) of Injectable Area

PURPOSE

This procedure defines requirements for monitoring of air quality relative to airborne particulates of injection manufacturing and packing area and delineates the responsibilities for assurance of compliance with this procedure. The room number, activity, and location number defined are hypothetical. Individual companies can apply the SOP as a model to develop their own SOPs.

RESPONSIBILITY

Area QAI (quality assurance inspector) is responsible for following the procedure. Production manager and quality assurance manager are responsible for SOP compliance.

PROCEDURE

1. Introduction

In the pharmaceutical industry, reduction and/or elimination of microbially nonviable particulates in the environmental air reduces the possibility of product contamination during processing. Particulate matter and microbial monitoring programs are necessary for areas where product quality can be affected by environmental conditions.

2. Airborne Particulate Cleanliness Class

Following are the airborne particulate cleanliness classes. Each class is defined by the maximum allowable number of particles equal to or greater than 0.5 and 5.0 μm in size per cubic foot of air. The class is considered met if the measured particle concentration per cubic foot is within the limits specified.

SOP No.: Val. 1700.20 Effective date: mm/dd/yyyy

Approved by:

Measured Particle Size (Micrometers)/Cubic Foot of Air

Class		0.5	5.0
100	ISO 5	100	—
1000	ISO 6	1,000	7
10,000	ISO 7	10,000	70
100,000	ISO 8	100,000	700

2.1 HEPA (high-efficiency particulate air) filter

HEPA filters are units or systems used to purify air entering controlled areas. This program includes air monitoring of nonviable particulate monitoring for all sites, where operations can be critically affected by particulates.

3. Cleanliness Classified Areas (at Rest and at Work, Dynamic) with Frequency of Monitoring Injectable Manufacturing Area

3.1 Classification (at rest, design criteria)

- Level I areas are classified as class 100 (LAF and LFH) areas where sterilized products and their contact parts may be exposed to the environment.
- Level II areas are classified as class 1000 areas and restricted to personnel who are certified to be gowned for the aseptic processing.
- Level III areas are classified as class 10,000 areas used to process product contact materials before they are sterilized or depyrogenated. Also, the rooms surrounding sterile core (level II) areas are classified as level III.
- Level IV areas are classified as class 100,000 areas; these include production corridors, air locks, packaging areas, etc.

Table 1 gives examples of level I, level II, level III, and level IV rooms.

3.2 Particulate testing (nonviable)

- Area QAI is responsible for performing the airborne particulate test in cleanliness classified areas for checking the particulate matter

SOP No.: Val. 1700.20   Effective date: mm/dd/yyyy

Approved by:

in environmental air. Test area, frequency, and limits are included in the attachments as follows:

■ Attachment no. 1700.20(A) — particulate monitoring injectable area
■ Attachment no. 1700.20(B) — nonconformance alert and investigational report

■ Particulate quantitative air sampling will be performed using device CI-500 innovation laser particle counter (serial no.: mentioned on the monitoring format) according to SOP (provide number)

■ A total of three readings will be carried out for the particles of 0.5 and 5.0 μm in the test area (stations mentioned in attachment no. 1700.20(C). Readings in the test station will be carried out from area of test station (enclosed sampling location drawing). If the particulate counter is transferred to a higher cleanliness class, it is decontaminated with 70% ethanol.

■ For each cleanliness class in the injectable preparation area (e.g., class 100, class 10,000, and class 100,000), alert limits will be half the number of particles allowed of 0.5 and 5.0 μm for each area. Area supervisors shall be immediately notified verbally of particles exceeding the alert limit (in any of the three readings) in any area for corrective action.

■ For sampling, the particle counter should be positioned at 1 m height above the floor level (i.e., working level).

■ For continuous monitoring in the filling area, particle counter probe will be positioned at a distance of not more than 12 in. from the filling nozzle (same identified position as used in the media fill runs for the vial and ampoule filling machine). Alert limits are as follows:

Class	ISO Class	Maximum Allowable Number (Action Limit) of Particles per Cubic Foot of Air		Alert Limit per Cubic Foot of Air	
		0.5 μm	5.0 μm	0.5 μm	5.0 μm
Class 100	ISO 5	100	—	50	—
Class 1,000	ISO 6	1,000	7	500	3.5
Class 10,000	ISO 7	10,000	70	5,000	35
Class 100,000	ISO 8	100,000	700	50,000	350

Table 1

Classifi-cation Level	Room No.	Activity	Cleanroom Class at Rest
IV	Provide room no.	Production corridor	100,000
I	Provide room no.	Syringe pass box (LAF)	100
III	Provide room no.	Syringe prep. (surrounding LFH)	10,000
I		Syringe prep. tunnel (LFH)	100
IV	Provide room no.	Change room I (entry side)	100,000
III	Provide room no.	Change room (sterile side)	10,000
IV	Provide room no.	Coding room	10,000
IV	Provide room no.	Air lock washing	100,000
IV	Provide room no.	Packaging	100,000
I	Provide room no.	Syringe filling (LAF)	100
II		Syringe filling (surrounding LAF)	100
I	Provide room no.	Unloading autoclave LAF	100
III		Unloading autoclave (surrounding LAF)	1000
IV	Provide room no.	Washing area for compounding vessels	100,000
IV	Provide room no.	Air lock for materials	10,000
IV	Provide room no.	Personnel air lock II	10,000
III	Provide room no.	Solution preparation	10,000
IV	Provide room no.	Vial capping	10,000
III	Provide room no.	Commodity washing and autoclave loading	10,000 10,000 100
IV	Provide room no.	Packaging	100,000
I	Provide room no.	Ampoule filling (LFH)	100
I		Vial filling (LFH)	100
II		Vial/ampoule filling (surrounding LAF)	100
		Tunnel outlet machine	100
I	Provide room no.	Loading freeze dryer (LAF)	100
II		Freeze dryer (surrounding LAF)	100
III	Provide room no.	Solution room	10,000
I		Solution room (pass box) (LFH)	100
I	Provide room no.	Sterilization tunnel	100
II		(LFH)	100 100

SOP No.: Val. 1700.20   **Effective date: mm/dd/yyyy**
                                              Approved by:

Cleanroom Class at Work	Location[a]	Frequency	Sampling Location ID No.[b]
100,000	Opposite F.D Tech. R	Once/	A
	Opposite pack. A	week	B
100	Under LFH of pass box	Each operating shift	C
10,000	Center of room	Each operating shift	D
100	Under LFH		E
100,000	Center of room	Once/week	F
10,000	Center of room	Each operating shift	G
100,000	Center of room	Twice/week	H
100,000	Center of room	Once/week	I
100,000	Center of room	Once/week	J
100	Under LAF	Each operating shift	K
1,000	Center of room		L
1,000	Under LAF	Each operating shift	M
10,000	Center of room		N
100,000	Center of room	Twice/week	O
100,000	Center of room	Twice/week	P
100,000	Center of room	Once/week	Q
100,000	Center of room (1)	Each operating shift	R
	Center of room (2)		S
10,000	Near capping machine	Each operating shift	T
100,000	Near washing tunnel	Each operating shift	U
100,000	Near autoclave		V
1,000	Near sink (under LFH)		W
100,000	Center of room	Once/week	X
100	Under LFH	Each operating shift	Y
100	Under LFH		Z
1,000	Center of room		AB
100	Under LFH		AC
100	Under LAF	Each operating shift	AD,AE,AF
1,000	Center of room		AG
10,000	Center of room	Each operating shift	AH
100	Under LFH		AI
00	Preheating zone (LFH)	Daily each operational	AJ
00	Heating zone (LFH)	shift	AK
00	Cooling zone (LFH)		AL

SOP No.: Val. 1700.20

Effective date: mm/dd/yyyy

Approved by:

Table 1 (continued)

[a] Location defined in Table 1 based on critical sampling site determination by grid technique.
[b] Refer to drawing for location identification (your facility).

Notes: Particulate monitoring during unloading freeze dryer area will be considered as level III; locations not in use will be monitored during the complete area monitoring once per week; R represents hypothetical room numbers.

■ If counts do exceed the class limit in any of the test reading, the QAI will immediately identify and record the cause of the higher counts. If no cause is apparent and the counts continue to exceed class criteria at the product opening, the QAI should inform QA manager and production manager to stop the production operation immediately and to carry out corrective actions.

■ A corrective action system (form attached, nonconformance alert and investigation report [nonviable]), attachment no. 1700.20(B) should be activated. This system contains provisions for the following:

■ Notification that particulate levels are exceeded

■ Review of associated product monitoring data

■ Corrective action taken to remedy the situation within an appropriate time frame

■ Verification of remedial action by data collection

Corrective actions shall address the problem generally and not merely resolve the counts associated with the specific sampling point.

Monthly data shall be collected for each cleanliness class and summarized for the management review.

SOP No.: Val. 1700.20   Effective date: mm/dd/yyyy

                                         Approved by:

REASONS FOR REVISION

Effective date: mm/dd/yyyy

- First time issued for your company, affiliates, and contract manu-facturers.

SOP No.: Val. 1700.20

Effective date: mm/dd/yyyy
Approved by:

Attachment No. 1700.20 (A)
PARTICULATE MONITORING

DATE: _____

FREQUENCY: _____

PRODUCTs / ACTIVITIES : _____

AREA	*Syringe Prep. Tunnel Room No. (Class 100) Reference sampling point		*Syringe Filling Room No. Class (100) Reference sampling point		*Syringe Filling Room No. (Class 100/1000) Reference sampling point		*Unloading under LAF Room No. (Class 100/1000) Reference sampling point	
SIZE	0.5 µ	5.0 µ	0.5 µ	5.0 µ	0.5 µ	5.0 µ	0.5 µ	5.0 µ
Reading 1								
Reading 2								
Reading 3								
AVERAGE								

AREA	*Loading freeze dryer Room No. (Class 100) Reference sampling point		*Loading freeze dryer Room No. Class (100/1000) Reference sampling point		*Solution Room Room No. (Class 100) Reference sampling point		*Sterilization tunnel Room No. (Class 100) Reference sampling point	
Reading 1								
Reading 2								
Reading 3								
AVERAGE								

AREA	*#Coding room Room No. (Class 10,000/100,000) Reference sampling point		*Unloading autoclave Room No. (Class 1000/10,000) Reference sampling point		#Material Airlock Room No. (Class 10,000/100,000) Reference sampling point		#Personnel Airlock Room No. (Class 10,000/100,000) Reference sampling point	
Reading 1								
Reading 2								
Reading 3								
AVERAGE								

AREA	*Commodity Washing Room No. (Class 100) Reference sampling point		*Solution room Room No. (Class 10,000) Reference sampling point		Production Corridor Room No. (Class 100,000) Reference sampling point		Production Corridor Room No. (Class 100,000) Reference sampling point	
Reading 1								
Reading 2								
Reading 3								
AVERAGE								

Remarks & comments: _____
_____
_____
_____

C.O.R = centre of the room. LFH = under laminar flow hood. LAF: Laminar flow

All limits per cubic foot of air sampled Corrective Action (s) in case of failure (NC and investigation report (n0n-viable)) a) Notification c) Corrective Action b) Review d) Verification by data collection	*Each operational shift #Twice per week Remaining all weekly

SOP No.: Val. 1700.20

Effective date: mm/dd/yyyy

Approved by:

SOP No.: 1700.20
Issued on: mm/dd/yyy
Revision No.: New

Device used: CI-500 Laser Particle Counter

Serial No.: _____

Calibration due: _____

*Ampoule filling Room No. (Class 100) Reference sampling point		*Vial filling (LFH) Room No. (Class 100) Reference sampling point		*Vial/Ampoules Room No. (Class 100/1000) Reference sampling point		*Loading freeze dryer Room No. (Class 100) Reference sampling point		*Loading freeze dryer Room No. (Class 100) Reference sampling point	
0.5 μ	5.0 μ	0.5 μ	5.0 μ	0.5 μ	5.0 μ	0.5 μ	5.0 μ	0.5 μ	5.0 μ

*Sterilization tunnel Room No. (Class 100) Reference sampling point		*Sterilization tunnel Room No. (Class 100) Reference sampling point		*HEPA b/w tunnel Room No. (Class 100) Reference sampling point		*Syringe Preparation (COR) Room No. (Class 10,000) Reference sampling point		*Change room (at work) Room No. (Class 10,000) Reference sampling point	

*Solution Preparation Room No. (Class 10,000/100,000) Reference sampling point		*Solution Preparation Room No. (Class 10,000/100,000) Reference sampling point		Vial capping (COR) Room No. (Class 10,000) Reference sampling point		* Commodity washing Room No. (Class 10,000/100,000) Reference sampling point		* Commodity washing Room No. (Class 10,000/100,000) Reference sampling point	

Change Room (at work) Room No. (Class 100,000) Reference sampling point		Air lock washing Room No. (Class 100,000) Reference sampling point		Packaging Room No. (Class 100,000) Reference sampling point		Washing Room No. (Class 100,000) Reference sampling point		Packaging Room No. (Class 100,000) Reference sampling point	

lass 100 (100 particles of 0.5m lass 1000 (1000 particles of 0.5m and 7.0 of 5.0m) lass 10,000 (10,000 particles of 0.5m and 7.0 of 5.0m) lass 100,000 (100,000 particles of 0.5m and 7.0 of 5.0m)	Done by:	Checked by:
	Q.A.I	QA Officer

SOP No.: Val. 1700.20

Effective date: mm/dd/yyyy

Approved by:

Attachment No. 1700.20(B)	**Issued on: mm/dd/yyyy**
PARTICULATE COUNT, NONCONFORMANCE ALERT,	**Revision No: New**
AND INVESTIGATION REPORT	

Date: _____ Location: _____ Room no.: _____
Product: _____ Batch no.: _____
Equipment used: _____ Calibrated on: _____
Calibration due on: _____

RESULTS

Area	Averages	
	0.5 μ	0.5 μ

Investigation

- No. of personnel: _____
- Activity: _____
- Materials handled inside the room: _____
- Unusual events during processing: _____
- Gowning used by the staff: _____
- Filter maintenance: _____
- Room cleaning records: _____
- Temperature of the area: _____
- Pressure differential of the area: _____
- Liquid-borne particulate of the product (filled): _____
- Visual inspections of the filled product: _____
- Sterility testing: _____
- Endotoxin test: _____
- Air sampling results (microbiological): _____
- Personnel hygiene: _____

Conclusion: _____
_____

Corrective action: _____
_____

| QAI | Production Manager | Quality Assurance Manager |

cc: Quality Assurance Manager/Production Manager
Original: QAI. In case of NC investigation, report to be raised by QAI.

SOP No.: Val. 1700.20   Effective date: mm/dd/yyyy

                                               Approved by:

Attachment No. 1700.20(C) PARTICULATE (NONVIABLE) SAMPLING POINTS LOCATION MAP	Issued on: mm/dd/yyyy Revision No: New

Provide scale-up drawing of your facility
(sampling location with I.D. number)

Drawn by: _____ Checked by: _____ Approved by: _____	Date: _____ Date: _____ Date: _____	Sampling point location map	Drawing no.: XYZ

| YOUR COMPANY |
| VALIDATION STANDARD OPERATING PROCEDURE |

SOP No.: Val. 1700.30

Effective date: mm/dd/yyyy

Approved by:

TITLE: QA Responsibilities in Injectable Area

AUTHOR:

_____
Name/Title/Department

_____
Signature/Date

CHECKED BY:

_____
Name/Title/Department

_____
Signature/Date

APPROVED BY:

_____
Name/Title/Department

_____
Signature/Date

REVISIONS:

No.	Section	Pages	Initials/Date

SOP No.: Val. 1700.30   Effective date: mm/dd/yyyy

                                                   Approved by:

SUBJECT: QA Responsibilities in Injectable Area

PURPOSE

The purpose is to provide the guidelines for the QA inspector responsible for parenteral facility.

RESPONSIBILITY

Area QAI (quality assurance inspector) is responsible for following the procedure. The quality assurance (QA) manager is responsible for SOP compliance.

PROCEDURE

The QAI will perform the following activities. In case of any deviation, the QAI will inform the QA manager and area manager to ensure that corrective actions are taken.

- QAI will check the particulate of the manufacturing area using laser particulate counter, per plant SOP (provide number).
- QAI will monitor the results of pressure differentials, temperature, and humidity obtained from the control room daily before starting the manufacturing process in the area for comparison of results with the standard provided for each area. They should be within limits.
- Whenever any operation such as manufacturing or filling starts, QAI will fill an inspection start-up checklist and give release for manufacturing or packaging. See attachment no. 1700.30(A), 1700.30(B), and 1700.30(C).
- QAI is responsible to give the line clearance and line release according to SOP (provide number) and countersign the formats. See attachment no. 1700.30(D) and 1700.30(E).
- When filling of the ampoules starts in the area, QAI will be informed. He will check the area/equipment clearance and then check the dose of the filled ampoules (by weight and volume), length of the ampoules, and sealing and also check that ampoules/vials are free from particles. See SOP (provide number). When the dose of the filled ampoules is found satisfactory, QAI will give release for filling

and coding. See attachment no. 1700.30(B). For entries of the readings, see attachment no. 1700.30(F).

■ After completion of visual inspection by manufacturing, QAI will perform (a) the review of printout of cold leak test carried out by production; and (b) the visual inspection. See SOP (provide number). He gives the release for packing if the results are satisfactory.

■ QAI will fill the optical inspection report attachment no. 1700.30(G) and 1700.30(H) for each batch optically inspected and for leakage.

■ When packaging of the batch starts, the production supervisor will request the area QAI for line clearance. The QAI will check the area and countersign the line clearance given by the supervisor. See attachment no. 1700.30(B).

■ QAI will perform the in-process audit checks hourly for the packaging of products according to SOP (provide number) and record the observations on form. See attachment no. 1700.30(I).

■ For sample size, see military standard 105-E and attachment no. 1700.30(J) and 1700.30(K).

REASONS FOR REVISION

Effective date: mm/dd/yyyy

■ First time issued for your company, affiliates, and contract manufacturers.

SOP No.: Val. 1700.30

Effective date: mm/dd/yyyy
Approved by:

Attachment No. 1700.30(A)

INSPECTION START-UP CHECK LIST

Issued date: mm/dd/yyyy
Revision no.:

Date: _____

Room/Area/Line: _____

Machine/Line: _____

Product to start: _____

Batch no.: _____

Code no.: _____

Expiry date: _____

Previous product: _____

Batch no.: _____

1. Differential pressure between room and corridor as required ( )
2. Room/area/line is clean ( )
3. Machine/line is clean ( )
4. No materials left from previous product or batch ( )
5. No documents left from previous product or batch ( )
6. Materials/bulk product to be used: ( )
   a) Available ( )
   b) Complete ( )
   c) Released by QC ( )
   d) Compared to planning requisition and found same ( )

Remarks _____

_____

_____

Batch is released to start by: _____
Signature and Date

Effective date: mm/dd/yyyy

Approved by:

Attachment No. 1700.30(B)

Issued date: mm/dd/yyyy
Revision no.:
Date: _____

QUALITY ASSURANCE DEPARTMENT
RELEASE
(MANUFACTURING)

Product: _____

Code no.:_____ Batch no.: _____

Area: _____ Machine: _____

Process: _____

Date: _____

Signature: _____

SOP No.: Val. 1700.30

Effective date: mm/dd/yyyy

Approved by:

Attachment No. 1700.30(C)

Issued date: mm/dd/yyyy
Revision no.:
Date: _____

QUALITY ASSURANCE DEPARTMENT
RELEASE
(PACKAGING)

Product: _____

Code no.: _____ Batch no.: _____

Line no.: _____ Consignee: _____

MFG. date: _____ Expiry date: _____

Date: _____

Signature: _____

SOP No.: Val. 1700.30

Effective date: mm/dd/yyyy
Approved by:

Attachment 1700.30(D) AREA/EQUIPMENT CLEARANCE PRODUCTION		SOP No.: 1700.30 Issued on: mm/dd/yyyy Revision No: New
___All related SOPs followed ___No traces of previous product ___Equipment/area cleaned per SOPs ___Correct material of product for which area equipment/clearance required		Product for which line clearance required Product name: _____ Batch no.: _____
Previous Product		*Present*
	Process: _____	
	Equipment/room no.: _____	
Previous product:		Time: _____
Batch no.:		Date: _____
Date:		Requested by: _____
Cleared by:	Checked by:	Checked by: _____
	Process: _____	
	Equipment/room no.: _____	
Previous product:		Time: _____
Batch no.:		Date: _____
Date:		Requested by: _____
Cleared by:	Checked by:	Checked by: _____
	Process: _____	
	Equipment/room no.: _____	
Previous product:		Time: _____
Batch no.:		Date: _____
Date:		Requested by: _____
Cleared by:	Checked by:	Checked by: _____
	Process: _____	
	Equipment/room no.: _____	
Previous product:		Time: _____
Batch no.:		Date: _____
Date:		Requested by: _____
Cleared by:	Checked by:	Checked by: _____

SOP No.: Val. 1700.30

Effective date: mm/dd/yyyy

Approved by:

Attachment No. 1700.30(E) **AREA/EQUIPMENT/LINE CLEARANCE** **PACKAGING**	SOP No.: 1700.30 Issued on: mm/dd/yyyy Revision No.: New

___Previously finished stock/component removed along with the previous product documents
___Equipment, area, belt cleaned
___Correct product and commodities available for which line clearance required
___All other requirements followed as mentioned in related SOPs

Previous Product on the Line

Product: _____
Code no.: _____
Batch no.: _____
Consignee: _____
Date: _____

Previous Product on the Line

Product:
Code no.:
Batch no.:
Consignee:
Date:

Product for Which Line Clearance Required

Product: _____
Code no.: _____
Batch no.: _____
Line no.: _____

Consignee:_____ Time: _____ Date: _____ Cleared by: (line boss)____ Checked by: (supervisor)__ Verified by: (QAI)_____	Number of units rejected during setup Number of units for rework	Consignee: _____ Time: _____ Date: _____ Cleared by: (line boss)_____ Checked by: (supervisor) _____ Verified by: (QAI) _____
Consignee:_____ Time: _____ Date: _____ Cleared by: (line boss)____ Checked by: (supervisor)__ Verified by: (QAI)_____	Number of units rejected during setup Number of units for rework	Consignee: _____ Time: _____ Date: _____ Cleared by: (line boss)_____ Checked by: (supervisor) _____ Verified by: (QAI) _____
Consignee:_____ Time: _____ Date: _____ Cleared by: (line boss)____ Checked by: (supervisor)__ Verified by: (QAI)_____	Number of units rejected during setup Number of units for rework	Consignee: _____ Time: _____ Date: _____ Cleared by: (line boss)_____ Checked by: (supervisor) _____ Verified by: (QAI) _____

SOP No.: Val. 1700.30

Effective date: mm/dd/yyyy

Approved by:

Attachment No. 1700.30(F)
QA IN-PROCESS X- AND R-CHARTS

SOP No.: 1700.30
Issued on: mm/dd/yyyy
Revision No.: New

Product:
Code no.:
Batch no.:

Process:
Batch size:

Machine:
Operator:
Room no.:

Target:
UCL:
UCL:

LCL: _____ (In-process limit)
LCL: _____ (Lab. limit)

Date/Time	Individual Fill Volume	Individual Fill Weight (g)	Average	Range
1				
2				
3				
4				
5				
6				
7				
8				
9				
10				

X-Chart

Limits	1	2	3	4	5	6	7	8	9	10

R-Chart

Limits	1	2	3	4	5	6	7	8	9	10

Description of Any Corrective Action
Ceramic printing

Inspection of Filled and Sealed Vials/Ampoules/Refilled Syringes

S. No.	Length	Sealing	Free from Particles	Breaking Ring
1				
2				
3				
4				
5				
6				
7				
8				
9				
10				

Batch
Average:

Signature (Quality Assurance Inspector)

SOP No.: Val. 1700.30　　　　　　　　**Effective date: mm/dd/yyyy**

　　　　　　　　　　　　　　　　　　　　Approved by:

Attachment No. 1700.30(G)

　　　　　　　　Issued date: mm/dd/yyyy

　　　　　　　　Revision no.:

　　　　　　　　Date: _____

From:　　Production Supervisor
To:　　　Quality Assurance Inspector
Subject:　Leak Test/Visual Inspection
Product:　_____
Code:　　_____
Batch no.:　_____

Sample Status	Signature Supervisor	Result (Passes/Fails)	Conclusion (OK/Repeat)	Signature (QAI)	Date
After leak test					
Leak test Repeat 1					
Leak test Repeat 2					
After optical Inspection					
Optical inspection Repeat 1					
Optical inspection Repeat 2					

In Case of Dispute

Remarks by system and inspection officers:

Remarks by operation and QA managers:

　　　　　　　　Signature/Date:

SOP No.: Val. 1700.30

Effective date: mm/dd/yyyy
Approved by:

Attachment No. 1700.30(H)
OPTICAL INSPECTION REPORT
IN-PROCESS CONTROL

Issued Date: mm/dd/yyyy
Revision No.: New

Product name: _____

Batch no.: _____

Code no.: _____

Sample size: _____

Optical Inspection for	First Sample	Second Sample	Remarks
Clarity of solution			
Particles			
Color (as specified)			
Fibers			
Glass pieces/glass dust			
Ampoule defect (sealing)			
Charring (burning of solution near the tip)			
Break ring color/color code			
Presence of OPC (one point cut)			
Printing (if applicable)			
Others			

Disposition: conforms/does not conform

Optical inspection by: _____ Date: _____

SOP No.: Val. 1700.30

Effective date: mm/dd/yyyy

Approved by:

SOP No.: 1700.30
Issued on: mm/dd/yyyy
Revision No.: New

Attachment No.1700.30(I)
Q.A. FINAL PACKAGING RECORD

Q.C.NO:
Product:
Code No.:
Batch No.:
No. of units:

Line No.:
Line boss:
Consignee:
Pack Size:
Batch Size:

Packaging started:
Packaging completed:
Mfg. date:
Expiry date:

Commodities used

S.No.	CRITICAL DEFECTS	Time / Sample size	1	2	3	4	5	6	7	8	9	10	Description of any corrective action and remarks
1	Wrong commodity												
2	Wrong batch no.:												
	Wrong expiry												
3	Wrong manufacturing												
	Wrong special stamping												
4	Incorrect no. of units in C. carton												
5	Foreign material												
6	Empty unit												
	Total defects												
	MAJOR DEFECTS	Sample size											
1	Missing component												
	(Dosing cup, leaflet, Cannula, etc.)												
2	Illegible B. No.												
	Illegible mfg. date												
	Illegible exp. date												
3	Others												
	Total defects												
	MINOR DEFECTS	Sample size											
1	Loose or torn labels/carton/box												
2	Double labels												
3	Printing smeared partly												
4	Crooked label												
5	Excessive shrinkage label												
6	Improper shrink wrap												
7	Other												
	Total defects	Sample size											
	SAMPLED QUANTITY/HOUR												TOTAL
	PHYSICAL / CHEMICAL												
	MICROBIOLOGICAL												
	GRAND TOTAL												
	SIGNATURE												RECEIVED IN Q.C. BY

QA0002A

SOP No.: Val. 1700.30 Effective date: mm/dd/yyyy

Approved by:

Attachment No. 1700.30(J)
MILITARY STANDARD 105-E
GENERAL INSPECTION LEVEL-I

Double Sampling Plan

Lot Size or Batch Size	Code Letter	Sample Size	Cumul. Sample Size	Critical Defects AQL 0.0%		Major Defects AQL 1.0%		Minor Defects AQL 2.5%	
				Acc.	Rej.	Acc.	Rej.	Acc.	Rej.
2–8	A	2	—	0	1	0	1	0	1
Single Sampling Plan									
9–15	A	2	—	0	1	0	1	0	1
Single Sampling Plan									
16–25	B	3	—	0	1	0	1	0	1
Single Sampling Plan									
26–50	C	5	—	0	1	0	1	0	1
Single Sampling Plan									
51–90	C	5	—	0	1	0	1	0	1
Single Sampling Plan									
91–150	D	8	—	0	1	0	1	0	1
Single Sampling Plan									
151–280	E	13	—	0	1	0	1	1	2
Single Sampling Plan									
281–500	F	20	—	0	1	0	1	1	2
Single Sampling Plan									

Notes: Acc. = acceptance number; Rej. = rejection number; Cumul. = cumulative sample.

If acceptance number exceeded on the first sample, but rej. number is not crossed, inspect another sample.

If rejection number exceeded on the first sample, no second sample is taken.

SOP No.: Val. 1700.30

Effective date: mm/dd/yyyy
Approved by:

Attachment No. 1700.30(K)
MILITARY STANDARD 105-E
GENERAL INSPECTION LEVEL-I

SOP No.: 1700.30
Issued on: mm/dd/yyyy
Revision No.: New

Double Sampling Plan

Lot Size or Batch Size	Code Letter	Sample Size	Cumul. Sample Size	Critical Defects AQL 0.0%		Major Defects AQL 1.0%		Minor Defects AQL 2.5%	
				Acc.	Rej.	Acc.	Rej.	Acc.	Rej.
501–1200	G	20	20	0	1	0	2	0	3
		20	40	0	1	1	2	3	4
1201–3200	H	32	32	0	1	0	2	1	4
		32	64	0	1	1	2	4	5
3201–10,000	J	50	50	0	1	0	3	2	5
		50	100	0	1	3	4	6	7
10,001–35,000	K	80	80	0	1	1	4	3	7
		80	160	0	1	4	5	8	9
35,001–150,000	L	125	125	0	1	2	5	5	9
		125	250	0	1	6	7	12	13
150,001–500,000	M	200	200	0	1	3	7	7	11
		200	400	0	1	8	9	18	19
500,001+	N	315	315	0	1	5	9	11	16
		315	630	0	1	12	13	26	27

| YOUR COMPANY |
| VALIDATION STANDARD OPERATING PROCEDURE |

SOP No.: Val. 1700.40

Effective date: mm/dd/yyyy

Approved by:

TITLE: Particulate Matter in Injectables, USP Criteria

AUTHOR: _____

Name/Title/Department

_____

Signature/Date

CHECKED BY: _____

Name/Title/Department

_____

Signature/Date

APPROVED BY: _____

Name/Title/Department

_____

Signature/Date

REVISIONS:

No.	Section	Pages	Initials/Date

SOP No.: Val. 1700.40

Effective date: mm/dd/yyyy

Approved by:

SUBJECT: Particulate Matter in Injectables, USP Criteria

PURPOSE

The purpose is to provide the USP criteria for the monitoring of liquid-borne particulate matter in injections (large- and small-volume parenterals).

RESPONSIBILITY

Area QAI (quality assurance inspector, injections) is responsible for the monitoring of liquid-borne particulate matter in the injection per the criteria mentioned in the SOP. The quality assurance (QA) manager is responsible for SOP compliance.

PROCEDURE

The test applies to large-volume injections labeled as containing more than 100 ml, unless otherwise specified in the individual monograph. It counts suspended particles that are solid or liquid. The test applies also to single- or multiple-dose, small-volume injections labeled as containing 100 ml or less that are in solution or in solution constituted from sterile solids, where a test for particulate matter is specified in the individual monograph. Injections packaged in prefilled syringes and cartridges are exempt from these requirements, as are products for which an individual monograph specifies that the label states that the product is to be used with a final filter.

1. Test Apparatus

The apparatus is an electronic, liquid-borne particle-counting system that uses a light-obscuration sensor with a suitable sample feeding device. It is the responsibility of those performing the test to ensure that the operating parameters of the instrumentation are appropriate to the required accuracy and precision of the test result.

Equipment: Climet CI 1000 particulate analyzer; Climet CI1020 batch sampler

Effective date: mm/dd/yyyy

Approved by:

2. Test Environment

Perform the test in an environment that does not contribute any significant amount of particulate matter. Specimens must be cleaned to the extent that any level of extraneous particles added has a negligible effect on the outcome of the test. The test specimen, glassware, closures, and other required equipment preferably are prepared in an environment protected by high efficiency particulate air (HEPA) filters. Nonshedding garments and powder-free gloves preferably are worn throughout the preparation of samples.

Cleanse glassware, closures, and other required equipment, preferably by immersing and scrubbing in warm, nonionic detergent solution. Rinse in flowing tap water and then rinse again in flowing filtered water. Organic solvents may also be used to facilitate cleaning.

3. Test Preparation and Procedure

For containers having volumes of less than 25 ml, test a solution pool of ten or more units. Single units of small-volume injections may be tested individually if the individual unit volume is 25 ml or greater.

Prepare the test specimens in the following sequence. Remove outer closures, sealing bands, and any loose or shedding paper labels. Rinse the exterior of containers with filtered distilled water, as described under "Test Environment," and dry, taking care to protect the containers from environmental contamination. Withdraw the contents of the containers in the normal or customary manner of use or as instructed in the package labeling. However, containers with removable stoppers may be sampled directly by removing the closure or, if test specimens are being pooled, by removing the closure and emptying the contents into a clean container.

3.1 Procedure

3.1.1 Liquid fill (contents of each unit are less than 25 ml)

Mix each unit by inverting it 20 times to resuspend any particles. Open and combine the contents of ten or more units in a cleaned container to obtain a volume of not less than 20 ml. Gently stir the contents of the container by hand-swirling or by mechanical means, taking care not to introduce air bubbles or contamination. Withdraw not less than three aliquot portions,

SOP No.: Val. 1700.40

Effective date: mm/dd/yyyy
Approved by:

each not less than 5 ml in volume, into the light obscuration counter sensor. Obtain the particle counts and discard the data from the first portion.

3.1.2 Liquid fill (contents of each unit are 25 ml or more and option of testing individual units is selected)

Mix one unit by inverting it 20 times. Remove the closure and insert the counter probe into the center of the solution. Gently agitate the contents of the unit by hand-swirling or by mechanical means. Withdraw not less than three aliquot portions, each not less than 5 ml in volume, into the light obscuration counter sensor. Obtain the particle counts and discard the data from the first portion.

3.1.3 Dry or lyophilized fill

Open the container, taking care not to contaminate the opening or cover. Constitute with a suitable volume of filtered water or with the appropriate filtered diluent if water is not suitable. Replace the closure and manually agitate the container to dissolve the drug. Allow to stand until the drug is completely dissolved. Prior to analysis, gently stir the contents of the containers by hand-swirling or by mechanical means, taking care not to introduce air bubbles or contamination. Pool or test individually the appropriate number of units and withdraw not less than three aliquot portions, each not less than 5 ml in volume, into the light obscuration counter sensor. Obtain the particle counts and discard the data from the first aliquot.

4. Acceptance Criteria

For *pooled samples (small-volume injections)*, average the counts from the two or more aliquot portions analyzed. Calculate the number of particles in each container by the formula

$$PV_t/V_a n$$

in which
  P is the average particle count obtained from the portions analyzed
  V_t is the volume of pooled sample, in milliliters
  V_a is the volume, in milliliters, of each portion analyzed
  n is the number of containers pooled

SOP No.: Val. 1700.40   Effective date: mm/dd/yyyy

Approved by:

For *individual samples* (*small-volume injections*), average the counts obtained for the 5-ml or greater aliquot portions from each separate unit analyzed and calculate the number of particles in each container by the formula

$$PV/V_a$$

in which
P is the average particle count obtained from the portions analyzed
V is the volume, in milliliters, of the tested unit
V_a is the volume, in milliliters, of each portion analyzed

See attachment no. 1700.40(B) for calculation and interpretation of results.

For *individual unit samples* (*large-volume injections*), average the counts obtained for the two or more 5-ml aliquot portions taken from the solution unit. Calculate the number of particles in each milliliter taken by the formula

$$P/V$$

in which
P is the average particle count for individual 5 ml or greater sample volume
V is the volume, in milliliters, of the portion taken

See attachment no. 1700.40(C) for calculation and interpretation of results.

5. Interpretation

The injection meets the requirements of the test if the average numbers of particles present in the units tested do not exceed the appropriate value listed in Table 1.

Table 1   Light Obscuration Test Particle Count

	$\geq 10\ \mu m$	$\geq 25\ \mu m$
Small-volume injections	6000	600 per container
Large-volume injections	25	3 per milliliter

6. Documentation

- Attachment 1700.40(A) for pooled sample/container
- Attachment 1700.40(B) for individual sample/container
- Attachment 1700.40(C) for individual sample/ml

REASONS FOR REVISION

Effective date: mm/dd/yyyy

- First time issued for your company, affiliates, and contract manufacturers.

SOP No.: Val. 1700.40

Effective date: mm/dd/yyyy

Approved by:

SOP No.: Val. 1700.40

Effective date: mm/dd/yyyy

Approved by:

ATTACHMENT No.1700.40(A)

LIQUID-BORNE PARTICULATE MATTER IN INJECTION
(For Pooled Sample / Container)

SOP No.: 1700-40
Issued on: mm/dd/yyyy
Revision No.: New

PRODUCT:

CODE No:

BATCH No:

PACK SIZE:

TEST APPARATUS
    Climet Cl 1000 particulate analyzer
    Climet Cl 1020 batch sampler

ACCEPTANCE CRITERIA
    The injection meets the requirements of the test if the average numbers of particles present
    in the units tested do not exceed the appropriate value listed in Table.

Light Obscuration Test Particulate Count

	_ 10 µm	_ 25 µm
Small-volume injections	6000	600 per container
Large-volume injections	25	3 per ml

POOLED SAMPLES
(Where the contents of each unit are less than 25 ml)

Run I

SIZE (µ)	COUNT
1.00	
5.00	
10.0	
15.0	
20.0	
25.0	

Run II

SIZE (µ)	COUNT
1.00	
5.00	
10.0	
15.0	
20.0	
25.0	

CALCULATION

The number of particles in each container

$$PV_t/V_a n$$

P = the average particles count obtained from the portions analyzed
V_t = the volume, in ml, of the tested unit
V_a = the volume, in ml, of each portion analyzed.

n = the number of containers pooled

For 10 µ   P
             V_t
             V_a
             n

For 25 µ   P
             V_t
             V_a
             n

RESULT:

_ 10 µm   =   nos.

_ 25 µm   =   nos.

DISPOSITION: CONFORM   /   DOESN'T CONFORM   ANALYZED BY: _____

REMARKS: _____   DATE: _____
_____   CHECKED BY: _____
                                                Date: _____

SOP No.: Val. 1700.40

Effective date: mm/dd/yyyy

Approved by:

SOP No.: Val. 1700.40

Effective date: mm/dd/yyyy

Approved by:

<table>
<tr><td>Attachment No. 1700.40 (B)
LIQUID-BORNE PARTICULATE MATTER IN INJECTION
(For individual Sample / Container)</td><td>SOP No.: 1700-40
Issued on: mm/dd/yyyy
Revision No.: New</td></tr>
</table>

PRODUCT: _____

CODE No.: _____

BATCH No.: _____

PACK SIZE: _____

TEST APPARATUS
    Climet CI 1000 particulate analyzer
    Climet CI 1020 batch sampler

ACCEPTANCE CRITERIA
    The injection meets the requirements of the test if the average numbers of particles present
    in the units tested do not exceed the appropriate value listed in Table.

Light Obscuration Test Particulate Count

	_ 10 µm	_ 25 µm
Small-volume injections	6000	600 per container
Large-volume injections	25	3 per ml

INDIVIDUAL SAMPLES
(Where the contents of each unit are more than 25 ml)

Run I

SIZE (µ)	COUNT
1.00	
5.00	
10.0	
15.0	
20.0	
25.0	

Run II

SIZE (µ)	COUNT
1.00	
5.00	
10.0	
15.0	
20.0	
25.0	

CALCULATION
The number of particles in each container

$$PV/V_a$$

P = the average particles count obtained from the portions analyzed
V = the volume, in ml, of the tested unit
v_a = the volume, in ml, of each portion analyzed.

For 10 µ   P
          V
          V_a

For 25 µ   P
          V
          V_a

RESULT:
  _ 10 µm   =   nos.

  _ 25 µm   =   nos.

DISPOSITION: CONFORM   /   DOESN'T CONFORM   ANALYZED BY: _____

REMARKS: _____   DATE: _____

_____   CHECKED BY: _____

          Date: _____

SOP No.: Val. 1700.40

Effective date: mm/dd/yyyy

Approved by:

SOP No.: Val. 1700.40

Effective date: mm/dd/yyyy

Approved by:

Attachment - No. 1700.40(C)

LIQUID-BORNE PARTICULATE MATTER IN INJECTION

(For individual Sample / ml)

SOP No.: 1700-40
Issued on: mm/dd/yyyy
Revision No.: New

PRODUCT.:
CODE No.:
BATCH No.:
PACK SIZE.:

TEST APPARATUS
    Climet Cl 1000 particulate analyzer
    Climet Cl 1020 batch sampler

ACCEPTANCE CRITERIA
    The injection meets the requirements of the test if the average numbers of particles present
    in the units tested do not exceed the appropriate value listed in Table.

Light Obscuration Test Particulate Count

	10 μm	25 μm
Small-volume injections	6000	600 per container
Large-volume injections	25	3 per ml

INDIVIDUAL SAMPLES .
(Large Volume Parenterals)

Run I

SIZE (μ)	COUNT
1.00	
5.00	
10.0	
15.0	
20.0	
25.0	

Run II

SIZE (μ)	COUNT
1.00	
5.00	
10.0	
15.0	
20.0	
25.0	

CALCULATION

The number of particles in each container

$$P/V$$

P = the average particles count obtained from the portions analyzed
V = the volume, in ml, of the tested unit

For 10 μ   P
           V

For 25 μ   P
           V

RESULT:

  10 μm =   nos.

  25 μm =   nos.

DISPOSITION: CONFORM / DOESN'T CONFORM

ANALYZED BY:

REMARKS:

DATE:

CHECKED BY:

Date:

YOUR COMPANY
VALIDATION STANDARD OPERATING PROCEDURE

SOP No.: Val. 1700.50   Effective date: mm/dd/yyyy
                                          Approved by:

TITLE: Visual Inspection of Lyophilized Product
                  Parenterals

AUTHOR: _____
                              Name/Title/Department

                  _____
                              Signature/Date

CHECKED BY: _____
                              Name/Title/Department

                  _____
                              Signature/Date

APPROVED BY: _____
                              Name/Title/Department

                  _____
                              Signature/Date

REVISIONS:

No.	Section	Pages	Initials/Date

SOP No.: Val. 1700.50   Effective date: mm/dd/yyyy

                                               Approved by:

SUBJECT: Visual Inspection of Lyophilized Product Parenterals

PURPOSE

The purpose is to provide a procedure for checking and confirming that lyophilized parenterals are passing the following test and can be packed:

- Reconstitution time
- Visual inspection

RESPONSIBILITY

It is the responsibility of the quality assurance (QA) inspector stationed in the parenterals area to follow the SOP. The QA manager is responsible for SOP compliance.

PROCEDURE

1. Equipment_

1.1 Optical checking hood

Divide into two portions; one should have white background and the other black. The illumination should be from a shielded 100-W bulb daylight-quality light source, which provides an intensity of illumination of not less than 100 and not more than 350 fc, at a point 25.4 cm (10 in.) from its source.

2. Procedure for Inspection of Caked Lyophilized Vials

Collect a representative sample per the single sampling plan (general level B) and inspect. Record the observation in the attached format. Following is the sampling plan.

SOP No.: Val. 1700.50   Effective date: mm/dd/yyyy
                                                Approved by:

2.1 Visual inspection of lyophilized product

Single Sample Plan for Normal Inspection (General Level B)

Lot Size	Sample Size	Critical (0.15%) Acc.	Rej.	Major (0.65%) Acc.	Rej.	Minor (2.50%) Acc.	Rej.
501–3200	80	0	1	1	2	5	6
3201–35,000	315	0	1	5	6	14	15
35,001–150,000	500	0	1	7	8	21	22
150,000–500,000	800	0	1	10	11	21	22

Notes: Acc. = acceptance numbers; Rej. = rejection numbers.

3. Procedure

Reconstitute the lyophilized caked powder by adding clear and particulate-free distilled water to the vials one by one. Amount of water to be used should be per the product reconstitution requirement prior to use. Note the reconstitution time for each vial from the addition of water to obtaining a clear solution. Reconstitution time should not be more than the specification limit. Record the reconstitution time on the attached format. Calculate the average. Test should be carried out on 20 vials. Shake to dissolve and check as described in the procedure for visual inspection of solutions.

3.1 Solutions

Hold three vials to be inspected and with a shaking motion invert the container and return it to the upright position under optical checking hood for 10 to 12 seconds in front of white background and then in black background to inspect for the following:

- Any particle in the solution that tends to float, hang, or slowly settle onwards, not any fine flocculent material
- The clarity of the solution: no haze, solution sparkling clear
- Color as specified

SOP No.: Val. 1700.50　　　　　　　Effective date: mm/dd/yyyy

Approved by:

4. Disposition

The result will be given on attachment no. 1700.50(A). If failure occurs, the batch quantity will be reinspected by production.

REASONS FOR REVISION

Effective date: mm/dd/yyyy

- First time issued for your company, affiliates, and contract manu-facturers.

SOP No.: Val. 1700.50   Effective date: mm/dd/yyyy

Approved by:

Attachment No. 1700.50(A)

Issued date: mm/dd/yyyy
Revision no.:

VISUAL INSPECTION OF LYOPHILIZED PRODUCT

Product name: _____

Batch no.: _____Code no.: _____

Batch size: _____Sample size: _____

Lyophilized Cake

Critical	Acc. Nos.:	Rej. Nos.	Results	Remarks
1.	Glass chip			
2.	Metal chip			
3.	Missing stoppers			
4.	Non-crimp			
5.	Glass defect — loss of integrity			
6.	Hazy — discolored			
7.	Low fill/high fill (qualitative)			
8.	Others			
Critical	**Acc. Nos.:**	**Rej. Nos.**	**Results**	**Remarks**
1.	Twisted stopper			
2.	Cracked glass (intensity interact)			
3.	Particulate matter			
4.	Inconsistent cake/melt back (lyophilized)			
5.	Empty vials			
6.	Heavy score lines			
Critical	**Acc. Nos.:**	**Rej. Nos.**	**Results**	**Remarks**
1.	Crimp defect cosmetic			
2.	Glass defect cosmetic			
3.	Foreign matter on exterior of vial			

SOP No.: Val. 1700.50

Effective date: mm/dd/yyyy
Approved by:

4.	Defective cap		
5.	Others		

Total Defective/Sample Size	Description
Critical: _____	
Major: _____	
Minor: _____	
Done by: _____ Date: _____	

Reconstitution time: Perform on 20 vials

1.		3.		5.		7.		9.		11.		13.		15.		17.		19.		Average
2.		4.		6.		8.		10.		12.		14.		16.		18.		20.			

Reconstituted

Critical	Acc. Nos.:	Rej. Nos.	Results	Remarks
1.	Hazy/discolored			
Critical	Acc. Nos.:	Rej. Nos.	Results	Remarks
1.	Particulate matter			

Total Defective/Sample Size	Description
Critical: _____	
Major: _____	
Done by: _____ Date: _____	

Disposition: Pass/Fail	
Done by: _____ Date: _____	
Checked by: _____ Date: _____	

| YOUR COMPANY |
| VALIDATION STANDARD OPERATING PROCEDURE |

SOP No.: Val. 1700.60   Effective date: mm/dd/yyyy

                                         Approved by:

TITLE: Microbiological Monitoring of Water

AUTHOR: _____

                        Name/Title/Department

           _____

                        Signature/Date

CHECKED BY: _____

                        Name/Title/Department

             _____

                        Signature/Date

APPROVED BY: _____

                        Name/Title/Department

             _____

                        Signature/Date

REVISIONS:

No.	Section	Pages	Initials/Date

SOP No.: Val. 1700.60

Effective date: mm/dd/yyyy

Approved by:

SUBJECT: Microbiological Monitoring of Water

PURPOSE

This SOP describes the microbiological monitoring program — sampling, testing, and interpretation of results. The objective of monitoring program is to provide sufficient information to control the microbiological quality of the water for its intended use. This SOP is intended for microbiological monitoring of water systems used in an injectable facility.

RESPONSIBILITY

It is the responsibility of following staff to follow the procedure. The quality assurance (QA) manager is responsible for SOP compliance.

1. Quality Control Department

1.1 Monitoring analyst

- Prepare the testing equipment and media and perform growth promotion on the used media.
- Assure that testing equipment, utilities and media are sterile and in proper condition during testing.
- Follow up the sampling schedule.
- Assure that sampling techniques are applied properly.
- Count and identify the microbial isolates.
- Document the used sampling techniques and all related details as well as the detected results.

1.2 Supervisor

- Check and interpret the results.
- Determine and report the out-of-specification results.
- Summarize the data.

1.3 QC manager

- Evaluate and approve the results.
- Review the trend analysis.

SOP No.: Val. 1700.60   **Effective date: mm/dd/yyyy**
                                                 Approved by:

■ Review the source of out-of-specification results.
■ Deviations from standard operating procedures should be noted and approved by QA manager.

2. Product Development Department

■ This department validates the applied methodology.

3. Quality Assurance Department

■ Assistant manager of QA Data reviews and follows the corrective actions.
■ QA manager recommends the remedial actions for OOS results.

4. Training

A formal personnel training program is required to minimize the risk of false positive or negative results.

4.1 Monitoring analyst

The training should include at least:

■ Instruction on the basic principles of aseptic techniques
■ Good sampling practices
■ Demonstration of water system and its potential sources of con-tamination
■ Instruction on the basic principles of microbiology, microbial phys-iology, disinfection and sanitation, media selection and preparation, taxonomy, and sterilization as required by the nature of personnel involvement in aseptic processing
■ Endotoxin basics, sources, and determination

4.2 Supervisor

■ Personnel involved in microbial identification will require special-ized training on required laboratory methods.
■ Additional training on the management of the water data collected will be needed.

- Knowledge and understanding of applicable operating procedures, especially those relating to corrective measures taken when an adverse drift in microbiological conditions is detected, will be necessary.
- Understanding of regulatory compliance policies and each individual's responsibilities with respect to good manufacturing practices (GMP) should be an integral part of the training program as well as training in conducting investigations and in analyzing data.

PROCEDURE

1. Preparation of Sampling Bottles

1.1 Wash the bottles carefully with WFI

- *WFI sampling bottles:* the samples shall be collected into bottles with caps depyrogenated at 250°C/30 minutes.
- *Purified water-sampling bottles:* the samples shall be collected into autoclavable wide-mouthed bottles with stoppers. Glass or polypropylene bottles may be used. Sterilize the bottles at 180°C/1 hour (dry heat) or autoclaved at 121°C/15 minutes.

1.2 Sample size

Size should be sufficient for all required tests, but not less than 100 ml/sample.

1.3 Sampling technique

1.3.1 Special precautions during sampling

- Follow up aseptic technique rules. Wear the proper garment, sterile gloves, and mask.
- Keep the sampling bottle away and do not open it until the sampling point will be ready.
- Be sure that the sampling point is in proper condition and under higher pressure.
- Avoid leaky fixtures that may carry contaminating bacteria from the outside surfaces into the sample.
- Disinfect the sampling point from inside and its surrounding with 70% ethyl alcohol or 3% H_2O_2 for 2 minutes.

SOP No.: Val. 1700.60   **Effective date: mm/dd/yyyy**

                                            Approved by:

1.3.2 Special precautions during for sampling

- *Pure steam:* pass the steam through an autoclavable condenser for 30 minutes and then collect the sample in a tightly closed container with a vent passage.
- *Bacterial endotoxin:* pass the water through a pyrogenic connection for about 1 minute and then collect the sample in a pyrogenic bottle. All materials coming in contact with test materials and reagents must be pyrogen free and careful technique is essential to prevent contamination with environmental endotoxin.

1.3.3 Multistep sampling methods

- Open the sampling point to maximum — about 30 seconds with pulse flushing (quick open and close) — to purge any dust and residual of the disinfectant.
- Remove the cap from the bottle immediately prior to obtaining a sample.
- The cap may be set on the top of a clean surface topside against the surface. The sample side of the cap should not come in contact with any surface, including the fingers or hand of the individual obtaining the sample.
- Adjust the sampling point to adequate level, open the sampling bottle under the water stream in a downward direction and then quickly invert the bottle upward to collect the sample. Fill the bottle, but not to overflowing. Leave sufficient space (1 in. or 2.5 cm) in the top of the bottle to facilitate mixing of the sample by shaking.
- Do not allow the bottle or the water in the bottle to come in contact with the source.
- Remove the bottle from the sample stream and place the cap on it as quickly as possible.

Samples must be properly labeled. Include at least the date, time of sampling, special notes, name of sampler, and sampling point.

SOP No.: Val. 1700.60   Effective date: mm/dd/yyyy
                                         Approved by:

Date:		Name of sampler:	
Time of sampling:		Sampling point:	
Special notes:			

Be sure that the stopper of the bottle is not mishandled or does not touch any source of contamination. Cover the neck of the bottle with new sterile wrapped material.

1.3.4 MicropreSure in-line filtration sampler

■ Prepare in-process water samples using the MicropreSure device.

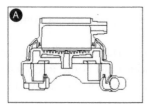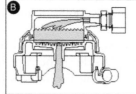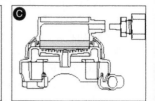

■ Flush the sanitary sampling valve port thoroughly, then close the valve.
■ Remove the yellow and blue stoppers and attach them to the base.

SOP No.: Val. 1700.60

Effective date: mm/dd/yyyy

Approved by:

■ Fit the MicropreSure outlet to the receiving graduated container or connect a length of tubing to the device outlet.

■ Insert the sampling valve port outlet into the MicropreSure inlet and, with a slight twisting movement, push the MicropreSure device onto the valve.

■ Close the sampling valve when the desired sample volume or the desired sampling time has been reached.

■ Gently pull the MicropreSure device away from the valve, using a rotating movement. Maintain the device in a horizontal position.

■ Place the yellow stopper over the MicropreSure device outlet and use the blue stopper to seal the device inlet.

■ Record the sample location, time, and volume on the side of the MicropreSure dome and send to the laboratory for processing.

SOP No.: Val. 1700.60

Effective date: mm/dd/yyyy

Approved by:

- Access the membrane by connecting the MSOpener™ to a vacuum source, and placing the MicropreSure in-line filtration sampler on top of the MSOpener.

- Incubate the membrane by lifting the membrane and transferring it onto a media plate

Sampling Points of WFI[a]

Location	Sampling Point	Activity	Frequency[b]
WFI plant	WFI inlet	Station	Each manufacturing day or at least twice a week
	WFI outlet		
Freeze dryer	WFI inlet		
	WFI outlet		
Room no.	WFI	Washing	Twice a week
Room no.	WFI	Solution preparation	Each manufacturing day or at least twice a week
Room no.	WFI	Washing machine	Each manufacturing day or at least twice a week
		Vessels machine	Each manufacturing day or at least twice a week

[a] Refer to attached drawing (provide number).
[b] It is appropriate to increase or decrease sampling based on the trend performance. The time and date shall be determined according to the activity.

SOP No.: Val. 1700.60   **Effective date: mm/dd/yyyy**

Approved by:

Pure Steam[a]

Sampling Point	Activity	Frequency[b]
Steam plant	Station	Weekly
Room no.	SIP	Weekly
Room no.	SIP	Weekly

[a] Refer to attached drawing (provide number).
[b] It is appropriate to increase or decrease sampling based on the trend performance. The time and date shall be determined according to the activity.

Sampling Points of DI Water[a]

Building	Sampling Point	Activity	Frequency[b]
Water treatment plant	DI water inlet DI water outlet	Station	Once a week
Room no.	DI water	Hand washing	Twice a week
Room no.	DI water	Washing	Once a week
Room no.	DI water	Solution preparation	Twice a week
Room no.	DI water	Washing machine Sink	Twice a week

[a] Refer to attached drawing (provide number).
[b] It is appropriate to increase or decrease sampling based on the trend performance. The time and date shall be determined according to the activity.

Feed Water and Drinking Water[a]

Building	Sampling Point	Activity	Frequency[b]
Water treatment plant	Feed water	Station	Once monthly
Water treatment plant	R.O.	Station	Once/3 months
Water treatment plant	Deionizer	Station	Once/3 months
Water treatment plant	After filtration	Station	Once/3 months
Water treatment plant	Chilled	Station	Once/3 months

[a] Refer to attached drawing (provide number).
[b] It is appropriate to increase or decrease sampling based on the trend performance. The time and date shall be determined according to the activity.

SOP No.: Val. 1700.60   Effective date: mm/dd/yyyy
                                         Approved by:

1.4 General instruction for water sample manipulation

- Deliver the sample to the laboratory as fast as possible; the maximum time is 2 hours. After that, the sample will be unacceptable for microbiological analysis.
- Maintain the quality of the sample through the transport to the laboratory — especially, the neck of the bottle must not mishandled.

2. Microcount Methodology Consideration

The method selected shall be capable of isolating the numbers and types of organisms that have been estimated significant relative to system control and product impact for each individual system. The recommended method is membrane filtration; pour plate and most probable number may be used per requirements.

Culture approaches for each type of water are further defined by the type of medium used in combination. This should be selected according to the monitoring needs presented by a specific water system as well as its ability to recover microorganisms that could have a detrimental effect on the product or process.

Cultivation on low nutrient media and incubation at 20 to 25°C for 5 to 7 days is recommended. The applied approach shall be documented and validated.

2.1 Media growth promotion

Media shall be able to support growth when inoculated with less than 100 CFUs of the challenge organisms per standard test method (provide reference number). The media shall be capable of supporting growth of indicator microorganisms and of environmental isolates from samples obtained through the monitoring program or their corresponding ATCC strains. The ability of the selected media to detect and quantitate these anaerobes or microaerophilic microorganisms shall be evaluated.

2.1.1 Controls

During routine monitoring, each preparation of media shall be tested for positive control (with 100 CFUs) and negative control (open during performing the test).

Effective date: mm/dd/yyyy

Approved by:

2.1.2 Methodology (according to the type of water)

■ Pure steam or water for injection
  ■ Total aerobic viable count: per standard test method (provide reference number) using membrane filtration method. Sample 100 ml shall be filtrated and placed on low-nutrient content media, at 20 to 22°C for 5 days and observed for another 2 days.
  ■ Bacterial endotoxin: kinetic test or gel clot limit standard test method (provide reference number)
■ Purified water: there are three types of purified water:
  ■ Class A purified water uses highly bacteria-critical products such as certain topicals, antacids, and inhalants as well as the deionizer water.
    ■ Total aerobic viable count: per standard test method (provide reference number) using membrane filtration method. Sample of 100 ml shall be filtrated and placed on low-nutrient content media, at 20 to 22°C for 5 days and observed for another 2 days.
    ■ *Pseudomonas*: sample of 100 ml shall be filtrated and placed the membrane filter into 50 ml of TSB incubated for 20 to 28 hours at 35 ± 2°C. Transfer into *Pseudomonas* agar F and P for 40 to 48 hours at 35 ± 2°C. If any characteristic colony is detected, identify using API and/or automated identification system
    ■ Enterobacteriaceae: sample of 100 ml shall be filtrated and placed the membrane filter into 50 ml of MacConkey broth, incubated for 20 to 28 hours at 35 ± 2°C. Transfer into EMB agar for 40 to 48 hours at 35 ± 2°C. If any characteristic colony is detected, identify using API and/or automated identification system
  ■ Class B purified water is used for rinsing. This is the same as purified water class A, except that the microbial limit is 100 CFUs/100 ml.
  ■ Class C purified water has minimum concerns associated with bacteria levels, such as washing water and laboratory usage.
    ■ Total aerobic viable count: pour 1 ml into plate containing 25 ml of tryptone glucose extract agar or TSA half concentration. Incubate at 22°C for 5 days.
    ■ *Pseudomonas*: pour 1 ml into plate containing 25 ml of *Pseudomonas* agar F or P. Incubate at 35 to 37°C for 2 days.

SOP No.: Val. 1700.60 Effective date: mm/dd/yyyy

Approved by:

■ Feed water, R.O. water, and drinking water. The drinking water depends on the local availability and is delivered by the municipal or other local public system or drawn from a private well or reservoir. This water serves as the starting material for most forms of water.

  ■ Total aerobic viable count: pour 1 ml into plate containing 25 ml of tryptone glucose extract agar or TSA half concentration. Incubate at 22°C for 5 days

  ■ *Pseudomonas*: pour 1 ml into plate containing 25 ml of *Pseudomonas* agar F or P. Incubate at 35 to 37°C for 2 days.

  ■ Coliform: using membrane filtration techniques, filtrate 100 ml. Place the membrane filter into Endo agar and identify using API and/or automated identification system.

  ■ Fecal streptococci: use direct inoculation of 1 ml into TSB; if there is turbidity, transfer into Azide dextrose broth or Azide blood agar and identify using API and/or automated identification system.

2.1.3 General instruction for water analysis

Analyze the sample as fast as possible. If there is any delay, keep it in a refrigerator for a maximum of 2 hours. Never incubate or keep the sample at room temperature. An appropriate level of control may be maintained by using data trending techniques and limiting specific contraindicated microorganisms.

2.2 Identification of microbial isolates

■ For general purposes: colony morphology, Gram staining, or microscopic morphology as a routine may be sufficient.

■ Phenotypical isolates from purified water shall be characterized. Biochemical testing (such as oxidase test, urease test, catalase test, citrate test, coagulase test, and indole test) and commercial test kits (such as API tests) and reagents may be used for conformation of some unique isolates.

■ Identification of isolates from critical sampling points such as water used for preparation and areas immediate to these critical sampling points such as water used for CIP should take precedence over identification of microorganisms from noncritical sampling points

SOP No.: Val. 1700.60   Effective date: mm/dd/yyyy
                                                Approved by:

■ Isolates from WFI shall be identified to species level (if possible) using API tests/automated identification system. This pertains to the detection of any isolate obtained from a sample that breaches the alert or action level.

■ Mold does not carry "special" status relative to bacteria. Any significant shifts in type or number require action — regardless of mold vs. bacteria.

Note: The organism recovered from production environments may be highly stressed due to physical factors, contact with chemicals, and thermal stress. It may be difficult to obtain typical biochemical reactions with these isolates. The databases for commercial test kits and ID systems are often designed for clinical isolates and may be incomplete with regard to industrial isolates. Thus, interpretation of such microbial data requires experienced judgment.

■ Identification methods should be verified, and ready-to-use kits should be qualified for their intended purpose. The methods used for identification of isolates should be verified using indicator microorganisms

■ The information gathered by an identification program can also be useful in the investigation of the source of contamination, especially when the action levels are exceeded.

3. Results

3.1 Result validity

The result shall be invalid if:

■ Negative control or positive control was unacceptable.
■ The used media or reagent is unacceptable.
■ Unusual events occurred during sampling.
■ The sample subjected to factor changing its original content such as extended transportation time, incubated at high temperature, or improper handling

Perform a laboratory technical investigation and retest the testing reagents. Repeat the test with more restrictive sampling and testing conditions. Perform investigational analysis of out-of-specification microbiological results. If the new sample is out of specification, the sampling point is rejected.

4. Trend Analysis

Trend: periodic printouts or tabulations of results for purified water systems. These printouts or data summaries should be reviewed. A trend analysis is used to facilitate decision-making for requalification of a controlled environment or for maintenance and sanitization schedules.

Interpretation of the significance of fluctuations in counts or a change in flora should be based on the experienced judgment of qualified personnel.

4.1 Limits

Limits are conservative measures designed to signal potential or actual drift from historical or design performance characteristics. They are not extensions of product specifications, but are intended to flag changes so that corrective action may be taken before product quality is adversely affected. *Not all situations require use of both alert and action limits.*

4.1.1 Alert levels

Alert levels are specific for a given facility and are established on the basis of a baseline developed under an environmental monitoring program. These levels are usually re-examined for appropriateness at an established frequency. When the historical data demonstrate improved conditions, these levels can be re-examined and changed to reflect the conditions.

Exceeding the alert level is not necessarily grounds for definitive corrective action, but it should at least prompt a follow-up investigation that could include sampling plan modifications.

4.1.2 Action levels

An action level in microbiological environmental monitoring is the level of microorganisms that, when exceeded, requires immediate follow-up and, if necessary, corrective action. The evaluation does not depend on the number of colonies only but also on the types of microbes isolated and the suspected hazard.

SOP No.: Val. 1700.60

Effective date: mm/dd/yyyy
Approved by:

Regarding microbiological results, for pure steam and water for injection, it is expected that they be essentially sterile. Because sampling frequently is performed in nonsterile areas and is not truly aseptic, occasional low-level counts due to sampling errors may occur. Agency policy is that less than 5 CFUs/100 ml is an acceptable action limit.

Type of Water	Alert Limit	Action Limit
Pure steam	≤1 CFU/100 ml Endotoxin: LT 0.06 USP EU/ml	≤5 CFUs/100 ml Endotoxin: LT 0.125 USP EU/ml
Water for injection	≤3 CFUs/100 ml Endotoxin: LT 0.125 USP EU/ml	≤5 CFUs/100 ml Endotoxin: LT 0.25 USP EU/ml
Purified water used for highly bacteria-critical products	≤10 CFUs/100 ml	≤20 CFUs/100 ml *Pseudomonas* absence in 100 ml Enterobacteriaceae absence in 100 ml
Purified water used for rinsing	≤50 CFUs/100 ml	≤100 CFUs/100 ml *Pseudomonas* absence in 100 ml Enterobacteriaceae absence in 100 ml
Purified water used for washing and lab work	≤25 CFUs/ml	≤100 CFUs/ml *Pseudomonas* absence in 1 ml
R.O. water Feed water Drinking water	NA	Total viable count: ≤500 CFUs/ml Coliform absence in 100 ml *Pseudomonas* absence in 1 ml Fecal streptococci absence in 1 ml

The data recovered from the sampling are reviewed for conformance to specifications. Maintenance and production management are notified immediately when action levels are exceeded. Results over action level (OAL) cause initiation of an investigation to identify the source of the contaminant. Recovered bacteria are identified to genus whenever possible. Sampling may be performed more frequently during the investigation or until counts fall back within limits. System sanitization will also be performed. The specifications for monitoring of water for injection are defined in manufacturing site SOPs.

4.1.3 Interpretation of data

Because microbiological test results from a water system are not usually obtained until after the drug product is manufactured, results exceeding limits shall be reviewed with regard to the drug product formulated from such water. Consideration with regard to the further processing or release of such a product will depend upon the specific contaminant, the process, and the end use of the product. Such situations are usually evaluated on a case-by-case basis.

4.1.4 Corrective action: per the attached flowchart (attachment no. 1700.60A and 1700.60B for excursions of WFI and DI water)

When action limits are exceeded, the QA manger will investigate the cause of the problem, take action to correct it, assess the impact of the microbial contamination on products manufactured with the water, and document the results of the investigation.

In addition to review of test results, summary data, investigation reports, and other data, the print of the system should be reviewed when conducting the actual physical inspection. As pointed out, an accurate description and print of the system is needed in order to demonstrate that the system is validated.

Maintenance and production management are notified immediately when action levels are exceeded. OAL results cause initiation of an investigation to identify the source of the contaminant. Recovered bacteria are identified to genus whenever possible. Sampling may be performed more frequently during the investigation or until counts fall back within limits. System sanitization will also be performed.

REASONS FOR REVISION

Effective date: mm/dd/yyyy

- First time issued for your company, affiliates, and contract manufacturers.

SOP No.: Val. 1700.60

Effective date: mm/dd/yyyy

Approved by:

Issued on: mm/dd/yyyy
Revision No.: New

Attachment No. 1700.60(A)

PURIFIED WATER BIOBURDEN EXCURSION INVESTIGATION REPORT

Date: _____ Time: _____   Sampling Point: _____

Used for: _____   Batch No.: _____

Alert limit	Action limit	Results

Trend: Isolated Excursion □   Frequent Excursion □

Attach the trend report of the point for the last 3 months

Comment upon the point during sampling:

Microbial identification:

Microbiologist

QA Investigation

Performed by: Reviewed by:

Water system temp. trends (if any): _____

Chemical trend reports: _____

Feed water pretreatment trend reports: _____

Unusual events during sampling day: _____

Other environmental monitoring data in the area: _____

Water system sanitization records: _____

Status of water pumps: _____

Status of main tank: _____

Status of vent filter: _____

Status of automatic valves: _____

Status of manual valves: _____

Status of sampling valves: _____

Status of sampling connection (if any): _____

SOP No.: Val. 1700.60

Effective date: mm/dd/yyyy

Approved by:

CIP of preparation tank: _____ CIP of holding tank: _____

Status of manual valves: _____

Comments: _____

_____

Corrective actions (performed by utilities engineer): _____

Conclusion: _____

_____

QA Manager

In cases of NC (nonconformance), investigation report to be raised by Quality Assurance Inspector.

SOP No.: Val. 1700.60   Effective date: mm/dd/yyyy

                                         Approved by:

                                         Issued on: mm/dd/yyyy
                                         Revision No.: New

Attachment No. 1700.60(B)

WFI BIOBURDEN EXCURSION INVESTIGATION REPORT

Date: _____Time: _____   Sampling Point: _____

Used for: _____   Batch No.: _____

Alert limit	Action limit	Results

Trend: Isolated Excursion □   Frequent Excursion □

Attach the trend report of the point for the last 3 months

Comment upon the point during sampling:

Microbial identification:

                                         Microbiologist

QA Investigation

     Performed by: Reviewed by:

WFI system temp. monitoring trends (if any): _____

Review operational parameters: _____

Chemical trend reports: _____

Feed water pretreatment trend reports: _____

Unusual events during sampling day: _____

Other environmental monitoring data in the area: _____

Water system sanitization records: _____

Status of water pumps 003P 001, 2, 3, and 4: _____

Status of heat exchangers: _____

Status of tank (03B 001): _____

Status of vent filter (03F 001): _____

Status of automatic valves: _____

Status of manual valves: _____

Status of sampling valves: _____

SOP No.: Val. 1700.60

Effective date: mm/dd/yyyy

Approved by:

Status of sampling connection (if any):_____ SIP of preparation tank: _____ SIP of mobile tank: _____ Status of manual valves: _____
Comments: _____ _____
Corrective actions (performed by utilities engineer): _____ _____ Conclusion: _____ _____

QA Manager

In cases of NC (nonconformance), investigation report to be raised by Quality Assurance Inspector.

YOUR COMPANY
VALIDATION STANDARD OPERATING PROCEDURE

SOP No.: Val. 1700.70   Effective date: mm/dd/yyyy

                                         Approved by:

TITLE: Monitoring of Personnel Hygiene

AUTHOR: _____

                        Name/Title/Department

              _____

                          Signature/Date

CHECKED BY: _____

                        Name/Title/Department

                 _____

                          Signature/Date

APPROVED BY: _____

                        Name/Title/Department

                  _____

                          Signature/Date

REVISIONS:

No.	Section	Pages	Initials/Date

SOP No.: Val. 1700.70   Effective date: mm/dd/yyyy

                                                  Approved by:

SUBJECT: Monitoring of Personnel Hygiene

PURPOSE

The purpose is to develop a procedure for monitoring of personnel hygiene (hands, cough, and clothes) in the production, packaging, and quality assurance staff.

RESPONSIBILITY

It is the responsibility of microbiologists to follow the procedure. Production and QA managers are responsible for SOP compliance.

PROCEDURE

This procedure is for staff working directly with the product or its utensils in production and packaging departments.

1. Frequency

- For sterile areas (class 100), the test is performed each shift of the production run.
- For nonsterile areas (class 10,000), the test is performed once per week.
- For nonsterile areas (class 100,000), the test is performed once every 3 months.

2. Material Used

- Swab for testing the clothes
- TSA-filled plates (the media contain neutralizing agents for disinfectants) for testing the hands

3. Method

3.1 Clothing swab

- Prepare the sterile swab with 3 ml of neutralizing solution to the applied disinfectant.
- Aseptically, take the sample from 25 cm of the clothes after work

SOP No.: Val. 1700.70   Effective date: mm/dd/yyyy
                                         Approved by:

- Under LAF, pour 3 ml of neutralizing solution into sterile plate with 25 ml of TSA.
- Incubate for 3 to 5 days at 32°C.
- Count the number of colonies per plate.
- Identify the detected microbes, using API tests.

3.2 Fingerprint

- Prepare sterile plate contain 25 ml of TSA with neutralizing agents to the applied disinfectant.
- Incubate for 2 days at 32°C for sterility assurance.
- Aseptically, let the staff being tested print each hand in one side of the plate.
- Mark the plate with staff name and I.D. number, date, time, and product run.
- Incubate for 3 to 5 days at 32°C.
- Count the number of colonies per plate.
- Identify the detected microbes, using API tests.

4. Acceptance Criteria

4.1 Cleaning hands

- For sterile area (class 100), the detected count for ten fingers is not more than 3 CFUs/gloves.
- For nonsterile area (class 10,000), the detected count for ten fingers is not more than 10 CFUs/gloves.
- For nonsterile area (class 100,000), the detect count for ten fingers is not more than 50 CFUs/gloves, absence of enterobacteriaceae.

4.2 Cleaning clothes

- For sterile area (class 100), the detected count for 25 cm is not more than 5 CFUs/swab.
- For nonsterile area (class 10,000), the detect count for 25 cm is not more than 10 CFUs/swab.
- For nonsterile area (class 100,000), the detected count for 25 cm is not more than 50 CFUs/swab, absence of enterobacteriaceae.

SOP No.: Val. 1700.70   Effective date: mm/dd/yyyy
                                         Approved by:

REASONS FOR REVISION

Effective date: mm/dd/yyyy

- First time issued for your company, affiliates, and contract manufacturers.

SOP No.: Val. 1700.70

Effective date: mm/dd/yyyy
Approved by:

Attachment No. 1700.70(A)

FINGER PRINTING MONITORING

I.D. No.	Name	Limit / 10 Fingers		Results	
		Alert	Action	Count	Microbe
		25 CFU	50 CFU		
		25 CFU	50 CFU		
		25 CFU	50 CFU		

GOWN MONITORING

I.D. No.	Name	Limit / Swab		Results	
		Alert	Action	Count	Microbe
		25 CFU	50 CFU		
		25 CFU	50 CFU		
		25 CFU	50 CFU		

Comments: _____

_____

Performed by: _____   Date: _____

Checked by: _____   Date: _____

Approved by: _____   Date: _____

YOUR COMPANY
VALIDATION STANDARD OPERATING PROCEDURE

SOP No.: Val. 1700.80

Effective date: mm/dd/yyyy
Approved by:

TITLE: Microbiological Environmental Monitoring
of Injectable Facility

AUTHOR: _____
Name/Title/Department

_____
Signature/Date

CHECKED BY: _____
Name/Title/Department

_____
Signature/Date

APPROVED BY: _____
Name/Title/Department

_____
Signature/Date

REVISIONS:

No.	Section	Pages	Initials/Date

SOP No.: Val. 1700.80

Effective date: mm/dd/yyyy

Approved by:

SUBJECT: Microbiological Environmental Monitoring of Injectable Facility

PURPOSE

This SOP describes a microbial monitoring program, testing, sampling, interpretation of results, and corrective actions related to the operations taking place in a controlled environment and auxiliary environments. The microbiological monitoring of air, facilities, equipment, and personnel aims to obtain representative estimates of bioburden of the environment and to detect an adverse drift in microbiological conditions in a timely manner that would allow for meaningful and effective corrective actions.

The procedure and contents described may vary from company to company. The similarity may be noticed due to the nature of the work, which is more or less similar in all companies. The author does not take any responsibility. Each company should validate its own procedure.

SCOPE

Manufacturing sterile areas

RESPONSIBILITY

The following are responsible for following the SOP. The quality assurance (QA) manager is responsible for SOP compliance.

1. Quality Control Department

1.1 Monitoring analyst

- Prepare the testing equipment and media and perform growth promotion on the used media.
- Assure that testing equipment, utilities and media are sterile and in proper condition during testing in the controlled area.
- Apply the proper behavior of cleanroom.
- Follow up the sampling schedule according to instruction of microbiology officer.
- Assure that sampling techniques are applied properly.
- Count and identify the microbial isolates.

SOP No.: Val. 1700.80   Effective date: mm/dd/yyyy
                                         Approved by:

- Observe and report any deviation from the proper aseptic techniques in cleanroom. See attachment no. 1700.80(L).
- Document the used sampling techniques and all related details as well as the detected results.

1.2 Supervisor

- Determine the sampling schedule according to the product requirement and the previous condition of the tested area.
- Check and interpret the results.
- Determine and report the out-of-specification results.
- Summarize the data.
- Suggest the required remedial actions to the minor out-of-specification results.

1.3 QC manager

- Evaluate and approve the results.
- Review the trend analysis.
- Review the source of out-of-specification results.
- Deviations from SOPs should be noted and approved by QC manager and QA manager.

2. Product Development

Manager validates the applied methodology.

3. Quality Assurance Department

3.1 QA inspector

- Observe the operation process and determine the kinds and times of interventions, label the produce items during these interventions and take separate representative samples.
- Observe and report any deviation from the proper aseptic techniques in cleanroom.
- Note and report any deviation of SOPs related to the sanitization program and good housekeeping practice.

SOP No.: Val. 1700.80   Effective date: mm/dd/yyyy

Approved by:

3.2 Assistant QA manager

- Review data.
- Monitor validation of the used equipment.
- Provide the environmental control monitoring parameter.
- Follow the corrective action.

3.3 QA manager

- Interpret the results of environmental control parameters such as flow rate and patterns, Rh, particulate maters, differential pressures, and temperature.
- Suggest the required remedial actions to the major and critical out-of-specification results.
- Follow up the corrective action.
- Determine the required training (or retraining) of the staff.

4. Training

The training of personnel responsible for the microbial monitoring program is critical when contamination of the clean working area could inadvertently occur during microbial sampling. A formal personnel training program is required to minimize this risk.

4.1 Monitoring analyst

The training should include at least:

- Instruction on the basic principles of aseptic processing and the relationship of manufacturing and handling procedures to potential sources of product contamination
- The importance of good personal hygiene and a careful attention to detail in the aseptic gowning procedure used by personnel entering the controlled environment
- Instruction on the basic principles of microbiology, microbial physiology, disinfection and sanitation, media selection and preparation, taxonomy, and sterilization as required by the nature of personnel involvement in aseptic processing
- Good sampling practices

SOP No.: Val. 1700.80   Effective date: mm/dd/yyyy

Approved by:

4.2 Supervisor

■ Personnel involved in microbial identification will require specialized training on required laboratory methods. Additional training on the management of the environmental data collected will need to take place.

■ It is necessary to know and understand applicable operating procedures, especially those relating to corrective measures taken when an adverse drift in microbiological conditions is detected.

■ Understanding of regulatory compliance policies and each individual's responsibilities with respect to good manufacturing practices (GMP) should be an integral part of the training program as well as training in conducting investigations and in analyzing data.

PROCEDURE

1. Media and Diluents Used

Total microbial count: TSA and standard plate count agar are generally used. Nutrient agar may be used for settling plate. Tryptone glucose extract agar, contact plate agar, brain–heart infusion agar, and lecithin agar may be used for contact plate or RODAC. Swab.

■ Swab transportation: tryptone saline, peptone water, enriched buffered gelatine, buffered saline, buffered gelatine, and brain–heart infusion are generally recommended media that may be used for swab transportation. For RCS + testing: agar strip TC (BIOTEST) or M air T air sampler for level I is used.

■ The detection and quantitation of yeast and moulds should be considered. Sabouraud dextrose agar is generally used. General mycological media, such as modified Sabouraud's agar or inhibitory mould agar, are acceptable. Other media that have been validated for promoting the growth of fungi, such as TSA, can be used. For RCS + testing: agar strip YM (BIOTEST) or M air T air sampler for level I is used.

■ Testing for obligatory anaerobes is performed routinely every 3 months. However, more frequent testing is indicated, especially when these organisms are identified in sterility testing.

■ Neutralizing agents: the preceding media may be supplemented with additives to overcome or to minimize the effects of sanitizing agents, or of antibiotics if used or processed in these environments.

SOP No.: Val. 1700.80 Effective date: mm/dd/yyyy
Approved by:

Table 1 Recommended Neutralizing Agents and Their Concentration

Type of Antimicrobial Agent	Inactivator	Concentration
Phenolics, quaternary ammonium compounds, chlorhexidine, tego compounds, and alcohols	Sodium lauryl sulfate	4 g/l
	Polysorbate 80 and lecithin	30 g/l and 3 g/l
	Egg yolk	5–50 ml/l
Halogens, iodine, chlorine, glutaraldehyde	Sodium thioglycolate	5 g/l
Hydrogen peroxide	Catalase	
Quaternary ammonium compounds	Egg yolk	5–50 ml/l
B-lactam antibiotic	Penicillinase enzyme	As required

According to the type of antimicrobial agent, the neutralizers shown in Table 1 should be added.

All microbiological isolation techniques used for the environmental monitoring shall be validated and, in addition, media shall be examined for sterility and for growth promotion as indicated next.

1.1 Media growth promotion

Media must be able to support growth when inoculated with less than 100 colony-forming units (CFUs) of the challenge organisms. The media shall be capable of supporting growth of indicator microorganisms and of environmental isolates from samples obtained through the monitoring program or their corresponding ATCC strains. The ability of the selected media to detect and quantitate these anaerobes or microaerophilic microorganisms shall be evaluated.

1.2 Controls

During routine monitoring, each preparation of media shall be tested for positive control (with 100 CFUs) and negative control (open during performing the test).

2. Sampling

The environment shall be sampled during normal operations to allow collection of meaningful data. Microbial sampling should be performed

SOP No.: Val. 1700.80   Effective date: mm/dd/yyyy
                                         Approved by:

when materials are in the area, processing activities are ongoing, and a full complement of operating personnel is on site.

3. Sampling Sites

The sampling plans should be dynamic with monitoring frequencies and sample plan locations adjusted based on trending performance. The intensity of sampling (routine or intensive) depends upon the activity inside the room and the trend performance. One room may be monitored intensively while the other is normally monitored.

Environmental monitoring is more critical for products that are aseptically processed (intensive monitoring) than for products that are processed and then terminally sterilized (normal monitoring). As manual interventions during operation increase and as the potential for personnel contact with the product increases, the number of sampling points increases.

Attachment nos. 1700.80(A) and 1700(B) show sampling sites for microbiological air monitoring and surface monitoring class 100 of the injectable facility. Attachment nos. 1700.80(C) and 1700.80(D) show sampling site microbiological air monitoring and surface monitoring of the sterility testing lab. Attachment nos. 1700.80(E) and 1700.80(F) show sampling sites for microbiological air monitoring and surface monitoring classes 10,000 and 100,000 of the injectable facility, respectively.

Sampling sites are selected because of their potential for product/container/closure contacts. Consideration should be given to the proximity to the product and whether air and surfaces might be in contact with a product or sensitive surfaces of container-closure systems. Such areas should be considered critical areas requiring more monitoring than non-product-contact areas.

3.1 Critical areas

Areas and surfaces in a controlled environment that are in direct contact with products, containers, or closures and the microbiological status of which can result in potential microbial contamination of the product/container/closure system should be identified. Once identified, these areas should be tested more frequently than non-product-contact areas or surfaces. Element that are likely critical product contact points may include: compressed air or nitrogen, room air, manufacturing equipment, tools, work surfaces storage containers, conveyors, gloved hands of personnel, and water.

SOP No.: Val. 1700.80

Effective date: mm/dd/yyyy
Approved by:

In a parenteral vial filling operation, areas of operation would typically include the container-closure supply, paths of opened containers, and other inanimate objects (e.g., fomites) that personnel routinely handle. Examples of non-product-contact elements may include: walls, floors, ceilings, doors, benches, chairs, personnel attire, waste containers, and test instruments.

3.2 Sampling sites for personnel hygiene (gowning)

Attachment nos. 1700.80(G) and 1700.80(H) describe the possible sampling sites. Samples are taken in the level I and level II areas during the beginning and concluding segments of each manufacturing operation.

The preceding is the recommended frequency; however, it is appropriate to increase or decrease sampling frequency according to the area performance. If the area is not used frequently, accompanying the manufacturing.

4. Test Methodologies

Each test method selected for routine monitoring should be validated. All handling should be performed in an aseptic way.

- Two days before testing: incubate in inverted position a sufficient number of RODAC plates (or contact plates), swabs, agar strips, and plates for fingerprints.
- Before testing
  - Examine all plates/strips for microbial growth under a strong light. Wipe their outer surfaces with disinfectant.
  - Inspect the unexposed plates that exhibit microbial growth.
  - Hold these suspicious plates in refrigerator until all results for that lot number of plates/strips have been read out. The plate can be submitted for identification, if a problem is encountered with organisms of the same colony morphology. Determine the suspected source of contamination.
- Labeling: prepare the required label for the specified tests with at least the following:
  - Area/room
  - Position of test
  - Time/date

- Test for
- Media used

4.1 Air monitoring

Settling plate, M air T air sampler, or slit sampling sampler are used for class 100, level I. Use operating procedure per plant SOP. RCS + (Biotest Hycon) air sampler is used for other classes. A specified sampler shall be used only for class 100 areas.

4.1.1 RCS + Sampler

- Prior to use: the rotor and the protective cap should be autoclaved daily at 121°C for 15 minutes or sterilized with dry heat at 180° for 2 hours. Alternatively, disinfection can be carried out using filtrated disinfectant; *however, the rotor must not be allowed to come into contact with alkaline solution.* The housing of the RCS + is made of polycarbonate and can be disinfected by spraying or wiping with a solution of 70% ethanol. The batteries must be charged.
- At the specified sites: adjust the tripod holder to be at the working level (about 1.5 m) and document activity level at site during test.
- To prevent contamination: aseptic technique should be strictly applied during handling of agar strips; all handling should be under LAF.

4.1.2 Insertion of agar strip

To insert the agar strip, first remove the protection cap by rotating it in a counterclockwise direction and then remove the rotor from the magnetic flange. Pull back the protective wrapper of the agar strip by approximate 4 cm and remove the strip by that. Slide the strip into the rotor with the culture medium facing inwards. The top of the strip marked with the batch number and type should be flushed with the guide plate of the rotor. Then replace the rotor onto the magnetic flange and replace the protective cap.

Use an infrared remote control to control the instrument as follows:

■ Switching the instrument on: the instrument is switched on by pressing the (RESET) key. The last sample volume or residual, unprocessed volume will appear in the display.

■ Setting the sample volume: the sample volume is adjusted using the keys [+] and [–]. A value one greater than the value shown in the display is selected using the [+] key and one less than the value shown in the display using the [–] key. The recommended sampling size is 500 or 200 l.

■ Starting the instrument: after the required volume has been set, the air sampler is started by pressing the [ON/OFF] key or using the remote control. Starting the instrument with the remote control is possible only after the instrument has been switched on.

■ Switching the instrument off: after the set volume of air has been sampled, the measuring process ends automatically. The value appears in the display, switched off by pressing the (RESET) key.

■ Removal of the agar strip: to remove the agar strip, first remove the protection cap and then the rotor. Then, with the half of the tab, pull the agar strip from the rotor and place it into the protective wrapper with the culture medium facing downwards. Seal the protection wrapper using one of the supplied slide seals or adhesive label to prevent the culture medium from drying only during incubation. Label the agar strip with the most relevant sample data to avoid problems of identification.

■ Incubate the strips up to 2 days at 32°C and 3 days at 22°C.

■ The colonies are counted by direct visual inspection after incubation in the sealed protective wrapper. The microbial count is to be stated with reference to the sampling volume. The microbial count (CFUs) corresponding to the selected sample should be recorded as CFUs/sample volume). For calculation of CFUs per volume, the following formula is applied.

$$CFU/m^3 = \frac{\text{Germ count} \times 1000}{\text{Volume of air (liter)}}$$

SOP No.: Val. 1700.80

Effective date: mm/dd/yyyy

Approved by:

4.1.3 Settling plates

- For each specified site, test for fungi (use 30 ml of SDA with 1% glycerine) and bacteria (use 30 ml of TSA with 1% glycerine).
- At working level, label the agar-filled petri dishes and simply open completely and expose to environmental conditions, collecting particles that settle on agar surface.
- After not more than 4 h, close the plate with the lid.
- Incubate TSA plates up to 3 days at 32°C and SDA plates up to 5 days at 22°C.

Air microbial monitoring should be performed in and around areas of high operator activity. It is not unusual to see settle plates and air sample locations well away from such areas. A typical example is where settle plates are located well to the rear of the filling machine where there is little or no operator activity. The same may be true for air sampling. It is important, therefore, to observe operator activity over a period of time and ensure that the monitoring sites are located so as to monitor operator activity.

4.2 Surface sampling

To minimize disruptions to critical operations, surface sampling should be performed at the stop or the end of operations. Surface sampling may be accomplished by the use of contact plates or by the swabbing method.

- Contact plates are used when sampling regular or flat surfaces. The test is performed with RODAC (replicate organisms detecting and counting) plates, or contact plates, where the convex agar surface rises above the brim of the plate.
- The swabbing method may be used for sampling of irregular surfaces, especially for equipment or discrete surface areas in locations difficult to reach (such as corners, crevices, and other nonflat areas). Swabbing is used to supplement contact plates for regular surfaces. A useful monitoring technique is to monitor the filling needles at the end of the filling session.

4.2.1 Contact plates

- Samples must be collected on dry surfaces.
- At the site of the area to be tested, carefully introduce the RODAC dishes and remove the lid.

SOP No.: Val. 1700.80   Effective date: mm/dd/yyyy

Approved by:

- Then carefully but firmly press (ensuring that pressure of about 500 g is distributed over the whole plate) the plate with a slight rolling movement onto the surface being examined and hold it for a few (2) minutes without moving it.
- Remove the plate and replace the lid immediately.
- The tested area is then to be cleaned and disinfected with 70% ethyl alcohol.

4.2.2 Swabbing

- The swab, generally composed of a stick with an absorbent extremity, is immersed in 3 ml of an appropriate diluent (tryptone saline with suitable neutralizer) before sampling.
- The area to be swabbed is defined using a sterile template of appropriate size. In general, it is in the range of 24 to 30 cm².
- The swab is then rinsed with vortex and the contents plated on TSA plates.
- The estimate of microbial count is done by direct inoculation on solid media or membrane filtration of the swab rinsing fluid specified nutrient agar.
- Incubate at 32°C for 2 days and at 22°C for more 3 days.
- Count the number of microbial colonies on each plate. The microbial estimates are reported per contact plate or per swab.
- Examine the types of colonies and submit the representative colonies for identification. Mold does not carry "special" status relative to bacteria. Any significant shifts in type or number require action — regardless of mold vs. bacteria.

5. Personnel Hygiene

5.1 Gowning

- Tested by the sterile swab or contact plate.
- For routine monitoring, choose places that have the potential for personnel contact with surface. For intensive monitoring, choose ten places at least.
- Aseptically, take the sample from 25 cm² of the clothes after work.
- Incubate for 2 days at 32°C and 3 days at 22°C.
- Count the number of colonies per plate.

SOP No.: Val. 1700.80

Effective date: mm/dd/yyyy

Approved by:

5.2 Fingerprint

- Prepare sterile plate containing 25 ml TSA with neutralizing agents to the applied disinfectant.
- Aseptically, let the under-test staff print each hand in one side of the plate; the press should be moderately placed (not too high to crush the media or too low to reduce test efficacy).
- Mark the plate with *staff name, identification number, date, time, and product run.*
- Incubate for 2 days at 32°C and 3 days at 22°C.
- Count the number of colonies per plate.

5.3 Incubation temperatures for environmental monitoring samples

There is no universal set of incubation conditions that will reliably detect all types of environmental microorganisms that could be present from a given sampling site at a given point in time. It follows that the purpose of using a defined set of incubation conditions for detecting environmental isolates is to establish whether any microbial changes or shifts are occurring within the manufacturing environment. Mesophilic incubation conditions (e.g., 20 to 35°C for 3 to 7 days) are suitable for recovery of microorganisms from normal ambient temperature manufacturing environments.

6. Identification of Microbial Isolates

All types of environmental monitoring-derived isolates shall be identified to the genus level and to the species level (if possible). Speciation or biocharacterization of specific or all organisms, as requested, should be undertaken according to the following:

- Colonial morphology of isolates obtained during predisinfection phase (baseline) of surfaces, air, and/or personnel
- Biocharacterization to category (i.e., Gram positive, Gram negative, yeast, mold, etc.)
- Automated microbial identification of computer-generated probabilities of identity — genus and species

Microbial monitoring may be not needed to identify and quantitate all microbial contaminants present in these controlled environments. Identification of isolates from critical areas and areas immediate to these critical

SOP No.: Val. 1700.80   Effective date: mm/dd/yyyy

Approved by:

areas should take precedence over identification of microorganisms from noncritical areas.

Identification methods should be verified, and ready-to-use kits should be qualified for their intended purpose. The methods used for identification of isolates should be verified using indicator microorganisms. A knowledge of the normal flora in controlled environments aids in determining the usual microbial flora anticipated for the facility being monitored, evaluating the effectiveness of the cleaning and sanitization procedures, methods, and agents; and recovery methods.

The information gathered by an identification program can also be useful in the investigation of the source of contamination, especially when the action levels are exceeded.

6.1 Evaluation

6.1.1 Interpretation of data

The following factors should be considered when data are interpreted:

- Contamination of a manufacturing environment may exist as discrete, nonhomogeneous, and dynamic entities.
- All of the viable microbes present at a site may not be detected due to:
  - Differences in optimal growth and test conditions among microorganisms
  - Range of colony overlap or antagonism (particularly if counts are high)
- Sampling techniques: the number of colony units formed may only reflect the number of particles holding microbes, not the total number of organisms.
- As a consequence of incubation delay for traditional test methods, control is rarely based on more than a retrospective review. By the time results are known, additional manufacturing and control steps have already been executed. Retests of identical conditions are not possible.
- Seasonal variation may be a factor.
- There are inherent limitations and testing for low levels of microorganisms, especially when only a limited portion of a whole population is analyzed. Limited or small sample volumes may not provide representative data, but a large number of samples may

partially compensate for this. The limit of detection of test methods should be considered.

- Sampling procedures may be subject to contamination.
- No single test method may be representative of production work cycles; sample sites may not be representative of all equipment activities in an area.
- Properties of contaminants may change once isolation has occurred.
- No single test method may completely characterize the microbial contamination in an area or on a surface.

6.1.2 Trend analysis

Trend analysis is data from a routine microbial environmental monitoring program that can be related to time, shift, facility, etc. This information is periodically evaluated to establish the status or pattern of that program to ascertain whether it is under adequate control. A trend analysis is used to facilitate decision-making for requalification of a controlled environment or for maintenance and sanitization schedules. Interpretation of the significance of fluctuations in counts or a change in flora should be based on the experienced judgment of qualified personnel. Each process datum must be evaluated as part of an overall monitoring program.

6.1.2.1 Limits

Limits are conservative measures designed to signal potential or actual drift from historical or design performance characteristics. They are not extensions of product specifications, but are intended to flag changes so that corrective action may be taken before product quality is adversely affected.

- Base limits on historical data.
- Calculate confidence limits by one of two methods:
    - Fit data to a recognized mathematical distribution (e.g., normal Poisson, binomial). When appropriate, transform the data (e.g. log 10 transformation). Calculate confidence limits.
    - If there is no mathematical fit, find the level below which 95% of the counts fall (for 95% confidence limit).
- Set your limits based on the confidence limits.

SOP No.: Val. 1700.80   Effective date: mm/dd/yyyy

Approved by:

- Perform continued trend analysis to see whether your limits remain appropriate.
- Watch for periodic spikes, even if averages stay within limits.

Not all situations require use of both alert and action limits.

6.1.2.2 Alert levels

Alert levels are specific for a given facility and are established on the basis of a baseline developed under an environmental monitoring program. These levels are usually re-examined for appropriateness at an established frequency. When the historical data demonstrate improved conditions, these levels can be re-examined and changed to reflect the condition (Table 2).

Table 2   Delineated Alert Limits

Class	Air	Surface	Fingerprint	Gowning
Class 100 level I	—	—	—	—
Class 100	1 CFU/m³	1 CFU/plate or swab	1 CFUs/ 10 fingers	2 CFUs/ plate
Potential product/ container contact areas	5 CFUs/m³	All sites: 3 CFUs/ plate or swab Floor: 5 CFUs/ plate or swab	5 CFUs/ 10 fingers	10 CFUs/ plate
Other support areas (class 100,000)	20 CFUs/m³	All sites: 15 CFUs/ plate or swab Floor: 15 CFUs/ plate or swab	10 CFUs/ 10 fingers	25 CFUs/ plate
Other support areas to aseptic processing areas but nonproduct contact (class 100,000 or lower)	25 CFUs/m³	All sites: 25 CFUs/ plate Floor: 50 CFUs/ plate	25 CFUs/ 10 fingers	25 CFUs/ plate

Note: It is appropriate to increase or decrease alert limit based on the trend performance.

SOP No.: Val. 1700.80

Effective date: mm/dd/yyyy

Approved by:

Exceeding the alert level is not necessarily grounds for definitive corrective action, but it should at least prompt a documented follow-up investigation that could include sampling plan modifications. However, the evaluation does not depend only on the number of colonies, but also on the types of microbes isolated and the suspected hazard from them, e.g., Gram-negative bacteria, anaerobic bacteria, fungi, or organisms not normally isolated.

6.1.2.3 Action levels

An action level in microbiological environmental monitoring is that level of microorganisms that, when exceeded, requires immediate follow-up and, if necessary, corrective action. The evaluation does not depend only on the number of colonies but also on the types of microbes isolated and the suspected hazard from them, e.g., Gram-negative bacteria, anaerobic bacteria, fungi, or organisms not normally isolated.

If an excursion occurs above an action level, as a minimum, one or more of the following action may include, but not be limited to, investigative action, corrective action, and/or informing responsible parties (Table 3).

6.1.3 Corrective actions

When action level excursions or frequent alert excursions are identified, a corrective action program, resolution deadline, and preventive plan shall be implemented. Risk analysis shall be performed to determine the probability of one or more causes of errors occurring, as well as to identify the potential consequences of excursions. (See attachment nos. 1700.80(I), 1700.80(J), and 1700.80(K), and 1700.80(L) for excursion of air, surface, personnel, and visual inspection report during the visit to plant by the microbiologist responsible.)

During the testing of the efficacy of the corrective actions, the production or QA areas may perform suitable professional remedial actions until the next monitoring result.

REASONS FOR REVISION

Effective date: mm/dd/yyyy

- First time issued for your company, affiliates, and contract manufacturers.

SOP No.: Val. 1700.80   Effective date: mm/dd/yyyy

                                                   Approved by:

Table 3   Delineated Action Limits

Class	Air	Surface	Fingerprint	Gown
Level I	1 CFU/m^3	1 CFU/plate or swab	1 CFU/10 fingers	1 CFU/ plate
Supporting areas immediately adjacent level	3 CFUs/m^3	3 CFUs/plate or swab	3 CFUs/10 fingers	5 CFUs/ plate
Potential product/ container contact areas	10 CFUs/m^3	All sites: 5 CFUs/ plate Floor: 10 CFUs/ plate	10 CFUs/ 10 fingers	20 CFUs/ plate
Other support areas (class 100,000)	35 CFUs/m^3	All sites: 25 CFUs/ plate Floor: 30 CFUs/ plate	20 CFUs/ 10 fingers	50 CFUs/ plate
Other support areas to aseptic processing areas but no product contact (class 100,000 or lower)	50 CFUs/m^3	All sites: 50 CFUs/ plate Floor: 100 CFUs/ plate	50 CFUs/ 10 fingers	50 CFUs/ plate

SOP No.: Val. 1700.80

Effective date: mm/dd/yyyy

Approved by:

Attachment No. 1700.80(A)

GUIDELINE OF NUMBER OF SAMPLING POINTS FOR AIR MONITORING CLASS 100 OF INJECTABLE FACILITY

Room	Activity/Works Area	Position for Routine Monitoring[a]	Position for Intensive Monitoring[a]	Frequency	Alert Limit	Action Limit
Room no.	Ampoule/vial filling			Each working shift or at least twice a week	—	1 CFU/4 h
					—	1 CFU/m³
					1 CFU/m³	3 CFUs/m³
Room no.	Loading freeze dryer/at rest				—	1 CFU/4 h
					—	1 CFU/m³
					1 CFU/m³	3 CFUs/m³
Room no.	Syringe filling				—	1 CFU/4h
					—	1 CFU/m³
					1 CFU/m³	3 CFUs/m³
Room no.	Unloading autoclave				—	1 CFU/4h
					—	1 CFU/m³
					1 CFU/m³	3 CFUs/m³

SOP No.: Val. 1700.80 **Effective date: mm/dd/yyyy**

Approved by:

Room	Area	Settle plate / Air sampler limit	Additional limit
Room no.	Pass box	1 CFU/4h	—
LAF path	Pass box	1 CFU/m³	—
Room no.	Gowning	1 CFU/4h	—
Room no.	Sterilization and depyrogenation tunnel	1 CFU/m³	1 CFU/m³
	Washing area	3 CFUs/m³	—
Room no.	LAF path	1 CFU/m³	1 CFU/m³
	Weighing area	3 CFUs/m³	1 CFU/m³
	LAF 2	1 CFU/4 h	—
Sampling room	LAF 1	3 CFUs/m³	1 CFU/m³
	LAF 2	3 CFUs/m³	1 CFU/m³

Notes: S = settled plate; R = RCS + air sampler; M = M air T air sampler.

[a] The sampling frequency and number of sampling positions can be decreased or increased according to the trend performance

SOP No.: Val. 1700.80   Effective date: mm/dd/yyyy

                                        Approved by:

Attachment No. 1700.80(B)

GUIDELINE OF NUMBER OF SAMPLING POINTS FOR SURFACE MONITORING CLASS 100 OF INJECTABLE FACILITIES

Activity	Room	Routine Monitoring[a]	Intensive Monitoring[a]	Frequency	Alert Limit	Action Limit
Ampoule/ vial filling	Room no.	Vial machine, ampoule machine, vial inlet, filling needles (nozzles), filling tank, ampoule inlet, conveyor, curtain, mobile LAF (wall, two shelves)	Vial filling machine (two positions), ampoule filling machine (two positions), each vial hopper, each curtain, vial inlet, filling needles (nozzles), filling tank, each control panel, each tool, table, storage containers, conveyor (four positions), ampoule inlet, mobile LAF (wall, each shelf)	Each working shift or at least twice a week	—	1 CFU
		Wall, floor, corner, control panel, mobile LAF (door handle and outer surface)	Wall, wall near machine, doorknob, above the door, floor beneath the machine, each control panel, floor below the door, chairs, belts of moving benches or chairs, waste containers, corners (especially unused), mobile LAF (door handle, glass, outer surface and movable belt)		1 CFU	3 CFUs
Loading/ unloading freeze dryer	Room no.	Inlet and outlet of conveyor, F. D. door, tray, curtain	Conveyor (five positions), tray, moving parts (two positions), each curtain, tray (five positions), F.D. door, inner floor		—	1 CFU

SOP No.: Val. 1700.80   **Effective date: mm/dd/yyyy**

Approved by:

			1 CFU	3 CFUs
Syringe filling	Room no.	Wall, floor, panel, corner	Wall, wall near machine, doorknob, above the door, floor beneath the F.D., floor below the door, chamber, control panel, each used tool, table, chairs, belts of moving benches or chairs, waste containers, corner (especially unused)	1 CFU	3 CFUs
		Machine surface, filling needles, filling tank, hopper	Machine, storage containers, machine surface, filling needles, filling tank (two positions), hopper, each curtain	—	1 CFU
		Wall, floor, control panel, corner, conveyor	Wall, wall near doorknob, above the door, floor beneath the machine, floor below the door, control panel, each tool, table, chairs, belts of moving benches or chairs, waste containers, corners (especially unused)	1 CFU	3 CFUs
Unloading autoclave	Room no.	Wall, floor, autoclave surface, control panel, movable table, corner	Wall, wall near autoclave, doorknob, above autoclave door, floor beneath autoclave, autoclave surface, control panel, each tool, table, chairs, belts of moving benches or chairs, waste containers, corners (especially unused)	1 CFU	3 CFUs
Syringe pass box	Room no.	Wall, floor	Wall, floor beneath door, each tool inside, corners	—	1 CFU
			Doorknob, door control unit	1 CFU	3 CFUs

SOP No.: Val. 1700.80

Effective date: mm/dd/yyyy

Approved by:

Activity	Room	Routine Monitoring[a]	Intensive Monitoring[a]	Frequency	Alert Limit	Action Limit
LAF path	LAF Path	Wall, floor	Wall, floor beneath the door, each tool inside, corners		—	1 CFU
			Doorknob, door control unit		1 CFU	3 CFUs
Change room	Room no.	Wall, floor	Wall, doorknob, floor beneath both doors, door control units, cupboard, mirror, corners		1 CFU	3 CFUs
Commodity washing	Room no.	Sink, curtain, wall, bench	Sink, curtain, wall (two positions), bench (two positions), glass, each vessel		1 CFU	3 CFUs
LAF	LAF (room no.)	Wall, floor	Wall, floor, each tool inside, corner		—	1 CFU
			Doorknob, door control unit		1 CFU	3 CFUs
Weighing	Weighing Area (room no.)	Curtain, wall, table, floor	Curtain, wall (two points), table, scalor, balance, floor (two points)		1 CFU	3 CFUs
Sampling	LAF1 sampling room	Wall, floor	Wall, curtain, plate, floor		1 CFU	3 CFUs
Sampling	LAF2 sampling room	LAF (screen, wall, floor)	LAF (screen, wall, left side, right side, electrical socket, floor)		—	1 CFU
		LAF surface, curtain, wall, floor	LAF surface, curtain, wall (two points), floor (two points), chair		1 CFU	3 CFUs

[a] The sampling frequency and number of sampling positions can be decreased or increased according to the trend performance.

SOP No.: Val. 1700.80

Effective date: mm/dd/yyyy

Approved by:

Attachment No. 1700.80(C)

GUIDELINE OF NUMBER OF SAMPLING POINTS FOR AIR MONITORING OF STERILITY TESTING ROOM

Room/ Activity	Position for Routine Monitoring[a]	Position for Intensive Monitoring[a]	Frequency	Alert Limit	Action Limit
Testing	Holten safe LAF:	Holten safe LAF:	Each operating day or at least twice/ week	—	1 CFU/m^3
	Near entrance, center	Near entrance, center, near left side		0.1 CFU/m^3	0.3 CFU/m^3
Incubation	Near entrance, near center	Near entrance, near center, near incubators		10 CFUs/m^3	20 CFUs/m^3
Change	Near entrance After bench	Near entrance Near lockers After bench, near mirror	Twice/ week	10 CFUs/m^3	20 CFUs/m^3

Notes: S = settled plate; R = RCS + air sampler.
[a] The sampling frequency and number of sampling positions can be decreased or increased according to the trend performance.

Effective date: mm/dd/yyyy

Approved by:

Attachment No. 1700.80(D)

GUIDELINE OF NUMBER OF SAMPLING POINTS FOR SURFACE MONITORING OF STERILITY TESTING ROOM

Room/ Activity	Position for Routine Monitoring[a]	Position for Intensive Monitoring[a]	Frequency	Alert Limit	Action Limit
Testing	Holten safe LAF (screen, wall, floor) steritest surface	Holten safe LAF: (electrical socket, screen, wall, floor, left side, right side) surface of steritest (two points)	Each operating day or at least twice/week	—	1 CFU
	Surface of LAF, door handle, wall, corner, floor	Surface of LAF, chair, door handle, doorknob wall (two points), corner, floor (two points)		1 CFU	3 CFUs
Incubation	Incubator (1), incubator (2), table, door handle, wall, corner, floor	Incubator (1), incubator (2), table, writing table, chair, door handle, door knob, wall (two points), corner, floor (two points)	Each operating day or at least twice/week	0.3 CFU

0.5 CFU | 0.5 CFU

10 CFUs |
| Changing | Wall, hanger, bench, floor, corner | Wall (two points), hanger, bench, mirror, door handle, floor (two points), corner | Twice/ week | 0.3 CFU

0.5 CFU | 0.5 CFU

10 CFUs |

[a] The sampling frequency and number of sampling positions can be decreased or increased according to the trend performance.

SOP No.: Val. 1700.80

Effective date: mm/dd/yyyy

Approved by:

Attachment No. 1700.80(E)

GUIDELINE OF NUMBER OF SAMPLING POINTS FOR AIR MONITORING CLASS 10,000 AND CLASS 100,000 OF INJECTABLE FACILITY

Activity	Room	Position for Routine Monitoring[a]	Position for Intensive Monitoring[a]	Frequency	Alert Limit	Action Limit
Production corridor	Provide room no.			Once/week	25 CFUs	50 CFUs
Syringe preparation	Provide room no.			Twice/week	0.5 CFU	10 CFUs
Change room	Provide room no.			Twice/week	20 CFUs	35 CFUs
Coding room	Provide room no.			Each operating day or at least twice/week	0.5 CFU	10 CFUs
Air lock washing	Provide room no.			Twice/week	25 CFUs	50 CFUs
Packaging	Provide room no.			Once/week	25 CFUs	50 CFUs
Washing area	Provide room no.			Once/week	25 CFUs	50 CFUs
Air lock for materials	Provide room no.			Twice/week	0.5 CFU	10 CFUs
Personnel air lock	Provide room no.			Twice/week	0.5 CFU	10 CFUs

SOP No.: Val. 1700.80

Effective date: mm/dd/yyyy

Approved by:

Activity	Room	Position for Routine Monitoring[a]	Position for Intensive Monitoring[a]	Frequency	Alert Limit	Action Limit
Solution preparation	Provide room no.			Each operating day or at least twice/ week	0.5 CFU	10 CFUs
LAF	Provide room no.			Each operating day or at least twice/ week	0.5 CFU	10 CFUs
Vial capping	Provide room no.			Twice/week	0.5 CFU	10 CFUs
Commodity washing	Provide room no.			Each operating day or at least twice/ week	0.5 CFU	10 CFUs
Packaging	Provide room no.			Once/ week	25 CFUs	50 CFU
Loading freeze dryer	Provide room no.			Each unloading	0.5 CFU	10 CFU
Solution room	Provide room no.			Each operating day or at least twice/ week	0.5 CFU	10 CFU
Personnel air lock	Provide room no.			Once/week	25 CFUs	50 CFU
Sampling room	Sampling room			Twice/week	25 CFUs	50 CFU

[a] The sampling frequency and number of sampling positions can be decreased increased according to the trend performance.

SOP No.: Val. 1700.80

Effective date: mm/dd/yyyy

Approved by:

Attachment No. 1700.80(F)

GUIDELINE OF NUMBER OF SAMPLING POINTS FOR SURFACE MONITORING CLASS 10,000 AND CLASS 100,000 OF INJECTABLE FACILITY

Activity	Room	Routine Monitoring[a]	Intensive Monitoring[a]	Frequency	Alert Limit	Action Limit
Production corridor	Provide room no.	Wall (two positions)	Wall (three positions), door handles, trays	Once /week	25 CFUs	50 CFU
		Floor (two positions)	Glass (two positions) Floor (three positions)		50 CFUs	100 CFU
Syringe preparation	Provide room no.	Wall Conveyor, curtain	Wall, floor, door handle, electrical box, panel conveyor, curtain	Twice /week	0.3 CFU 0.1 CFU	0.5 CFUs 0.3 CFUs
		Floor	Floor, corner		0.5 CFU	10 CFUs
Change room	Provide room no.	Wall	Wall, sink, gown shelf, drier	Twice /week	15 CFUs	25 CFUs
		Floor	Floor (two positions), corner		15 CFUs	30 CFUs

SOP No.: Val. 1700.80

Effective date: mm/dd/yyyy

Approved by:

Activity	Room	Routine Monitoring[a]	Intensive Monitoring[a]	Frequency	Alert Limit	Action Limit
Coding room	Provide room no.	Wall, machine	Wall, machine, table, trolley	Each operating day or at least twice/week	0.3 CFU	0.5 CFU
		Floor	Floor (two positions), corner		0.5 CFU	10 CFUs
Air lock washing	Provide room no.	Wall	Wall (two positions), doors handles	Once /week	25 CFUs	50 CFUs
		Floor	Floor (two positions), corner		50 CFUs	100 CFUs
Packaging	Provide room no.	Wall	Wall (two positions), machine, table	Once /week	25 CFUs	50 CFUs
		Floor	Floor (two positions), corner		50 CFUs	100 CFUs
Washing area	Provide room no.	Wall	Wall, gown hanger, table	Once /week	25 CFUs	50 CFUs
		Floor	Floor (two positions), drain, corner		50 CFUs	100 CFUs

[a] The sampling frequency and number of sampling positions can be decreased or increased according to the trend performance.

SOP No.: Val. 1700.80

Effective date: mm/dd/yyyy

Approved by:

Attachment No. 1700.80(F)

Activity	Room	Routine Monitoring[a]	Intensive Monitoring[a]	Frequency	Alert Limit	Action Limit
Air lock for material	Provide room no.	Wall	Wall (two positions), door handles	Once /week	0.3 CFU	0.5 CFU
		Floor	Floor (two positions), corner		0.5 CFU	10 CFUs
Personnel air lock	Provide room no.	Wall	Wall (two positions), sink, shelf	Once /week	0.3 CFU	0.5 CFU
		Floor	Floor (two positions), corner		0.5 CFU	10 CFUs
Solution preparation	Provide room no.	Wall, prep. tank, mobile tank, table	Wall (two positions), prep. tank, mob. tank, table, panel, balance, tools	Each operating day or at least twice/ week	0.3 CFU	0.5 CFU
		Homogenize Floor	Homogenize Floor (two positions), corner		0.5 CFU	10 CFUs
AF	Provide room no.	Wall, table, cage	Wall (two points), table, cage, door handle	Each operating day or at least twice /week	0.3 CFU	0.5 CFU
		Floor, corner	Floor (three points), corner		0.5 CFU	10 CFUs

SOP No.: Val. 1700.80

Effective date: mm/dd/yyyy

Approved by:

Activity	Room	Routine Monitoring[a]	Intensive Monitoring[a]	Frequency	Alert Limit	Action Limit
Vial capping	Provide room no.	Wall, hopper, machine surface, conveyor, panel	Wall (two positions), hopper, machine surface, conveyor, panel, glass, hopper	Twice /week	0.3 CFU	0.5 CFU
		Floor	Floor (two positions), corner		05 CFU	10 CFUs
Commodity washing	Provide room no.	Wall, washing machine, table, electric panel	Wall (two positions), washing machine, electric panel sink, autoclave door, table, trolley	Each operating day or at least twice /week	0.3 CFU	0.5 CFU
		Floor	Floor (two positions), corner		0.5 CFU	10 CFU
Packaging	Provide room no.	Wall, machine surface	Wall (two positions), panel, machine surface, machine tray,	Once /week	25 CFUs	50 CFU
		Floor	Floor (two positions), corner		50 CFUs	100 CFU

[a] The sampling frequency and number of sampling positions can be decreased c increased according to the trend performance.

SOP No.: Val. 1700.80

Effective date: mm/dd/yyyy

Approved by:

Attachment No. 1700.80(F)

Activity	Room	Routine Monitoring[a]	Intensive Monitoring[a]	Frequency	Alert Limit	Action Limit
Loading freeze dryer	Provide room no.	Wall, corner, conveyor	Wall, conveyor, moving belts, door knob, glass	During unloading process	0.3 CFU	0.5 CFU
		Floor	Floor, corner		0.5 CFU	10 CFUs
Solution room	Provide room no.	Wall, table, mobile tank (top dish, surface)	Wall (two positions), table, filter cover, LAF control unit, mobile tank (top dish, surface, nitrogen hose, silicon hose)	Each operating day or at least twice /week	0.3 CFU	0.5 CFU
		Floor	Floor (two positions), corner		0.5 CFU	10 CFUs
Sampling room	Sampling room	Wall	Wall, drier, sink, refrigerator, scalor	Once /week	25 CFUs	50 CFUs
		Floor	Floor (two points), corner		50 CFUs	100 CFUs
Personnel air lock	Provide room no.	Wall	Wall (two points), shelves	Once /week	25 CFUs	50 CFUs
		Floor	Floor (two points), corner		50 CFUs	100 CFUs

Effective date: mm/dd/yyyy

Approved by:

Attachment No. 1700.80(G)

THE POSSIBLE SAMPLING POINT OF THE GOWNING SUITES

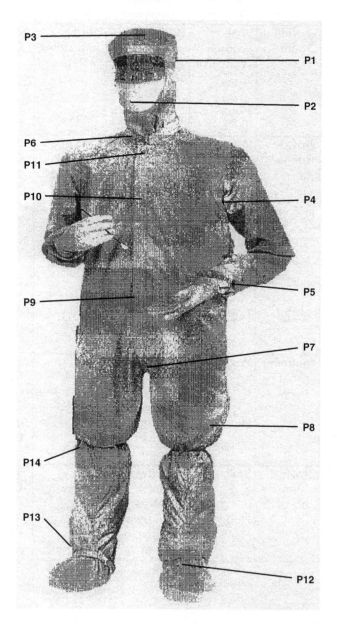

SOP No.: Val. 1700.80

Effective date: mm/dd/yyyy
Approved by:

Attachment No. 1700.80(H)

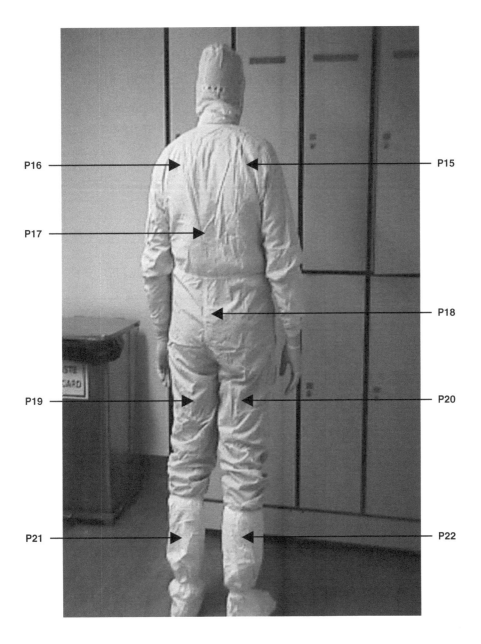

SOP No.: Val. 1700.80

Effective date: mm/dd/yyyy
Approved by:

Attachment No. 1700.80(I)

ENVIRONMENTAL EXCURSION INVESTIGATION REPORT: MICROBIAL AIR MONITORING OF INJECTABLE PLANT

Microbial Air Monitoring of Injectable Plant

Date: _____ Time: _____ Room No.: _____

Product: _____ Batch No.: _____

Cleaner Name: _____ ID No.: _____

ALERT LIMIT	ACTION LIMIT	RESULTS

Trend: Isolated Excursion ☐   Frequent Excursion ☐

Microbial Identification: _____

Microbiological Section Head

QA Investigation

Performed by: Reviewed by:

Area environmental monitoring trend: _____

Materials handled inside the room: _____

HEPA filter status: _____

Unusual events during processing: _____

Method of sampling: _____

Other environmental monitoring data in the area: _____

Sanitization records: _____

RH/temperature of the area: _____

Pressure differential of the area: _____

Source of particulate count: _____

Personnel hygiene: _____

Visual inspections: _____

SOP No.: Val. 1700.80

Effective date: mm/dd/yyyy

Approved by:

Comments: _____

_____

Corrective actions (performed by production): _____

_____

Conclusion: _____

_____   _____

QC Manager   QA Manager

cc: QA File/Area Manager/QC Microbiology

SOP No.: Val. 1700.80

Effective date: mm/dd/yyyy

Approved by:

Attachment No. 1700.80(J)

ENVIRONMENTAL EXCURSION INVESTIGATION REPORT: SURFACE MONITORING OF INJECTABLE FACILITY

Surface Monitoring of Injectable Facility

Date: _____ Time: _____ Room No.: _____

Product: _____ Batch No.: _____

Cleaner Name: _____ ID No.: _____

ALERT LIMIT	ACTION LIMIT	RESULTS

Trend: Isolated Excursion ☐   Frequent Excursion ☐

Microbial Identification: _____

Microbiological Section Head

QA Investigation

Performed by: Reviewed by:

Area environmental monitoring trend:_____

Cleaner training record: _____

Cleaner glove and gowning monitor: _____

Unusual events during processing: _____

Method of sampling:_____

Other environmental monitoring data in the area: _____

Sanitization records: _____

Microbiological status of used disinfectant: _____

Microbiological status of preparation water: _____

Holding time of the disinfectant: _____

Status of cleaning tools: _____

Visual inspections:_____

SOP No.: Val. 1700.80   Effective date: mm/dd/yyyy
                                         Approved by:

Comments: _____

_____

Corrective actions (performed by production): _____

_____

Conclusion: _____

_____

_____   _____
   QC Manager   QA Manager
cc: QA File/Area Manager/QC Microbiology

SOP No.: Val. 1700.80

Effective date: mm/dd/yyyy

Approved by:

Attachment No. 1700.80(K)

ENVIRONMENTAL EXCURSION INVESTIGATION REPORT: PERSONNEL HYGIENE MONITORING OF STERILE PRODUCT PREPARATION

Personnel Hygiene Monitoring of Sterile Product Preparation

Date: _____ Time: _____ Room No.: _____

Product: _____ Batch No.: _____

Cleaner Name: _____ ID No.: _____

ALERT LIMIT	ACTION LIMIT	RESULTS

Trend: Isolated Excursion ☐   Frequent Excursion ☐

Microbial Identification: _____

Microbiological Section Head

QA Investigation

Performed by: _____   Reviewed by: _____

Operator training record: _____

Operator glove and gowning trend: _____

Unusual events during processing: _____

Sanitization records: _____

Gowning sterilization cycle: _____

Operator aseptic technique: _____

Comments: _____

_____

Corrective actions (performed by production): _____

_____

Conclusion: _____

_____

_____   _____

  QC Manager   QA Manager

cc: QA File/Area Manager/QC Microbiology

Written by:

Plant:

SOP No.: 1700.80
Issued On: mm/dd/yyyy
Revision No. New

Attachment No. 1700.80(L)

QUALITY CONTROL DEPARTMENT
Microbological Laboratories
VISUAL INSPECTION REPORT

Date:

Activity:

Observation/Deviation	Result	Room	Operator ID No.	Description of Deviation	Nature of Citation			Recommendation
					Minor	Major	Critical	
Personnel Hygiene								
Sick operator (flu or cold)	Yes No							
Improper gowning	Yes No							
Dirty gowning	Yes No							
Long nail	Yes No							
Wearing jewelry	Yes No							
Personnel Behavior								
Shouting	Yes No							
Running/fast movement	Yes No							
Crowding	Yes No							
Touching/sitting on floor	Yes No							

SOP No.: Val. 1700.80   **Effective date: mm/dd/yyyy**

                                                      Approved by:

Cleaning		
Dust/Dirt on the area	Yes	No
Unclean equipments	Yes	No
Dirty cages/pallets	Yes	No
Absence of hand disinfectant	Yes	No
Absence of alcohol	Yes	No
Others		
_____	Yes	No
_____	Yes	No
_____	Yes	No
_____	Yes	No

Prepared by: _____

Checked by: _____

QC Manager

YOUR COMPANY
VALIDATION STANDARD OPERATING PROCEDURE

SOP No.: Val. 1700.90

Effective date: mm/dd/yyyy

Approved by:

TITLE: Chemical and Physical Monitoring of DI, WFI, and RO Waters

AUTHOR:
_____
Name/Title/Department

_____
Signature/Date

CHECKED BY:
_____
Name/Title/Department

_____
Signature/Date

APPROVED BY:
_____
Name/Title/Department

_____
Signature/Date

REVISIONS:

No.	Section	Pages	Initials/Date

SOP No.: Val. 1700.90   **Effective date: mm/dd/yyyy**

Approved by:

SUBJECT: Chemical and Physical Monitoring of DI, WFI, and RO Waters

PURPOSE

This procedure defines the requirements for the monitoring of chemical and physical testing of water and the interpretation of the results related to the water system used in the injectable manufacturing plant.

RESPONSIBILITY

It will be the responsibility of the chemical lab analyst to follow the procedure. Quality assurance manager is responsible for SOP compliance.

PROCEDURE

- Use suitable sample containers (500 ml) that have been scrupulously cleaned of organic residue.
- Flush water for 1 minute prior to collecting samples.
- Remove container stopper just before sample collecting.
- Rinse the container twice using sampling water.
- Collect the water sample with minimal headspace.
- Immediately stopper the container after sampling.
- Label sample with information showing sample location, date, and time, then sign for "sampled by."

Notes:

- In case of noncompliance, copy of reports should be sent to QA manager.
- Use extreme caution when obtaining samples for TOC analysis.
- Water samples can be easily contaminated during the process of sampling and transportation to the testing facility.
- Collect the test solution in a tight container with minimal headspace, and test in a timely manner to minimize the impact of organic contamination from the closure and container.

SOP No.: Val. 1700.90   Effective date: mm/dd/yyyy

Approved by:

1. Sampling Locations and Frequencies

Sample locations and number of samples shall be decided based on the system in place.

Injectable Plant

Sampling Points	Activity	Frequency
Provide location	DIW tank	Once weekly
Provide location	WFI tank	Once weekly
Provide location	WFI tank	Once weekly
Provide room no.	WFI and DIW washing	Once weekly
Provide room no.	WFI and DIW solution preparation	Every working day
Provide room no.	WFI and DIW (equipment washing)	Every working day
Provide room no.	DIW (steam sterilizer)	As required

Note: It is appropriate to increase or decrease samples based on the trend performance; the time and date should be determined according to the activity.

2. Water Testing

■ All water samples should be tested per current USP (pH, conductivity, and TOC). See attachment no. 1700.90(A).
■ If TOC is not available, the samples are tested for pH, ammonia, calcium, carbon dioxide, chloride, sulphate, and oxidizable substances. See attachment no. 1700.90(B).
■ For chilled water and water for injection, see attachment no. 1700.90(C).

REASONS FOR REVISION

Effective date: mm/dd/yyyy

■ First time issued for your company, affiliates, and contract manufacturers.

SOP No.: Val. 1700.90   **Effective date: mm/dd/yyyy**

                                      Approved by:

Attachment No. 1700.90(A)

QUALITY CONTROL DEPARTMENT WATER ANALYTICAL REPORT

Plant: _____   Sampled by: _____

Sample: _____   Sampled on: _____

Location: _____

Analysis	Limits	Results
Description	A clear, colorless, odorless liquid	
pH	5–7	
Conductivity	N.M.T. 1.3 µS/cm at 25°C	
T.O.C.	N.M.T. 500 ppb	

Analyzed by: _____   Date: _____

Analyzed on: _____   Date: _____

Checked by: _____   Date: _____

                                     Quality Control Manager

SOP No.: Val. 1700.90 | Effective date: mm/dd/yyyy
Approved by:

Attachment No. 1700.90(B)

QUALITY CONTROL DEPARTMENT WATER ANALYTICAL REPORT

Plant: _____   Sampled by: _____

Sample: _____   Sampled on: _____

Location: _____

Analysis	Limits	Results
Description	A clear, colorless, odorless liquid	
• pH	5.0–7.0	
• Chloride	L.T 0.5 ppm	
• Sulphate	Passes test	
• Ammonia	NMT 0.3 ppm	
• Calcium	Passes test	
• Carbon dioxide	Passes test	
• Oxidizable substance	Passes test	

Analyzed by: _____   Date: _____

Analyzed on: _____   Date: _____

Checked by: _____   Date: _____

Quality Control Manager

SOP No.: Val. 1700.90

Effective date: mm/dd/yyyy

Approved by:

Attachment No. 1700.90(C)

CHEMICAL MONITORING OF WATER — WATER TESTING REPORT

Frequency: 2 Weeks　　　　Chiller Water (HVAC and Process)

Analysis	Limits	Results
pH	9.5–10.5	
TDS	For information	
Conductivity	For information	
Nitrite as NO$_2$	1200 ppm	
Total alkalinity	1000 ppm	

Frequency: Weekly　　　　Sample: Water for Injection

Analysis	Limits	Results
Description	A clear, colorless, odorless liquid (USP 24 req.)	
• pH	5.0–7.0	
• Chloride	Passes test	
• Sulphate	Passes test	
• Limit of ammonia	NMT 0.3 ppm	
• Calcium	Passes test	
• Carbon dioxide	Passes test	
• Heavy metals	Passes test	
• Oxidizable substance	Passes test	
• Total solid	NMT 0.001%	
• Conductivity	NMT 20 µS/cm	
Evaluation	Conform:	Not conform:

Analyzed by: _____　　Date: _____

Analyzed on: _____　　Date: _____

Checked by: _____　　Date: _____

Quality Control Manager

YOUR COMPANY
VALIDATION STANDARD OPERATING PROCEDURE

SOP No.: Val. 1700.100   Effective date: mm/dd/yyyy

Approved by:

TITLE: Validation of Sterility Test

AUTHOR: _____

Name/Title/Department

_____

Signature/Date

CHECKED BY: _____

Name/Title/Department

_____

Signature/Date

APPROVED BY: _____

Name/Title/Department

_____

Signature/Date

REVISIONS:

No.	Section	Pages	Initials/Date

SOP No.: Val. 1700.100   Effective date: mm/dd/yyyy

Approved by:

SUBJECT: Validation of Sterility Test

PURPOSE

The purpose is to develop a guideline to validate the efficacy of the sterility test method for a specific product or material. The similarity of the validation approach with the other pharmaceutical manufacturers shall be considered coincidental due to the similarity of operations and the nature of the work.

RESPONSIBILITY

It is the responsibility of concerned microbiologist to follow the procedure. The quality control (QC) manager is responsible for SOP compliance.

PROCEDURE

1. Frequency

The validation is performed:

- When the test for sterility must be carried out on reformulated or new product
- Whenever there is a change in the experimental conditions of the test
- If aseptically processed products have a higher rate of invalid results because this may be indicative of sterility problems not identified during validation

The validation may be performed simultaneously with the sterility test of the product being examined, but before the results of this test are interpreted.

2. Test Method Validation

- Before tests for sterility for any product are initially carried out, it is necessary to demonstrate the validity of the test method used by recovery of a small number of microorganisms in the presence of the product.
- Validation should mimic the test proper in every detail, such as in the volumes of media used and quantities and dilutions of product

and diluents; the approach depends on the method of test and details are given in each section. It may be performed concurrently with the actual test for sterility, but should be confirmed as successful before the results of the sterility test are interpreted.

■ All validation procedures should be carried out by personnel responsible for the routine testing of the product and should be done for each facility manufacturing that product.

■ Testing associated with the sterility test but requiring the use of live microorganisms (e.g., validation, stasis testing) should be carried out in laboratory facilities completely separate from the cleanroom.

2.1 Testing for antimicrobial activity

The items to be tested for sterility should be tested for antimicrobial activity during the product development stages, if this is possible. If they are found to have such activity, preparatory or test procedures will need to be modified to neutralize this activity. If all items are found to be free of such activity when first tested or after modification of procedures, application of the test for antimicrobial activity to every sample is not necessary.

To demonstrate that the mixture does not manifest antimicrobial activity, carry out the test as described previously up to the incubation step and add an inoculum of viable cells of the specified aerobic bacteria, anaerobic bacteria, and fungi.

2.1.1 Interpretation

Growth of each of the added microorganisms should be apparent within 48 hours. If conspicuous growth does not occur within 3 days for bacteria and 5 days for fungi, the test procedure is not valid; the article is bacteriostatic or fungistatic and must be modified (e.g., by using additional washes, using antagonists to the antimicrobial agent such as polysorbate 80, lecithin, azolectin, or β-lactamase, or other procedure) until conspicuous growth does occur when tests as previously described are carried out.

2.2 Bacteriostasis and fungistasis test validation

The test methodology should be validated by inoculation with 10 to 10 CFUs of challenge organism strains to the media/product container at the beginning of the test incubation period.

SOP No.: Val. 1700.100   Effective date: mm/dd/yyyy
Approved by:

The challenge inoculum should be verified by concurrent viable plate counts. The preferred validation method involves addition of challenge organisms directly to the product prior to direct inoculation or membrane filtration. However, when this is not practical due to inhibition or irreversible binding by the product, the challenge organisms should be added directly to the media containing the product, in the case of direct test methodology, or to the last rinse solution if membrane filtration methodology is used.

Microorganisms for Use in Growth Promotion, Validation, and Stasis Tests

Microorganism		Incubation Conditions	
Species	Suitable Strain	Temperature (°C)	Maximum Duration
Type: anaerobic bacteria		30–25	3 days
Clostridium sporogenes	ATCC 19404 CIP 79.3		
Type: aerobic bacteria		30–35	3 days
Staphylococcus aureus	ATCC6538 CIP 53.156 NCTC 7447 NCIMB 9518		
Bacillus subtilis	ATCC 6633 CIP 52.62 NCIB 8054		
Pseudomonas aeruginosa	ATCC 9027 NCIMB 8626 CIP 82.118		
Type: fungi		20–25	5 days
Candida albicans	ATCC 10231 IP 48.72 ATCC 2091 IP 1180.79		
Aspergillus niger	ATCC 16404		

Periodically, strains of microorganisms collected from the manufacturing environment should be used as challenge organisms.

2.2.1 Interpretation

The test is declared invalid if validation challenge organisms do not show clearly visible growth of bacteria within 3 days and fungi within 5 days in the test media-containing product. In most cases, unless the sterile product causes turbidity in the media, visual recovery times should be comparable to those of the growth promotion test. Records of validation and/or revalidation tests should be maintained.

2.3 Recovery of injured microorganisms

The test method should detect any microbial growth, especially injured microbes that may have actually survived — for example, due to improper disinfection techniques. Several artificial injuring methods — such as sonication for 2 minutes, heating to sublethal temperature, sublethal UV exposure, or sublethal disinfectant exposure — can be applied.

2.4 Stasis test

It is recommended that a stasis test be performed when antibiotics, inherently antimicrobial, or preserved products are tested. The stasis test can identify problems with dehydrated commercial media that were not apparent when a validation test was conducted at the beginning of the incubation period. It is necessary to demonstrate that growth-promoting qualities of media are retained and that preservative inhibitors remain stable for the full test period.

The test is performed by inoculation of 10 to 100 CFUs of challenge organisms directly to representative test containers of media containing product, which do not display any signs of contamination at the end of the test incubation period.

2.4.1 Interpretation

Stasis test challenge organisms should show clearly visible growth in the test medium within 3 days for bacteria and 5 days for fungi; otherwise the test is invalid. Records of stasis tests should be maintained.

SOP No.: Val. 1700.100

Effective date: mm/dd/yyyy
Approved by:

2.5 Testing for residual antimicrobial activity

To validate the test method, carry out the test procedures up to the final wash procedure of the membrane. To the final wash, add an inoculum of viable cells of the specified aerobic bacteria, anaerobic bacteria, and fungi. The number of microorganisms is to be between 10 and 100 CFUs.

After the final wash with the added microorganisms has been passed through the filter, incubate one filter disc in thioglycolate medium at 32 ± 2°C and one in TSB at 22 ± 2°C.

Growth of each of the added microorganisms should be apparent within 48 hours. If conspicuous growth does not occur within 3 days for bacteria and 5 days for fungi, the test procedure is not valid and must be modified (e.g., by using additional washes, using antagonists to the antimicrobial agent, or other procedures) until conspicuous growth does occur when tests as previously described are carried out.

2.4.1 Interpretation

If the membrane is found to be free of such antimicrobial activity when first tested or after modification of procedures, application of the test to every sample is not necessary.

Tests for sterility are to be carried out by trained personnel using techniques and equipment that minimize the risks of accidental microbial contamination of the tests and of the testing environment.

2.6 Media sterility test

All media should be preincubated for 14 days at appropriate test temperatures to demonstrate sterility prior to use. Alternatively, this control test may be conducted concurrently with the product sterility test. Media sterility testing may involve a representative portion or 100% of the batch.

2.7 Negative test controls

Negative controls for SteriDILUTOR, antibiotic membrane filter, diluting fluid sterility, and all other components of the sterility test should be conducted.

SOP No.: Val. 1700.100

Effective date: mm/dd/yyyy

Approved by:

2.8 Negative product controls

Negative product controls, which are similar in type and packaging to the actual product under test, should be included in each test session. These controls facilitate the interpretation of test results, particularly when used to declare a test invalid because of contamination in the negative product controls.

A minimum of ten negative product control containers may be adequate to simulate manipulations by the operator during a membrane filtration test. An equivalent number of samples to the test samples may be necessary to simulate the manipulations of the product by the operator during a direct inoculation test.

The negative control contamination rate should be calculated and recorded. The results of negative product control tests facilitate the interpretation of sterility test results, particularly when used to declare a test invalid because of contamination in the negative product controls.

During each working session (i.e., that uninterrupted period of time in which a sample or group of samples is tested) in which sterility testing is carried out, at least ten negative product control containers should be tested. For a direct inoculation test, these controls should be tested when possible at regular intervals during the test session.

A negative product control is usually a terminally sterilized item of undoubted sterility — that is, it has been subjected to the equivalent of two sterilization cycles by autoclaving or by dry heat sterilization. A negative control should be similar in type and container (or packaging if a device) to the product under test. The essential element of the negative control is that the manipulations involved in testing the control should be similar to those involved in testing the product. There should be similar risks of introducing contamination in the control and product tests.

A suitable negative product control for an aqueous product could be distilled water in a similar container. A negative product control for testing an ointment could be a container of liquid paraffin or ointment base that has been sterilized by dry heat; pouring the liquid paraffin from a container would be adequate to simulate squeezing of ointment from a tube.

When a retest is being carried out in the working session, these simulated negative controls should be processed concurrently with that retest.

The negative control contamination rate should be calculated and recorded. In order to derive the maximum information from the result of sterility tests, it is essential that the level of contamination detected in negative control tests be minimal.

SOP No.: Val. 1700.100   Effective date: mm/dd/yyyy

Approved by:

2.9 Positive test controls

All positive control tests in this section use viable challenge microorganisms. These tests should be conducted in a laboratory environment separate from the aseptic area where the product is tested.

2.9.1 Incubation period

■ All test containers should be incubated at temperatures specified by the pharmacopeial method for each test media for at least 14 days, regardless of whether filtration or direct inoculation test methodology is used.

■ The temperature of incubators should be monitored and there should be records of calibration of the temperature monitoring devices.

■ Test containers should be inspected at intervals during the incubation period and these observations recorded.

■ If the product produces a suspension, flocculation, or deposit in the media, after 14 days, suitable portions (e.g., 2 to 5%) of the contents of the containers should be transferred to fresh media under cleanroom conditions and reincubated for a further 7 days.

■ If any of these deviations may have compromised the integrity of the sterility test, it would be consistent with cGMP not to proceed with the test.

2.9.2 Observation and interpretation of results

At intervals during the incubation period and at its conclusion, examine the media for macroscopic evidence of microbial growth when the material being tested renders the medium turbid so that the presence or absence of microbial growth cannot be determined readily by visual examination. The test may be considered invalid only when one or more of the following conditions are fulfilled:

■ The data of the microbiological monitoring of the sterility testing facility show a fault.

■ A review of the testing procedure used during the test in question reveals a fault.

■ Microbial growth is found in the negative controls.

■ After determination of the identity of the microorganisms isolated from the test the growth of this species or these species may be ascribed unequivocally to faults with respect to the material and/or the technique used in conducting the sterility test procedure.

If the test is declared to be invalid, it is repeated with the same number of units as in the original test. If no evidence of microbial growth is found in the repeat test, the product examined complies with the test for sterility. If microbial growth is found in the repeat test, the product examined does not comply with the test for sterility.

2.10 Sterility test validation: per manufacturer validation protocol

2.10.1 Sterility testing operators

Aseptic techniques used should be reviewed periodically to ensure that departures from aseptic practices do not develop. Personnel should undergo periodic aseptic technique training (SOP [provide number]), particularly when problems are detected (during the course of routine environmental and negative control monitoring) or when operators perform the test infrequently.

The operator's testing technique should be monitored during every test session by use of negative product controls. The examination of test and control containers during and at the end of the incubation period should be included as part of the operator training. Statistics show that skilled operators working under the prescribed conditions can achieve a level as low as one contamination in 5000 control inoculations (0.02%).

Personnel training should be documented and records maintained.

2.11 Environmental control monitoring

A concurrent intensive monitoring of air, surface, and personnel hygiene should be performed during all validation procedures within the aseptic area. For documentation use Attachment No. 1700.100(A). For template validation report refer to Attachment No. 1700.100(B).

REASONS FOR REVISION

Effective date: mm/dd/yyyy

- First time issued for your company, affiliates, and contract manufacturers.

SOP No.: Val. 1700.100

Effective date: mm/dd/yyyy
Approved by:

Attachment No. 1700.100(A)

MICROBIOLOGY ANALYSIS RECORD
STERITEST MEMBRANE FILTRATION METHOD

Product/Bulk/Raw Material Name: _____   Code No. : _____   _____

Batch/Lot No.: _____   _____   Analytical No.: _____   _____

Test Method: _____   _____   Validation Reference: _____

Sampled on: _____   _____   Sampling Remarks: _   _____

Analyzed on: _____   Analyzed by: _____

Inspection of sample units for cracks, defective crimps, and cloudy solution: ☐ _____
Sample exterior surfaces sprayed with sterile 70% EPA ☐ 3% H₂O₂-----☐--- for ≡ 10 min prior to test.

Exterior surfaces of steritest and all items relating to the test, sanitized with sterile 70% EPA ☐ 3% H₂O₂-----
Type of canister: TTHA LA2 10 ☐ TTGS LA2 10 ☐ TTHV AB2 10 ☐ TTHA PC2 10 ☐ TLHV SL2 10 ☐ Lot No. __
Sample preparation _____ / Reconstituted with 10 ml of sterile _____ Using _____
No. of tested samples: _____ Tested as: 1 group ☐   3 groups ☐   Individual samples
Holding time before filtration _____ Sampling preparation remarks _____

Type of rinsing fluid: A ☐   Fluid D ☐   Lot No. _____ Sterility Reference: _____

Step	*Pump speed*
Pre-wet the membranes with 25 mL of sterile rinse solution Product testing	
Clear the product from the tubing and canisters	
Rinse the tubing, canisters, and membranes	

Additives to the diluent   Penicillinase   Other neutralizing agent

Add Media to the Canisters

Media used	*Lot No.*	*Sterility Ref.*	*Pump speed*	*Negative control*		
Fluid Thioglycolate				Conform	Not conform	
TSB				Conform	Not conform	

Under LAF, with sterile scissors, cut the tubing approximately 2 cm above the clamp, fold over and insert the tubing into the top air vents of the canisters

Accidental events during test: _____

Negative control using _____

Negative product control using _____

Incubation: Incubator Memmert K896-0002 at 32°C Incubator Precision WB 83326658 at 22°C

SOP No.: Val. 1700.100

Effective date: mm/dd/yyyy

Approved by:

Attachment No. 1700.100(A)

MICROBIOLOGY ANALYSIS
STERILITY TEST RESULTS

Result of environmental microbiological monitoring

Monitoring	Results			Microbial Identification
	Position 1	Position 2	Position 3	
RCS+ of sterility room				
Settle plates in LAFH				
Surface monitoring of LAFH				
Surface monitoring of sterility room				
Operators' gloved hand plates				
Operators' gowns				

Results of sterility test

Test	3 days	7 days	14 days	21 days
TSB				
Fluid Thioglycolate				
Negative control				
Product negative control				

*T = turbidity   C = clear

Identification of the detected growth: _____

_____

Stasis test within 3 days

Organism	TSB		Fluid Thioglycolate	
	Tested with about 100 cfu of each organism			
B. subtilis ATCC 6633	Growth	No growth	Growth	No growth
S. aureus ATCC 6538P	Growth	No growth	Growth	No growth
C. albicans ATCC 10231	Growth	No growth	Growth	No growth
Asp. niger ATCC 16404	Growth	No growth	Growth	No growth
Environmental isolate	Growth	No growth	Growth	No growth

Evaluation _____ **Conform**   **Invalid test**   **Should be repeated**   **Not conform**

Final Comment: _____

_____

Performed by: _____ Date: _____

Checked by: _____ Date: _____

Approved by: _____ Date: _____

QC Manager

SOP No.: Val. 1700.100   Effective date: mm/dd/yyyy

                                          Approved by:

Attachment No. 1700.100(B)

MICROBIOLOGY VALIDATION REPORT
STERILITY TEST OF _____

1.0 Purpose

To validate sterility test method no. _____ membrane filtration for estimation for sterility test. The recovery of pharmacopoeial and in-house organisms by filtration method has been tested and to validate that quantity of rinsing fluid is suitable to neutralize antimicrobial activity of injectable product. The methodology shall not provide an opportunity for false negative.

2.0 Scope

Microbial recovery studies of membrane filtration method challenged with less than 100 CFU of each organism listed in USP-25, EP-2002 and in-house microbial isolates. Two lots of (Product Name) (Batch Numbers) have been validated in triplicate for each organism. Validation mimicked the test proper in every detail, such as in the volumes of media used, quantities and dilutions of product and diluents.

3.0 References

3.1 USP 25 Sterility test

3.2 USP 25 <1227> Validation of Microbial Recovery from pharmacopoeial articles

3.3 Quality System Regulation, Code of Federal Regulations, 21 CFR 820 (1996)

3.4 Parenteral Drug Association, Technical Report No. 21, Bioburden Recovery Validation, Vol. 44, No. S3 (PDA, 1990)

3.5 APP SOP 0110070001

3.6 SOP QCS 102

4.0 Conclusion

4.1 (Product Name) can be tested for sterility by membrane filtration with 3 x 300 rinsing fluid A after incubation of 14 days and more than one week.

4.2 The neutralization procedure showed an effective neutralization for the antibiotic effect (if applicable).

Checked by: _____ Date: _____

Approved by: _____ Date: _____

SOP No.: Val. 1700.100

Effective date: mm/dd/yyyy

Approved by:

Attachment No. 1700.100(B)

MICROBIOLOGY VALIDATION REPORT
STERILITY TEST OF (PRODUCT NAME)
COMPLEMENTARY TEST

Complementary Tests	Procedure	Results
Millipore steritest validation	Steritest system validation protocol # TCPU 0 VG 01 Rev. 5	Steritest system is valid
Environmental control monitoring	SOP No._____	All result of active air sampling; settle plates; RODAC plates, swabs; and operator's gloved hand plates within the acceptance limit
Sterility of antibiotic membrane filter	STM No._____	Lot No. F9AM31473 is sterile (incubated for 21 days)
Sterility of Sterisolutest	STM No._____	Lot No. R7AM22431 is sterile (incubated for 21 days)
Media sterility	STM No._____	Fluid thioglycolate Millipore Lot No. H2CN11088 B is sterile (incubated for 21 days) TSB Millipore Lot No. H2CN11505B is sterile; growth appears in all inoculated media containers within 2 days
Count of the challenge inoculum	STM No._____	The inoculum count of all challenge organisms in the range between 10 and 100 CFU
Media growth promotion test	STM No._____	Fluid thioglycolate Millipore Lot No. H1PN04885 A is sterile (incubated for 21 days)
Sterility test	STM No._____	Sterile, and pass stasis test after incubation period
Sterility test	STM No._____	Sterile, and pass stasis test after incubation period
Negative control (sterile placebo)	STM No._____	Sterile (incubated for 21 days), performed by two analysts (Hamdy & Amal)
Recovery of injured microorganisms	SOP No._____	Conform to the acceptance criteria
Testing for residual antimicrobial activity	SOP No._____	Holding 1h & then rinsing 3 times with 300 ml fluid A nullifies the residual antimicrobial activity of the product
Validation of aseptic working technique when performing sterility test	SOP No._____	Two analysts passed the acceptance criteria

SOP No.: Val. 1700.100   Effective date: mm/dd/yyyy
                                          Approved by:

Attachment No. 1700.100(B)

MICROBIOLOGY VALIDATION REPORT
STERILITY TEST OF (PRODUCT NAME)
TEST FOR BACTERIOSTASIS AND FUNGISTASIS
FINAL REPORT

The filter is pre-wetted with diluent before filtration. 21 vial (Product Name) is reconstituted with WFI and held for 1 h. In triplicate, the reconstituted sample is filtered through the membrane filter and washed carefully by 3 x 300-mL portions of diluting-fluid A. Inoculate the final 100-mL portion with less than 100 CFU of each challenge microorganism, and pass through the filter. This filter vessel is then filled with the appreciated medium and incubated for recovery. A concurrent viable count has been carried out when performing any of these tests using membrane filtration method and Miles-Misra control method as a check that the working dilution has been correctly prepared and calculated. After incubation, morphological characteristics of the detected growth obtained from both methods have been checked.

Acceptance criteria: The test media are satisfactory if visual evidence of growth appears in all inoculated media containers within 5 days of incubation.

Batch No. _____

Challenge Microorganism	Initial count		FTB (32°C) showed turbidity after			TSB (22°C) showed turbidity after		
Pseud. aeruginosa ATCC 9027	65	Aerobically	48 h	48 h	48 h	48 h	48 h	48 h
E. coli ATCC 8739	58	Aerobically	48 h	48 h	48 h	48 h	48 h	48 h
Bacillus subtilis ATCC 6633	76	Aerobically	48 h	48 h	48 h	48 h	48 h	48 h
Staphylococcus aureus ATCC 6538	85	Aerobically	48 h	48 h	48 h	48 h	48 h	48 h
Enterobacter colcacae isolate from purified water	55	Aerobically	48 h	48 h	48 h	48 h	48 h	48 h
Micrococcus luteus environmentally isolated	58	Aerobically	48 h	48 h	48 h	48 h	48 h	48 h
Staph. epidermis environmentally isolated	49	Aerobically	48 h	48 h	48 h	48 h	48 h	48 h
Cl. sporogenes ATCC 11437	59	An-Aerobically	48 h	48 h	48 h	48 h	48 h	48 h
Candida albicans ATCC 10231	33	Aerobically	48 h	48 h	48 h	48 h	48 h	48 h
Aspergillus niger ATCC 16404	58	Aerobically	48 h	48 h	48 h	48 h	48 h	48 h

All tested organisms showed conform results for (Product Name) using membrane filtration technique and rinsed with 3 x 300 ml of fluid A.

SOP No.: Val. 1700.100
Effective date: mm/dd/yyyy
Approved by:

Attachment No. 1700.100(B)

MICROBIOLOGY VALIDATION REPORT
STERILITY TEST OF (PRODUCT NAME)
TEST FOR BACTERIOSTASIS AND FUNGISTASIS
FINAL REPORT

Batch No. _____

Challenge Microorganism	Initial count		FTB (32°C) showed turbidity after			TSB (22°C) showed turbidity after		
Pseud. aeruginosa ATCC 9027	65	Aerobically	48 h	48 h	48 h	48 h	48 h	48 h
E. coli ATCC 8739	58	Aerobically	48 h	48 h	48 h	48 h	48 h	48 h
Bacillus subtilis ATCC 6633	76	Aerobically	48 h	48 h	48 h	48 h	48 h	48 h
Staphylococcus aureus ATCC 6538	85	Aerobically	48 h	48 h	48 h	48 h	48 h	48 h
Enterobacter colcacae isolate from purified water	55	Aerobically	48 h	48 h	48 h	48 h	48 h	48 h
Micrococcus luteus environmentally isolated	58	Aerobically	48 h	48 h	48 h	48 h	48 h	48 h
Staph. epidermis environmentally isolated	49	Aerobically	48 h	48 h	48 h	48 h	48 h	48 h
Cl. sporogenes ATCC 11437	59	An-Aerobically	48 h	48 h	48 h	48 h	48 h	48 h
Candida albicans ATCC 10231	33	Aerobically	48 h	48 h	48 h	48 h	48 h	48 h
Aspergillus niger ATCC 16404	58	Aerobically	48 h	48 h	48 h	48 h	48 h	48 h

All tested organisms showed conform results for (Product Name) using membrane filtration technique and rinsed with 3 x 300 ml of fluid A.

YOUR COMPANY
VALIDATION STANDARD OPERATING PROCEDURE

SOP No.: Val. 1700.110

Effective date: mm/dd/yyyy
Approved by:

TITLE: Visual Inspection of Small-Volume Parenterals

AUTHOR: _____
Name/Title/Department

_____
Signature/Date

CHECKED BY: _____
Name/Title/Department

_____
Signature/Date

APPROVED BY: _____
Name/Title/Department

_____
Signature/Date

REVISIONS:

No.	Section	Pages	Initials/Date

SOP No.: Val. 1700.110

Effective date: mm/dd/yyyy

Approved by:

SUBJECT: Visual Inspection of Small-Volume Parenterals

PURPOSE

The purpose is to provide criteria for accepting or rejecting small-volume parenterals during visual inspection.

RESPONSIBILITY

It is the responsibility of the inspection operator to perform visual inspection. Production supervisor and sterile production manager are responsible for SOP compliance.

DOCUMENTATION

■ Document the results of the visual inspection process of ampoules in attachment no. 1700.110(A).
■ Document the results of the visual inspection process of lyophilized vials in attachment no. 1700.110(B).
■ Document the results of the visual inspection process of liquid-filled vials in attachment no 1700.110(C).

PROCEDURE

1. Liquid Form Products

1.1 Solution defects

■ Clarity: The solution should be sparkling clear; there should be no turbidity or haze.
■ Particles: The solution should be free from visible particles (e.g., glass pieces/glass dust, fibers, other particles). *Note:* During inspection, particles float, hang, or slowly settle downwards.
■ Color: Color of the solution should be as specified in the MFM.
■ Volume variation: empty, underfilled, or overfilled ampoules/vials should be removed by visual comparison to a correctly filled ampoule/vial.

SOP No.: Val. 1700.110   Effective date: mm/dd/yyyy
                                                Approved by:

1.2 Ampoule defects

- Ampoule height: this should be visually compared to a correctly sealed ampoule.
- Bad sealing: sealing should be smooth and rounded in shape. If one of the following defects is found, the ampoule should be rejected:
  - Dome bubbling
  - Flat or concave dome
  - Sharp protrusions
- Charring (burning of solution near the tip): ampoules should be free from charring.
- Bad coding: ampoules should have the correct color ring code (if specified) with the correct number of continuous rings.
- Broken: ampoules should not be broken and should not have any cracks.

1.3 Vial defects

- Broken: vials should not be broken and should not have any cracks.
- Air bubbles: vials should be free from any glass air bubbles (glass vials manufacturing defect).
- Faulty crimping: vials should be free from faulty crimping) (Figure 1[A]):
  - Lower cap edge is crimped downwards along the bottle neck (Figure 1[B]).
  - Lower edge of the cap is not crimped inwards completely (Figure 1[C]).

2. Lyophilized Products

2.1 Cake defects

- Volume: the cake should be of the correct volume (cake height should be appropriate to the correct volume filled, not showing variation of low or high fill).
- Appearance: the cake should have the correct appearance and color (i.e., no back melting or pockets of moisture or particulate matter).

SOP No.: Val. 1700.110   Effective date: mm/dd/yyyy

Approved by:

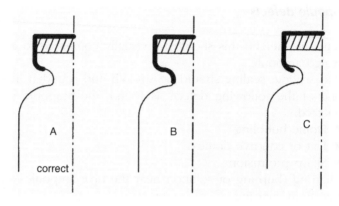

Figure 1

2.2 Vial defects (see Section 1.3)

Note 1: Maximum allowable reject rate is 2%. Investigate if rejects exceed this limit. If the limit is exceeded, refer to SOP no. (provide number)
Note 2: For samples of defects, see the photos in attachment no. 1700.110(D).

REASONS FOR REVISION

Effective date: mm/dd/yyyy

■ First time issued for your company, affiliates, and contract manu-facturers.

SOP No.: Val. 1700.110

Effective date: mm/dd/yyyy

Approved by:

Attachment No. 1700.110(A)

VISUAL INSPECTION REPORT: AMPOULES

Product Name: _____ Batch No.: _____ Qty. (after leaker): _____ Date: _____

Operator	Solution Defects								Ampoule Defects					Rejects/ Operator
	Clarity	Color	Particles			Volume Variation		Ampoule Height	Bad Sealing	Charring	Bad Coding	Broken	Other	
			Glass	Fibers	Other	Low	High							
												Total Rejects		

A) Total rejects = _____ ampoules
B) Total inspected = _____ ampoules
% Rejects = A/B × 100 = _____%
(Maximum allowable reject rate of 2%; investigate if reject exceeds limit; refer to SOP No. PGS-059)

Done by _____

Checked by _____

Line boss _____

Sterile area supervisor _____

SOP No.: Val. 1700.110

Effective date: mm/dd/yyyy

Approved by:

Attachment No. 1700.110(B)

VISUAL INSPECTION REPORT: LYOPHILIZED VIALS

Product Name: _____ Batch No.: _____ Qty. (lyophilized): _____ Date: _____

Operator	Cake Defects					Vial Defects			Total Rejects
	Color	Back Melting	Pockets of Moisture	Volume Variation (Cake Height)		Bad Crimping	Air Bubbles (Glass Defect)	Broken or Cracked	Other	
				Low	High					
										Total rejects

Done by _____   Checked by _____

Line boss _____   Sterile area supervisor _____

A) Total rejects = _____ vials
B) Total inspected = _____ vials
% Rejects = A/B × 100 = _____ %
(Maximum allowable reject rate of 2%; investigate if reject exceeds limit; refer to SOP No. PGS-059)

SOP No.: Val. 1700.110

Effective date: mm/dd/yyyy

Approved by:

Attachment No. 1700.110(C)

VISUAL INSPECTION REPORT: LIQUD-FILLED VIALS

Product Name: _____ Batch No.: _____ Qty. (after filling): _____ Date: _____

Operator	Solution Defects					Volume Variation		Vial Defects				Rejects/ Operator
	Clarity	Color	Glass	Fibers	Other	Low	High	Bad Crimping	Air Bubble (Glass Defect)	Broken or Cracked	Other	
			Particles									
											Total rejects	

A) Total rejects = _____ vials

B) Total inspected = _____ vials

% Rejects = A/B × 100 = _____ %

(Maximum allowable reject rate of 2%; investigate if reject exceeds limit; refer to SOP No. PGS-059)

Done by _____ Checked by _____

Line boss _____ Sterile area supervisor _____

SOP No.: Val. 1700.110   Effective date: mm/dd/yyyy
                                          Approved by:

Attachment No. 1700.110(D)

VISUAL INSPECTION: SAMPLES OF DEFECTS

White particle.
Concentrated light is sent through the container bottom. Looking from the
other side of the container, the operator, looking from the side detects
light reflections scattered by the particle against a dark background.

Black particle.
Light is sent through the container side. Looking from the other side of the
container, the operator looking from the other side of the container detects
shadows created by the particle.

SOP No.: Val. 1700.110   Effective date: mm/dd/yyyy
                                           Approved by:

Bad crimping.

Inspection of cap, neck function.

Freeze-dried cake: good surface.

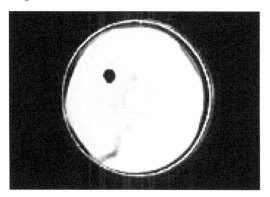

SOP No.: Val. 1700.110

Effective date: mm/dd/yyyy

Approved by:

Freeze-dried cake: bottom defect.

Back melting.
Bad sealing. Light is sent through the container side. Looking from the other side of the container, the operator detects shadows created by the particle

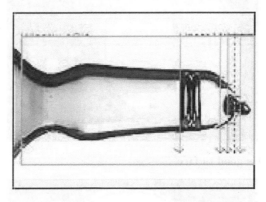

YOUR COMPANY
VALIDATION STANDARD OPERATING PROCEDURE

SOP No.: Val. 1700.120

Effective date: mm/dd/yyyy

Approved by:

TITLE: Fill Volume/Weight and Other Checks for Parenteral Products during Filling

AUTHOR: _____
Name/Title/Department

_____
Signature/Date

CHECKED BY: _____
Name/Title/Department

_____
Signature/Date

APPROVED BY: _____
Name/Title/Department

_____
Signature/Date

REVISIONS:

No.	Section	Pages	Initials/Date

SOP No.: Val. 1700.120   Effective date: mm/dd/yyyy

Approved by:

SUBJECT: Fill Volume/Weight and Other Checks for Parenteral Products during Filling

PURPOSE

The purpose is to provide a procedure for the fill check (by weight and/or by volume) for liquid products (ampoules/vials) and for powder by weight and prefilled syringes.

RESPONSIBILITY

It is the responsibility of the area QA inspector to carry out the testing per the described procedure. The quality assurance (QA) manager will be responsible for SOP compliance.

PRECAUTIONS

- Make sure that no air space is left on top of syringe while transferring; also liquid should not touch the walls of cylinder.
- Cylinders and syringes should be completely dry before use.

PROCEDURE

1. **Equipment Required**

- One syringe each of 2, 5, and 10 ml fitted with 21-gauge needle of NMT 1 in. in length
- 400-ml beakers
- Ethanol
- WFI
- Only 10-ml cylinder
- Balance

2. **General Procedure**

Fill volume check for liquid/lyophilized parenterals

- Take eight samples from the machine.
- Place a dry cylinder on calibrated balance and tare.
- Place a syringe in the cylinder and tare.
- Remove the syringe from the cylinder.

SOP No.: Val. 1700.120   **Effective date: mm/dd/yyyy**

Approved by:

- Take up the content of first containers in the dry hypodermic syringe of a rated capacity not exceeding three times the volumes to be measured.
- Expel any air bubbles from the syringes and needle.
- Then place the filled syringe back on the cylinder.
- Note the weight in grams on the control chart.
- Repeat the same procedure for the remaining seven ampoules.
- Take out the average of the eight ampoules or vials or prefilled syringes (by weight) and write in the average column.
- Find out the range by subtracting the lowest weight from highest weight (eight ampoules checked) and note in the range column.
- Plot the average and range for each hourly check.
- Each hour, also check on three ampoules the length of the ampoules, sealing, particles, break ring presence, or any printing and note down in X and R charts for parenterals.
- Immediately inform the supervisor in case of any adjustment required for volume or any physical defects.
- Control limits for the average chart will be drawn on the basis of specific gravity done by QC at the time of bulk sample analysis prior to filling.

3. Exceptions

For volumes of 10 ml or more, the content of containers (ampoules) holding more than 10 ml will be directly transferred to the graduated cylinders without using the syringe.

4. Excess volume

Labeled Size (ml)	Mobile Liquids (ml)	Theoretical	Actual (–1%)
1.0	0.10 ml	1.1 ml	1.09 ml
2.0	0.15 ml	2.15 ml	2.12 ml
3.0	0.2 ml	3.15 ml	3.12 ml
4.0	0.3 ml	4.3 ml	4.25 ml
5.0	0.3 ml	5.3 ml	5.25 ml
10.0	0.5 ml	10.5 ml	10.40 ml
20.0	0.6 ml		

5. Prefilled Syringes

Prefilled syringes will be monitored for volume check by directly dispensing the content into a tared calibrated cylinder of appropriate capacity, reading the volume, and noting down the weight in the average and range chart.

6. Documentation

See attachment no. 1700.120.

REASON FOR REVISION

Effective date: mm/dd/yyyy

■ First time issued for your company, affiliates, and contract manufacturers.

SOP No.: Val. 1700.120

Effective date: mm/dd/yyyy

Approved by:

Attachment No. 1700.120(A)

QA INPROCESS X- AND R-CHARTS

SOP No. 1700.120
Issued on: mm/dd/yyyy
Revision No.: New

Product:
Code No.:
Batch No.:

Process:
Batch Size:

Machine:
Operator:
Room No.:

Target:
UCL:
UCL:

LCL: _____ (Inprocess Limit)
LCL: _____ (Lab.Limit)

X-CHART

R-CHART

LIMITS

LIMITS

DATE/TIME | INDIVIDUAL FILL VOLUME | INDIVIDUAL FILL WEIGHT(gm) | AVERAGE | RANGE

INSPECTION OF FILLED AND SEALED VIALS/AMPOULES/REFILLED SYRINGES

S. No. | Length | Sealing | Free from particles | Breaking ring

Description of any corrective action

ceramic printing

BATCH AVERAGE:

Signature (Quality Assurance Inspector)

| YOUR COMPANY |
| VALIDATION STANDARD OPERATING PROCEDURE |

SOP No.: Val. 1700.130

Effective date: mm/dd/yyyy

Approved by:

TITLE: Preservative Efficacy Test

AUTHOR: _____
Name/Title/Department

_____
Signature/Date

CHECKED BY: _____
Name/Title/Department

_____
Signature/Date

APPROVED BY: _____
Name/Title/Department

_____
Signature/Date

REVISIONS:

No.	Section	Pages	Initials/Date

SOP No.: Val. 1700.130   **Effective date: mm/dd/yyyy**

                                               Approved by:

SUBJECT: Preservative Efficacy Test

PURPOSE

1. Objective

The objective is to provide standard procedure for performing the preservative efficacy test for nonsterile and sterile dosage forms.

1.1 Product categories under preservative efficacy test (Table 1)

Table 1   Product Categories

Category	Product Description
Category I	Ia. Injections, other parenterals including emulsions, otic, sterile nasal products, and ophthalmic products made with aqueous bases or vehicles
	Ib. Topically used products made with aqueous bases or vehicles, nonsterile nasal products, and emulsions, including those applied to mucous membrane.
	Ic. Oral products made with aqueous bases or vehicles.
Category II	All preserved dosage forms listed under category I made with nonaqueous (anhydrous) bases or vehicles.

2. Materials

2.1 Equipment

- Laminar air flow
- Autoclave
- Water bath
- Incubator at 376°C
- Dry heat sterilizer
- Refrigerator
- Gas burner
- Petri dishes (20 × 100 mm)
- Micropipette
- pH Meter

SOP No.: Val. 1700.130   Effective date: mm/dd/yyyy

Approved by:

- Spectrophotometer
- Vortex mixer
- Glassware
- Colony counter

2.2 Reagent

2.2.1 Harvesting fluid-1 (JUSP sterile saline TS)

- Sodium chloride: 9.0 g
- Distilled water: q.s. 1000 ml
- Sterilize in the autoclave at 121°C for 15 minutes

2.2.1 Harvesting fluid 2

- Sodium chloride: 9.0 g
- Polysorbate 80 (w/v): 0.5 g
- Distilled water q.s.: 1000 ml
- Sterilize in the autoclave at 121°C for 15 minutes

2.2.3 Diluent

- Sodium chloride: 9.0 g
- Polysorbate 80 (w/v): 0.5 g
- Peptone 49: 1.0 g
- Distilled water q.s.: 1000 ml
- Sterilize in the autoclave at 121°C for 15 minutes

2.3 Media: perform growth promotion test for all media used in preservative efficacy test

2.3.1 Tryptone soya agar (TSA)

- Pancreatic digest of casein: 15.0 g
- Papaic digest of soya bean: 5.0 g
- Sodium chloride: 5.0 g
- Agar: 15.0 g

SOP No.: Val. 1700.130   **Effective date: mm/dd/yyyy**
                                                    Approved by:

- Distilled water q.s.: up to 1000 ml
- Final pH: 7.3 to 0.2 (after sterilization) at 25°C
- Suspend 40 g in 1 l distilled water or deionized water
- Mix and boil to create solution
- Sterilize in the autoclave at 121°C for 15 minutes

2.3.2 Sabouraud dextrose agar (SDA)

- Peptone: 10.0 g
- Dextrose: 40.0 g
- Agar: 15.0 g
- Distilled water q.s.: up to 1000 ml
- Final pH 5.6 to 0.2 (after sterilization)
- Suspend 65 g in 1 l distilled water or deionized water
- Mix and boil to create solution
- Sterilize in the autoclave at 121°C for 15 minutes

2.4 Test organisms

- *Aspergillus niger* (ATCC no. 16404)
- *Candida albicans* (ATCC no. 10231)
- *Escherichia coli* (ATCC no. 8739)
- *Pseudomonas aeruginosa* (ATCC no. 9027)
- *Staphylococcus aureus* (ATCC no. 6538)

Note: Other microorganisms may be included in the test on an optional basis, if they represent likely contaminants to the preparation. *Zygosaccharomyces rouxii* may be used for oral preparations containing a high concentration of sugar.

- The viable microorganism used in the test must not be more than five passages removed from the original ATCC culture.
- One passage is defined as the transfer of organisms from an established culture to fresh medium and count all transfers.
- Resuscitate the cultures received from the ATCC according to the ATCC directions.

3. Method

3.1 Preparation of inoculum

- Inoculate the surface of tryptone soya agar slant for bacteria and Sabouraud dextrose agar slant for fungi from recently revived stock culture of each of the test microorganisms.
- Incubate the bacterial cultures at 32.5 to 2.5°C for 18 to 24 hours.
- Incubate the culture of *C. albicans* at 22.5 to 2.5°C for 48 hours and the culture of *A. niger* at 22.5 to 2.5°C for 6 to 10 days or until good sporulation is obtained.
- Harvest the growth of bacteria and *C. albicans* with 10 ml of sterile harvesting fluid 1 and the growth of *A. niger* with 10 ml of harvesting fluid 2.
- Dilute the suspension with respective sterile fluid to reduce the microbial count to about 1 × 108 colony-forming units (CFUs) per liter.
- Determine the number of colony-forming units per liter in each suspension by plate count method, using the conditions of media and microbial recovery incubation times listed in Table 3 to confirm the initial CFUs per ml estimate.
- Use the bacterial and yeast suspensions within 24 hours of harvesting.
- Use the fungal suspension up to 7 days of harvest if stored under refrigeration.

3.2 Procedure

3.2.1 Challenging formulated product

- Conduct the test in original product container or in sterile, capped containers of suitable size into which a sufficient volume of product, at least 20 ml, has been transferred.
- Inoculate each product container with one of the prepared and standardized inoculum suspensions and mix well.
- Use the volume of suspension between 0.5 and 1.0% of the volume of the product so that each liter of product after inoculation contains between 1×10^5 to 1×10^6 cells of the test microorganisms.

SOP No.: Val. 1700.130
Effective date: mm/dd/yyyy
Approved by:

Table 2 Sampling Intervals per Product Category

Product Category	Sampling Intervals				
	6 hours	4 hours	7 days	14 days	28 days
Category Ia	+	+	+	+	+
Category Ib	+	+	+	+	+
Category Ic	+	+	+	+	+
Category II	+	+	+	+	+

Note: "+" indicates sampling intervals.

■ Estimate the initial concentration of viable microorganisms in each test preparation based on the concentration of microorganisms in each of the standardized inoculum suspensions.
■ Incubate the inoculated product containers at 22.5 to 2.5°C.
■ Sample each container at the appropriate intervals specified in Table 2.

3.2.2 Standard plate count method

■ Preparation of sample
   1. Transfer aseptically 1 ml, accurately measured, of sample from each inoculated product container to a dilution bottle containing 9 ml of sterile diluent and mix well. This is 10:1 dilution.
   2. Transfer aseptically 1 ml, accurately measured, from 10:1 dilution to a second dilution blank containing 9 ml of sterile diluent and mix well. This is 10:2 dilution.
   3. Continue tenfold dilution up to 10-'s or suitable as required.
■ Total aerobic bacterial count
   1. Pipette 1 ml aliquot from each dilution to duplicate sets of sterile petri dishes.
   2. Promptly pour into each petri dish about 15 to 20 ml of sterile melted tryptone soya agar medium previously melted and cooled to approximately 45°C.
   3. Cover the dishes and mix the sample well with agar by tilting and rotating the petri dishes.
   4. Pour one plate with uninoculated agar medium and uninoculated diluent as negative control.
   5. After the agar has solidified, invert the dishes and incubate them at 32.5 to 2.5°C for 3 to 5 days.

Table 3   Culture Conditions for Inoculum Preparation and Microbial Recovery

Organism	Suitable Medium	Incubation Temperature	Incubation Time	Microbial Recovery Time
Escherichia coli (ATCC no. 8739)	TSA	32.5–2.5°C	18–24 h	3–5 days
Pseudomonas aeruginosa (ATCC no. 9027)	TSA	32.5–2.5°C	19–24 h	3–5 days
Staphylococcus aureus (ATCC no. 6538)	TSA	32.5–2.5°C	18–24 h	3–5 days
Candida albicans (ATCC no. 10231)	SDA	22.5–2.5°C	44–52 h	3–5 days
Aspergillus niger (ATCC no. 16404)	SDA	22.5–2.5°C	6–10 day	3–5 days

Note: TSA = tryptone soya agar; SDA = Sabouraud dextrose agar.

■ Total molds and yeasts count
1. Pipette 1 ml aliquot from each dilution to duplicate sets of sterile petri dishes.
2. Promptly pour into the petri dishes about 15 to 20 ml of sterile melted Sabouraud dextrose agar medium previously melted and cooled to approximately 45°C.
3. Cover the dishes and mix the sample with agar carefully by tilting and rotating the petri dishes.
4. Pour one plate with uninoculated agar medium and uninoculated diluent as negative control.
5. After the agar has solidified, invert the dishes and incubate them at the conditions listed in Table 3.

4. Acceptance Criteria

The criteria for evaluation of antimicrobial activity are given in Table in terms of the log reduction in the number of viable microorganism using as baseline the value obtained for the inoculum. "No increase" defined as not more than 0.5 log unit higher than the previous valu obtained.

SOP No.: Val. 1700.130

Effective date: mm/dd/yyyy

Approved by:

Table 4 Acceptance Criteria for Preservative Efficacy

	Required Log Reduction from the Initial Calculated Count at Interval			
Product Category	*Organism*	*7 days*	*14 days*	*28 days*
Category Ia	Bacteria	1	3	NI
	Fungi	NI	NI	NI
Category Ib	Bacteria	—	2	NI
	Fungi	—	NI	NI
Category Ic	Bacteria	—	1	NI
	Fungi	—	NI	NI
Category II	Bacteria and fungi	—	NI	NI

Note: NI = no increase.

REASONS FOR REVISION

Effective date: mm/dd/yyyy

■ First time issued for your company, affiliates, and contract manufacturers.

YOUR COMPANY
VALIDATION STANDARD OPERATING PROCEDURE

SOP No.: Val. 1700.140　　　　　　　　　　　Effective date: mm/dd/yyyy

Approved by:

TITLE:　　　Disinfectant Validation

AUTHOR:　　　_____

Name/Title/Department

_____

Signature/Date

CHECKED BY:　　　_____

Name/Title/Department

_____

Signature/Date

APPROVED BY:　　　_____

Name/Title/Department

_____

Signature/Date

REVISIONS:

No.	Section	Pages	Initials/Date

SOP No.: Val. 1700.140

Effective date: mm/dd/yyyy
Approved by:

SUBJECT: Disinfectant Validation

PURPOSE

The purpose is to assure that the recommended disinfectant has the acceptable relative standard of the antimicrobial activity using the membrane filtration technique and surface testing technique.

RESPONSIBILITY

It is the responsibility of all analysts (microbiologists) to follow the procedure. The quality control (QC) manager is responsible for SOP compliance.

PROCEDURE

1. Frequency

- The *membrane filtration technique* is used once, prior to the introduction of a new disinfectant within the production department.
- The *surface testing technique* is used prior to any changes in the recommended procedure for evaluating its effectiveness on surfaces to be treated and demonstrating activity against contamination for various contact times.

2. Equipment

- Incubator 22 ± 2°C and 32 ± 2°C
- Standard loop
- Membrane filtration unit
- Vortex
- Cultures
  - *Staphylococcus aureus* ATCC 6538
  - *Candida albicans* ATCC 10231
  - *Bacillus subtilis* ATCC 6633
  - *Clostridium sporogenes* ATCC 11437
  - *Pseudomonas aeruginosa* ATCC 9027
  - *Escherichia coli* ATCC 8739

SOP No.: Val. 1700.140   Effective date: mm/dd/yyyy

Approved by:

- *Aspergillus niger* ATCC 16404
- Organism(s) recovered from environment (The criteria for selection of environmental isolates include frequency of occurrence and representation from a broad spectrum of organisms.)
- Sterile distilled water
- Sterile screw-cap test tubes
- Sterile buffer (dilution blanks)
- Poured, sterile soybean casein digest agar (SCDA) petri plates
- Poured, sterile potato dextrose agar (PDA)
- Sterile pipettes: 25-, 10-, and 1-ml sizes
- Sterile forceps
- Sterile membrane filtration unit with 47-mm diameter
- Pore size: 0.45 μm
- Solvent-resistant membrane
- Sterile, empty petri plates
- Sterile, empty containers of ≈200-ml volume
- Sterile Ca. alignate swabs

3. Method for Preparation of Cell Suspension of *S. aureus*

- Every 4 months, open a new ampoule and subculture to TSB. The incubation conditions for this organism are 24 hours at 32 ± 2°C. Subculture from the TSB to (TSA) slopes (stock slopes) and concurrently plate onto a TSA plate to check for purity.
- Every month, subculture from the stock slope to a fresh TSA slope.
- Every week, subculture from the monthly slope to a TSA plate. If the culture is pure, subculture from the slope into 10 ml of TSB and incubate for 24 hours at 32 ± 2°C. This culture is used to prepare the working dilution as follows:
  - Assuming that the 24-hour culture contains 1×10^6 CFUs/ml carry out sufficient serial dilutions in 0.1% peptone saline to arrive at approximately 1000 CFUs/ml.
  - Prepare about 100 ml of this dilution, which will be the working dilution of surface test.
  - Carry out a viable count on the working dilution on TSA plates.
  - Store the 100 ml of the working dilution at 2 to 8°C. Use as needed but do not keep longer than a week.

SOP No.: Val. 1700.140   Effective date: mm/dd/yyyy

Approved by:

4. Preparation of Spore Suspensions of *B. subtilis* Stock Suspension

- Preparation of the stock suspension may be carried out every 12 months.
- Open ampoule and subculture into SCD and incubate at 32°C ± 2 for 24 hours.
- Inoculate five 45-ml sporulation agar slopes (in 100-ml medical flats) with approximately 1.0 ml of the 24-hour broth culture and incubate at 32 ± 2°C for 5 days. Concurrently, plate the 24-hour broth culture onto TSA to check for purity. Incubate at 32 ± 2°C overnight. The next day, if it is pure, discard; otherwise purify it.
- Check spore production after 5 days by spore stain. If the percentage of cells sporing is less than 70 to 80%, continue incubation. When a 70 to 80% spore yield is achieved, wash off the growth from the flats with 20 ml of sterile normal saline and dispense into sterile centrifuge tubes.
- Centrifuge at 1500 rpm for 20 minutes. Decant (and discard) the supernatant liquid.
- Resuspend the sediment in 10 ml of fresh sterile normal saline and spin again; repeat this process three times.
- After the third wash, decant off the supernatant liquid except for approximately 1 ml. Resuspend the spores in 2 ml of normal saline.
- Heat the spore suspension at 56°C for 30 minutes to kill the vegetative cells.
- Carry out a viable count on TSA using peptone saline as diluent, incubating at 32 ± 2°C for 24 hours. The final preparation should contain approximately 10^6 spores/ml.
- Prepare about 100 ml of this dilution, which will be the working dilution.
- Carry out a viable count on the working dilution for surface test on TSA plates to arrive at approximately 1000 CFUs/ml; incubate the plates at 32 ± 2°C for 24 hours.

5. Method for Preparation of Cell Suspension of *E. coli*

- Every 4 months open a new ampoule and subculture to TSB. The incubation conditions for this organism are 24 hours at 32 ± 2°C. Subculture from the TSB to (TSA) slopes (stock slopes) and concurrently plate onto a TSA plate to check for purity.
- Every month, subculture from the stock slope to a fresh TSA slope.

SOP No.: Val. 1700.140
Effective date: mm/dd/yyyy
Approved by:

■ Every week, subculture from the monthly slope to a TSA plate. If the culture is pure, subculture from the slope into 10 ml of TSB and incubate for 24 hours at 32 ± 2°C. This culture is used to prepare the working dilution as follows:
  ■ Assuming that the 24-hour culture contains 1×10^6 CFUs/ml, carry out sufficient serial dilutions in 0.1% peptone saline to arrive at approximately 1000 CFUs/ml for surface test
  ■ Prepare about 100 ml of this dilution, which will be the working dilution.
  ■ Carry out a viable count on the working dilution for surface test on TSA plates to arrive at approximately 1000 CFUs/ml; incubate the plates at 32 ± 2°C for 24 hours.

6. **Method for Preparation of Cell Suspension of *P. aeruginosa***

■ Every 4 months, open a new ampoule and subculture to TSB. The incubation conditions for this organism are 24 hours at 32 ± 2°C. Subculture from the TSB to (TSA) slopes (stock slopes) and concurrently plate on to a TSA plate to check for purity.
■ Every month, subculture from the stock slope to a fresh TSA slope
■ Every week, subculture from the monthly slope to a TSA plate. If the culture is pure, subculture from the slope into 10 ml of TSB and incubate for 24 hours at 32 ± 2°C. This culture is used to prepare the working dilution as follows:
  ■ Assuming that the 24-hour culture contains 1×10^6 CFUs/ml, carry out sufficient serial dilutions in 0.1% peptone saline to arrive at approximately 1000 CFUs/ml.
  ■ Prepare about 100 ml of this dilution, which will be the working dilution of surface test.
  ■ Carry out a viable count on the working dilution on TSA plates
  ■ Store the 100 ml of the working dilution at 2 to 8°C. Use as needed but do not keep longer than a week.

7. **Preparation of Spore Suspension of *C. sporogenes* Stock Suspension**

■ Preparation of the stock suspension may be carried out every 1 months.
■ Open ampoule and inoculate into reinforced clostridial medium (RCM). Incubate under anaerobic conditions at 32°C for 48 hours

SOP No.: Val. 1700.140 **Effective date: mm/dd/yyyy**

Approved by:

- Subculture about 3 to 5 ml of the broth onto each of five 45-ml solid RCM agar slopes (in 100-ml medical flats) and incubate anaerobically at 32°C for until 70 to 80% of the population is sporing (approximately 2 weeks). Spore production should be checked every few days by spore stain.
- When a 70 to 80% spore yield is achieved, wash off the growth from the flats with 20 ml of sterile normal saline and dispense into sterile McCartney bottles or centrifuge tubes.
- Centrifuge at 1500 rpm for 20 minutes. Decant (and discard) the supernatant liquid.
- Resuspend the sediment in 10 ml of fresh sterile normal saline and spin again; repeat this process three times.
- After the third wash, decant off the supernatant liquid except for approximately 1 ml. Resuspend the spores in 2 ml of normal saline.
- Heat the spore suspension at 56°C for 30 minutes to kill the vegetative cells.
- Carry out a viable count on TSA using peptone saline as diluent, incubating anaerobically at 32 to 37°C for 24 to 48 hours. The final preparation should contain approximately 10^8 spores/ml.
- Carry out sufficient serial dilutions in 0.1% peptone saline to arrive at approximately 1000 CFUs/ml.
- Prepare about 100 ml of this dilution, which will be the working dilution of surface test.

8. Method for Preparation of Cell Suspension of *C. albicans*

- Every 4 months, open a new ampoule and subculture to Sabouraud dextrose broth (SDB). Unless otherwise stated, the incubation conditions for this organism are 24 to 48 hours at 30°C. Subculture from the SDB to Sabouraud dextrose agar (SDA) slopes (stock slopes) and concurrently plate onto an SDA plate to check for purity. Prepare sufficient stock slopes to last 4 months.
- Every month, subculture from the stock slope to a fresh SDA slope.
- Every week, subculture from the monthly slope to an SDA plate. If the culture is pure, subculture from the slope into 10 ml of SCD and incubate for 24 hours at 30°C.
- This culture is used to prepare the working dilution, as follows:

- Assuming that the 24-hour culture contains 1×10^6 CFUs/ml, carry out sufficient serial dilutions in 0.1% peptone saline to arrive at approximately 1000 CFUs/ml;
- Prepare about 100 ml of this dilution, which will be the working dilution of surface test.
- Carry out a viable count on the working dilution on SDA plates and incubate at 30°C for 24 hours.
- Store the 100 ml of the working dilution at 2 to 8°C. Use as needed but do not keep longer than a week.

9. Method for Preparation of Cell Suspension of *A. niger*

- Every 4 months, open a new ampoule and subculture to Sabouraud dextrose broth (SDB). Unless otherwise stated, the incubation conditions for this organism are 24 to 48 hours at 25°C. Subculture from the SDB to Sabouraud dextrose agar (SDA) slopes (stock slopes) and concurrently plate onto an SDA plate to check for purity. Prepare sufficient stock slopes to last 4 months.
- Every month subculture from the stock slope to a fresh SDA slope.
- Every week, subculture from the monthly slope to an SDA plate. If the culture is pure, subculture from the slope into 10 ml of SCD and incubate for 24 hours at 30°C.
- This culture is used to prepare the working dilution, as follows:
  - Assuming that the 24-hour culture contains 1×10^6 CFUs/ml carry out sufficient serial dilutions in 0.1% peptone saline to arrive at approximately 1000 CFUs/ml.
  - Prepare about 100 ml of this dilution, which will be the working dilution of surface test.
  - Carry out a viable count on the working dilution on SDA plate and incubate at 30°C for 24 hours.
  - Store the 100 ml of the working dilution at 2 to 8°C. Use as needed but do not keep longer than a week.

Periodically, the strains referred to here should be supplemented by strains of microorganisms collected from the manufacturing environment

From the results of the viable count, calculate the volume of each organism that contains between 10 and 100 CFUs and use this for validation, growth promotion, and stasis testing. A concurrent viable cou

SOP No.: Val. 1700.140   **Effective date: mm/dd/yyyy**

Approved by:

should be carried out when performing any of these tests as a check that the working dilution has been correctly prepared and calculated.

Store the 100 ml of the working dilution at 2 to 8°C. Use as needed but do not keep longer than a week.

- Sample preparation:
  - Dilute the tested sanitizer to the use dilution (according to the recommendations of the manufacturer).
  - Prepare further dilutions of the use dilution (10^{-1} or 10^{-2}).

10. Membrane Filtration Technique

- Perform the entire operation in a laminar flow hood with aseptic technique.
- In triplicate, pipette 10 ml of each of the dilutions into separate sterile test tubes (changing pipettes after each transfer).
- For each challenge organism, provide appropriate numbers of tubes for each dilution and time intervals.
- Using a calibrated pipette, inoculate each separate complement of dilution tubes with different challenge microorganisms, using an inoculum volume of 10^6 CFUs. Shake well and let stand at room temperature.
- For positive controls, inoculate appropriate duplicate dilution blanks with challenge microorganisms.
- Contact time: assay at 5-, 10-, and 15-minute intervals at room temperature.
- At the conclusion of each challenge time interval, pass the contents of each tube through a separate membrane filter unit.
- Neutralization of the sanitizing agent: wash each membrane with appropriate quantity of the appropriate neutralizing solution.
- Count of the survivors: remove each membrane from its filter unit and plate face up on the surface of appropriate poured agar plate.
- Positive controls must be tested last.
- Incubate plates at 32 ± 2°C for 48 to 72 hours or at 22 ± 2°C for 3 to 5 days.
- Examine each day for signs of growth of the inocula; count and record.
- Growth promotion controls: +ve control membranes must show confluent growth of each of the challenge microorganisms.

10.1 Acceptance criteria

The recommended disinfectant solution must be able to establish a six-log reduction of each of the inoculated microorganisms within 5 minutes. For shelf-life determination, perform membrane filtration technique to conform to the manufacturer shelf life data.

11. Surface Testing Technique

This type of testing generally involves inoculation of a coupon (chosen to simulate cleanroom surfaces, e.g., stainless steel, polycarbonate, etc.) with the challenging organism.

- Clean each coupon with detergent and rinse well with distilled water.
- For each challenge organism, provide appropriate numbers of coupons for each dilution and time interval.
- Using a swab, inoculate each coupon with different challenge microorganisms, using the volume of the working dilution that contains approximately 1000 CFUs/cm².
- Allow drying for 30 minutes at room temperature.

12. Application Method (Spraying, Wiping, Full Immersion)

- Spraying: disinfect using the trigger spray system. All surfaces of the inoculated items should be sprayed at distance of 30 cm. The item should be allowed to stand for 2 minutes for excess disinfectant to drain off before processing.
- Wiping: a wipe should be placed over the top of the item, which is then gripped through the wipe and lifted. The remaining accessible surfaces are disinfected using parallel overlapping wipes from a second wipe. A fresh surface of the wipe is used for each pass. The second wipe is then discarded and the item gripped in the free hand through a third wipe.

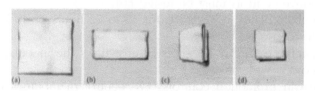

Folding of a sealed-edge wiper into quarters.

SOP No.: Val. 1700.140   Effective date: mm/dd/yyyy

Approved by:

- Spraying then wiping: the item is sprayed as described previously, but is not given the 2-minute drainage time. Instead, the item is immediately wiped following the same procedure.
- Wiping then spraying: the item is wiped as detailed earlier and then immediately sprayed, following the same procedure as before. The item is allowed to stand for 2 minutes for excess disinfectant to drain off before processing.
- After drying, immerse two strips in sanitizer solution to be tested.
- Contact time: allow contact for 30 seconds or 1, 2, 3, 4, 5, 10, or 15 minutes at room temperature.
- Interfering substance: addition of specific interfering material depends upon the application site of the disinfectant. Commonly, organic material, hard water, dried-on organisms, or the presence of heavy proteinaceous soils can be used.
- Sampling: contact plates and swabs shall be used.
- Swab each strip thoroughly with a premoistened Ca alginate swab.
- Neutralization of the sanitizing agent: neutralizing agents may be used to neutralize the activity of antimicrobial agents. They may be added to buffered Na Cl–peptone solution pH 7, preferably before sterilization. A typical neutralizing fluid has the following composition:

Polysorbate 80	30 g/l
Lecithin (egg)	3 g/l
Histidine HCl	1 g/l
Peptone (meat or casein)	1 g/l
NaCl	4.3 g/l
H_2PO_4	3.6 g/l
$H(PO_4)_2$	7.2 g/l

Table 1

Type of Antimicrobial Agent	Inactivator	Concentration	Comment
Phenolics	Sodium lauryl sulfate	4 g/l	Add after sterilization of buffered NaCl–peptone solution pH 7
	Polysorbate 80 and lecithin	30 g/l and 3 g/l	
	Egg yolk	5–50 ml/l	
Organomercurials	Sodium thioglycolate	0.5–5 g/l	
Halogens	Sodium thioglycolate	5 g/l	
Quaternary ammonium compounds	Egg yolk	5–50 ml/l	Add after sterilization of buffered NaCl–peptone solution pH 7

If the solution has insufficient neutralizing capacity, the concentration of polysorbate 80 or lecithin may be increased. Alternatively, according to the type of antimicrobial agent, the neutralizers listed in Table 1 should be added.

- Counting: use pour plate technique.
- Media: use PDA for *C. albicans* and *A. niger* and SCDA for all others.
- Repeat the preceding for each test organism.
- For positive control, use sterile DI water instead of sanitizer solution.
- For negative control, use sterile strips.

12.1 Acceptance criteria

The recommended disinfectant solution must be able to establish a three log reduction of each of the inoculated microorganisms within the contact time.

- *In situ* evaluation: establish the efficacy of the sanitizing agent against environmental bioburden, where operator technique, dried on organisms, and the presence of heavy proteinaceous soils may present a greater challenge to the disinfectant.

SOP No.: Val. 1700.140   Effective date: mm/dd/yyyy

Approved by:

- This test is not regular practice but shall be used if a resistant microorganism is detected, for cleaning techniques and personnel practices trouble-shooting, or for a specific disinfectant.
- Strain: choose the isolated environmental organism.
- Inoculum: use a microorganism concentration of approximately 1000 CFUs per cm^2.
- Experimental test conditions: the *in situ* evaluation should take place under worst-case conditions (e.g., at the end of a preventative maintenance shut-down).
- Contact time: allow the contact for recommended contact time at room temperature.
- Sampling: contact plates and swabs shall be used.
- Swab each strip thoroughly with a premoistened Ca alginate swab.
- Neutralize the sanitizing agent as mentioned earlier in surface technique.
- Counting: use pour plate technique.
- Media: use PDA for *C. albicans* and *A. niger*, SCDA for all others.
- Repeat the preceding for each test organism.
- For positive control, use sterile DI water instead of sanitizer solution.
- For negative control, use sterile strips.

13. Documentation

See attachment nos. 1700.140(A) to 1700.140(M) to report the results.

REASONS FOR REVISION

Effective date: mm/dd/yyyy

- First time issued for your company, affiliates, and contract manufacturers.

SOP No.: Val. 1700.140   Effective date: mm/dd/yyyy
                                                Approved by:

Attachment No. 1700.140(A)

DISINFECTANT VALIDATION RESULTS

Disinfectant: _____   Lot No.: _____

Supplier: _____

Analyzed on: (provide date)   Analyzed by: (provide name)

Tested for: Disinfectant Efficacy

The test has been carried out at the predetermined concentration of
manufacturer (____%)

Filtration Technique

Organism	Acceptance Criteria	Result
Pseudomonas aeruginosa ATTCC 9027	Six-log reduction within 5 minutes	
Escherichia coli ATCC 8739	Six-log reduction within 5 minutes	
Bacillus subtilis ATCC 6633	Six-log reduction within 5 minutes	
Staphylococcus aureus ATCC 6538	Six-log reduction within 5 minutes	
Clostridium sporogenes ATCC 11437	Six-log reduction within 5 minutes	
Candida albicans ATCC 10231	Six-log reduction within 5 minutes	
Aspergillus niger ATCC 16404	Six-log reduction within 5 minutes	
Environmental isolates from wall (micrococcus lutes)	Six-log reduction within 5 minutes	
Environmental isolates from operator's fingerprint (Bacillus spp.)	Six-log reduction within 5 minutes	
Environmental isolates from floor (Aspergillus)	Six-log reduction within 5 minutes	
Environmental isolates from glass window (Staphylococcus epidermis)	Six-log reduction within 5 minutes	

Analyzed by: Microbiologist   Checked by: QC Manager

Signature/date: _____   Signature/date: _____

SOP No.: Val. 1700.140

Effective date: mm/dd/yyyy
Approved by:

Attachment No. 1700.140(B)

DISINFECTANT VALIDATION RESULTS

Disinfectant: _____ Lot No.: _____

Supplier: _____

Analyzed on: (provide date)   Analyzed by: (provide name)

Tested for: Disinfectant Efficacy

The test has been carried out at the predetermined concentration of
manufacturer (_____%)

Surface Technique
Efficacy Test on Wall (Sandwich Panel) (Coupons)

Organism	Acceptance Criteria	Result
Pseudomonas aeruginosa ATTCC 9027	Three-log reduction within 5 minutes	
Escherichia coli ATCC 8739	Three-log reduction within 5 minutes	
Bacillus subtilis ATCC 6633	Three-log reduction within 5 minutes	
Staphylococcus aureus ATCC 6538	Three-log reduction within 5 minutes	
Clostridium sporogenes ATCC 11437	Three-log reduction within 5 minutes	
Candida albicans ATCC 10231	Three-log reduction within 5 minutes	
Aspergillus niger ATCC 16404	Three-log reduction within 5 minutes	
Environmental isolates from wall (micrococcus lutes)	Three-log reduction within 5 minutes	
Environmental isolates from operator's fingerprint (*Bacillus* spp.)	Three-log reduction within 5 minutes	
Environmental isolates from floor (*Aspergillus*)	Three-log reduction within 5 minutes	
Environmental isolates from glass window (*Staphylococcus* epidermis)	Three-log reduction within 5 minutes	

Analyzed by: Microbiologist   Checked by: QC Manager

Signature/date: _____   Signature/date: _____

SOP No.: Val. 1700.140

Effective date: mm/dd/yyyy

Approved by:

Attachment No. 1700.140(C)

DISINFECTANT VALIDATION RESULTS

Disinfectant: _____ Lot No.: _____

Supplier: _____

Analyzed on: (provide date)   Analyzed by: (provide name)

Tested for: Disinfectant Efficacy

The test has been carried out at the predetermined concentration of
manufacturer (____%)

Efficacy Test on Ceiling (Cleanroom Design) (Coupons)

Organism	Acceptance Criteria	Result
Pseudomonas aeruginosa ATTCC 9027	Three-log reduction within 5 minutes	
Escherichia coli ATCC 8739	Three-log reduction within 5 minutes	
Bacillus subtilis ATCC 6633	Three-log reduction within 5 minutes	
Staphylococcus aureus ATCC 6538	Three-log reduction within 5 minutes	
Clostridium sporogenes ATCC 11437	Three-log reduction within 5 minutes	
Candida albicans ATCC 10231	Three-log reduction within 5 minutes	
Aspergillus niger ATCC 16404	Three-log reduction within 5 minutes	
Environmental isolates from wall (micrococcus lutes)	Three-log reduction within 5 minutes	
Environmental isolates from operator's fingerprint (Bacillus spp.)	Three-log reduction within 5 minutes	
Environmental isolates from floor (Aspergillus)	Three-log reduction within 5 minutes	
Environmental isolates from glass window (Staphylococcus epidermis)	Three-log reduction within 5 minutes	

Analyzed by: Microbiologist   Checked by: QC Manager

Signature/date: _____   Signature/date: _____

SOP No.: Val. 1700.140

Effective date: mm/dd/yyyy

Approved by:

Attachment No. 1700.140(D)

DISINFECTANT VALIDATION RESULTS

Disinfectant: _____ Lot No.: _____

Supplier: _____

Analyzed on: (provide date) Analyzed by: (provide name)

Tested for: Disinfectant Efficacy

The test has been carried out at the predetermined concentration of manufacturer (_____%)

Efficacy Test Doors (Galvanized Steel Finished with White Powder) (Coupons)

Organism	Acceptance Criteria	Result
Pseudomonas aeruginosa ATTCC 9027	Three-log reduction within 5 minutes	
Escherichia coli ATCC 8739	Three-log reduction within 5 minutes	
Bacillus subtilis ATCC 6633	Three-log reduction within 5 minutes	
Staphylococcus aureus ATCC 6538	Three-log reduction within 5 minutes	
Clostridium sporogenes ATCC 11437	Three-log reduction within 5 minutes	
Candida albicans ATCC 10231	Three-log reduction within 5 minutes	
Aspergillus niger ATCC 16404	Three-log reduction within 5 minutes	
Environmental isolates from wall (micrococcus lutes)	Three-log reduction within 5 minutes	
Environmental isolates from operator's fingerprint (Bacillus spp.)	Three-log reduction within 5 minutes	
Environmental isolates from floor (Aspergillus)	Three-log reduction within 5 minutes	
Environmental isolates from glass window (Staphylococcus epidermis)	Three-log reduction within 5 minutes	

Analyzed by: Microbiologist   Checked by: QC Manager

Signature/date: _____   Signature/date: _____

SOP No.: Val. 1700.140   Effective date: mm/dd/yyyy

Approved by:

Attachment No. 1700.140(E)

DISINFECTANT VALIDATION RESULTS

Disinfectant: _____ Lot No.: _____

Supplier: _____

Analyzed on: (provide date)   Analyzed by: (provide name)

Tested for: Disinfectant Efficacy

The test has been carried out at the predetermined concentration of manufacturer (_____%)

Efficacy Test on Glass Windows (Coupons)

Organism	Acceptance Criteria	Result
Pseudomonas aeruginosa ATTCC 9027	Three-log reduction within 5 minutes	
Escherichia coli ATCC 8739	Three-log reduction within 5 minutes	
Bacillus subtilis ATCC 6633	Three-log reduction within 5 minutes	
Staphylococcus aureus ATCC 6538	Three-log reduction within 5 minutes	
Clostridium sporogenes ATCC 11437	Three-log reduction within 5 minutes	
Candida albicans ATCC 10231	Three-log reduction within 5 minutes	
Aspergillus niger ATCC 16404	Three-log reduction within 5 minutes	
Environmental isolates from wall (micrococcus lutes)	Three-log reduction within 5 minutes	
Environmental isolates from operator's fingerprint (Bacillus spp.)	Three-log reduction within 5 minutes	
Environmental isolates from floor (Aspergillus)	Three-log reduction within 5 minutes	
Environmental isolates from glass window (Staphylococcus epidermis)	Three-log reduction within 5 minutes	

Analyzed by: Microbiologist   Checked by: QC Manager

Signature/date: _____   Signature/date: _____

SOP No.: Val. 1700.140   Effective date: mm/dd/yyyy

                                                  Approved by:

Attachment No. 1700.140(F)

DISINFECTANT VALIDATION RESULTS

Disinfectant: _____   Lot No.: _____

Supplier: _____

Analyzed on: (provide date)   Analyzed by: (provide name)

Tested for: Disinfectant Efficacy

The test has been carried out at the predetermined concentration of manufacturer (_____%)

Efficacy Test on Concrete Walls Finished with Antifungal Oil Paint (Coupons)

Organism	Acceptance Criteria	Result
Pseudomonas aeruginosa ATTCC 9027	Three-log reduction within 5 minutes	
Escherichia coli ATCC 8739	Three-log reduction within 5 minutes	
Bacillus subtilis ATCC 6633	Three-log reduction within 5 minutes	
Staphylococcus aureus ATCC 6538	Three-log reduction within 5 minutes	
Clostridium sporogenes ATCC 11437	Three-log reduction within 5 minutes	
Candida albicans ATCC 10231	Three-log reduction within 5 minutes	
Aspergillus niger ATCC 16404	Three-log reduction within 5 minutes	
Environmental isolates from wall (microccocus lutes)	Three-log reduction within 5 minutes	
Environmental isolates from operator's fingerprint (Bacillus spp.)	Three-log reduction within 5 minutes	
Environmental isolates from floor (Aspergillus)	Three-log reduction within 5 minutes	
Environmental isolates from glass window (Staphylococcus epidermis)	Three-log reduction within 5 minutes	

Analyzed by: Microbiologist   Checked by: QC Manager

Signature/date: _____   Signature/date: _____

SOP No.: Val. 1700.140   Effective date: mm/dd/yyyy
                                          Approved by:

Attachment No. 1700.140(G)

DISINFECTANT VALIDATION RESULTS

Disinfectant: _____ Lot No.: _____

Supplier: _____

Analyzed on: (provide date)   Analyzed by: (provide name)

Tested for: Disinfectant Efficacy

The test has been carried out at the predetermined concentration of
manufacturer (____%)

Efficacy Test on Floor (Epoxy) (Coupons)

Organism	Acceptance Criteria	Result
Pseudomonas aeruginosa ATTCC 9027	Three-log reduction within 5 minutes	
Escherichia coli ATCC 8739	Three-log reduction within 5 minutes	
Bacillus subtilis ATCC 6633	Three-log reduction within 5 minutes	
Staphylococcus aureus ATCC 6538	Three-log reduction within 5 minutes	
Clostridium sporogenes ATCC 11437	Three-log reduction within 5 minutes	
Candida albicans ATCC 10231	Three-log reduction within 5 minutes	
Aspergillus niger ATCC 16404	Three-log reduction within 5 minutes	
Environmental isolates from wall (micrococcus lutes)	Three-log reduction within 5 minutes	
Environmental isolates from operator's fingerprint (Bacillus spp.)	Three-log reduction within 5 minutes	
Environmental isolates from floor (Aspergillus)	Three-log reduction within 5 minutes	
Environmental isolates from glass window (Staphylococcus epidermis)	Three-log reduction within 5 minutes	

Analyzed by: Microbiologist   Checked by: QC Manager

Signature/date: _____   Signature/date: _____

SOP No.: Val. 1700.140   Effective date: mm/dd/yyyy

Approved by:

Attachment No. 1700.140(H)

DISINFECTANT VALIDATION RESULTS

Disinfectant: _____   Lot No.: _____

Supplier: _____

Analyzed on: (provide date)   Analyzed by: (provide name)

Tested for: Disinfectant Efficacy

The test has been carried out at the predetermined concentration of manufacturer (_____%)

Efficacy Test on Machines (Product Contact Stainless Steel 316) (Coupons)

Organism	Acceptance Criteria	Result
Pseudomonas aeruginosa ATTCC 9027	Three-log reduction within 5 minutes	
Escherichia coli ATCC 8739	Three-log reduction within 5 minutes	
Bacillus subtilis ATCC 6633	Three-log reduction within 5 minutes	
Staphylococcus aureus ATCC 6538	Three-log reduction within 5 minutes	
Clostridium sporogenes ATCC 11437	Three-log reduction within 5 minutes	
Candida albicans ATCC 10231	Three-log reduction within 5 minutes	
Aspergillus niger ATCC 16404	Three-log reduction within 5 minutes	
Environmental isolates from wall (micrococcus lutes)	Three-log reduction within 5 minutes	
Environmental isolates from operator's fingerprint (*Bacillus* spp.)	Three-log reduction within 5 minutes	
Environmental isolates from floor (*Aspergillus*)	Three-log reduction within 5 minutes	
Environmental isolates from glass window (*Staphylococcus* epidermis)	Three-log reduction within 5 minutes	

Analyzed by: Microbiologist   Checked by: QC Manager

Signature/date: _____   Signature/date: _____

SOP No.: Val. 1700.140   Effective date: mm/dd/yyyy
                                          Approved by:

Attachment No. 1700.140(I)

DISINFECTANT VALIDATION RESULTS

Disinfectant: _____ Lot No.: _____

Supplier: _____

Analyzed on: (provide date)   Analyzed by: (provide name)

Tested for: Disinfectant Efficacy

The test has been carried out at the predetermined concentration of
  manufacturer (_____%)

Efficacy Test on Manufacturing Vessels (Non-Product-Contact Stainless Steel 304) (Coupons)

Organism	Acceptance Criteria	Result
Pseudomonas aeruginosa ATTCC 9027	Three-log reduction within 5 minutes	
Escherichia coli ATCC 8739	Three-log reduction within 5 minutes	
Bacillus subtilis ATCC 6633	Three-log reduction within 5 minutes	
Staphylococcus aureus ATCC 6538	Three-log reduction within 5 minutes	
Clostridium sporogenes ATCC 11437	Three-log reduction within 5 minutes	
Candida albicans ATCC 10231	Three-log reduction within 5 minutes	
Aspergillus niger ATCC 16404	Three-log reduction within 5 minutes	
Environmental isolates from wall (micrococcus lutes)	Three-log reduction within 5 minutes	
Environmental isolates from operator's fingerprint (*Bacillus* spp.)	Three-log reduction within 5 minutes	
Environmental isolates from floor (*Aspergillus*)	Three-log reduction within 5 minutes	
Environmental isolates from glass window (*Staphylococcus* epidermis)	Three-log reduction within 5 minutes	

Analyzed by: Microbiologist   Checked by: QC Manager

Signature/date: _____   Signature/date: _____

SOP No.: Val. 1700.140

Effective date: mm/dd/yyyy

Approved by:

Attachment No. 1700.140(J)

DISINFECTANT VALIDATION RESULTS

Disinfectant: _____ Lot No.: _____

Supplier: _____

Analyzed on: (provide date)   Analyzed by: (provide name)

Tested for: Disinfectant Efficacy

The test has been carried out at the predetermined concentration of manufacturer (____%)

Efficacy Test on Washing Basin (Stainless Steel 304) (Coupons)

Organism	Acceptance Criteria	Result
Pseudomonas aeruginosa ATTCC 9027	Three-log reduction within 5 minutes	
Escherichia coli ATCC 8739	Three-log reduction within 5 minutes	
Bacillus subtilis ATCC 6633	Three-log reduction within 5 minutes	
Staphylococcus aureus ATCC 6538	Three-log reduction within 5 minutes	
Clostridium sporogenes ATCC 11437	Three-log reduction within 5 minutes	
Candida albicans ATCC 10231	Three-log reduction within 5 minutes	
Aspergillus niger ATCC 16404	Three-log reduction within 5 minutes	
Environmental isolates from wall (micrococcus lutes)	Three-log reduction within 5 minutes	
Environmental isolates from operator's fingerprint (Bacillus spp.)	Three-log reduction within 5 minutes	
Environmental isolates from floor (Aspergillus)	Three-log reduction within 5 minutes	
Environmental isolates from glass window (Staphylococcus epidermis)	Three-log reduction within 5 minutes	

Analyzed by: Microbiologist   Checked by: QC Manager

Signature/date: _____   Signature/date: _____

Attachment No. 1700.140(K)

DISINFECTANT VALIDATION RESULTS

Disinfectant: _____ Lot No.: _____

Supplier: _____

Analyzed on: (provide date) Analyzed by: (provide name)

Tested for: Disinfectant Efficacy

The test has been carried out at the predetermined concentration of manufacturer (_____%)

Efficacy Test on Bench (Stainless Steel 304) (Coupons)

Organism	Acceptance Criteria	Result
Pseudomonas aeruginosa ATTCC 9027	Three-log reduction within 5 minutes	
Escherichia coli ATCC 8739	Three-log reduction within 5 minutes	
Bacillus subtilis ATCC 6633	Three-log reduction within 5 minutes	
Staphylococcus aureus ATCC 6538	Three-log reduction within 5 minutes	
Clostridium sporogenes ATCC 11437	Three-log reduction within 5 minutes	
Candida albicans ATCC 10231	Three-log reduction within 5 minutes	
Aspergillus niger ATCC 16404	Three-log reduction within 5 minutes	
Environmental isolates from wall (*micrococcus lutes*)	Three-log reduction within 5 minutes	
Environmental isolates from operator's fingerprint (*Bacillus* spp.)	Three-log reduction within 5 minutes	
Environmental isolates from floor (*Aspergillus*)	Three-log reduction within 5 minutes	
Environmental isolates from glass window (*Staphylococcus* epidermis)	Three-log reduction within 5 minutes	

Analyzed by: Microbiologist   Checked by: QC Manager

Signature/date: _____   Signature/date: _____

SOP No.: Val. 1700.140

Effective date: mm/dd/yyyy

Approved by:

Attachment No. 1700.140(L)

DISINFECTANT VALIDATION RESULTS

Disinfectant: _____ Lot No.: _____

Supplier: _____

Analyzed on: (provide date) Analyzed by: (provide name)

Tested for: Disinfectant Efficacy

The test has been carried out at the predetermined concentration of manufacturer (____%)

Efficacy Test on Tables (Stainless Steel 304) (Coupons)

Organism	Acceptance Criteria	Result
Pseudomonas aeruginosa ATTCC 9027	Three-log reduction within 5 minutes	
Escherichia coli ATCC 8739	Three-log reduction within 5 minutes	
Bacillus subtilis ATCC 6633	Three-log reduction within 5 minutes	
Staphylococcus aureus ATCC 6538	Three-log reduction within 5 minutes	
Clostridium sporogenes ATCC 11437	Three-log reduction within 5 minutes	
Candida albicans ATCC 10231	Three-log reduction within 5 minutes	
Aspergillus niger ATCC 16404	Three-log reduction within 5 minutes	
Environmental isolates from wall (micrococcus lutes)	Three-log reduction within 5 minutes	
Environmental isolates from operator's fingerprint (Bacillus spp.)	Three-log reduction within 5 minutes	
Environmental isolates from floor (Aspergillus)	Three-log reduction within 5 minutes	
Environmental isolates from glass window (Staphylococcus epidermis)	Three-log reduction within 5 minutes	

Analyzed by: Microbiologist Checked by: QC Manager

Signature/date: _____ Signature/date: _____

SOP No.: Val. 1700.140   Effective date: mm/dd/yyyy
                                                 Approved by:

Attachment No. 1700.140(M)

DISINFECTANT VALIDATION RESULTS

Disinfectant: _____ Lot No.: _____

Supplier: _____

Analyzed on: (provide date)   Analyzed by: (provide name)

Tested for: Disinfectant Efficacy

The test has been carried out at the predetermined concentration of
manufacturer (____%)

Efficacy Test on Doors/Lockers (Melamin) (Coupons)

Organism	Acceptance Criteria	Result
Pseudomonas aeruginosa ATTCC 9027	Three-log reduction within 5 minutes	
Escherichia coli ATCC 8739	Three-log reduction within 5 minutes	
Bacillus subtilis ATCC 6633	Three-log reduction within 5 minutes	
Staphylococcus aureus ATCC 6538	Three-log reduction within 5 minutes	
Clostridium sporogenes ATCC 11437	Three-log reduction within 5 minutes	
Candida albicans ATCC 10231	Three-log reduction within 5 minutes	
Aspergillus niger ATCC 16404	Three-log reduction within 5 minutes	
Environmental isolates from wall (micrococcus lutes)	Three-log reduction within 5 minutes	
Environmental isolates from operator's fingerprint (Bacillus spp.)	Three-log reduction within 5 minutes	
Environmental isolates from floor (Aspergillus)	Three-log reduction within 5 minutes	
Environmental isolates from glass window (Staphylococcus epidermis)	Three-log reduction within 5 minutes	

Analyzed by: Microbiologist   Checked by: QC Manager

Signature/date: _____   Signature/date: _____

SECTION

VAL 1800.00

YOUR COMPANY
VALIDATION STANDARD OPERATING PROCEDURE

SOP No.: Val. 1800.10   Effective date: mm/dd/yyyy

Approved by:

TITLE: Process Simulation (Media Fill) Test

AUTHOR: _____

                     Name/Title/Department

             _____

                         Signature/Date

CHECKED BY: _____

                     Name/Title/Department

             _____

                         Signature/Date

APPROVED BY: _____

                     Name/Title/Department

             _____

                         Signature/Date

REVISIONS:

No.	Section	Pages	Initials/Date

SOP No.: Val. 1800.10   Effective date: mm/dd/yyyy

Approved by:

SUBJECT: Process Simulation (Media Fill) Test

PURPOSE

This procedure describes methods and procedures for the conduct of process simulation tests of sterile products, including media fill procedures, media selection, fill volume, incubation, time and temperature, inspection of filled volume, documentation, interpretation of results, and possible corrective action required where its main purposes are to

- Demonstrate control over the aseptic operations
- Qualify aseptic processing personnel
- Qualify aseptic techniques
- Comply with cGMP requirements

RESPONSIBILITY

It is the responsibility of all concerned departmental managers to follow the procedure. The quality assurance (QA) manager is responsible for SOP compliance.

Manufacturing departments will be responsible for the following:

- Raise media fill request form to all concerned departments with schedule (attachment no. 1800.10).
- Perform media fill per manufacturing direction and protocols.
- Provide operators involved in process simulation.
- Indicate responsible supervisory personnel.
- Provide the material to be used.
- Clean the line after the run and prepare the area for real work.
- Schedule the validation runs.
- Prepare media.

The engineering and maintenance departments are responsible for the following:

- Calibrate process equipment instrumentation on a regularly scheduled basis and after repairs.
- Complete major repairs/renovations prior to validation runs if any

SOP No.: Val. 1800.10   Effective date: mm/dd/yyyy

Approved by:

Quality control is responsible for the following:

- Test the microbial contamination and inspect vials.
- Monitor area for viable count.
- Test water.
- Test media.
- Review any major repairs.
- Review calibration status.

Quality assurance is responsible for the following:

- Initiate protocols and review with concerned departments.
- Inspect the process.
- Review batch documents.
- Compile final report.

The validation group is responsible for the following:

- Monitor completeness, accuracy, technical excellence, and applicability.
- Schedule the validation runs (in conjunction with manufacturing department).
- Conduct the validation runs, including recording of all data, etc.
- Review data and accept validation run.
- Schedule revalidation.

The QC manager, QA manager, and production manager will evaluate and approve the results in case of failure and determine the remedial actions.

1. Frequency

1.1 Initial validation

Three media fill runs are required for the following (this list is not all inclusive):

- New filling room or machine
- Major room modification (such as wall or ceiling reconfigurations)

SOP No.: Val. 1800.10 Effective date: mm/dd/yyyy

Approved by:

- New equipment or new machines that have product contact or affect product flow and are not an exact equivalent of the original equipment
- Major mechanical or line configuration changes (such as new equipment additions or major preventive maintenance)
- Major HVAC changes that may affect airflow patterns related to the critical filling area (class 100) (such as addition/removal of HEPA housings)

Revalidation should take place:

- Twice per year, once per each shift (every 6 months)
- Additional tests in response to adverse trends or failures in the ongoing monitoring of the facilities or process, such as (1) continued critical area environmental monitoring results above the action levels; or (2) an increased incidence of product sterility tests failures or to evaluate changes to procedures and practices, such as
  - Exceeding the 6-month requalification interval
  - A media fill failure with a conclusive nonpersonnel assignable cause (determined during the investigation process)
  - New product/process evaluation (number of runs to be determined on a case-by-case basis)

One media fill run is required for the following (this list is not all inclusive):

- Major modification to areas adjoining filling rooms (such as expanded core area)
- Minor filling assignment (such as a different turntable)
- Filling areas that have not been in operation for a period of time not to exceed 6 months

Each of the following worst-case challenge conditions will be performed on each line once per year:

- Vial/ampoule filling speed challenge/fill volume:
  - The machine will be run at optimal speed of normal production. See manufacturing directions.
  - Fill volume is of approximately 50% of vial/ampoule capacity

SOP No.: Val. 1800.10   Effective date: mm/dd/yyyy

Approved by:

The following list details requirements to be followed for all aseptic process simulations:

- A total of three people is to be in the filling room (operator, microbiologist, and technician or maintenance) during the filling of the entire fill or additional personnel should be identified and documented.
- All activity performed during filling (or lyophilization) shall be per related SOPs.
- All personnel scheduled to work in the aseptic area must participate in at least one media fill.
- New hires must participate in the next scheduled media fill after receiving pertinent cleanroom training per appropriate SOPs.
- Each semiannual requalification run must be of a duration greater than or equivalent to a routine fill and is to occur using first and second filling shifts. The duration of biannual requalifications is maximum 24 hours.
- During each media fill, all major and minor interventions defined will be performed once. The time and the personnel performing the intervention will be documented per approved manufacturing direction.
- Media fill operation shall be monitored for the operator's behavior (the usual and unusual intervention) through fixed cameras.
- If multiple sizes of the same container/closure configuration are aseptically filled, selective sizes may be used for media fills. Containers with the widest diameter openings and slowest line speed should be included in the media-filling regimen and may be representative of worst-case conditions. However, the small containers may be representative of these conditions because of lack of container stability in the line operations.
- The duration of the run should be sufficient to cover all manipulations normally performed in actual processing and all production shifts for line/product/container combinations.

2. Environmental Monitoring

Environmental monitoring is to be performed as required per current SOPs for the following:

SOP No.: Val. 1800.10   Effective date: mm/dd/yyyy

Approved by:

- Airborne bioburden will be analyzed per SOP (provide number).
- Airborne particulate will be analyzed per SOP (provide number).
- Surface bioburden will be analyzed per SOP (provide number). In addition, surface samples of the room will be analyzed per the weekly requirements of this SOP.
- Personnel bioburden will be analyzed per SOP (provide number). In addition, QC/microbiology will perform audit monitoring of all media fill participants, including themselves.

3. Data Review

- All data review will be completed within 21 days after the completion of the simulation of the process.
- QA will review the batch record and make the final summary report.

4. Summary Report/Documentation

For each media fill qualification, a summary report that will include the following information will be written and reviewed:

- The reason for the media fill
- The date, time duration, and lot number of the media fill
- The filling room (re)qualified
- The names of the participants in the media fill
- The number and size of vials and ampoules filled as well as the size of closure used
- The target volume of media used
- The machine speed of the vial/ampoule filling
- The simulation(s) performed
- Identification of filling room used
- Container/closure type and size
- Filter lot and catalogue number
- Type of media filled
- Number of units rejected at inspection and reason
- Number of units incubated
- Number of units positive
- Incubation time and temperature for each group of units incubated and whether any group of units is subjected to two different temperatures during the incubation

- Procedures used to simulate any steps of a normal production fill
- Microbiological monitoring data obtained during the media-fill set-up and run
- Growth promotion results of the media removed from filled containers
- Identification of the microorganisms from any positive units and investigation of any contamination events observed during media fills
- Length of time media were stored in holding tank prior to filtration
- Length of time taken to fill all containers
- Results of the requalifications as outlined in the acceptance criteria of the SOP
- Verification of media sterility
- In-process inspection
- Media fill unit inspection
- Management review

PROCEDURE

1. Solutions

1.1 Media to be used

Soybean casein digest broth shall be used for normal media runs, but thioglycolate broth shall be used for the detection of anaerobic organisms, especially when a filtered nitrogen purge is used to ensure anaerobic conditions. Prepare the media according to manufacturing direction. Adjust pH; the medium is sterilized by 0.2 μ. Sterile filtration technique is used.

1.2 Manufacturing methods/process simulation

The manufacturing methods/process simulation include the following steps (see individual MFM [manufacturing formula and method] for media fill run):

- Filling equipment preparation
- Media preparation
- Sterilization
- Sterile filtration

- Washing of commodity depyrogenation
- Vial stoppering, unloading, and scaling

Media addition
Filling
Lyophilized products
Vial transport and loading of freeze dryer
Vial incubation in FD chamber under vacuum

SOP No.: Val. 1800.10   Effective date: mm/dd/yyyy

Approved by:

The filling operation of nonlyophilized product shall be completed in 12 hours (two shifts) as follows:

Table 1 Nonlyophilized Products

Runs	Filling 12 Hours		Break
Run I	Shift 1/day 1	Shift 2/day 1	3–4
Run II	Shift 1/day 2	Shift 2/day 2	3–4
Run III	Shift 1/day 3	Shift 2/day 3	3–4

1.3 Verification of medium sterility

■ Perform sterility test for the bulk media after filtration and bioburden count before filteration.
■ Aseptically, QA inspector shall take samples (each 100 ml of media) at the beginning each 2-hour period and near the end of the process.

1.4 Filling operations

■ The containers and closures shall be cleaned and sterilized per current SOPs.
■ The filling machine is operated at the predetermined and validated fill rate for the container size utilized.
■ During filling, all major and minor interventions will be performed as defined in approved manufacturing direction.
■ Power shutdown is to be simulated by stopping the laminar flow unit and the main air handling unit (AHU) for 30 seconds.
■ The containers are sealed and the medium-filled units are collected in sequentially numbered trays or boxes (notified to the filling time).
■ The filled units should be briefly inverted and swirled after filling to assure closure contact with the medium.
■ Increase the size of filling crew to more than the number necessary to fill the batch (maximum of three people).
■ All routine activities that take place on the filling line should be part of the test, i.e., weight adjustments, replacement of container

SOP No.: Val. 1800.10   Effective date: mm/dd/yyyy

                                                   Approved by:

addition of components, change of filling needle, installation of machine parts, installation of sterile filter, etc.

2. Lyophilized Products

2.1 Compounding and filling operations

Compounding takes place as described in the previous section. Filling operations take place as mentioned earlier; however, the containers are to be partially stoppered. Then, they will be transferred through the dedicated conveyors to the automatic loading/unloading systems (ALUSs).

The filling operation of lyophilized products shall be accomplished in 24 hours as follows:

Table 2 Lyophilized Products

Runs	Filling Loading	Unloading
Run I	Shift 1/day 1/1 break Shift 1/day 2/1 break	Day 3
Run II	Shift 1/day 1/1 break Shift 1/day 2/1 break	Day 3
Run III	Shift 1/day 1/1 break Shift 1/day 2/1 break	Day 3

2.2 Lyophilization simulation

The methods employed for lyophilization process simulation testing generally are similar to those used for solution fills with the addition of the transport and freeze-drying steps. However, it should focus on loading and sealing activities, which are presumed to be the greatest source of potential contamination. The test is conducted according to the process details and description in the manufacturing direction.

Containers are filled with medium, and stoppers are partially inserted. The containers are loaded into the lyophilizer. A partial vacuum is drawn on the chamber and this level is held for a predetermined time. The vacuum must not be so low as to permit the medium in the containers to boil out. The chamber is then vented and the stoppers are seated within the chamber. The stopper units are removed from the aseptic processing area and sealed.

SOP No.: Val. 1800.10   Effective date: mm/dd/yyyy

Approved by:

Anaerobic condition. There may be need for sterile inert gas to break the vacuum on the chamber and remain in the container after sealing. The use of anaerobic medium (e.g., alternative fluid thioglycolate medium) would be appropriate when the presence of anaerobic organisms has been confirmed in environmental monitoring or, more likely, during end product sterility testing. When anaerobic organisms have not been detected in the environmental monitoring or sterility testing, lyophilizer process simulation tests should utilize TSB and air.

Lyophilizer media fills will consist of a simulation of all filling and handling procedures that would be required when processing a lyophilized product, such as steam sterilization of lyophilizer 12 hours (minimum) before loading the first tray of media, introduction of the vials into the lyophilizer, holding vacuum for a minimum of 24 hours, and subsequent releasing of vacuum and activation of the stoppering mechanism.

3. Powders

Selection of placebo powder. The chosen material must be easily sterilizable, dispersible, or dissolvable in the chosen medium. The principal sterile placebo materials (irradiated in a final container) are irradiated lactose, mannitol, polyethylene glycol 6000, and sodium chloride. The material should pass the solubility testing at the desired concentration with suitable amount and time of agitation.

3.1 Compounding operations

- A quantity of an appropriate sterilized placebo powder is blended with sterile excipients prior to filling (if needed), in a manner similar to the production process being simulated.
- The medium is passed through the run as though it were an actual product batch, and all routine procedures used in manufacture of a batch are performed.
- Once the medium has been processed, it is held for a period of time at least equal to that for aseptically produced materials.
- Any of aseptic manipulations performed during and at the end of the hold period should be simulated hold times and product recalculation.

SOP No.: Val. 1800.10   Effective date: mm/dd/yyyy

Approved by:

3.2 Filling operations

- The containers, and closures are cleaned and sterilized using SOPs.
- The filling machine is operated at the predetermined fill rate for the container size utilized, as well as at the fastest speed (maximization) and the slowest speed (handling difficulty).
- The containers are sealed and the medium-filled units are collected in sequentially numbered trays or boxes (notified to the filling time).
- All routine activities that take place on the filling line should be a part of the test, i.e., weight adjustments, replenishment of containers, addition of components, change of filling agar, installation of machine parts, etc.
- Increase the size of filling crew to more than the number necessary to fill the batch.

3.3 Powder reconstitution

Before incubation of the vials, powder showed reconstituted with adapted media (TSB or thioglycolate broth) using aseptic technique under laminar flow. The reconstitution volume is according to the volume described in original formula. A strict environmental monitoring should be followed through this step.

3.3.1 Media-filled vial/ampoule analysis

Media-filled vials will be incubated in an upright position at two temperatures for a minimum of 14 days. The vials/ampoules will be inspected for growth at two separate times, once at a minimum of 7 days of incubation at $22.5 \pm 2.5°C$ and again at the end of the 14 days of incubation at $32.5 \pm 2.5°C$. Record of these inspections will be documented in QC notebook. Any vials found positive for growth will be investigated as detailed in the acceptance criteria section of the SOP.

3.3.2 Growth promotion of media-filled vials

- Growth promotion of the media-filled vials will be performed on representative vials of each media-fill run after the completion of the 14-day incubation period per instructions given in standard test method (provide number). The media shall be challenged after being subjected to the same condition of the other media fill units.

■ Growth promotion ability of the medium in final filled containers must be demonstrated using ten randomly selected containers for each challenge organism (more than 100 microorganisms per test container).

■ The units used for growth testing must be subject to the same processing steps (e.g., cleaning, depyrogenation, sterilization, filtration, filling, lyophilization, reconstition) up to the point at which they are placed into incubation.

■ In case of growth-promotion controls, growth must be observed in at least 95% of the test containers for all the challenge microorganisms.

■ If no growth is observed in all of the ten challenged containers, a second-stage test must be conducted to rule out the laboratory. In the second-stage test, both containers must support growth.

3.3.3 Negative control

■ Media from the bulk media container is incubated at 32.5 ± 2.5°C for 14 days as negative control for liquid media fill to check its sterility.

■ 500 containers filled only with the reconstitution medium serve as negative control for powder fill run.

■ No growth should be demonstrated in negative controls to indicate the acceptability of media fill. If growth is observed, the discrepancy should be investigated and documented.

3.3.4 Incubation and inspection of media-filled units

■ Leaking or damaged media-fill evaluation units shall be removed and a record made of such removal, following processing and prior to incubation of the media.

■ Media-filled evaluation units shall be incubated for a minimum of 14 days.

■ Incubation temperatures shall be appropriate for the specific growth requirements of microorganisms that are anticipated in the aseptic filling area. *Note:* Environmental monitoring data can assist

SOP No.: Val. 1800.10

Effective date: mm/dd/yyyy

Approved by:

in identifying the optimum incubation temperatures. Frequently used temperature ranges for incubation are 22.5 ± 2.5°C or 32.5 ± 2.5°C.

■ Media-filled units shall be stored or manipulated to allow contact of the media with all product contact surfaces in the unit.

■ After completion of the incubation period, the media-filled containers shall be visually inspected for the presence of microbial growth.

■ Microorganisms present in contaminated units shall be identified to help determine the likely source of the contamination. *Note:* Media-filled units should be chronologically identified within the batch to help identify the time at which a contaminated unit was filled.

4. Acceptance Criteria

A minimum of 4800 units will be filled and incubated. For multiple shift fills, at least 3000 vials/ampoules are to be filled per shift. Table 3 shows the acceptance criteria for initial performance qualification of an aseptic processing line. Table 4 shows the acceptance criteria for requalification of an aseptic processing line. Table 5 shows alert and action levels when a 0.1% contamination rate is attained for large numbers of media-filled units, i.e., when one elects to media-fill more than 3000 units.

Successful USP growth promotion test results must be provided for the media utilized in the media fill.

Any contaminated units (not just failures) will be investigated. This investigation should include, but is not limited to, the following:

■ Any recovered microorganisms shall be investigated related to the reason for them and their possible origin.

■ Examine contaminated vials for a breach of integrity.

■ Identify positive vial growth to at least the genus level and the species level if possible.

■ Examine the activities occurring during the filling of the vials in the tray where the positive vial was found.

■ Review sanitization records.

■ Review environmental monitoring data including personnel monitoring data and training records. Alert and action level results will also be documented per (provide SOP number).

Table 3 Initial Qualification Alert and Action Levels

No. Batch Size (No. Units)	Minimum No. Media Fill	Media Fill Alert Level and Required Action	Media Fill Action Level
4800–7500	Three fills	One contaminated unit in any run: investigate cause, conduct one additional run (repeat performance qualification if the additional run fails)	Two contaminated units in a single run or one each in two runs: investigate cause; repeat initial qualification media fills
7501–10,000	Three fills	One contaminated unit in any run: investigate cause; conduct one additional run (repeat performance qualification if the additional run fails)	Two contaminated units in a single run or one each in two runs: investigate cause; repeat initial qualification media fills
10,001+	Three fills	One contaminated unit in any run: investigate cause; conduct one additional run (repeat performance qualification if the additional run fails)	Two contaminated units in a single run or one each in two runs: investigate cause; repeat initial qualification media fills

Notes: Any recovered microorganisms shall be investigated related to the reason for them and their possible origin.

A sufficient number of units should be media filled to permit most interventions and worst-case conditions that may occur during production conditions.

SOP No.: Val. 1800.10

Effective date: mm/dd/yyyy

Approved by:

Table 4 Periodic Requalification — Media Fill Acceptance Criteria

Production Batch Size (No. Units)	No. Media Fills	Minimum No. Units per Media Fill	Media-Fill Alert Level	Media-Fill Action Level
≤500	≥3	Max. batch size[a]	One contaminated unit in any run: cease requalification; investigate cause; repeat periodic requalification	Two contaminated units in a single run: cease requalification; investigate cause; repeat initial qualification per Table 3
≥500–2999	1	Max.[a] batch size	As above Repeat the media-fill run	See Table 3 for maximum action level values
≥3000	1	3,000	Repeat the media-fill run if the alert values in Table 5 are exceeded	See Table 5 for maximum action level values

[a] Fewer than 3,000 units would be filled in any one run, so the 95% confidence limit table cannot be directly applied; however, on an empirical accumulative basis, no positive units in each of the individual media-fill runs would suggest a low contamination level. See Table 5.

5. Media-Fill Runs Exceeding Action Levels

5.1 Investigation

When media-fill action levels are exceeded, an investigation shall be conducted and documented regarding the cause. If action levels are exceeded, there shall be a prompt review of all appropriate records relating to aseptic production between the current media fill and the last successful one.

Table 5 Alert and Action Levels for Large Numbers of Media-Filled Units

No. Units in Single Media-Fill Test	Alert Level[a] (No. Contaminated Units in a Single Media-Fill Run)	Action Level[b] (No. Contaminated Units in a Single Media-Fill Run)
3,000	Not applicable	1
4,750	1	2
6,300	1	3
7,760	1	4
9,160	1	5
10,520	2	6
11,850	2	7
13,150	3	8
14,440	3	9
15,710	4	10
16,970	4	11

[a] This alert level is based on selection of a 0.05% contamination rate.
[b] Contamination rate of >0.1% at a 95% confidence level.

The investigation should include, but not be limited to, consideration of the following:

- Microbial environmental monitoring data
- Particulate monitoring data
- Personnel monitoring data (finger impressions, etc.)
- Sterilization cycles for media, commodities, and equipment
- HEPA filter evaluation (airborne particulate levels, smoke-challeng testing, velocity measurements, etc.)
- Room airflow patterns and pressures
- Operator technique and training
- Unusual events that occurred during the media fill
- Storage conditions of sterile commodities
- Identification of contaminants as a clue to the source of th contamination
- Housekeeping procedures and training
- Calibration of sterilization equipment
- Pre- and postfilter integrity test data and/or filter housing assemb

- Product and/or process defects and/or limitation of inspectional processes
- Documented disqualification of samples for obvious reasons prior to final reading

5.2 Corrective actions

- Media-fill tests that exceed action levels in Table 3 shall require action.
- Decisions on whether or not to take action against product being held and/or distributed shall be based upon an evaluation of all the information available and shall be documented.
- All products produced on a line following the media fill shall be quarantined until a successful resolution of the media fill has occurred.
- The materials and equipment used for manufacturing this batch will be documented in the batch record.

5.3 Failure investigation and corrective actions

A contaminated container should be examined carefully for any breach in the container system. All positives (from integral containers) should be identified to at least genus and to species whenever possible. The identification of contaminant should be compared to the database of the organisms recently identified. The biochemical profile of the contaminant can then be compared to that of microorganisms obtained from the sterility tests and bioburden and environmental monitoring programs, in order to help identify the potential sources of the contaminant.

- *If media-fill contaminant is same as sterility test contaminant:* increase media-fill vial quantities and routine filling environmental monitoring to identify the source of contamination. Review environmental data obtained during line set up.
- *If media-fill contaminant is same as media fills environmental contaminant:* increase routine environmental monitoring to determine whether the contamination potential exists during routine filling operation.
- *If media-fill contaminant is same as routine environmental contaminant:* increase media fill environmental monitoring (in the same location) to confirm the contaminant source.

SOP No.: Val. 1800.10

Effective date: mm/dd/yyyy

Approved by:

- *If sterility test contaminant is same as media fill environmental contaminant:* increase routine environmental monitoring (in the same location) and number of media-fill vials to conform.
- *If sterility test contaminant is same as routine environmental contaminant:* the sterility test is voided. Investigate sterility test procedures and room sanitation/sterilization methods to eliminate cause.
- *If media-fill environmental contaminant is same as routine environmental contaminant:* increase the number of media-fill vials in media fill to determine the product risk potential. Review monitoring technique for possible problem. Review personnel practices, gowning, sanitation, and sterilization.
- *If media-fill environmental contaminant is same as sterility test contaminant and if routine environmental contaminant is same as sterility test contaminant:* check environmental monitoring methods and techniques closely for problems. Review personnel practices, gowning, sanitation, and sterilization.
- *If the failure repeated represent a potential for product concern:* a corrective action system should be activated. This system should contain provision for the following:
  - Critical systems (HVAC, compressed air/gas, water, steam) should be reviewed for documented changes.
  - Calibration records should be checked.
  - All HEPA filters in the filling area should be inspected and rectified, if warranted.
  - Training records for all individuals (production, maintenance cleaning) involved in the fill should be reviewed to assure proper training was provided.
  - The root cause is assignable; corrective action needs to be taken and documented.
  - If three consecutive runs over action levels occur, a media-fill failure investigation and corrective action report must be issued
  - For template microbiology laboratory reports refer to Attachment No. 1800.10(A).

REASONS FOR REVISION

Effective date: mm/dd/yyyy

- First time issued for your company, affiliates, and contract manufacturers.

SOP No.: Val. 1800.10 Effective date: mm/dd/yyyy

Approved by:

Issued Date: mm/dd/yyyy
Revision No.:

Attachment No. 1800.10

MEDIA FILL REQUEST FORM

From: Production Manager

To: List Vials/Ampoules

Date: _____ Size: _____

Subject: Initial Qualification/Requalification

The media fill is scheduled from_____ to _____ to comply with the SOP Val-1800.10 for the following reason:

Check applicable box

Three media fill runs are required for the following (this list is not all inclusive)

1. New filling room or machine ☐

2. Major room modification (such as wall or ceiling reconfigurations) ☐

3. New equipment or new machines that have product contact or affect product flow and are not an exact equivalent of the original equipment ☐

4. Major mechanical or line configuration changes (such as new equipment additions or major preventive maintenance). ☐

5. Major HVAC changes that may affect airflow patterns related to the critical filling area (class 100) (such as addition/removal of HEPA housings) ☐

6. Exceeding the 6-month requalification interval ☐

7. A media fill failure with a conclusive nonpersonnel assignable cause (determined during the investigation process) ☐

8. New product/process evaluation (number of runs to be determined on a case-by-case basis) ☐

SOP No.: Val. 1800.10

Effective date: mm/dd/yyyy

Approved by:

*Check
applicable
box*

One media fill run is required for the following (this list is not all inclusive)

1. Major modification to areas adjoining filling rooms (such as ☐ expanded core area)

2. Minor filling assignment (such as a different turntable) ☐

3. Filling areas that have not been in operation for a period of ☐ time not to exceed 6 months

4. Any other reason: _____ ☐

Each of the following worst-case challenge conditions will be performed on each line once per year

Large vial/slow speed challenge/fill volume:

The largest vial size with the largest opening run on a particular ☐ filling line

The machine run at a speed lower than routine production ☐

Fill volume of approximately 50% of vial capacity ☐

Small vial/fast speed challenge/fill volume:

The smallest vial size with the smallest run on a particular filling ☐ line

The machine run at a speed faster than routine production ☐

Fill volume of approximately 50% of vial capacity ☐

No. of units _____ MFM No. _____ ☐

List: QA Manager/QC Manager/Validation Manager/Microbiologist

SOP No.: Val. 1800.10

Effective date: mm/dd/yyyy
Approved by:

Attachment No. 1800.10(A)

Process Simulation
Media Fill Test
Microbiological Report
Summary of Results

1.0 Purpose

To validate the aseptic technique for _____ mL vials

2.0 Scope

Validation of aseptic technique for _____ mL vials. Media fill batch No. _____ has been investigated with monitoring of water, personnel and microbiological environmental monitoring on: (Date).

3.0 References

3.1 USP27 <1116> media fill
3.2 ISO/TC 198 N171
3.3 QAS – 269
3.4 STM MC 020
3.5 STM MC-0043
3.6 MFM VMF 004

4.0 Conclusion:

4.1 No positive ampoule was detected.
4.2 Microbiological environmental monitoring during the full operation showed conform results.
4.3 Water samples showed conform results.
4.4 Compressed air and nitrogen samples showed conform results.

Performed by: _____ Date: _____

Checked by: _____ Date: _____

Approved by: _____ Date: _____

SOP No.: Val. 1800.10

Effective date: mm/dd/yyyy

Approved by:

Attachment No. 1800.10(A)

Process Simulation
Media Fill Test
Microbiological Report
Summary of Results

Media Fill Procedures : Aseptically filled filtrated media
Batch (media) volume : _____ L
Type of container / closure to be used : _____ mL vials
No. of units to be filled : _____ vials
Used media : Tryptone Soya Broth (TSB)
Volume of medium filled into the vials : _____ mL
Incubation time / temperature : 21 days (14 days / 32°C & 7 days / 22°C)
Starting from _____ up to _____

SOP No.: Val. 1800.10

Effective date: mm/dd/yyyy

Approved by:

Attachment No. 1800.10(A)

Process Simulation Media Fill Test Microbiological Report Summary of Results

WATER FOR INJECTION REPORTS

Sample Date: _____   Time: _____

Sampling Remarks: Normal as per plant SOP No._____

Standard Test Method (STM): _____, Membrane filtration method, using filtration unit Milliflex sensor II

Funnel: Milliflex 100 ml funnel 0.22 μm Lot No. _____, EXP. Date: _____

Holder: Milliflex Solid media Cassette Lot No. _____, EXP. Date: _____

TGEA Media for TMC: 48 h pre-incubated showed negative growth

Test filtrate: 100 ml. The media with filter incubated at 22°C for 5 days, in cold incubator Precision WB 83326658

Positive Control: Inoculated with *E. coli* showed growth after 48 h

Negative Control: Filtrated sterile WFI under the same conditions of the test

Result

Sampling Point ID No.	Sampling Point	Room No.	TMC	
	Inlet source		Zero	CFU/100 ml
	Out source		Zero	CFU/100 ml
	Preparation room		Zero	CFU/100 ml
	Washing room		Zero	CFU/100 ml
	Machine parts washing		Zero	CFU/100 ml
	Washing machine		Zero	CFU/100 ml

Action Limit: 5 CFU/100 ml

SOP No.: Val. 1800.10

Effective date: mm/dd/yyyy

Approved by:

Attachment No. 1800.10(A)

Process Simulation Media Fill Test Microbiological Report Summary of Results

DEIONIZED WATER REPORTS

Sample Date: _____   Time: _____

Sampling Remarks: Normal as per SOP No. _____

Standard Test Method: STM MC 042, Membrane filtration method, using filtration unit Milliflex sensor II

Funnel: Milliflex 100 ml funnel 0.22 μm Lot No. _____, EXP. Date: _____

Holder: Milliflex Solid media Cassette Lot No. _____, EXP. Date: _____

TGEA Media for TMC: 48 h pre-incubated showed negative growth

Test filtrate: 100 ml. The media with filter incubated at 22°C for 5 days, in cold incubator Precision WB 83326658

Positive Control: Inoculated with *P. aeruginosa* showed growth after 48 h

Negative Control: Filtrated sterile DI under the same conditions of the test

Result

Sampling Point ID No.	Sampling Point	TMC	
	Inlet source	Zero	CFU/100 ml
	Out source	Zero	CFU/100 ml
	Preparation room	Zero	CFU/100 ml
	Washing room	Zero	CFU/100 ml
	Machine parts washing	Zero	CFU/100 ml
	Washing machine	Zero	CFU/100 ml

Action Limit: 100 CFU/ml

SOP No.: Val. 1800.10   Effective date: mm/dd/yyyy

                                            Approved by:

Attachment No. 1800.10(A)

Process Simulation
Media Fill Test
Microbiological Report
Summary of Results

STERILITY TEST FOR MEDIA AFTER FILTRATION

Standard Test Method: according to USP & CFR, filtration method

Filtration unit used: Steritest Lot No. _____

Fluid A: Lot No. _____, EXP. _____

TSB Media: Difico, Lot No. _____, EXP. _____, 48 h pre-incubated showed negative growth

Fluid thioglycolate broth: MILLIPORE, Lot No._____, EXP. _____, 48 h pre-incubated showed negative growth

Sample: _____ sampled from the same media used for media fill

Positive Control: Inoculated, with *E. coli for FTB* and *B. subtilis for TSB,* showed growth after 48 h

Negative Control: Filtrated sterile WFI under the same conditions of the test

Result:

Sampling			
Date	Time	TSA	FTB
		No turbidity after 21 days	No turbidity after 21 days
		No turbidity after 21 days	No turbidity after 21 days
		No turbidity after 21 days	No turbidity after 21 days
Negative Control		No turbidity after 21 days	No turbidity after 21 days
Positive control		Turbidity after 48 h	Turbidity after 48 h

Comment: The media used for media fill are sterile.

Attachment No. 1800.10(A)

Process Simulation
Media Fill Test
Microbiological Report
Summary of Results

STERILITY TEST FOR RUBBER STOPPER AND GLASS VIALS

Standard Test Method: according to USP & CFR, direct method

Sampling Date: _____

TSB Media: Difico, Lot No. _____, EXP. _____, 48 h pre-incubated showed negative growth

Fluid thioglycolate broth: MILLIPORE, Lot No._____, EXP. _____, 48 h pre-incubated showed negative growth

Sample: 100 rubber stopper (_____mm, supplier lot No. _____) randomly sampled from the same sterilization run use for media fill

25 glass vials (Tubular, Type I, Flint, size 10 ml) randomly sampled from the same sterilization run use for media fill

Positive Control: Inoculated, with *P. aeruginosa for FTB and C. albicans for TSB* showed growth after 48 h

Negative Control: Filtrated sterile WFI under the same conditions of the test

Result

Sampling	TSA	FTB
Rubber stopper	No turbidity after 21 days	No turbidity after 21 days
Glass vial	No turbidity after 21 days	No turbidity after 21 days
Negative control	No turbidity after 21 days	No turbidity after 21 days
Positive control	Turbidity after 48 h	Turbidity after 48 h

Comment: The vials and stopper used for media fill are sterile.

SOP No.: Val. 1800.10

Effective date: mm/dd/yyyy

Approved by:

Attachment No. 1800.10(A)

Process Simulation Media Fill Test Microbiological Report Summary of Results

GROWTH PROMOTION OF USED MEDIA

Medium: Fluid Thioglycolate medium, Lot No. _____ Exp. Date _____

Physical appearance: light beige, homogenous granules

PH: _____

Method of sterilization: 0.2 μm filtration

Test method: As per Standard Test Method No._____

Organism	Limits	Results
P. aeruginosa ATCC 9027	Turbidity during 3 days	Turbid
E.coli ATCC .14169	Turbidity during 3 days	Turbid
Bacillus subtilis ATCC 6633	Turbidity during 3 days	Turbid
Staph. aureus ATCC 6538	Turbidity during 3 days	Turbid
Candida albicans ATCC 10231	Turbidity during 3 days	Turbid
Aspergillus niger ATCC 16404	Turbidity during 3 days	Turbid
Clostridium sporogenes ATCC 19404	Turbidity during 3 days	Turbid
Staph epidermis isolated from floor	Turbidity during 3 days	Turbid
Micrococcus luteus isolated from wall	Turbidity during 3 days	Turbid

Comment: The lot of media is acceptable for media fill.

SOP No.: Val. 1800.10   Effective date: mm/dd/yyyy
                                          Approved by:

Attachment No. 1800.10(A)

Process Simulation
Media Fill Test
Microbiological Report
Summary of Results

GROWTH PROMOTION OF FILLED VIALS

Medium: Fluid Thioglycolate medium filled in 3-ml glass vials

Physical appearance: Very light amber, clear solution

PH: _____

Method of sterilization: 0.2-µm filtration

Test method: As per STM No.: _____

Sample: 100 CFU of each organism has been inoculated into 10 vials of the same media fill run.

Organism	1st	2nd	3rd	4th	5th	6th	7th	8th	9th	10th
P. aeruginosa ATCC 9027	T	T	T	T	T	T	T	T	T	T
E.coli ATCC .14169	T	T	T	T	T	T	T	T	T	T
Bacillus subtilis ATCC 6633	T	T	T	T	T	T	T	T	T	T
Staph. aureus ATCC 6538	T	T	T	T	T	T	T	T	T	T
Candida albicans ATCC 10231	T	T	T	T	T	T	T	T	T	T
Aspergillus niger ATCC 16404	T	T	T	T	T	T	T	T	T	T
Clostridium sporogenes ATCC 19404	T	T	T	T	T	T	T	T	T	T
Staph epidermis isolated from floor	T	T	T	T	T	T	T	T	T	T
Micrococcus luteus isolated from wall	T	T	T	T	T	T	T	T	T	T

T = turbidity during 3 days

Comment: The prepared media and the handling procedure are acceptable for media fill.

SOP No.: Val. 1800.10   Effective date: mm/dd/yyyy
                                         Approved by:

Attachment No. 1800.10(A)

Process Simulation
Media Fill Test
Microbiological Report
Summary of Results

HYGIENE MONITORING FOR PERSONNEL PARTICIPATING

The processing rooms other than clean room have been monitored according to SOP No._____

Date: _____   Media preparation Room No. _____, Ampoule sterilization Room No._____ Class 10,000

Name	ID No.	Fingerprint	Clothes			
			Head cover	Forearm	Knee	Shoe cover
		1 CFU	0 CFU	0 CFU	1 CFU	1 CFU
		0 CFU	0 CFU	1 CFU	1 CFU	1 CFU
		1 CFU	0 CFU	0 CFU	0 CFU	0 CFU

Action Limit: 10 CFU   Action Limit: 20 CFU

For Clean room, all staff may access the area that has been involved and be monitored for gowning practices.

Date: _____   Class 100

Staff has had the gowning practices performed and enter the clean room. The test has been carried out during de-gowning.

Name	ID No.	Finger print	Clothes			
			Head cover	Forearm	Knee	Shoe cover
		0 CFU	0 CFU	0 CFU	0 CFU	0 CFU
		0 CFU	0 CFU	0 CFU	0 CFU	0 CFU
		0 CFU	0 CFU	0 CFU	0 CFU	0 CFU
		0 CFU	0 CFU	0 CFU	0 CFU	0 CFU
		0 CFU	0 CFU	0 CFU	0 CFU	0 CFU

Action Limit: 3 CFU   Action Limit: 5 CFU

Attachment No. 1800.10(A)

Process Simulation
Media Fill Test
Microbiological Report
Summary of Results

ENVIRONMENTAL MONITORING

The monitoring has been performed according to SOP QCS No._____

1- Clean room No. _____ has been monitored _____ at rest and during filling to demonstrate the influence of (upright vials, load stoppers, clean vial from line, fill needle adjustment, jammed stopper head, fill pump manifold adjustment, bleeding of sterilization filter, movement of stalled vials) interventions on the area microbiological profile.

On _____, Room No. _____ has been monitored during power shutdown intervention.

2- Changing room No. _____ and autoclave unloading room No. _____ have been monitored on _____.

4- Capping room No. _____ and room No. _____ have been monitored on _____.

5- The processing rooms No. _____ and room No. _____ have been monitored on _____.

6- All service rooms have been monitored on _____.

Comments:
The environmental monitoring results were found acceptable as per plant SOP.

SOP No.: Val. 1800.10   Effective date: mm/dd/yyyy
                                         Approved by:

Attachment No. 1800.10(A)

Process Simulation
Media Fill Test
Microbiological Report
Summary of Results

VISUAL INSPECTION OF VIALS
(MICROBIOLOGICAL INVESTIGATION)

Date: _____

Tray No.	No. of Samples on Each Tray	Observation	Inspected by	Sign	Microbial Isolates
1		Clear			
2		Clear			
3		Clear			
4		Clear			
5		Clear			
6		Clear			
7		Clear			
8		Clear			
9		Clear			
10		Clear			
11		Clear			
12		Clear			
13		Clear			
14		Clear			
15		Clear			
16		Clear			
17		Clear			
18		Clear			
19		Clear			
20		Clear			

SOP No.: Val. 1800.10   **Effective date: mm/dd/yyyy**

Approved by:

Tray No.	No. of Samples on Each Tray	Observation	Inspected by	Sign	Microbial Isolates
21		Clear			
22		Clear			
23		Clear			
24		Clear			
25		Clear			
26		Clear			
27		Clear			
28		Clear			
29		Clear			
30		Clear			
31		Clear			
32		Clear			
33		Clear			
34		Clear			
35		Clear			
36		Clear			
37		Clear			
38		Clear			
39		Clear			
40		Clear			
41		Clear			
42		Clear			
43		Clear			
44		Clear			
45		Clear			
46		Clear			
47		Clear			
48		Clear			
49		Clear			
50		Clear			
51		Clear			
52		Clear			
53		Clear			

SOP No.: Val. 1800.10   **Effective date: mm/dd/yyyy**

                                             Approved by:

Tray No.	No. of Samples on Each Tray	Observation	Inspected by	Sign	Microbial Isolates
54		Clear			
55		Clear			
56		Clear			
57		Clear			
58		Clear			
59		Clear			
60		Clear			
61		Clear			
62		Clear			
63		Clear			
64		Clear			
65		Clear			
66		Clear			
67		Clear			
68		Clear			
69		Clear			
70		Clear			
71		Clear			
72		Clear			
73		Clear			
74		Clear			
75		Clear			
76		Clear			
77		Clear			
78		Clear			
79		Clear			
80		Clear			
81		Clear			
82		Clear			
83		Clear			
84		Clear			
85		Clear			
86		Clear			

SOP No.: Val. 1800.10

Effective date: mm/dd/yyyy

Approved by:

Tray No.	No. of Samples on Each Tray	Observation	Inspected by	Sign	Microbial Isolates
87		Clear			
88		Clear			
89		Clear			
90		Clear			
91		Clear			
92		Clear			
93		Clear			
94		Clear			
95		Clear			
96		Clear			

Note: Prepare same for each monitoring day (visual inspection on 7th, 4th and 21st days).

SOP No.: Val. 1800.10   Effective date: mm/dd/yyyy

Approved by:

Attachment No. 1800.10(A)
Process Simulation
Media Fill Test
Microbiological Report
Summary of Results

Total no. of filled vials = _____ vials

Damaged vials during filling = _____

No. of vials rejected to visual inspection = _____

No. of vials for inprocess control = _____

No. of incubated units	No. of positive units after 7 days	No. of positive units after 14 days	No. of positive units after 21 days
	0	0	0

Damaged containers were not considered in the evaluation (acceptance) of the aseptic processing.

Acceptance Criteria

The % of contamination = $\dfrac{\text{No. of vials with microbial growth} \times 100}{\text{No. of vials filled} - \text{No. of damaged vials}}$

The acceptable number of contaminated vial is zero/run.

All vials showed no turbidity.

YOUR COMPANY
VALIDATION STANDARD OPERATING PROCEDURE

SOP No.: Val. 1800.20

Effective date: mm/dd/yyyy

Approved by:

TITLE: **Media-Fill Microbiological Examination**

AUTHOR: _____

Name/Title/Department

_____

Signature/Date

CHECKED BY: _____

Name/Title/Department

_____

Signature/Date

APPROVED BY: _____

Name/Title/Department

_____

Signature/Date

REVISIONS:

No.	Section	Pages	Initials/Date

SOP No.: Val. 1800.20

Effective date: mm/dd/yyyy

Approved by:

SUBJECT: Media-Fill Microbiological Examination

PURPOSE

This procedure describes standard test method for media-fill microbiological examination.

RESPONSIBILITY

It is the responsibilty of the microbiologist to follow the procedure. The quality control (QC) manager is responsible for SOP compliance.

PROCEDURE

1. Media to Be Used

The used batch of media shall be tested for growth promotion to verify its validity for media-fill usage.

- Tryptic soya broth (soybean-casein digest medium) for normal media fill run
- Fluid thioglycolate medium for detection of anaerobic growth
- Irradiated sucrose or sodium carbonate (for powder fill run)

Media are prepared per manufacturer instruction. Adjust the pH. The sterilization procedure is conducted per related MFM (manufacturing formula and method) for media fill run.

. Test Conditions (per SOP Val 1800.10)

- During a media-fill run, all equipment, personnel, procedures, etc. must be the same as if it were a normal production run.
- For presterilized liquid media-fill run, the membrane filter holder without the membrane must be in line during the test.
- Monitoring the environment for viable and nonviable particles shall be conducted during each media fill per SOP (provide number) the same as if it were a normal production run.
- During each media fill, all major and minor defined interventions will be microbiologically monitored.

SOP No.: Val. 1800.20   Effective date: mm/dd/yyyy

Approved by:

3. Environmental Monitoring

In time, environmental monitoring shall be performed as required per current SOPs for the following:

- Airborne bioburden will be monitored per SOP (number).
- Airborne particulate will be monitored per SOP (number).
- Surface bioburden will be monitored per SOP (number).
- Personnel bioburden will be monitored per SOP (number).
- In addition, QA/QC microbiology will perform audit monitoring of all media-fill participants, including themselves.
- If required, nitrogen gas and compressed air should be monitored for microbiological quality.
- Water used will be monitored per SOP (provide number).
- The process simulation tests should utilize TSB. The use of anaerobic medium (e.g., alternative fluid thioglycolate medium) would be appropriate when the presence of anaerobic organisms has been confirmed in environmental monitoring or, more likely, during end product sterility testing. Oxygen-sensitive products sparged and blanketed with nitrogen create an environment that favors the growth of microaerophillic bacteria, such as *Proplonibacterium acnes*
- Follow the related MFM for media preparation, filling, and further procedures.
- Prepare media per requirement depending upon commodity to be filled. The bulk media shall be monitored for growth promotion test.
- Verification of medium sterility: sterilized media will be monitored for sterility during the holding time. QA inspector shall take sample (each 100 ml of media) at the beginning and at appropriate interval periods.
- Perform growth promotion test using medium exposed to the actual process and product contacting, to detect the potential of growth inhibition due to various factors (e.g., extractables from filter, polymer hoses, and residuals from disinfectants/cleaning agents
- Process simulation units should be grouped and marked for date and time to permit the error investigation. If found, video taping of process simulations may be useful as an investigative tool the evaluation of positive units and/or aseptic training efficacy The use of videotaping is optional.

SOP No.: Val. 1800.20   Effective date: mm/dd/yyyy

Approved by:

■ Incubation conditions shall be suitable for recovery of bioburden and environmental isolates at 20 to 35°C. Microbiology department shall provide a rationale for the selection of their specific incubation conditions according to the trend analysis of area environmental control and presterilization bioburden isolates. The temperature chosen shall be based upon its ability to recover microorganisms normally found environmentally or in the product bioburden. This same panel of microorganisms shall be used in growth testing the media-filled containers.

■ The selected temperature shall be controlled and monitored continuously throughout the incubation period.

■ Process simulation tests shall be incubated for a minimum of 14 days up to 21 days.

■ Process simulation units shall not be required to be inverted at some point during the incubation period. All filled units shall be sufficiently manipulated to assure the contact of all sterile surfaces by the growth media prior to incubation. Momentary inversion of test units shall be surfaces of the container/closure system.

■ Reconciliation requirements of process simulation units shall be equivalent to the requirement for production. The target will be 100% reconciliation/accountability of all filled units. Any excursion must be investigated and documented; however, a variance is not an automatic invalidation of a process simulation test. Process simulation testing shall simulate normal production as closely as possible because its purpose is to assess the potential of contamination in units representative of normal production.

■ Nonintegral (i.e., damaged) units identified during procedurally performed sorting shall not be incubated. The reason for discarding these units shall be documented.

■ If the process includes a "leak test," this evaluation shall be performed for the process simulation units prior to incubation. Failing "leak test" units shall not be incubated.

■ Excluding cosmetic defects, all units that would be procedurally discarded during production shall not be incubated. The reason for discarding shall be documented. It is important that the procedures for excluding units be sufficiently specific to ensure consistency between process simulation runs and normal operations.

■ Incubation of nonintegral units would introduce a risk for microbial contamination (e.g., during transport to the microbiology lab or

SOP No.: Val. 1800.20   Effective date: mm/dd/yyyy

Approved by:

incubation room), which does not exist in routine production. Because such conditions would be unrealistic case rather than worst case, they are not acceptable.

5. Growth Promotion Control

- Growth promotion batch of media before release to media fill.
- Growth promotion of sterilized media during filling is most meaningful if performed concurrent with incubation of the media-fill units. Growth promotion at the end of incubation may not detect any interaction between contaminants and containers that may mask growth inhibition.
- Growth promotion ability of the medium in final filled containers must be demonstrated using ten randomly selected containers for each challenge organism (more than 100 microorganisms per test container). The unit used for growth promotion test shall be subjected to the same processing steps as media fill units (e.g. cleaning, depyrogenation, sterilization, filtration, filling) up to the incubation step.
- Growth promotion shall be performed per standard test method (provide number), where USP growth promotion test organisms and representative panel of microorganisms normally found environmentally or in the product bioburden shall be used in growth testing the medium-filled containers.

6. Visual Inspection

- Every week, all media-fill units shall be visually inspected by trained microbiologist for the presence of microbial turbidity.
- The containers should be compared to a known sterile container because some microbial growth shows up as a faint haze, which is difficult to detect unless there is a control container to compare against. Personnel should be trained for this task.

7. Results

- The run being examined passes the media fill test when:
  - No growth should be demonstrated in negative controls indicate the acceptability of media fill.

SOP No.: Val. 1800.20   Effective date: mm/dd/yyyy

Approved by:

- Growth promotion controls' growth must be observed in at least 95% of the test containers for all the challenge microorganisms.
- All media-fill units showed no detected growth.
- Laboratory investigation:
- If growth is observed in negative controls, the discrepancy should be investigated and documented.
- If growth promotion controls for all the challenge microorganisms showed less than 95% growth is observed in all of the ten challenged containers, a second-stage test must be conducted to rule out the laboratory. In the second-stage test, both containers must support growth.
- The conditions under which a process simulation be invalidated:
- Failure of growth promotion of media, providing there are no positive units in the process simulation
- Failure of the physical conditions in the aseptic processing area (e.g., power outage, failure of HEPA units)
- Failure of the operators to follow proper procedure not permitted in normal production
- Process simulations can be aborted for any reason that, according to procedure, would also lead to discontinuation of a production batch. However, clear documentation of the event(s) that caused discontinuation of the process simulation shall be performed and maintained.
- In case of growth detection in any media fill unit (not just failure), at least the following points should be included:
- Record the details of the unit that showed positive growth for QA investigation.
- Examine the unit for a breach of its integrity.
- Identify the growth up to species at least.
- Correlate the detected organism with the clean area microbiological profile to detect the possible source of contamination
- The biochemical profile of the contaminant can then be compared to that of microorganisms obtained from the sterility tests and bioburden and environmental monitoring programs, in order to help identify the potential sources of the contaminant.
- If media-fill contaminant is same as sterility test contaminant, increase media-fill vial quantities and routine filling environmental monitoring to identify the source of contamination. Review environmental data obtained during line set up.

Effective date: mm/dd/yyyy

Approved by:

- If media-fill contaminant is same as media fill's environmental contaminant, increase routine environmental monitoring to determine whether the contamination potential exists during routine filling operation.
- If media-fill contaminant is same as routine environmental contaminant, increase media-fill environmental monitoring (in the same location) to confirm the contaminant source.
- If media-fill environmental contaminant is same as routine environmental contaminant,
  - Increase the number of media-fill vials in media fill to determine the product risk potential.
  - Review monitoring technique for possible problem.
  - Review personnel practices, gowning, sanitation, and sterilization.
- If media fill environmental contaminant same as sterility test contaminant and if routine environmental contaminant is same as sterility test contaminant
  - Check environmental monitoring methods and techniques closely for problems.
  - Review personnel practices, gowning, sanitation, and sterilization.
- The impact of growth detection on the products shall be considered according to but not be limited to the following:
  - Level of contamination
  - Identification of contaminants
  - Review of preservatives in product, if any
  - Product sterility test results
  - Results of investigation conducted
  - Implications of the detected growth on other filling lines and products
  - Microbial environmental control data

8. Acceptance Criteria

Damaged containers shall not be considered in the evaluation (acceptance of the aseptic processing capability of the process. Use Attachment 1800.10(A) for reporting the media visual inspection results.

9. Corrective Action

In case of media-fill failure, apply SOP (provide number), media-fill failure investigation.

SOP No.: Val. 1800.20   **Effective date: mm/dd/yyyy**

                                               Approved by:

REASONS FOR REVISION

Effective date: mm/dd/yyyy

- First time issued for your company, affiliates, and contract manufacturers.

YOUR COMPANY
VALIDATION STANDARD OPERATING PROCEDURE

SOP No.: Val. 1800.30   Effective date: mm/dd/yyyy

                                           Approved by:

TITLE: Process Simulation (Media Fill) Test Protocol

AUTHOR: _____

                                Name/Title/Department

                    _____

                                   Signature/Date

CHECKED BY: _____

                                Name/Title/Department

                            _____

                                   Signature/Date

APPROVED BY: _____

                                Name/Title/Department

                                   Signature/Date

REVISIONS:

No.	Section	Pages	Initials/Date

SOP No.: Val. 1800.30   **Effective date: mm/dd/yyyy**
                                         Approved by:

SUBJECT: Process Simulation (Media Fill) Test Protocol

PURPOSE

The purpose is to describe the format and contents of the process simulation (media fill) test protocol.

RESPONSIBILITY

It is the responsibility of validation team members to follow the procedures. The quality assurance (QA) manager is responsible for SOP compliance.

PROCEDURE

Written by:	Signature and Date
Validation Manager	_____
Checked by:	Signature and Date
Microbiologist	_____
Reviewed by:	Signature and Date
Production Manager	_____
Authorized by:	Signature and Date
Quality Assurance Manger	_____

. Purpose

ˈo demonstrate the control of aseptic operations maintained by well-ained personnel, defined procedures, and appropriately designed equipent and facilities. Media filling in conjunction with comprehensive environmental monitoring will be conducted three times to demonstrate that the aseptic processing of lyophilized powder is functioning as intended.

. Background

ˈ ensure the sterility during the aseptic processing, it was decided to mulate three media-fill runs on A-ml vials. The study performed will rve following two purposes:

SOP No.: Val. 1800.30   Effective date: mm/dd/yyyy
Approved by:

■ Sterility assurance
■ Generation of data for an ANDA submission

3. Procedure

3.1 Lyophilization

■ Lyophilizer media fills will consist of a simulation of all filling and handling procedures that would be required when processing a lyophilized product, such as steam sterilization of lyophilizer 24 hours (minimum) before loading the first shelf of media, introduction of the vials into the lyophilizer, holding vacuum for a minimum of 24 hours, and subsequent releasing of vacuum and activation of the stoppering mechanism.

■ During each media fill, all major and minor interventions defined in plant SOP will be performed at least once. The time and the personnel performing the intervention will be documented on attachment no. 1800.30(C). QC will be requested to monitor air surfaces, and personnel for bioburden during each intervention.

3.2 Environmental monitoring

Environmental monitoring is to be performed as required per current SOP for the following:

■ Airborne bioburden will be analyzed per SOP.
■ Airborne particulate will be analyzed per SOP.
■ Surface bioburden will be analyzed per SOP. In addition, surface samples of the room will be analyzed per the weekly requirement of this SOP.
■ Personnel bioburden will be analyzed per SOP. In addition QC/Microbiology will perform audit monitoring of all media-fill participants, including themselves.

3.3 Manufacturing methods/process simulation

Prepare all necessary filling equipment for the filling machine.

3.4 Media preparation

Record the equipment and settings used in compounding as indicated the manufacturing direction.

SOP No.: Val. 1800.30

Effective date: mm/dd/yyyy
Approved by:

3.5 Media addition

■ At specified time intervals after the addition of the media, collect approximately 50 ml and observe the bulk solution for completeness of dissolution.

■ Take a 50-ml sample for in-process to check for appearance and pH; make entries in the manufacturing direction.

3.6 Sterilization

Review and verify the autoclave printout for the following sterilization cycles:

■ Rubber stoppers
■ Filtration assembly/vial machine parts
■ TSB media

3.7 Sterile filtration

■ Set up for media prefiltration and final filtration bubble point test per SOP.

■ Review and verify the true flow integrity test printouts for prefiltration and final filtration, before and after.

3.8 Washing

■ Review and verify the washing parameters per the validated cycle: dry heat sterilization/depyrogenation.

■ Review and verify the tunnel recorder printout to assure complete sterilization and depyrogenation of the vials per the current validated cycle.

3.9 Filling

■ Set up the filling machine for the media fill configuration of XX.0 ml fill in an X-ml tubing vial.

■ Review and verify the fill volume/weight (X and R Chart) and filling machine speed.

■ Start the filling of the product and then load the lyophilizer on line using the automatic loading system.

■ For filling/loading and unloading, see Table 1.

Effective date: mm/dd/yyyy
                                                      Approved by:

Table 1

Runs	Filling Loading	Unloading
Run 1	Shift 1/Day 1/1 break Shift 1/Day 2/1 break	Day 3
Run 2	Shift 1/Day 1/1 break Shift 1/Day 2/1 break	Day 3
Run 3	Shift 1/Day 1/1 break Shift 1/Day 2/1 break	Day 3

■ Filling and loading will be validated for 24 hours' duration.
■ Vials will be kept in the lyophilize chamber for 24 hours before unloading.
■ Each filling operation shall be monitored for air and surface, interventions, and personnel bioburden.
■ At the end of filling the product solution, perform a bubble point integrity test of the final filter.
■ The first two shelves will be completely loaded with the product vials (3200 on each).

3.10 Lyophilization (vacuum and temperature simulation)

■ Follow the lyophilization instructions; set loading temperature chamber pressure, and duration of vacuum; and record all required documentation on the MFM.
■ Review and verify the CIP/SIP cycle, per the validated cycle.
■ Review and verify chamber leak test records and make entries on the MFM.
■ After each media run, collect 100-ml sample from the CIP rinse water using the drain sample valve. Send the sample to QC for analysis. Results will be used for validation of CIP cycle for media fill simulation. See plant SOP.
■ Record the actual values of vacuum and set temperature process parameters on MFM.
■ Unload the shelves as one and two.

Note: The media-filled vials will have a light yellowish-brown, clear solution.

SOP No.: Val. 1800.30

Effective date: mm/dd/yyyy

Approved by:

3.11 Sampling

After completion of crimping, perform visual inspection and after reconciliation send vials to QC for incubator. Record the quantity on MFM.

4. Media-Filled Vial Analysis

Media-filled vials will be incubated in an upright position at 30 to 35°C for a minimum of 14 days. The vials will be inspected for growth at two separate times, once at a minimum of 7 days of incubation and again at the end of the 14 days of incubation. Record of these inspections will be documented in QC registers. Any vials found positive for growth will be investigated as detailed in the acceptance criteria section of the SOP.

5. Growth Promotion of Media-Filled Vials

Growth promotion of the media-filled vials will be performed on representative vials of each media-fill run after the completion of the 14-day incubation period per instructions given in plant SOP.

5.1 Media selection and growth support

- Verification of growth promotion of media used in specific media-fill runs shall be conducted following the run.
- The incubation temperature shall be the same as that used for the media-filled units.
- The media selected for media-fill runs shall be capable of growing a wide spectrum of microorganisms and of supporting microbiological recovery and growth of low numbers of microorganisms, i.e., 100 CFUs/unit or less.

.2 Incubation and inspection of media-filled units

- Leaking or damaged media-fill evaluation units shall be removed, and a record made of such removal, following processing and prior to incubation of the media.
- Media-filled evaluation units shall be incubated for a minimum of 14 days.

SOP No.: Val. 1800.30   Effective date: mm/dd/yyyy

                                                  Approved by:

■ Incubation temperatures shall be appropriate for the specific growth requirements of microorganisms that are anticipated in the aseptic filling area. Note: Environmental monitoring data can assist in identifying the optimum incubation temperatures. Frequently used temperatures ranges for incubation are 20 to 25°C and 30 to 35°C, or 28 to 32°C.

■ Media-filled units shall be stored or manipulated to allow contact of the media with all product contact surfaces in the unit.

■ After completion of the incubation period, the media-filled containers shall be visually inspected for the presence of microbial growth.

■ Microorganisms present in contaminated units shall be identified to help determine the likely source of the contamination. Note: Media-filled units should be chronologically identified within the batch to help identify the time at which a contaminated unit was filled.

6. Acceptance Criteria

■ A minimum of 6400 vials will be filled actually and incubated. For multiple shift fills (24-hour media fills), at least 200 vials are to be filled per shift.

■ The alert and action levels for number of positive media-filled vial are listed in Table 2.

Table 2   Alert and Action Levels

No. Batch Size (No. Units)	Minimum No. Media Fill	Media-Fill Alert Level and Required Action	Media-Fill Action Level
4800–7500	Three fills	One contaminated unit in any run: investigate cause, conduct one additional run (repeat performance qualification if the additional run fails)	Two contaminated units in a single run or one each in two runs: investigate cause; repeat initial qualification media fills

SOP No.: Val. 1800.30   Effective date: mm/dd/yyyy

Approved by:

- Successful USP growth promotion test results must be provided for the media utilized in the media fill.
- Any contaminated unit(s) (not just failures) will be investigated. This investigation should include, but is not limited to, the following:
  - Examine contaminated vial(s) for a breach of integrity.
  - Identify positive vial growth to at least the genus level and species level if possible.
  - Examine the activities occurring during the filling of the vials in the tray where the positive vial was found.
  - Review sanitization records.
  - Review environmental monitoring data, including personnel monitoring data and training records.
- Alert and action level results will also be documented per plant SOP.

7. Data Review

- All data review will be completed within 21 days after the completion of the simulation of the press.
- QA will review the batch record and make the final summary report.

7.1 Manufacturing records review

- Temperature and pressure differential records (historical trend event log)
- MFM with proper entries and signatures
- Sterilization records for machine parts
  - Filter/pressure vessel
  - Gowns
- Ampoule/vial washing sterilization records
- Seal/stopper sterilization records
- Area/equipment clearance (manufacturing)
- Printout for leak test performed in autoclave
- Lyophilizer sterilization record and vacuum record
- Reconciliation record

7.2 Quality assurance (in process)

- Air sampling record
- Surface monitoring
- Glove prints
- Water testing reports
- Process simulation media-fill test report

7.3 Other document

- Certificate of media-fill run

8. Summary Report

For each media fill qualification, a summary report will be written and reviewed, which will include the following information:

- The reason for the media fill
- The date, time duration, and lot number of the media fill
- The filling room (re)qualified
- The names of the participants in the media fill
- The number and size of vials and ampoules filled as well as the size of closure used
- The target volume of media used
- The machine speed of the vial/ampoule filling
- The simulation(s) performed
- Identification of filling room used
- Container/closure type and size
- Filter lot and catalogue number
- Type of media filled
- Number of units rejected at inspection and reason
- Number of units incubated
- Number of units positive
- Incubation time and temperature for each group of units incubated and whether any group of units is subjected to two different temperatures during the incubation
- Procedures used to simulate any steps of a normal production

SOP No.: Val. 1800.30   Effective date: mm/dd/yyyy
                                          Approved by:

- Microbiological monitoring data obtained during the media-fill set-up and run
- Growth promotion results of the media removed from filled containers
- Identification of the microorganisms from any positive units and investigation of any contamination events observed during media fills
- Length of time media were stored in holding tank prior to filtration
- Length of time taken to fill all containers
- Results of the requalifications as outlined in the acceptance criteria of the SOP
- Verification of media sterility
- In-process inspection
- Media fill unit inspection
- Management review

9. Media-Fill Runs Exceeding Action Levels

9.1 Investigation

When media-fill action levels are exceeded, an investigation shall be conducted and documented regarding the cause. If action levels are exceeded, there shall be a prompt review of all appropriate records relating to aseptic production between the current media fill and the last successful one.

The investigation should include, but not be limited to, consideration of the following:

- Microbial environmental monitoring data
- Particulate monitoring data
- Personnel monitoring data (finger impressions, etc.)
- Sterilization cycles for media, commodities, and equipment
- HEPA filter evaluation (airborne particulate levels, smoke-challenge testing, velocity measurements, etc.)
- Room airflow patterns and pressures
- Operator technique and training
- Unusual events that occurred during the media fill
- Storage conditions of sterile commodities

SOP No.: Val. 1800.30   Effective date: mm/dd/yyyy
                                         Approved by:

- Identification of contaminants as a clue to the source of the contamination
- Housekeeping procedures and training
- Calibration of sterilization equipment
- Pre- and postfilter integrity test data and/or filter housing assembly
- Product and/or process defects and/or limitation of inspectional processes
- Documented disqualification of samples for obvious reasons prior to final reading

10. Corrective Action

- Media-fill tests that exceed action levels shall require action as described in Table 2.
- Decisions on whether or not to take action against product being held and/or distributed shall be based upon an evaluation of all the information available and shall be documented.
- All products produced on a line following the media fill shall be quarantined until a successful resolution of the media fill has occurred.
- The materials and equipment used for manufacturing this batch will be documented in the batch record.

REASONS FOR REVISION

Effective date: mm/dd/yyyy

- First time issued for your company, affiliates, and contract manufacturers.

SOP No.: Val. 1800.30 Effective date: mm/dd/yyyy
Approved by:

Attachment No. 1800.30(A)

FINAL REPORT

SOP No.: Val. 1800.30

Effective date: mm/dd/yyyy

Approved by:

Attachment No. 1800.30(B)

COMPOUNDING OPERATION DATA SHEET

Batch No.: _____ MFM No: _____

Compounding Tank ID: _____

Recorder by: _____ Date: _____

Media Dissolution Monitoring

Time of Monitoring	Elapsed Mixing Time (Minutes) Completion of Media Addition	Visual Observation
	15	
	30	
	60	
	75	
	90	

Recorded by: _____ Date: _____

Checked by: _____ Date: _____

SOP No.: Val. 1800.30

Effective date: mm/dd/yyyy

Approved by:

Attachment No. 1800.30(C)

Room No.: _____

Batch No.: _____

M. Direction No.: _____

Date: _____

INTERVENTIONS SIMULATIONS RECORD

Intervention Category	Intervention Description	Time (Hours)	Air Sampling				Surface Sampling			
			Alert Limit	Action Limit	Results	Remarks	Alert Limit	Action Limit	Results	Remarks
Fill pump replacement	Major									
Fill needle replacement	Major									
Input/output tubing replacement	Major									
Open back panel for individual pump adjustment	Minor									
Upright vials	Minor									
Load vials/stoppers	Minor									
Clear vials from line	Minor									
Fill needle adjustment	Minor									
Jammed stopper head	Minor									
Fill pump manifold adjustment	Minor									
Bleeding of sterilizing filter	Minor									
Movement of stalled vials	Minor									
Employee break	Minor									

Performed by: _____ Date: _____

Checked by: _____ Date: _____

YOUR COMPANY
VALIDATION STANDARD OPERATING PROCEDURE

SOP No.: Val. 1800.40

Effective date: mm/dd/yyyy

Approved by:

TITLE: Media-Fill Run Report

AUTHOR: _____

Name/Title/Department

_____

Signature/Date

CHECKED BY: _____

Name/Title/Department

_____

Signature/Date

APPROVED BY: _____

Name/Title/Department

_____

Signature/Date

REVISIONS:

No.	Section	Pages	Initials/Date

SOP No.: Val. 1800.40   Effective date: mm/dd/yyyy

Approved by:

SUBJECT: Media-Fill Run Report

PURPOSE

The purpose is to describe the contents of the process simulation (media fill) test protocol.

RESPONSIBILITY

It is the responsibility of validation team members to follow the procedures. The quality assurance (QA) manager is responsible for SOP compliance.

PROCEDURE

Written by:	Signature and Date
Validation Manager	_____
Checked by:	Signature and Date
Microbiologist	_____
Reviewed by:	Signature and Date
Production Manager	_____
Authorized by:	Signature and Date
Quality Assurance Manger	_____

MEDIA FILL SUMMARY REPORT

Compliance Certificate

This is to certify that the vial filling line (A-ml vials) including freeze dryer has qualified for the media-fill requirements for current USP monograph and ISO 13408-1:1998 (E). All manufacturing, QA, and QC documents for batch numbers 001, 002, and 003 were reviewed and found in compliance. On the basis of the batch record and media-fill summary report, the vial filling line (including freeze drying) is approved for regular use. A detailed summary report is attached.

SOP No.: Val. 1800.40

Effective date: mm/dd/yyyy

Approved by:

Introduction

The purpose of the present media fill study is to conduct the process simulation test in A-ml vials to demonstrate that solution preparation, vial preparation, system for cleaning, washing, and sterilization/depyrogenation of glass containers, filling equipment, laminar air flow, air locks, autoclaves, dry heat tunnel, freeze drying, closure sealing machine, conveyors, and the steps in which the process is being carried out will produce sterile products. The evaluation includes environment as well as the personnel working in the area. All the equipment is validated; calibrations are maintained in compliance with the cGMP requirement.

Background

To ensure the sterility during the aseptic processing, it was decided to simulate three media-fill runs on A-ml vials because the B-ml vial size will be bracketed between A- and C-ml vial size. Moreover, the details are as follows:

Existing Available Sizes

Size/Description	A ml	B ml	C ml
Fill volume	__ ml	__ ml	__ ml
Closure dia	__ mm	__ mm	__ mm
Line speed	__ vials/min	__ vials/min	__ vials/min

For each run, more than _____ vials were used. The details of each run are as follows:

B. No.	Media Prep.	Filling and Loading	Unloading from FD
001	mm/dd/yyyy	mm/dd/yyyy	mm/dd/yyyy
002	mm/dd/yyyy	mm/dd/yyyy	mm/dd/yyyy
003	mm/dd/yyyy	mm/dd/yyyy	mm/dd/yyyy

SOP No.: Val. 1800.40   **Effective date: mm/dd/yyyy**

                                                  Approved by:

PROCESS SIMULATION

Following is the summary of some essential processes for media-fill simulation in place at ABC Pharmaceutical Industries.

1. Equipment Description

Following validated equipment was used for the solution preparation, vial preparation, sterilization, filling, and freeze-drying during the media-fill simulation of 15-ml vials.

1.1 Solution preparation

- Preparation vessel: S/No.
- Mobile holding tank: S/No.
- pH meter: S/No.
- Weighing balances: S/No.
- Integrity tester: S/No.
- N_2 generator: S/No.
- Prefiltration assembly: S/No.

1.2 Vial preparation

- Vial washing machine: S/No.
- Hot air sterilization tunnel: S/No.

.3 Vial filling and lyophilization

- Vial filling and closing machine: S/No.
- Automatic loading/unloading system: S/No.
- Freeze dryer: S/No.
- Vial capping machine: S/No.

.4 Sterilization

- Autoclave: S/No.
- Final filtration assembly: S/No.

1.5 Vial washing

The machine parts of A-ml vials were set per SOP and the vials were washed according to the validated cycle per SOP with the following controlled parameters:

Parameters	Set Parameters	Actual Run 1	Run 2	Run 3
Deionized water pressure	Minimum ___bar			
Water for injection pressure	Minimum ___bar			
Compressed air pressure	Minimum ___bar			
Machine speed	___ vials/min			

1.6 Vial sterilization

The A-ml vials used were sterilized and depyrogenated at ___°C according to the validated cycle prior to the filling operation for run 1, run 2, and run 3.

Study No./Batch No.	Set Temperature	Actual Temperature
1/001	___°C	___°C
2/002	___°C	___°C
3/003	___°C	___°C

1.7 Stopper sterilization

The rubber stoppers were sterilized at ___°C for ___ minutes according to the validated cycles for each run.

Study No./Batch No.	Date	Set Temperature	Sterilization Time
1/001	mm/dd/yyyy	122°C	20 minutes
2/002	mm/dd/yyyy	122°C	20 minutes
3/003	mm/dd/yyyy	122°C	20 minutes

1.8 Filtration assembly/machine parts sterilization/gowning sterilization

The filtration assembly, gowning, and vial filling machine parts we[re] sterilized at _____°C for _____ minutes according to the validated cy[cle] prior to run 1, run 2, and run 3.

SOP No.: Val. 1800.40

Effective date: mm/dd/yyyy

Approved by:

Study No./Batch No.	Date	Set Temperature	Sterilization Time
1/001	mm/dd/yyyy	___°C	___ minutes
2/002	mm/dd/yyyy	___°C	___ minutes
3/003	mm/dd/yyyy	___°C	___ minutes

1.9 Media preparation

Media for the consecutive three runs were prepared under aseptic conditions per manufacturing direction.

Study No./Batch No.	Manufactured on	Quantity
1/001	mm/dd/yyyy	L
2/002	mm/dd/yyyy	L
3/003	mm/dd/yyyy	L

1.10 Presterile filtration

The forward flow integrity test (using WFI as wetting liquid) was performed before and after prefiltration according to SOP and per MFM. Forward flow results for presterile filtration of media fill run 1, run 2, and run 3 are found satisfactory.

Study No./ Batch No.	Forward Flow Rate			Pressure	
	Specs.	Before	After	Specs.	Actual
1/001	Max. ___ ml per minute			NMT ___ bar	
2/002	Max. ___ ml per minute			NMT ___ bar	
3/003	Max. ___ ml per minute			NMT ___ bar	

.11 Final sterile filtration

erformed the forward flow integrity test (using WFI as wetting liquid) efore and after final sterile filtration according to SOP and per MFM. orward flow results for final sterile filtration of media fill run 1, run 2, nd run 3 are found satisfactory.

Study No./ Batch No.	Forward Flow Rate			Pressure	
	Specs.	Before	After	Specs.	Actual
1/001	Max. ___ ml per minute			NMT ___ bar	
2/002	Max. ___ ml per minute			NMT ___ bar	
3/003	Max. ___ ml per minute			NMT ___ bar	

　　　　　　　　　　Effective date: mm/dd/yyyy

Approved by:

1.12 Sterile filling

The previously sterilized machine parts were assembled and the vial filling machine set up according to SOP. Filling machine was operated according to SOP and fill volume was adjusted as specified in manufacturing direction. Filling operation covered a 24-hour period (around ___ of the vials were filled on day 1 and remaining batch [around ___ vials] were filled on day 2). The fill volume was checked by area QAI per SOP. Results for filling of media fill run 1, run 2, and run 3 are found within the specified limit.

Study No./	Filling Speed (Vials/min)		Filling Quantity		Fill Volume	
Batch No.	Set	Actual	Day 1	Day 2	Specs.	Actual
1/001					g	
2/002					g	
3/003					g	

1.13 Interventions during media fill

The following interventions were categorized to major and minor and simulated per approved MFM and SOPs during media fill. Details are:

Power shutdown	Major	Jammed stopper head	Minor
Fill needle replacement	Major	Fill pump manifold	Minor
Upright vial line	Minor	adjustment	
stoppage		Bleeding of sterilizing	Minor
Load stoppers	Minor	filter	
Kept vials and rubber	Minor	Movement of stalled vials	Minor
stoppers		Maximum personnel	Minor
Fill needle adjustment	Minor	Filter change during	Minor
		process	

1.14 Freeze dryer SIP and process simulation

The CIP and SIP cycle was performed according to the SOP and proce was simulated according to the media fill batch manufacturing direction The CIP, SIP, leak test, and process results of the freeze dryer for mec fill run 1, run 2, and run 3 are found satisfactory.

SOP No.: Val. 1800.40

Effective date: mm/dd/yyyy
Approved by:

Sterilization

Parameters	Specification	Batch 001	Batch 002	Batch 003
Sterilization time	___ Minutes			
Sterilization temp.	___°C			
Drying cooling	___ Hours			
Chamber wall cool down to	___°C			
Shelves cool down to	___°C			

Process

Study No./ Batch No.	Lyophilization Temperature (°C)		Target Vacuum (mbar)		Vacuum Duration (mbar)	
	Set	Actual	Set	Actual	Set	Actual
1/001	___°C		___mbar		___Hours	
2/002	___°C		___mbar		___Hours	
3/003	___°C		___mbar		___Hours	

1.15 Environmental monitoring: viable count records

The viable count records of room air quality during media-filling run 1, run 2, and run 3 were reviewed and found satisfactory, in accordance with the SOP. All the results were found within the alert limit. The microbes were identified and represented the normal flora, indicating no new microbe introduced during the aseptic processing.

1.16 Air monitoring records review

The following critical rooms were tested at different locations on a daily basis during three media-fill runs.

Room No.	Activity	No. Tests Performed		
		Run 1	Run 2	Run 3
	Changing room			
	Unloading autoclave			
	Compounding			
	Capping			
	Ampoule washing			
	Vial filling room			
	Freeze dryer			
	Solution room			
	Packaging			
	Total			

1.17 Room cleaning and sanitation records

The room cleaning and sanitation records were checked and found completed for three runs.

1.18 Surface monitoring records review

The following critical areas were monitored on a daily basis during three media-fill runs.

Room No.	Activity	No. Tests Performed		
		Run 1	Run 2	Run 3
	Changing room			
	Unloading autoclave			
	Compounding			
	Capping			
	Ampoule washing			
	Vial filling room			
	Freeze dryer			
	Solution room			
	Packaging			
	Total			

SOP No.: Val. 1800.40   Effective date: mm/dd/yyyy

Approved by:

1.19 Nonviable count records

The environmental monitoring test results of nonviable count in solution preparation, change room vial/ampoule washing and loading, commodity washing (near autoclave), vial filling, loading/unloading (freeze dryer) capping room were found in compliance with the acceptance criteria for class 10,000 and class 100, respectively, per SOP during run 1, run 2, and run 3. Critical room nonviable counts follow. The instrument used was Climet CI-500. All results were found within the specification.

Run 1, Run 2, and Run 3

	Preparation Room		Washing Room		Autoclaving Room	
Area/Class	0.5 μm	0.5 μm	0.5 μm	0.5 μm	0.5 μm	0.5 μm
LFH (100)						
Room (10,000)						

	Filling Room		Lyop. Loading		Lyop. Unloading	
Area/Class	0.5 μm	0.5 μm	0.5 μm	0.5 μm	0.5 μm	0.5 μm
LFH (100)						
Room (10,000)						

1.20 Personnel monitoring

Personnel monitoring was performed three times a day; which personnel worked in processing rooms and cleanrooms has been monitored according to SOP. The results of personnel monitoring were reviewed for the fingerprints and garments and found satisfactory during run 1, run 2, and run 3.

Media Fill Run	Date	Activity	Identification No.	Staff Involved
Run 1, Run 2, and Run 3		Media preparation; vial washing/stoppers sterilization; filling and loading		
		Vial washing/sterilization; filling and loading		
		Vial unloading		

1.21 Room temperature, humidity, and air pressure

The room temperature, humidity, and air pressure differential records were checked using the calibrated sensors for three consecutive runs and found within tolerance.

Area	Preparation Room Max	Preparation Room Min	Washing Room Max	Washing Room Min	Filling Room Max	Filling Room Min	Lyophilization Loading Max	Lyophilization Loading Min	Lyophilization Unloading Max	Lyophilization Unloading Min
Rh: Actual										
Limit	_%	_%	_%	_%	_%	_%	_%	_%	_%	_%
Temp.: Actual										
Limit	_°C	_°C	_°C	_°C	_°C	_°C	_°C	_°C	_°C	_°C
P. Diff.: Actual										
Limit	_ Pa	_ Pa	_ Pa	_ Pa	_ Pa	_ Pa	_ Pa	_ Pa	_ Pa	_ Pa

1.22 DOP test records of HEPA filters and revalidation records

Although the results of nonviable particles were found within limits, th DOP test records were still checked and found satisfactory. The air velocit records were reviewed in each critical room during run 1, run 2, and ru 3 and found within tolerance. See SOP (number).

Area/ Limits	Preparation Room Result	Preparation Room Valid up to	Washing Room Result	Washing Room Valid up to	Autoclave Room Result	Autoclave Room Valid up to	Filling Room Result	Filling Room Valid up to	Freeze Drying Room Result	Freeze Drying Room Valid up to	Capping Room Result	Capping Room Va up
LFH												
Filter no.												
Room												
Filter no.												

SOP No.: Val. 1800.40

Effective date: mm/dd/yyyy

Approved by:

1.23 Water testing

During all three media-fill runs, the following tests were also performed by microbiological laboratory for microbiological assurance.

1.23.1 Water for injection

Water for injection was tested microbiologically for the following and results are found in compliance.

Sampling Point	Action Limit	Result (CFU/100 ml)		
		Run 1	Run 2	Run 3
Inlet source	10 CFUs/100 ml			
Outlet source	10 CFUs/100 ml			
Solution preparation	10 CFUs/100 ml			
Utensil washing	10 CFUs/100 ml			
Hand washing	10 CFUs/100 ml			
Vial washing	10 CFUs/100 ml			

1.23.2 Deionized water

Deionized water was tested microbiologically for the following and results are found in compliance.

Sampling Point	Action Limit	Result (CFU/100 ml)		
		Run 1	Run 2	Run 3
Inlet source	100 CFUs/100 ml			
Outlet source	100 CFUs/100 ml			
Solution preparation	100 CFUs/100 ml			
Utensil washing	100 CFUs/100 ml			
Hand washing	100 CFUs/100 ml			
Vial washing	100 CFUs/100 ml			

1.23.3 CIP water

CIP water was tested microbiologically after completion of cleaning of the following equipment and results were found in compliance.

		Result (CFU/100 ml)		
Sampling Point	Action Limit	Run 1	Run 2	Run 3
Manufacturing vessel	10 CFUs/100 ml			
Mobile tank	10 CFUs/100 ml			

1.24 Nitrogen and compressed air sterility

Nitrogen and compressed air sterility was checked for the following points and found satisfactory.

Nitrogen		Compressed Air	
Activity	Room No.	Activity	Room No.
Solution preparation			
Freeze dryer			
Vial filling			

1.25 Operator training records

The operators are trained twice in a year on aseptic processing techniques. The training results of operator evaluation were found satisfactory.

SOP No.: Val. 1800.40   Effective date: mm/dd/yyyy

Approved by:

1.26 Growth-promotion test

Growth-promotion test was performed per standard test method (STM) media on ten media-filled vials. All results are found in compliance.

		Results					
		Run 1		Run 2		Run 3	
Organism	Limit	Used Media	10 Vials	Used Media	10 Vials	Used Media	10 Vials
P. aeruginosa ATTC 9027	Turbid during 3 days						
E. coli ATTC 14169	Turbid during 3 days						
Bacillus ATCC 6633	Turbid during 3 days						
S. aureus ATCC 6538	Turbid during 3 days						
C. albicans ATCC 10231	Turbid during 3 days						
A. niger ATCC 1604	Turbid during 3 days						
C. sporogenes ATCC 19404	Turbid during 3 days						

1.27 Unusual events occurring during media fill runs

Further review of media fill records indicated no unusual intervention observed during media filling operations.

1.28 Postmedia fill run viable and nonviable count results

The postmedia-fill run results for viable and nonviable count after cleaning and sanitation did not show any deviation and the results for the 3 consecutive days were found satisfactory and within alert limits.

1.29 Results

The results of media-fill run 1, run 2, and run 3 did not show any positive vials after 14 days and 21 days. The result meets the USP requirement and ISO 13408-1:1998(E).

Runs	No. Vials Incubated	14 Days	21 Days
Run 1/001			
Run 2/002			
Run 3/003			

1.30 Conclusion

All documents of batch numbers 001, 002, and 003 (process simulation of 24-hour duration) of media-fill study 1, study 2, and study 3 for aseptically filled A-ml vials were reviewed and found to meet the media-fill qualification requirements per current USP monograph and ISO 13408-1:1998 (E), including process simulation, environmental and personnel monitoring results.

1.31 Documentation

Complete documentation and actual results are available in the plant media-fill file.

REASONS FOR REVISION

Effective date: mm/dd/yyyy

■ First time issued for your company, affiliates, and contract manufacturers.

SECTION

VAL 1900.00

YOUR COMPANY
VALIDATION STANDARD OPERATING PROCEDURE

SOP No.: Val. 1900.10   Effective date: mm/dd/yyyy

                                               Approved by:

TITLE: Determination of Components Bioburden before Sterilization

AUTHOR: _____

                                Name/Title/Department

                       _____

                                     Signature/Date

CHECKED BY: _____

                                Name/Title/Department

                       _____

                                     Signature/Date

APPROVED BY: _____

                                Name/Title/Department

                       _____

                                     Signature/Date

REVISIONS:

No.	Section	Pages	Initials/Date

SOP No.: Val. 1900.10   Effective date: mm/dd/yyyy

Approved by:

SUBJECT: Determination of Components Bioburden before Sterilization

PURPOSE

The purpose is to evaluate the bioburden, spore bioburden, and endotoxin present on rubber stoppers and unprocessed glass vials. The similarity of contents and equipment may be coincidental as the similar and common inventory are used by the generic manufacturers.

RESPONSIBILITY

It is the responsibility of concerned microbiologists to follow the procedure. The quality control (QC) manager is responsible for SOP compliance

PROCEDURE

- These procedures apply to all vials and all stoppers used for sterile products.
- Each lot of each type of vials is to be tested per the following procedures.
- Each lot of each type of stoppers is to be tested before and after sterilization per the procedures described here.

1. Definitions

- Prewashed, siliconized rubber stoppers
- LAL = limulus amebocyte lysate
- Stopper type: stoppers made from the same compound and manufactured by the same process
- Vial type: vials made from the same compound and manufacture by the same process

2. Equipment/Materials

- Test samples (60 vials or stoppers of one size, style, manufacture and lot number)
- Sterile media
  - Trypticase soy agar (TSA) in bottles and prepoured in steri petri dishes

SOP No.: Val. 1900.10   **Effective date: mm/dd/yyyy**
                                              Approved by:

- 0.1% Tween 80 in WFI in appropriately sized containers
- Sterile filtration units, 47-mm size, with filtration manifold
- Sterile 0.45 μm, 47 mm, mixed esters of cellulose (MEC), tortuous path filter membranes
- Sterile forceps
- Vacuum pump
- Laminar flow hood
- Water bath set at 45 to 55°C
- Incubator set at 30° to 35°C
- Mechanical shaker
- Sterile surgical gloves
- Self-contained anaerobe jar
- Bunsen burner
- Sterile, pyrogen-free WFT
- Depyrogenated soda-lime (flint) 10 × 75 test tubes
- Depyrogenated borosilicate 13 × 100 mm test tubes
- Test tube rack
- Sterile, pyrogen-free syringes
- Sonicator
- Endotoxin standard (source: Associates of Cape Cod
- Limulue amebcyte lysate (LAL), sensitivity 0.125 EU/ml
- Vortex mixer
- Calibrated thermometer with range above 100°C
- Water bath with temperature range to 100°C

3. Bioburden Method

1. Sampling according to SOP (provide number)
2. Under a laminar flow hood, aseptically place 20 stoppers of the size and mold number to be tested into a sterile jar containing 50 ml of 0.1% Tween solution such that all stoppers are completely immersed (100 ml 0.1% Tween for 28-mm stoppers).
3. Allow stoppers to soak for 1 hour undisturbed. The soaking is intended to facilitate greater recovery of any bioburden present on the stoppers.
4. Shake container for approximately 30 minutes at "low" speed setting on mechanical shaker.
5. Vortex just before testing.

6. Don a fresh pair of sterile gloves. Sanitize all interior hood surfaces with a QC-approved antimicrobial agent:

7. Spray the surface of the hood with a QC-approved antimicrobial agent until visibly wet.

8. Wipe all wet surfaces with a sterile, lint-free towel. Allow surfaces to dry.

9. Turn on the pump and Bunsen burner or lab fuel. Don a fresh pair of sterile gloves and a pair of sterile gloves.

10. Using a QC-approved antimicrobial agent, sanitize the exterior surfaces of the sample container by spraying with the agent until visibly wet or wiping the surfaces with a sterile, lint-free, agent-soaked towel. Put all the samples to be tested under the hood. Allow surfaces to dry before testing.

11. Aseptically unwrap four sterile filter funnels and place in filtration manifold.

12. For each of the four filter funnels:

13. Holding the funnel portion of the filtration assembly with one hand, carefully lift the funnel from the base.

14. Using a sterile forceps (flame the tips of the forceps and cool to room temperature before using), take a presterilized 47-mm, 0.45-μm, mixed esters of cellulose (MEC) tortuous path membrane filter (hereafter referred to as membrane filter) and place filter, grid side up, in the filter holder.

15. Aseptically decant the Tween solution, splitting the total volume present equally among each of four sterile membrane filter holders

16. Apply vacuum to each of the filter holders and draw the Tween solution.

17. Turn off vacuum supply.

18. Pour an additional 10 ml (approximately) of sterile 0.1% Tween solution into each filter using aseptic technique. This rinse is intended to wash the interior walls of the filtration units and carry any bacteria adhering to these surfaces onto the surface of the membrane filter.

19. Turn on vacuum supply and filter the contents of the filter holder. Turn off vacuum supply.

20. Using sterile forceps (or forceps whose tips have been flamed and then cooled to room temperature), carefully remove each membrane filter and place onto prepoured TSA plate.

SOP No.: Val. 1900.10   **Effective date: mm/dd/yyyy**

                                                         Approved by:

21. Prepare each of two negative controls by filtering 50 ml of sterile 0.1% Tween solution. Place filter on sterile prepoured TSA plates as described previously.
22. Invert all plates. Incubate two of the test sample plates and one negative control plate at 30 to 35°C for no less than 72 hours.
23. Incubate the remaining two plates and the remaining negative control plate anaerobically in anaerobic jar. Incubate at 30 to 35°C for no less than 72 hours.
24. Following incubation, enumerate the colonies and record.
25. Calculations are done separately for aerobic and anaerobic plates.
26. Report these calculations with one decimal place, rounding up if the second decimal place is 5 or higher.
27. Record all data.
28. Sample size will be 20 vials per test: follow the procedures described in steps 2 through 27, vial size permitting. (Small vials will be immersed.)
29. For 20-cc or larger vials, fill each vial about halfway with rinsing fluid instead of immersing as described earlier. Then proceed as follows:
    - After filling, place depyrogenated foil and parafilm onto each vial and place onto a mechanical shaker.
    - Shake approximately 30 minutes at "low" speed setting on mechanical shaker.
    - Put all containers into a single sterile container.
    - Follow steps 6 through 23.

. Identification of Isolates

1. Identify a representative of each morphologic type.
2. Assign each isolate to be identified by number to permit easy tracking and documentation.
3. Streak any anaerobes recovered onto TSA and incubate aerobically to determine whether these are true (obligate) anaerobes.
4. Perform resistance testing on all Gram-positive spore formers. Submit any organism surviving the resistance test for D-value determination.
5. Summarize stopper and vial data separately.

SOP No.: Val. 1900.10   Effective date: mm/dd/yyyy

Approved by:

5. Spore Bioburdon Method

5.1 Analysis of stoppers

1. Follow the first three steps in section 3 of the Bioburdon method for sampling.
2. Heat-shock the rinses at 98 to 100°C by immersing in a water bath. Fill a separate container with the same amount of Tween solution used to rinse the stoppers and immerse with the test units. Include a thermometer in the extra container. Once 98°C is reached on the thermometer, begin timing the exposure period of 20 minutes with a calibrated timer.
3. After heat shocking, remove the rinse samples and place in cool water
4. After cooling, shake container on mechanical shaker at slow setting for 30 minutes. Sonicate for 2 minutes just prior to testing.
5. Transfer the samples to a laminar flow hood.
6. Continue testing as described in bioburden method, steps 7 through 19, using aseptic technique.
7. Incubate two plates and a negative at 30 to 35°C. The remaining two plates and a negative incubate at 93 to 60°C with parafilm to prevent media from drying.
8. Enumerate the colonies on all plates according to bioburden method, steps 21 and 22. Record all data.

5.2 Analysis of vials

9. Follow steps 1 through 3 in bioburden method for sampling and preparation of vials.
   - ▪ Small vials will be immersed. Sample size will be 20 vials per test
   - ▪ For 20-cc or larger vials, fill each vial about halfway with rinsing fluid instead of immersing as described previously. Place a sterile stopper and crimp seal onto each vial.
10. Follow steps 2 through 5 in the section about stopper bioburden method for heat shocking, cooling, shaking, and sonicating sample (one container containing 20 small vials 20 half-filled vials).
11. If testing 20-cc or larger vials, pool contents of the 20 half-filled vials into a single sterile container. Vortex.
12. Continue testing as described in spore bioburden method steps through 8.

SOP No.: Val. 1900.10   Effective date: mm/dd/yyyy

Approved by:

5.3 Resistance testing of spore formers

13. If any Gram-positive rods survive the heat shock, further testing is required.
14. Subculture isolates on TSA incubated at 30 to 53°C and identify each morphological type of colony.
15. Perform resistance testing on all Gram-positive spore formers. Submit any organism surviving the resistance test for D-value determination.
16. Summarize stopper and vial data separately.

6. Endotoxin Test Method

6.1 Endotoxin standard preparation

1. Using a sterile, nonpyrogenic pipette tip and pipettor, aseptically pipette 5.0 ml of pyrogen-free water into the endotoxin vial.
2. Vortex vial intermittently for not less than 20 minutes. Water should come in contact with all internal surfaces, including the stopper during reconstitution. Reconstituted stock can be stored for 14 days at 2 to 8°C. Never freeze stock solution. Label stock with initials and date prepared. Vortex refrigerated reconstituted stocks 5 minutes before use.
3. Determine endotoxin potency (EU/ng) from certificate of analysis (C of A). Calculate endotoxin units per milliliter (EU/ml), based on C of A.
4. When making dilutions, vortex each tube 1 minute between each dilution step. Do not freeze endotoxin dilutions. Do not store dilutions.

.2 Preparation of pyrotell

5. Remove vials of reagent from freezer, tear off metal seal, and gently remove the stopper.
6. Using a sterile, nonpyrogenic disposable syringe and needle, inject 5.0 ml of pyrogen-free water to reconstitute the reagent. Reconstitution is more rapid if reagents are at room temperature.
7. Gently swirl the vial to dissolve the contents completely. Mixing too vigorously can cause excessive foaming, which can lead to loss of reagent sensitivity. Keep reagent in the refrigerator after reconstitution until needed.

SOP No.: Val. 1900.10   Effective date: mm/dd/yyyy

Approved by:

Note: Following reconstitution, reagent should be used immediately. Any reagent remaining after reconstitution is stable at 2 to 8°C for 24 hours or for up to 1 month at −20°C or below. Reconstituted reagent can be frozen–thawed only once. Reconstituted reagent should be slightly opalescent. Yellow color indicates contamination or denaturation. Label vial with initials, date reconstituted, and expiration.

6.3 Analysis of stoppers

8. Wear sterile gloves and a clean lab coat.
9. Sampling according to SOP (provide number).
10. On laminar flow bench, open the bag to expose the stoppers. Using sterile forceps, randomly select 20 stoppers, transferring them to a sterile, nonpyrogenic sample container. Seal the container after the transfer of stoppers is complete.
11. Place the container of stoppers on a laminar flow bench along with a container of sterile, nonpyrogenic water for injection (WFI)
12. Don a fresh pair of sterile gloves.
13. Using a sterile, nonpyrogenic pipette, add 100 ml for 28-mm stoppers. Recap the container.
14. Gently mix the stoppers in the container to ensure that all surface come in contact with the water and sonicate for 30 minutes.
15. Return the container to the laminar flow hood. Open the containe and, using an automatic pipettor fitted with sterile, nonpyrogeni tips, pipette 0.1 ml of the water into each of two sterile pyroger free test tubes.
16. Controls:
    ■ Positive controls: pipette duplicate 0.1-ml aliquots of fresh prepared endotoxin standard solution, bracketing the labele lysate sensitivity, into separate 10 × 75 mm pyrogen-free te tubes.
    ■ Negative controls: pipette 0.1-ml sterile pyrogen-free water in two 10 × 75 mm test tubes.
17. Add 0.1 ml of reconstituted reagent to each tube prepared in and 17.
18. Gently mix the contents but avoid foaming.
19. Immediately place all the tubes in an upright position in a 37 1°C water bath and incubate undisturbed for 60 ± 2 minutes. Recc the thermometer temperature and number.

20. After incubation, inspect all tubes for the presence or absence of a firm gel. Gently but firmly pick up each tube by its top. Invert the tube 180° without shaking or jerking the tube and hold it there. A firm gel that adheres to the bottom of the test tube in this inverted position is scored as positive. If the contents of the tube slide down the tube, score it as negative.
21. Record results.
22. In the event of a negative test, the stoppers meet the acceptance criteria for endotoxin.

6.3.1 Phase II — confirmation

23. If one or both of the duplicate tubes yields a positive endotoxic response, repeat steps 15 through 21 using the original container of 20 stoppers. Record results.
24. If none of the quadruplicate tubes gel, the stoppers meet the acceptance criteria for endotoxin.
25. If any of the four reaction tubes shows a firm gel, proceed to phase III.

6.3.2 Phase III — individual stopper

26. Resample an additional 20 stoppers according to steps 8 through 14.

6.3.3 Pyroburden determination

27. Transfer individual stoppers to separate sterile nonpyrogenic containers.
28. Add only 10 ml of WFI to each container.
29. Repeat steps 15 through 21. Record results.
30. If no gel is obtained from any reaction tube, the stoppers meet the acceptance criteria.
31. If any reaction tube shows a firm gel, the stoppers do not meet the acceptance criteria.
32. In the event the results do not meet the acceptance criteria, an investigation will be initiated and QC and production management will be notified immediately. The vendor will be put on alert status

for that stopper and every incoming lot number of the type of stopper must be tested and shown to be free of endotoxin prior to use in production. The investigation must be completed and the vendor must be recertified by testing stoppers from five manufacturer's lot numbers for endotoxin before the vendor can be removed from the alert list for that stopper.

6.4 Analysis of vials

33. Wearing sterile gloves and a clean lab coat, collect 20 random samples of the vials to be tested. Place in a sterile container for transport to the microbiology lab.

6.4.1 Phase I — screening

34. Put on a new pair of sterile gloves.
35. Working under a laminar flow hood, add sterile pyrogen-free WFI to the vials to be tested. The following amounts are added:

WFI	Vial Size
10 ml	30 ml
5 ml	20 ml
1 ml	15 ml

36. Cover with depyrogenated foil.
37. Gently swirl the vials to ensure that all surfaces come in contact with the water and sonicate for 3 minutes.
38. Pool water from all 20 vials into one sterile pyrogen-free container. Vortex to mix. Pipette two aliquots of 0.1 ml each into labeled 10 × 75 mm sterile pyrogen-free glass tubes.
39. Follow steps 17 through 21 in endotoxin test method. Record results.
40. In the event of a negative test, vials meet the acceptance criteria for endotoxin.

SOP No.: Val. 1900.10   Effective date: mm/dd/yyyy

Approved by:

6.4.2 Phase II — confirmation

41. If one or both of the duplicate tubes yields a positive endotoxic response, proceed to phase II testing. Repeat steps 38 and 39. Record results.
42. If none of the quadruplicate tubes gel, the vials meet the acceptance criteria.
43. If any of four-reaction tubes gel, proceed to phase III.

6.4.3 Phase III — individual vial pyroburden determination

44. Resample an additional 20 vials from the same lot, following steps 34 through 40.
45. Do not pool samples, but test 20 samples individually.
46. Pipette duplicate 0.1-ml aliquots from each vial into labeled 10 × 75 mm sterile pyrogen-free glass tubes. Follow steps 10 through 14 in the stopper section of the SOP. Record results.
47. Evaluation is the same as for the stoppers; see steps 30 through 32.

Limit

Test	Ready-to-Sterilize Stoppers	Unprocessed Vials
Bioburden	NMT 25 CFUs/stopper	NMT 100 CFUs/vial
Spore bioburden	Spore bioburden data are collected to screen for heat-resistant organisms. Organisms surviving heat-resistance testing are to be submitted to the terminal sterilization laboratory for D-value analysis. D-values are then compared to established models for the component sterilization process.	
Endotoxin	NMT 2.5 EUs/stopper	NMT 10 EUs/vial

Note: NMT = not more than.

REASONS FOR REVISION

Effective date: mm/dd/yyyy

■ First time issued for your company, affiliates, and contract manufacturers.

YOUR COMPANY
VALIDATION STANDARD OPERATING PROCEDURE

SOP No.: Val. 1900.20   Effective date: mm/dd/yyyy

Approved by:

TITLE: Sterility Test Results Failure Investigation

AUTHOR: _____

Name/Title/Department

_____

Signature/Date

CHECKED BY: _____

Name/Title/Department

_____

Signature/Date

APPROVED BY: _____

Name/Title/Department

_____

Signature/Date

REVISIONS:

No.	Section	Pages	Initials/Date

SOP No.: Val. 1900.20

Effective date: mm/dd/yyyy

Approved by:

SUBJECT: Sterility Test Results Failure Investigation

PURPOSE

The purpose is to describe the procedure for the investigation of actions in case of sterility positive test results.

RESPONSIBILITY

It is the responsibility of the microbiologist and analyst to follow the procedure. The quality assurance (QA) manager is responsible for SOP compliance.

PROCEDURE

- The product batch with "sterility positive result" will be put on "hold."
- SOP (provide number) will be referred to for out-of-specification results.
- Team comprising micro lab in charge, QC manager, production manager, and validation manager will be responsible for investigation, which will include, but not be limited to:
  - Review sterility test method and handling of samples.
  - Microbial environmental monitoring data (air, surfaces, water)
  - Particulate monitoring data review
  - Personnel monitoring data (finger impressions, etc.)
  - Sterilization cycles for commodities and equipment
  - HEPA filter and LAF evaluation (airborne particulate levels, smoke-challenge testing, velocity measurements, etc.)
  - Room air flow patterns and pressures
  - Operator technique and training
  - Unusual events that occurred during the manufacturing
  - Storage conditions of sterile commodities (holding time)
  - Identification of contaminants as a clue to the source of the contamination
  - Housekeeping procedures and training
  - Calibration of sterilization equipment
  - Pre- and postfilter integrity test data and/or filter housing assembly

Effective date: mm/dd/yyyy

Approved by:

- Product and/or process defects and/or limitation of inspectional processes
- Documented disqualification of sample for obvious reasons prior to final reading
- Media fill records review (previous)
- All products that have been produced following this specific "sterility positive result" batch will be quarantined until a successful resolution of the problem and review of sterility results of subsequent batches.
- For corrective actions, see SOP, Val 1800.10: Process Simulation (Media Fill) Test.

REASONS FOR REVISION

Effective date: mm/dd/yyyy

- First time issued for your company, affiliates, and contract manufacturers.

YOUR COMPANY
VALIDATION STANDARD OPERATING PROCEDURE

SOP No.: Val. 1900.30   Effective date: mm/dd/yyyy

Approved by:

TITLE: Bacterial Endotoxin Determination in WFI, In-Process, and Finished Products

AUTHOR: _____

Name/Title/Department

_____

Signature/Date

CHECKED BY: _____

Name/Title/Department

_____

Signature/Date

APPROVED BY: _____

Name/Title/Department

_____

Signature/Date

REVISIONS:

No.	Section	Pages	Initials/Date

SOP No.: Val. 1900.30   Effective date: mm/dd/yyyy

Approved by:

SUBJECT: Bacterial Endotoxin Determination in WFI, In-Process, and Finished Products

PURPOSE

OBJECTIVE

The purpose is to establish a standard test method for endotoxin concentration monitoring of in-process or finished product by LAL gel formation method.

RESPONSIBILITY

It is the responsibility of the microbiologist concerned to follow the procedure. The quality control (QC) manager is responsible for SOP compliance.

1. Test Principle

The determination of the endotoxin with limulus amebocyte lysate (LAL) is based on gel formation of a mixture consisting of a solution of endotoxin of Gram-negative bacteria with a solution of lysate. The extent and speed of the reaction depend on the endotoxin concentration, pH, and temperature. The reaction requires the presence of certain cations, a proclotting enzymes system, and clottable protein, which are produced by lysate.

2. Requirements

- LAL test tube (10 × 75 mm pyrogen free)
- 8-ml test tube (USP type I)
- Disposable syringe
- Micropipette 100 µl, 1000 µl
- Disposable tips for micropipette (sterile/nonpyrogenic)
- Washed depyrogenated vials 20 to 50 ml for pooling product samples
- Test tube stand

SOP No.: Val. 1900.30   **Effective date: mm/dd/yyyy**

Approved by:

- Test tube shaker
- Dry bath incubator/water bath 37 ± 1°C
- Laminar air flow hood
- pH meter
- Calibrated timer
- Vortex
- Sterile parafilm
- Hot air oven 180 to 250°C

3. Reagents

- LAL reagent (pyrotell)
- Reference (RSE) or control (CSE) standard endotoxin
- LAL water (water for injection)
- Pyrosol LAL buffer
- Sterile nonpyrogenic 0.1 N hydrochloric acid LAL and 0.1 N sodium hydroxide LAL

Note 1: All materials coming in contact with test materials and reagents must be pyrogen free. Careful techniques are essential to prevent contamination with environmental endotoxin.

. Sampling Technique

.1 In-process product

- Samples should be collected aseptically in nonpyrogenic containers. Reused depyrogenate glassware or sterile, disposable polystyrene plastics are recommended to minimize adsorption of endotoxin.
- A sample (100 ml) of prefiltered solution will be taken in a sterile pyrogen-free container compounding.
- The sample is to be tested immediately upon receipt if the test is a requirement for batch release. If the test is not a requirement of batch release, the sample may stored at 2 to 8°C for up to 24 hours; however, the sample shall then be vortexed for 30 minutes prior to test.

4.2 Aseptically filled product

■ Three samples will be randomly taken from each product lot to be tested. The first sample should be taken in the beginning, the second in the middle, and the third at the end of the production run.

■ If the fill volume of an end product is less than 5 ml, a minimum of ten samples will be taken for endotoxin testing.

■ All samples must be properly labeled, showing the name of the product, lot number, and beginning, middle, or end of run. They should be stored upright and at room temperature while in the microbiology department.

4.3 Terminally sterilized

■ Target fill < 5 ml: the minimum number of samples required, pe. lot, is not less than ten. An approximately equal increment o samples is to be collected from each autoclave load of the lot.

■ Target fill > 5 ml: the minimum number of samples required, pe lot, is not less than three. The minimum number of samples fron each autoclave load is not less than one.

5. Preparation of Sample

■ For finished product, the test dilution should be approximately on half the MVD or, in cases in which this number is less than DROI the approximate midpoint between the MVD and the DROIE.

■ Check what sensitivity of reagent is to be used and what is use to reconstitute the reagent. Unless otherwise noted, use 0.06 U! EU/ml sensitivity reagent and WFI for reconstitution of the reage to be used

■ Using aseptic technique, open the required number of vials, vc texing each prior to removing an aliquot. Pool the samples ase tically in a sterile, pyrogen-free vial. For in-process sample, tre the bulk solution as a pooled sample.

■ If the fill volume of a product is 10 ml or more, take an equ volume of the product aseptically from each of the vials to tested and pool these samples in a sterile, pyrogen-free vial.

■ In cases of freeze-dried (lyophilized) products, reconstitute w the appropriate volume of sterile nonpyrogenic WFI before pool the samples.

SOP No.: Val. 1900.30 Effective date: mm/dd/yyyy

Approved by:

- In cases of terminally sterilized products, pool a minimum of one sample from each autoclave load.
- In cases of empty sterile vials, aseptically inject 2.5 ml of sterile nonpyrogenic WFI into each vial before pooling the sample.
- Vortex samples prior to pooling.

5.1 Preparation of LAL reagent (pyrotell)

5.1.1 Reconstitution

- Gently tap the vial/tube to cause loose LAL to fall to the bottom of the vial. Remove the crimp seal and break the vacuum by lifting the gray stopper.
- Remove and discard the stopper. Do not inject through or reuse the stopper. Do not contaminate the mouth of the vial and do not inject through or reuse the stopper. A small amount of LAL left on the stopper will not affect the test.
- Cover the vial with parafilm.
- For a single test vial, reconstitute with 0.2 ml of the test sample during the test procedure.
- For a multitest vial, reconstitute with 1.0, 2.0, or 5.0 ml of LAL reagent water as indicated on the vial label.
- Before use, gently mix the content of the vial to ensure homogeneity.
- Label the vial with initials, date reconstituted, and expiration.

Note 2: Mixing too vigorously may cause excessive foaming, which can cause a loss of sensitivity. Immediately after reconstitution, reagent should be used. Any reagent remaining shall be stored for 24 hours at 2 to 8°C (or for 1 to 3 months at −20°C but frozen once only).

Note 3: Reconstituted reagent should be slightly opalescent. Yellow color indicates contamination or denaturation.

Note 4: The USP inhabitation/enhancement test must be repeated on one lot of the product if the lysate manufacturer is changed. The recommended source is Pyroquant Diagostik.

Note 5: Before using LAL reagent or pyrosol, make sure the sensitivity of the lysate specified by the manufacturer is confirmed.

SOP No.: Val. 1900.30

Effective date: mm/dd/yyyy

Approved by:

5.2 Preparation of control standard endotoxin (CSE)

- Reconstitute and store the CSE according to manufacturer instructions. The potency of CSE specified by the manufacturer should be confirmed. Constitute the entire contents of one vial of CSE with LAL reagent water (LRW) and mix intermittently for 30 min, using a vortex mixer. Water should come in contact with all internal surfaces, including stopper sometime during reconstitution.
- Label the endotoxin vial with potency, initials, and preparation date.
- Use this concentrate for making appreciate serial dilutions. Use 13 × 100 mm pyrogen-free culture tubes.
- Prepare a series of twofold dilutions of CSE to give concentrations of 2λ, 1λ, 0.5λ, and 0.25λ, where λ is the labeled sensitivity of the LAL reagent in endotoxin units per milliliter. Vortex each tube for 1 minute between each dilution step — e.g., if λ = 0.06 EU/ml the endotoxin concentration to use in the assay shall be 0.25 EU/ml 0.125 EU/ml, 0.06 EU/ml, and 0.015 EU/ml concentration.
- Label standards with endotoxin concentration in EU/ml, initials and preparation date.
- Do not store dilutions. Use only on the day prepared.

5.2.1 Positive product control

- Positive product control is inhibitory control and consists of the specimen or dilutions of the specimen, per STM 063, to which standard endotoxin is added.
- The final concentration of the added endotoxin in the test specimen should be 2λ, e.g., the product to be tested using 0.06 USP EU/n
- Aseptically pipette 100 µl of the stock solution of endotoxin containing 1.2 USP EUs/ml into 900 µl of the tested product dilutio This gives a concentration of 0.12 USP EU/ml.

5.2.2 Negative control

- Take two 10 × 75 mm sterile, pyrogen-free glass test tubes.
- Pipette 0.1 ml of pyrogen-free water into the bottom of each tube
- If using pyrosol, pipette 0.1 ml pyrosol buffer into two addition 10 × 75 mm tubes.
- Mark these tubes "negative."

SOP No.: Val. 1900.30

Effective date: mm/dd/yyyy
Approved by:

6. Procedure

6.1 Gelation method (limit test)

- Perform the test in duplicate using a dilution not exceeding the maximum valid dilution of the preparation being examined.
- Before dispensing the sample to test tube, if necessary, per STM 063, take 1.0 ml of the diluted sample in a clean beaker and add the same amount (1:1) of LAL reagent being used, then check the pH of the sample. Adjust the pH of the sample in the range of 6.0 to 8.0 using HCl or NaOH or buffer. Only 10% or less additional dilution is acceptable.
- Arrange the 10 × 75 mm sterile, pyrogen-free glass tubes per the following table.
- Label tube with the dilution.

Solution	Endotoxin Concentration/Solution to Which Endotoxin Is Added	No. Replicates
A	None/dilute sample solution	2
B	2/Diluted sample solution	2
C	2/LAL reagent water	2
D	None/LAL reagent water	2

6.2 Semiquantitative gelation method

Prepare the following solutions:

- Two independent replicate solutions of the preparation being examined at the dilution with which the test for interfering factors was completed. Use water BET to make two independent dilution series of four tubes containing the preparation being examined at concentrations of 1, 1/2, 1/4, and 1/8 relative to the dilution with which the test for interfering factors was completed.
- Two series of four tubes of water BET containing endotoxin standard at a concentration of 2λ, λ, $1/2\lambda$, and $1/4\lambda$, respectively.
- Two independent replicate solutions of the preparation being examined at the dilution with which the test for interfering factors was completed and endotoxin standard BRP at a concentration of 2λ.

- Water BET (as a negative control). For each dilution, use separate pipette/syringe.
- After making the dilution mark the LAL tubes with the concentration of the sample with the dilution factor.
- Label tube with the dilution.

Solution	Endotoxin Concentration/Solution to Which Endotoxin Is Added	Dilution Factor	Initial Endotoxin Concentration	No. Replicates
A	None/dilute sample	1	—	2
	solution	2		2
		4		2
		8		2
B	2 /Diluted sample solution	1	2λ	2
C	2 /LAL reagent water	1	2λ	2
		2	1λ	
		4	0.5λ	
		8	0.25λ	
D	None/LAL reagent water	—	—	2

7. Incubation

- Cap the tube, swirl each tube gently to mix.
- Incubate the test tube in dry bath incubator or noncirculating wate bath, for 60 ± 2 minutes at $37 \pm 1°C$. Place the reaction mixture without vibration and minimizing loss of water by evaporatior Record the temperature and the thermometer number.

Note 6: Strict adherence to time and temperature is essential. The reactic begins when LAL is added to the test sample but does not proceed at a optimum rate until the mixture reaches 37°C. If large numbers of tub are tested in parallel, the tests should be batched and started at interva that permit the reading of each within the time limit.

Note 7: The gel-forming reaction is delicate and may be irreversil terminated, so extreme care must be exercised to prevent agitation vibration of the reaction tubes during incubation period.

SOP No.: Val. 1900.30

Effective date: mm/dd/yyyy

Approved by:

- After incubation, all tubes are visually inspected for the presence or absence of gel. Gently but firmly pick up each tube by its top in a single smooth motion.
- Do not wipe it. Do not shake or jerk the tube.
- Turn the tube upside down and hold it there.
- A positive result is indicated by the formation of a firm gel that does not disintegrate when the receptacle is gently inverted.
- While negative result is characterized by the absence of such a gel or formation of a viscous gel, which does not remain when inverted, record it as negative.

8. Recording Results

- Record in time the result in the log book.
- For semiquantitative gelation method, the geometric mean of the endotoxin concentration = antilog ($\Sigma e/f$), where Σe is the sum of the log endpoint concentrations of the dilution series used, and f is the number of replicate test tubes.
- To determine the endotoxin concentration of solution A, calculate the endpoint concentration for each replicate series of dilution factor by λ. The endotoxin concentration in the sample is the geometric mean of the endotoxin concentration of the replicates.
- For limit test: calculate the concentration of endotoxin in the test solution as follows:

$$\text{Endotoxin unit/ml} = \text{dilution factor of last end point (D.F.)} \times \\ \text{LAL reagent sensitivity}$$

Acceptance Criteria

The preparation being examined passes the test if the endotoxin concentration of each of all tested samples is less than the endotoxin limit concentration specified in the specification and the test meet the following three conditions:

- The results of the negative controls show acceptable nongel formation.
- The results of positive product controls or the endotoxin standard show the endpoint concentration to be within plus or minus twofold dilutions from the label claim sensitivity.
- All of the endotoxin standards form firm gels.

The test does not meet parameters:

- If the positive product control or endotoxin standard does not show the endpoint concentration to be within plus or minus twofold dilutions from the label claim sensitivity of the LAL reagent or if any negative control shows a gel clot endpoint.
- If all of the endotoxin standards do not form firm gels, reagent has deteriorated or the test has not been conducted properly.
- If negative controls show firm gel, water, glassware, or reagent is contaminated.

Perform a laboratory technical investigation and retest the testing reagents. Repeat the test with more restrict conditions.

The sample does not meet the specification:

- If the endotoxin concentration of one of the two solutions is lower and the other one is higher than this limit, repeat the test.
- If sample and positive control tubes form firm gel, this indicate the sample is pyrogenic. Resample with more careful technique and retest.

Apply SOP for investigational analysis of out-of-specification microbological results. If the new sample is pyrogenic, the batch is rejected.

REASONS FOR REVISION

Effective date: mm/dd/yyyy

- First time issued for your company, affiliates, and contract manufacturers.

| YOUR COMPANY |
| VALIDATION STANDARD OPERATING PROCEDURE |

SOP No.: Val. 1900.40

Effective date: mm/dd/yyyy

Approved by:

TITLE: Monitoring the Bioburden, Spore Bioburden, and Endotoxin Present on Stoppers and Unprocessed Vials

AUTHOR: _____

Name/Title/Department

_____

Signature/Date

CHECKED BY: _____

Name/Title/Department

_____

Signature/Date

APPROVED BY: _____

Name/Title/Department

_____

Signature/Date

REVISIONS:

No.	Section	Pages	Initials/Date

SOP No.: Val. 1900.40

Effective date: mm/dd/yyyy

Approved by:

SUBJECT: Monitoring the Bioburden, Spore Bioburden, and Endotoxin Present on Stoppers and Unprocessed Vials

PURPOSE

The purpose is to evaluate the bioburden, spore bioburden, and endotoxin present on rubber stoppers and unprocessed glass vials. The similarity of the contents and procedure description shall be considered coincidental due to the similarity of generic methods.

RESPONSIBILITY

It is the responsibility of the microbiologist to follow the procedure. The quality control (QC) manager is responsible for SOP compliance.

PROCEDURE

1. **Applicability**

These procedures apply to all vials and all stoppers used for sterile products

2. **Frequency**

   ■ Each lot of each type of vial to be tested per the procedure described here.
   ■ Each lot of each type of stoppers to be tested before and after sterilization per the procedures described here.

3. **Definitions**

   ■ Prewashed, siliconized rubber stoppers
   ■ LAL: limulus amebocyte lysate
   ■ Stopper type: stoppers made from the same compound and manufactured by the same process
   ■ Vial type: vials made from the same compound and manufactured by the same process

4. **Equipment/Materials**

   ■ Test samples (60 vials or stoppers of one size, style, manufacturer and lot number)

SOP No.: Val. 1900.40　　　　　　　　**Effective date: mm/dd/yyyy**

　　　　　　　　　　　　　　　　　　　　Approved by:

- Sterile media:
  - Trypticase soy agar (TSA), in bottles and prepoured in sterile petri dishes
  - 0.1% Tween 80 in WFI in appropriately sized containers
- Sterile filtration units, 47-mm size, with filtration manifold
- Sterile 0.45-μm, 47-mm, mixed esters of cellulose (MEC), tortuous path filter membranes
- Sterile forceps
- Vacuum pump
- Laminar flow hood
- Water bath set at 45 to 55°C
- Incubator set at 30 to 35°C
- Mechanical shaker
- Sterile surgical gloves
- Self-contained anaerobe jar
- Bunsen burner
- Sterile, pyrogen-free WFT
- Depyrogenated soda-lime (flint) 10 × 75 test tubes
- Depyrogenated borosilicate 13 × 100 mm test tubes
- Test tube rack
- Sterile, pyrogen-free syringes
- Sonicator
- Endotoxin standard (source — Associates of Cape Cod)
- Limulue amebcyte lysate (LAL), sensitivity 0.125 EU/ml
- Vortex mixer
- Calibrated thermometer with range above 100°C
- Water bath with temperature range to 100°C

5. Bioburden Method

1. Sampling according to SOP (provide number):
   - Under a laminar flow hood, aseptically place 20 stoppers of the size and mold number to be tested into a sterile jar containing 50 ml of 0.1% Tween solution so that all stoppers are completely immersed (100 ml 0.1% Tween for 28-mm stoppers).
   - Allow stoppers to soak for 1 hour undisturbed. The soaking is intended to facilitate greater recovery of any bioburden present on the stoppers.

- Shake container for approximately 30 minutes at "low" speed setting on mechanical shaker.
- Vortex just before testing.
2. Don a fresh pair of sterile gloves. Sanitize all interior hood surfaces with a QC-approved antimicrobial agent:
   - Spray the surface of the hood with a QC-approved antimicrobial agent until visibly wet.
   - Wipe all wet surfaces with a sterile, lint-free towel. Allow surfaces to dry.
3. Turn on the pump and bunsen burner or lab fuel. Don a fresh pair of sterile gloves.
4. Using a QC-approved antimicrobial agent, sanitize the exterior surfaces of the sample container by spraying with the agent until visibly wet or wiping the surfaces with a sterile, lint-free, agent-soaked towel. Put all the samples to be tested under the hood; allow surfaces to dry before testing.
5. Aseptically unwrap four sterile filter funnels and place in filtration manifold.
6. For each of the four filter funnels:
   - Holding the funnel portion of the filtration assembly with one hand, carefully lift the funnel from the base.
   - Using a sterile forceps (flame the tips of the forceps and cool to room temperature before using), take a presterilized 47-mm 0.45-µm, mixed esters of cellulose (MEC), tortuous path membrane filter (hereafter referred to as membrane filter) and plac filter, grid side up, in the filter holder.
7. Aseptically decant the Tween solution, splitting the total volum present equally among each of four sterile membrane filter holder.
8. Apply vacuum to each of the filter holders and draw the Twee solution.
9. Turn off vacuum supply.
10. Pour an additional 10 ml (approximately) of sterile 0.1% Twee solution into each filter using aseptic techniques. This rinse intended to wash the interior walls of the filtration units and car any bacteria adhering to these surfaces onto the surface of th membrane filter.
11. Turn on vacuum supply and filter the contents of the filter hold Turn off vacuum supply.

12. Using sterile forceps (or forceps whose tips have been flamed and then cooled to room temperature), carefully remove each membrane filter and place onto prepoured TSA plate.

13. Prepare each of two negative controls by filtering 50 ml of sterile 0.1% Tween solution. Place filter on sterile prepoured TSA plates as described earlier.

14. Invert all plates. Incubate two of the test sample plates and one negative control plate at 30 to 35°C for no less than 72 hours.

15. Incubate the remaining two plates and the remaining negative control plate anaerobically in anaerobic jar. Incubate at 30 to 35°C for no less than 72 hours.

16. Following incubation, enumerate the colonies and record.

17. Calculations are done separately for aerobic and anaerobic plates.

18. Report these calculations with one decimal place, rounding up if the second decimal place is 5 or higher.

19. Record all data

20. Sample size will be 20 vials per test: follow the procedures described in steps 2 through 19, vial size permitting. (Small vials will be immersed.)

21. For 20-cc or larger vials, fill each vial about halfway with rinsing fluid instead of immersing as described earlier. Then proceed as follows:
    ■ After filling, place depyrogenated foil and parafilm onto each vial and place onto a mechanical shaker.
    ■ Shake approximately 30 minutes at "low" speed setting on mechanical shaker.
    ■ Put all containers into a single sterile container.
    ■ Follow steps 2 through 19.

1 Identification of isolates

1. Identify a representative of each morphologic type.

2. Assign each isolate to be identified by number to permit easy tracking and documentation.

3. Streak any anaerobes recovered onto TSA and incubate aerobically to determine whether these are true (obligate) anaerobes.

4. Perform resistance testing on all Gram-positive spore formers. Submit any organism surviving the resistance test for D-value determination.

5. Summarize stopper and vial data separately.

SOP No.: Val. 1900.40

Effective date: mm/dd/yyyy

Approved by:

6. Spore Bioburden Method

6.1 Analysis of stoppers

1. Follow the first three steps in section 3 of Bioburden method for sampling and preparation of stoppers.
2. Heat-shock the rinses at 98 to 100°C by immersing in a water bath. Fill a separate container with the same amount of Tween solution used to rinse the stoppers and immerse with the test units. Include a thermometer in the extra container. Once 98°C is reached on the thermometer, begin timing the exposure period of 20 minutes with a calibrated timer.
3. After heat shocking, remove the rinse samples and place in cool water.
4. After cooling, shake container on mechanical shaker at slow setting for 30 minutes. Sonicate for 2 minutes just prior to testing.
5. Transfer the samples to a laminar flow hood.
6. Continue testing as described in bioburden method, steps 7 through 19, using aseptic technique.
7. Incubate two plates and a negative at 30 to 35°C. The remaining two plates and a negative incubate at 93 to 60°C with parafilm to prevent media from drying.
8. Enumerate the colonies on all plates according to bioburden method, step 21. Record all data.

6.2 Analysis of vials

9. Follow steps 1 through 3 in bioburden method for sampling and preparation of vials.
   - Small vials will be immersed. Sample size will be 20 vials per test
   - For 20-cc or larger vials, fill each vial about halfway with rinsing fluid instead of immersing as described previously. Place a sterile stopper and crimp seal onto each vial.
10. Follow steps 2 through 5 in the section about stopper bioburden method for heat shocking, cooling, shaking, and sonicating sample (one container containing 20 small vials 20 half-filled vials).
11. If testing 20-cc or larger vials, pool contents of the 20 half-filled vials into a single sterile container. Vortex.
12. Continue testing as described in spore bioburden method steps through 8.

SOP No.: Val. 1900.40 Effective date: mm/dd/yyyy

Approved by:

6.3 Resistance testing of sporeformers

13. If any Gram-positive rods survive the heat shock, further testing is required.
14. Subculture isolates on TSA incubated at 30 to 53°C and identify each morphological type of colony.
15. Perform resistance testing on all Gram-positive spore formers. Submit any organism surviving the resistance test for D-value determination.
16. Summarize stopper and vial data separately.

7. Endotoxin Test Method

7.1 Endotoxin standard preparation

1. Using a sterile, nonpyrogenic pipette tip and pipettor, aseptically pipette 5.0 ml of pyrogen-free water into the endotoxin vial.
2. Vortex vial intermittently for not less than 20 minutes. Water should come in contact with all internal surfaces, including the stopper during reconstitution. Reconstituted stock can be stored for 14 days at 2 to 8°C. Never freeze stock solution. Label stock with initials and date prepared. Vortex refrigerated reconstituted stocks 5 minutes before use.
3. Determine endotoxin potency (EU/ng) from certificate of analysis (C of A). Calculate endotoxin units per milliliter (EU/ml), based on C of A.
4. When making dilutions, vortex each tube 1 minute between each dilution step. Do not freeze endotoxin dilutions. Do not store dilutions.

.2 Preparation of pyrotell

5. Remove vials of reagent from freezer, tear off metal seal, and gently remove the stopper.
6. Using a sterile, nonpyrogenic disposable syringe and needle, inject 5.0 ml of pyrogen-free water to reconstitute the reagent. Reconstitution is more rapid if reagents are at room temperature.
7. Gently swirl the vial to dissolve the contents completely. Mixing too vigorously can cause excessive foaming, which can lead to loss of reagent sensitivity. Keep reagent in the refrigerator after reconstitution until needed.

SOP No.: Val. 1900.40

Effective date: mm/dd/yyyy

Approved by:

Note: Following reconstitution, reagent should be used immediately. Any reagent remaining after reconstitution is stable at 2 to 8°C for 24 hours or for up to 1 month at −20°C or below. Reconstituted reagent can be frozen–thawed only once. Reconstituted reagent should be slightly opalescent. Yellow color indicates contamination or denaturation. Label vial with initials, date reconstituted, and expiration.

7.3 Analysis of stoppers

8. Wear sterile gloves and a clean lab coat.
9. Sampling according to SOP (provide number).
10. On laminar flow bench, open the bag to expose the stoppers. Using sterile forceps, randomly select 20 stoppers, transferring then to a sterile, nonpyrogenic sample container. Seal the container afte the transfer of stoppers is complete.
11. Place the container of stoppers on a laminar flow bench alon; with a container of sterile, nonpyrogenic water for injection (WFI)
12. Don a fresh pair of sterile gloves.
13. Using a sterile, nonpyrogenic pipette, add 100 ml for 28-mn stoppers. Recap the container.
14. Gently mix the stoppers in the container to ensure that all surface come in contact with the water and sonicate for 30 minutes.
15. Return the container to the laminar flow hood. Open the containe and, using an automatic pipettor fitted with sterile, nonpyrogen tips, pipette 0.1 ml of the water into each of two sterile pyroge free test tubes.
16. Controls:
    ■ Positive controls: pipette duplicate 0.1-ml aliquots of fresh prepared endotoxin standard solution, bracketing the labele lysate sensitivity, into separate 10 × 75 mm pyrogen-free t tubes.
    ■ Negative controls: pipette 0.1-ml sterile pyrogen-free water ir two 10 × 75 mm test tubes.
17. Add 0.1 ml of reconstituted reagent to each tube prepared in ste 16 and 17.
18. Gently mix the contents but avoid foaming.
19. Immediately place all the tubes in an upright position in a 37 1°C water bath and incubate undisturbed for 60 ± 2 minutes. Recc the thermometer temperature and number.

SOP No.: Val. 1900.40   Effective date: mm/dd/yyyy
                                                Approved by:

20. After incubation, inspect all tubes for the presence or absence of a firm gel. Gently but firmly pick up each tube by its top. Invert the tube 180° without shaking or jerking the tube and hold it there. A firm gel that adheres to the bottom of the test tube in this inverted position is scored as positive. If the contents of the tube slide down the tube, score it as negative.
21. Record results.
22. In the event of a negative test, the stoppers meet the acceptance criteria for endotoxin.

7.3.1 Phase II — confirmation

23. If one or both of the duplicate tubes yields a positive endotoxic response, repeat steps 15 through 21 using the original container of 20 stoppers. Record results.
24. If none of the quadruplicate tubes gel, the stoppers meet the acceptance criteria for endotoxin.
25. If any of the four reaction tubes shows a firm gel, proceed to phase III.

7.3.2 Phase III — individual stopper

26. Resample an additional 20 stoppers. Number according to steps 8 through 13.

7.3.3 Pyroburden determination

27. Transfer individual stoppers to separate sterile nonpyrogenic containers.
28. Add only 10 ml of WFI to each container.
29. Repeat steps 15 through 21. Record results.
30. If no gel is obtained from any reaction tube, the stoppers meet the acceptance criteria.
31. If any reaction tube shows a firm gel, the stoppers do not meet the acceptance criteria.
32. In the event the results do not meet the acceptance criteria, an investigation will be initiated and QC and production management will be notified immediately. The vendor will be put on alert status

for that stopper and every incoming lot number of the type of stopper must be tested and shown to be free of endotoxin prior to use in production. The investigation must be completed and the vendor must be recertified by testing stoppers from five manufacturer's lot numbers for endotoxin before the vendor can be removed from the alert list for that stopper.

7.4 Analysis of vials

33. Wearing sterile gloves and a clean lab coat, collect 20 random samples of the vials to be tested. Place in a sterile container for transport to the microbiology lab.

7.4.1 Phase I — screening

34. Put on a new pair of sterile gloves.
35. Working under a laminar flow hood, add sterile pyrogen-free WFI to the vials to be tested. The following amounts are added:

WFI	Vial Size
10 ml	30 ml
5 ml	20 ml
1 ml	15 ml

36. Cover with depyrogenated foil.
37. Gently swirl the vials to ensure that all surfaces come in contact with the water and sonicate for 3 minutes.
38. Pool water from all 20 vials into one sterile pyrogen-free container. Vortex to mix. Pipette two aliquots of 0.1 ml each into labeled 10 × 75 mm sterile pyrogen-free glass tubes.
39. Follow steps 17 through 21 in endotoxin test method. Record results.
40. In the event of a negative test, vials meet the acceptance criteria for endotoxin.

SOP No.: Val. 1900.40   Effective date: mm/dd/yyyy

Approved by:

7.4.2 Phase II — confirmation

41. If one or both of the duplicate tubes yields a positive endotoxic response, proceed to phase II testing. Repeat steps 38 and 39. Record results.
42. If none of the quadruplicate tubes gel, the vials meet the acceptance criteria.
43. If any of four-reaction tubes gel, proceed to phase III.

7.4.3 Phase III — individual vial pyroburden determination

44. Resample an additional 20 vials from the same lot, following steps 34 through 40.
45. Do not pool samples, but test the 20 samples individually.
46. Pipette duplicate 0.1-ml aliquots from each vial into labeled 10 × 75 mm sterile pyrogen-free glass tubes. Follow steps 10 through 14 in the stopper section of the SOP. Record results.
47. Evaluation is the same as for the stoppers; see steps 30 through 32.

Limit

Test	Ready-to-Sterilize Stoppers	Unprocessed Vials
Bioburden	NMT 25 CFUs/stopper	NMT 100 CFUs/vial
Spore bioburden	Spore bioburden data are collected to screen for heat-resistant organisms. Organisms surviving heat resistance testing are to be submitted to the terminal sterilization laboratory for D-value analysis. D-values are then compared to established models for the component sterilization process.	
Endotoxin	NMT 2.5 EUs/stopper	NMT 10 EUs/vial

Note: NMT = not more than.

REASONS FOR REVISION

Effective date: mm/dd/yyyy

- First time issued for your company, affiliates, and contract manufacturers.

SECTION

VAL 2000.00

YOUR COMPANY
VALIDATION STANDARD OPERATING PROCEDURE

SOP No.: Val. 2000.10

Effective date: mm/dd/yyyy

Approved by:

TITLE: Technical Training

AUTHOR:

_____

Name/Title/Department

_____

Signature/Date

CHECKED BY:

_____

Name/Title/Department

_____

Signature/Date

APPROVED BY:

_____

Name/Title/Department

_____

Signature/Date

REVISIONS:

No.	Section	Pages	Initials/Date

SOP No.: Val. 2000.10

Effective date: mm/dd/yyyy

Approved by:

SUBJECT: Technical Training

PURPOSE

The purpose is to describe the training needs, courses titles, and the frequency of the training to ensure the staff is acquainted (familiar) with their jobs and GMP requirements.

RESPONSIBILITY

All employees working inside the plant are responsible to follow the procedure. All departmental managers are responsible for the SOP compliance. The following departments will be subjected to training as applicable to their operations:

1. Quality assurance
2. Quality control
3. Maintenance
4. Manufacturing
5. Packaging
6. Stores
7. Calibration lab
8. Shipping
9. Purchasing
10. Utilities (HVAC, gases, water)
11. Registration
12. Shipping

PROCEDURE

■ The manager concerned will identify the training needs and w inform the QA coordinator to provide the training sheet.
■ Frequency of training courses: the following courses to be co ducted at the time of joining ABC Pharmaceutical Industries a every 5 years after that. If the training is not conducted earlier, should be considered as initial from the first time conducted.

SOP No.: Val. 2000.10

Effective date: mm/dd/yyyy
Approved by:

- All equipment operating procedures — initial and once every 5 years
- All system-related SOPs — initial and once/year
- All equipment and system SOPs to be read after revision regardless of time frame
- SOP reading per day should be controlled to ensure understanding of the procedure and to avoid overburdening employees. Moreover, each department should ensure that when recording the SOP reading, a staff member should mention the following details:
  - Exact date (mm/dd/yyyy)
  - SOP number along with the revision number
- Alternatively, as a part of training, SOP validation/training records are also acceptable.
- The following training will be provided to all supervisors, technicians, analysts, workers, and QA staff working in sterile and non-sterile areas. The type of training can be selected from the topics; also, training course materials can be searched from the Internet — for instance, courses provided by companies such as Savant, Micron, etc.

Training Course	Frequency
Documentation skills and techniques	Initial, then once/year
GMP training of workers	Initial, then once/year
Now ... wash your hands	Initial, then once/year
Preventive contamination	Initial, then once/year
The GMP of personal hygiene	Initial, then once/year
Understanding GLP	Initial, then once/year
Understanding GMP	Initial, then once/year
Your first days at work with GMP	Initial, then once/year

- The following additional training should be provided to production and packaging staff, supervisors, technician, workers, and QA staff working in sterile areas:

SOP No.: Val. 2000.10   Effective date: mm/dd/yyyy

Approved by:

Training Course	Frequency
Behavior in the cleanroom	Initial, then once/year
Cleanroom clothing — design and performance	Initial, then once/year
Cleanroom clothing — service cycle	Initial, then once/year
Cleaning of production rooms and equipment	Initial, then once/year
Equipment cleaning	Initial, then once/year
GMP of dressing for sterile production	Initial, then once/year
Personal cleanliness in the cleanroom	Initial, then once/year
Planning work areas in cleanrooms	Initial, then once/year
Preparing to clean the room	Initial, then once/year
Robbing for the clean room	Initial, then once/year
Safety in clean room	Initial, then once/year
The GMP of cleaning and disinfecting cleanrooms	Initial, then once/year
Understanding sterile productions	Initial, then once/year
Working under controlled conditions	Initial, then once/year

■ The following training is to be conducted for laboratory analyst as applicable:

Training Course	Frequency
Introduction of UV visible spectroscopy	Initial, then once/5 years
Basic microbiology	Initial, then once/5 years
Column selection high performance liquid chromatography	Initial, then once/5 years
Electrical safety in the laboratory	Initial, then once/5 years
Flammable and explosive	Initial, then once/5 years
Instrumentation of HPLC (slide show and discussion)	Initial, then once/5 years
Laboratory ergonomics	Initial, then once/5 years
Laboratory hoods	Initial, then once/5 years
Orientation to laboratory safety	Initial, then once/5 years
Planning for laboratory emergencies	Initial, then once/5 years
Principles of HPLC	Initial, then once/5 years
Principle and practice	Initial, then once/5 years
Principles of IR quantitative	Initial, then once/5 years
Qualitative and quantitative (slide show and discussion)	Initial, then once/5 years

SOP No.: Val. 2000.10

Effective date: mm/dd/yyyy

Approved by:

Training Course	Frequency
Safe handling of laboratory glassware	Initial, then once/5 years
Safety showers and eye washes	Initial, then once/5 years
Technique of solid sample handling for IR spectroscopy	Initial, then once/5 years
OSHA formaldehyde standards	Initial, then once/5 years
Trouble-shooting liquid chromatography	Initial, then once/5 years

- All departmental managers shall update this SOP if additional training courses are available.
- Outside training conducted and received by the employees will be kept in a separate file by QA
- Technical training needs will be evaluated once in a year.
- Training that is conducted will be documented.

REASONS FOR REVISION

Effective date: mm/dd/yyyy

- First time issued for your company, affiliates, and contract manufacturers.

SECTION

VAL 2100.00

YOUR COMPANY
VALIDATION STANDARD OPERATING PROCEDURE

SOP No.: Val. 2100.10

Effective date: mm/dd/yyyy

Approved by:

TITLE: Environmental Performance Test Procedure

AUTHOR: _____

Name/Title/Department

_____

Signature/Date

CHECKED BY: _____

Name/Title/Department

_____

Signature/Date

APPROVED BY: _____

Name/Title/Department

_____

Signature/Date

REVISIONS:

No.	Section	Pages	Initials/Date

SOP No.: Val. 2100.10

Effective date: mm/dd/yyyy
Approved by:

SUBJECT: Environmental Performance Test Procedure

PURPOSE

The purpose is to describe the environmental performance test criteria.

RESPONSIBILITY

It is the responsibility of validation team members to follow the procedures. The quality assurance (QA) manager is responsible for SOP compliance.

PROCEDURE

Controlled environments should generally be subjected to the following set of performance tests related to environmental control:

- Airflow, volume, and distribution tests
- Temperature test
- Humidity test
- Airborne particle count test
- Main air supply and make-up supply volume and reserve capaci
- Lighting level test
- Microbial level assessment test

The preceding tests are required for the final certification and validatic of the environment (where applicable). The instruments used for the tests should be calibrated; reports should be available. The qualificatic shall include conditions at rest, dynamic, and or under stress (as applicable

1. At-Rest and Dynamic Testing

Tests performed under "at-rest" conditions to serve as baseline informati are needed to determine the degree by which the environmental para eters are affected by the process after they are repeated at dynam conditions (simulated fully operational conditions). After this analys review procedures, equipment, methods, etc. and change if necessa Performance tests executed at dynamic conditions are the only way obtain a clean representation of the prevailing environmental conditior

SOP No.: Val. 2100.10   Effective date: mm/dd/yyyy

Approved by:

2. Stress Testing

Testing at stress condition is performed to determine the ability of the system to remain stable at all times during operational conditions defined as continuity. Create stress conditions to determine the span of control for an individual system or rooms. Execute temperature, particle counting, etc. Verify obtained results with adequate operational conditions and determine whether the system is acceptable; if it is not, then alert systems are set to report the unacceptable condition. Determine the ability of the system to recover after an unacceptable limit has been reached. Stress testing and alert systems for hardware and software are an important part of the scope of work for the validation team.

3. Reporting Forms

The protocols define the procedures to be used to verify the performance of qualified equipment. As part of validation, the results obtained should be carefully recorded and compared with the design conditions. Deviations or diversions contrary to the specified levels determine the suitability of the controlled environment, so the reporting form represents the document for certification or acceptance of the system. The reporting form should show the following information:

- Date
- Name of person performing the test
- Location of the test
- Testing equipment with serial numbers
- Calibration dates
- Temperature (when applicable)
- Humidity (when applicable)
- Air velocity (when applicable)
- Design conditions
- Actual conditions
- Signatures of those involved in the test
- Diagrams showing test locations

Expand the list of reporting information as required.

SOP No.: Val. 2100.10

Effective date: mm/dd/yyyy

Approved by:

REASONS FOR REVISION

Effective date: mm/dd/yyyy

- First time issued for your company, affiliates, and contract manufacturer.

YOUR COMPANY
VALIDATION STANDARD OPERATING PROCEDURE

SOP No.: Val. 2100.20

Effective date: mm/dd/yyyy

Approved by:

TITLE: HEPA Filters Leak Test (DOP) Procedure

AUTHOR: _____
                              Name/Title/Department

              _____
                                  Signature/Date

CHECKED BY: _____
                              Name/Title/Department

              _____
                                  Signature/Date

APPROVED BY: _____
                              Name/Title/Department

              _____
                                  Signature/Date

REVISIONS:

No.	Section	Pages	Initials/Date

SOP No.: Val. 2100.20

Effective date: mm/dd/yyyy

Approved by:

SUBJECT: HEPA Filters Leak Test (DOP) Procedure

PURPOSE

The purpose is to check the integrity for HEPA filters by leak test to assure that they are not damaged during installation or operation.

RESPONSIBILITY

It is the responsibility of validation team members to follow the procedures. The quality assurance (QA) manager is responsible for SOP compliance.

PROCEDURE

1. Equipment and Material

- DOP aerosol generator equipped with Laskin-type nozzles, as described in USA Standard N-5.11
- Aerosol photometer
- DOP (dioctyl phalate)
- Nitrogen gas cylinder (nitrogen purity ≥99.9%)
- Pressure regulator

2. Method

- This test should be performed by trained or certified personnel who introduce DOP aerosol upstream of the filter through a test port and search for leaks downstream with an aerosol photometer
- Remove air diffusers and airflow laminators before performing the test (whatever is applicable) to get access to the filters.
- Introduce the challenge aerosol upstream from the HEPA filter following the generator operating manual.
- A challenge aerosol concentration of 10 to 20 μg/l of air is satisfactory
- In upstream position (gain adjustment), adjust the concentration immediately upstream from the filter to read 100% on the photometer. If the upstream concentration of 100% cannot be reached increase generator output.

SOP No.: Val. 2100.20   Effective date: mm/dd/yyyy
                                         Approved by:

- Adjust the zero of the calibrated photometer according to operating manual.
- Change into the test position (downstream) and scan the total surface of the filter and the edges. Hold the probe approximately 2.5 cm from the surface of the filter with a scanning speed of not more than 3 ft/minute (scan test).

3. Acceptance Criteria

- HEPA filters 99.997%: challenge aerosol penetration is lower or equal to 0.003% of the upstream concentration.
- HEPA filters 99.99%: challenge aerosol penetration is lower or equal to 0.01% of the upstream concentration.
- HEPA filters 99.97%: challenge aerosol penetration is lower or equal to 0.03% of the upstream concentration.
- HEPA filters 95.00%: challenge aerosol penetration is lower or equal to 5.0% of the upstream concentration.
- Use silicon sealant to seal the leaks (Dow or 3M weatherstrip adhesive). Recheck the filter for leaks after the repair and document.
- The filter repair is limited to 5% of the filter area; if it is more, filter must be rejected and a new one installed.

Documentation

Make entries in attachment no. 2100.20.

REASONS FOR REVISION

Effective date: mm/dd/yyyy

- First time issued for your company, affiliates, and contract manufacturers.

SOP No.: Val. 2100.20

Effective date: mm/dd/yyyy
Approved by:

Attachment No. 2100.20

HEPA FILTER LEAK (DOP) TEST RESULTS
Company Name

SOP No. 2100.20
Issue Date: mm/dd/yyyy
Revision No.: New

HEPA filter serial no: _____   System ID: _____

HEPA filter type: _____   Dimension mm × mm × mm: _____

HEPA filter efficiency: _____

Location: _____

Test date: _____

Std. equipment used and no.: _____   Std. calib. valid up to:_____

Scale of photometer: _____

Penetration (%)

Acceptance criteria:

- For HEPA filter of 99.997% and ≤ 0.003%
- For HEPA filter of 99.99% and more ≤ 0.01%
- For HEPA filter of 99.97% ≤ 0.03%
- For HEPA filter of 95% ≤ 5%
- For HEPA filter of 99.999% ≤ 0.001%
- For HEPA filter of 99.995 ≤ 0.005

Remarks:

_____

_____

Conclusion:

_____

_____

Performed by: _____   Date: _____

Checked by: _____   Date: _____

YOUR COMPANY
VALIDATION STANDARD OPERATING PROCEDURE

SOP No.: Val. 2100.30   Effective date: mm/dd/yyyy

                                          Approved by:

TITLE: Temperature Control Test Procedure

AUTHOR: _____

                      Name/Title/Department

               _____

                        Signature/Date

CHECKED BY: _____

                      Name/Title/Department

               _____

                        Signature/Date

APPROVED BY: _____

                      Name/Title/Department

               _____

                        Signature/Date

REVISIONS:

No.	Section	Pages	Initials/Date

SOP No.: Val. 2100.30

Effective date: mm/dd/yyyy

Approved by:

SUBJECT: Temperature Control Test Procedure

PURPOSE

The purpose is to demonstrate the capability of the HVAC system to control temperature.

RESPONSIBILITY

It is the responsibility of validation team members to follow the procedures The quality assurance (QA) manager is responsible for SOP compliance

PROCEDURE

- Complete the air balancing of the area.
- Measure the temperature at specified worst locations under "at-rest" and "dynamic" conditions.
- Operate the system 6 hours prior to the test.

1. Test Apparatus

- Calibrated Dickson temperature and humidity recorder or equivale
- Calibrated thermo-hygro recorder or equivalent
- Calibrated data logger

Note: Select test apparatus from 1.1 to 1.3 for data collection.

1.1 Dickson chart recorder

- Place the Dickson recorder in particular room after fixing ne chart on it, for 3 days' continuous monitoring.
- After completion of 3 days, remove the chart from recorder.
- Transfer the recorded data in a working format.
- Keep the chart in a file for reference.

SOP No.: Val. 2100.30   Effective date: mm/dd/yyyy

                                       Approved by:

1.2 Thermo-hygro meter

- Place the thermo-hygro meter in the specific location after date, time intervals, setting printing options for 3 days.
- After completion of 3 days, take printout from hygro meter.
- Note down the reading in a working format.
- Keep printout in a file for reference.

1.3 Data logger

- Place the data logger in a particular location after loading the start time and end time through computer software.
- After completion of 3 days of data, remove from the area and download the data by computer software and print.
- Note down the data in a working format.
- Keep the printout in a file for reference.

2. Documentation

Make entries in attachment no. 2100.30.

3. Acceptance Criteria

The room temperature should be in compliance with the room requirements, per validation master plan, at rest and dynamic conditions.

REASONS FOR REVISION

Effective date: mm/dd/yyyy

- First time issued for your company, affiliates, and contract manufacturers.

SOP No.: Val. 2100.30

Effective date: mm/dd/yyyy

Approved by:

Attachment No. 2100.30

COMPANY NAME

Issue Date: mm/dd/yyyy
Revision No.: New

Equipment type: _____

Equipment serial no.: _____

Equipment calibrated on: _____

Next calibration due: _____

Test starting date: _____

Test completion date: _____

Location/area: _____

		Condition		Temperature		Within Tolerance	
Date	Time	Rest	Dynamic	Maximum	Minimum	Yes	No

Remarks:

_____

_____

Conclusion:

_____

_____

Performed by: _____ Date: _____

Checked by: _____ Date: _____

YOUR COMPANY
VALIDATION STANDARD OPERATING PROCEDURE

SOP No.: Val. 2100.40

Effective date: mm/dd/yyyy

Approved by:

TITLE: Airflow and Uniformity Test Procedure

AUTHOR: _____

Name/Title/Department

_____

Signature/Date

CHECKED BY: _____

Name/Title/Department

_____

Signature/Date

APPROVED BY: _____

Name/Title/Department

_____

Signature/Date

REVISIONS:

No.	Section	Pages	Initials/Date

SOP No.: Val. 2100.40

Effective date: mm/dd/yyyy

Approved by:

SUBJECT: Airflow and Uniformity Test Procedure

PURPOSE

The purpose is to demonstrate that the air system is balanced and capable of delivering sufficient air volumes to maintain a minimum cross-sectional velocity under the absolute terminal/filter modules measured 6 in. downstream of the filters.

RESPONSIBILITY

It is the responsibility of validation team members to follow the procedures. The quality assurance (QA) manager is responsible for SOP compliance.

PROCEDURE

1. Equipment

Hot-wire anemometer and stand

2. Method

- Specify the location of absolute terminal filters modules installed.
- Draw a grid on the floor as indicated in the room diagram.
- Measure and document the velocity at the center of each grid the specified above the obstacle or a surface 40 in. above the floor.
- Allow no objects within 10 ft of the anemometer, except for built-in equipment. Document the number of people during the "at-rest" testing.
- Take the measurement for a minimum of 15 seconds.
- Record the pressure readings (in inches) from the manometer connected to the module's plenum.

3. Acceptance Criteria

Average measured clean air velocity shall be according to designed standard specified in the validation master plan at 6 in. downstream from filter face. Velocity differences within the same plenum should be no more than 25%.

SOP No.: Val. 2100.40

Effective date: mm/dd/yyyy
Approved by:

4. Documentation

Make entries in attachment no. 2100.40.

REASONS FOR REVISION

Effective date: mm/dd/yyyy

▪ First time issued for your company, affiliates, and contract manufacturers.

Effective date: mm/dd/yyyy

Approved by:

Attachment No. 2100.40

AIR FLOW UNIFORMITY TEST PROCEDURE
Company Name

Issue Date: mm/dd/yyyy
Revision No.: New

Equipment name/type: _____

Equipment serial no.: _____

Equipment calibrated on: _____

Next calibration due on: _____

Test performed on: _____

Area	Room No.	Reference Drawing No.	Filter Serial No.	Grid No.	Height of Sampling	Velocity	Within Tolerance	
							Yes	No

Acceptance criteria:

Average measured clean air velocity shall be according to designed standard specified in the validation master plan at 6 in. downstream from the filter face. Velocity differences within the same plenum should be no more than 25%.

Remarks:

_____

_____

Conclusion:

_____

_____

Performed by: _____ Date: _____

Checked by: _____ Date: _____

YOUR COMPANY
VALIDATION STANDARD OPERATING PROCEDURE

SOP No.: Val. 2100.50   Effective date: mm/dd/yyyy
                                         Approved by:

TITLE: Pressure Control Test Procedure

AUTHOR: _____
                        Name/Title/Department

              _____
                          Signature/Date

CHECKED BY: _____
                        Name/Title/Department

              _____
                          Signature/Date

APPROVED BY: _____
                        Name/Title/Department

              _____
                          Signature/Date

REVISIONS:

No.	Section	Pages	Initials/Date

SOP No.: Val. 2100.50   Effective date: mm/dd/yyyy

Approved by:

SUBJECT: Pressure Control Test Procedure

PURPOSE

The purpose is to demonstrate the capability of the system to control pressure levels within the specified limits.

RESPONSIBILITY

It is the responsibility of validation team members to follow the procedures. The quality assurance (QA) manager is responsible for SOP compliance.

PROCEDURE

1. Equipment

Pressure gauge with resolution of 0.01 in. of water

2. Method

- All HVAC and laminar flow systems are to be in continuous oper ation when performing these tests at least for a period of 6 hours
- To avoid unexpected changes in pressure and to establish a base line, all doors in the sterile facility must be closed and no traff is to be allowed through the facility during the test.
- Pressure readings are taken with the high- and low-pressure tubin at the each room.

3. Documentation

Make entries in attachment no. 2100.50.

4. Acceptance Criteria

- Pressure differentials should be as indicated in the design con tions at all times under static conditions.
- Pressure differentials should be maintained as indicated in t design conditions under standard simulated operating conditio

SOP No.: Val. 2100.50   **Effective date: mm/dd/yyyy**
                                             Approved by:

- Pressure differentials should be above 0.02 in. at the primary environments when stress conditions occur.
- The system will not be acceptable if, at any time during normal dynamic, static, or stress conditions, the pressure in the primary environments becomes less than zero or negative.

REASONS FOR REVISION

Effective date: mm/dd/yyyy

- First time issued for your company, affiliates, and contract manufacturers.

SOP No.: Val. 2100.50

Effective date: mm/dd/yyyy
Approved by:

Attachment No. 2100.50

COMPANY NAME

Issue Date: mm/dd/yyyy
Revision No.: New

Equipment type: _____

Equipment serial no.: _____

Equipment calibrated on: _____

Next calibration due on: _____

Room no./location/area: _____

Date	Time	Location	Tolerance Limit	Pressure

Remarks:

_____

_____

Conclusion:

_____

_____

Performed by: _____   Date: _____

Checked by: _____   Date: _____

YOUR COMPANY
VALIDATION STANDARD OPERATING PROCEDURE

SOP No.: Val. 2100.60   Effective date: mm/dd/yyyy

                                         Approved by:

TITLE: Particulate Count Test Procedure

AUTHOR: _____

                        Name/Title/Department

              _____

                        Signature/Date

CHECKED BY: _____

                        Name/Title/Department

              _____

                        Signature/Date

APPROVED BY: _____

                        Name/Title/Department

              _____

                        Signature/Date

REVISIONS:

No.	Section	Pages	Initials/Date

SOP No.: Val. 2100.60

Effective date: mm/dd/yyyy

Approved by:

SUBJECT: Particulate Count Test Procedure

PURPOSE

The purpose is to establish that, at critical work locations within clean-rooms, a count of less than 100 particles per cubic foot of air, 0.5 μm in diameter or larger, is maintained.

RESPONSIBILITY

It is the responsibility of validation team members to follow the procedures. The quality assurance (QA) manager is responsible for SOP compliance.

PROCEDURE

1. Equipment

CI-500 laser particulate counter

2. Method

- Complete the HEPA filter leak tests and air velocity tests at each location and room, respectively
- Perform the following tests with at-rest condition, without operational personnel, and equipment switched off.
  - Count particles greater than or equal to 0. 5 μm in diameter heights of 40 in. in the center of each grid.
  - If the particle count in the 0.5-μm range is less than 50 per cubic foot of air, four additional counts at this location are taken to place these particle counts within a 50% confidence interval
- Repeat the second step with operator present and the fill equipment running. If at any time there is a deviation from accepted parameters, the components of the systems in operation shall be reviewed, repaired, or adjusted until the desired conditions are achieved.

3. Documentation

Make entries in attachment no. 2100.60.

SOP No.: Val. 2100.60

Effective date: mm/dd/yyyy

Approved by:

4. Acceptance Criteria

■ The air system can be considered validated when the results of three consecutive sets of tests are within accepted operational parameters.

■ At any of the designated critical locations, a critical location being where any sterilized product or material is exposed to the working environment, the particulate count shall not exceed 100 particles 0.5-μm in diameter and larger per cubic foot of air.

■ The same test should be repeated at ancillary environments. Ancillary environment shall not exceed particle count of 10,000 and 100,000 particles 0.5-μm in diameter and larger per cubic foot of air in order to be considered acceptable by current regulations, in cascade order respectively.

Particle Matter — Action Levels

| | Maximum Allowable Number of Particles (Per Cubic Foot of Air) | | | |
| | Action Limit | | Alert Limit | |
Class	0.5 μm	5.0 μm	0.5 μm	5.0 μm
Class 100	100	0	50	0
Class 1,000	1,000	7	500	4
Class 10,000	10,000	70	5,000	35
Class 100,000	100,000	700	50,000	350

REASONS FOR REVISION

Effective date: mm/dd/yyyy

■ First time issued for your company, affiliates, and contract manufacturers.

SOP No.: Val. 2100.60

Effective date: mm/dd/yyyy
Approved by:

Attachment No. 2100.60

PARTICULATE COUNT TEST PROCEDURE

Issue Date: mm/dd/yyyy
Revision No.: New

Equipment name or type: _____

Equipment serial no.: _____

Equipment calibrated on: _____

Next calibration due on: _____

Room No.	Area	Location	Class	Frequency	5.0 µm			5.0 µm		
					Avg.	Max.	Min.	Avg.	Max.	Min.

Acceptance Criteria

	Max. Allowable Number of Particles (per Cubic Foot of Air)			
	Action Limit		Alert Limit	
Class	0.5 µm	5.0 µm	0.5 µm	5.0 µm
100	100	0	50	0
1,000	1,000	7	500	4
10,000	10,000	70	5,000	35
100,000	100,000	700	50,000	350

Remarks:

_____

_____

Conclusion:

_____

_____

Performed by: _____   Date: _____

Checked by: _____   Date: _____

YOUR COMPANY
VALIDATION STANDARD OPERATING PROCEDURE

SOP No.: Val. 2100.70   Effective date: mm/dd/yyyy
                                               Approved by:

TITLE: Humidity Control Test Procedure

AUTHOR: _____
                        Name/Title/Department

                 _____
                        Signature/Date

CHECKED BY: _____
                        Name/Title/Department

                 _____
                        Signature/Date

APPROVED BY: _____
                        Name/Title/Department

                 _____
                        Signature/Date

REVISIONS:

No.	Section	Pages	Initials/Date

SOP No.: Val. 2100.70   Effective date: mm/dd/yyyy

Approved by:

SUBJECT: Humidity Control Test Procedure

PURPOSE

The purpose is to demonstrate the capability of the air handling system to control humidity at the specified level for each room.

RESPONSIBILITY

It is the responsibility of validation team members to follow the procedures. The quality assurance (QA) manager is responsible for SOP compliance.

PROCEDURE

1. Equipment

- Calibrated Dickson relative humidity and temperature recorder
- Psychrometer (dry-bulb and wet-bulb thermometers)

2. Method

- Complete the air-balancing procedures.
- Operate the system for at least 6 hours prior to the start of the te.
- Measure and record humidities for the specified locations for eve room under "at-rest" and dynamic conditions.
- Measure and record the humidity at 1-hour intervals for the peri of 8 hours/day (at least 3 days) at each of the indicated locatio for each room (see attached reporting form).

3. Documentation:

Make entries in attachment no. 2100.70.

4. Acceptance Criteria

The relative humidity at each grid point shall be within the specified lev and tolerance limits indicated for each room in the validation master p If these levels are attained, the system is accepted.

SOP No.: Val. 2100.70   **Effective date: mm/dd/yyyy**
                                          Approved by:

REASONS FOR REVISION

Effective date: mm/dd/yyyy

■ First time issued for your company, affiliates, and contract manufacturers.

SOP No.: Val. 2100.70

Effective date: mm/dd/yyyy

Approved by:

Attachment No. 2100.70

HUMIDITY CONTROL TEST PROCEDURE
Company Name

Issue Date: mm/dd/yyyy
Revision No.: New

Equipment type: _____

Equipment serial no.: _____

Equipment calibrated on: _____

Next calibration due on: _____

Test starting date: _____

Test completion date: _____

				Humidity		Within Tolerance	
Date	Time	Area	Room No.	Maximum	Minimum	Yes	No

Remarks:

_____

_____

Conclusion:

_____

_____

Performed by: _____ Date: _____

Checked by: _____ Date: _____

| YOUR COMPANY |
| VALIDATION STANDARD OPERATING PROCEDURE |

SOP No.: Val. 2100.80 — Effective date: mm/dd/yyyy

Approved by:

TITLE: Recovery Test Procedure

AUTHOR: _____

Name/Title/Department

_____

Signature/Date

CHECKED BY: _____

Name/Title/Department

_____

Signature/Date

APPROVED BY: _____

Name/Title/Department

_____

Signature/Date

REVISIONS:

No.	Section	Pages	Initials/Date

SOP No.: Val. 2100.80   Effective date: mm/dd/yyyy
                                         Approved by:

SUBJECT: Recovery Test Procedure

PURPOSE

The purpose is to determine the capabilities of the system to recover from internally generated contamination.

RESPONSIBILITY

It is the responsibility of validation team members to follow the procedures. The quality assurance (QA) manager is responsible for SOP compliance.

PROCEDURE

1. Equipment

- Visual smoke generator
- Particle counter

2. Method

- Generate smoke for 1 to 2 minutes and shut off the smoke tub at a predesignated location.
- After 2 minutes, advance the sample tube of the particle counter t a point directly under the smoke source and at the level of the wor zone. Record the particle count. If it is not 100/ft^3 or less, repe: the test with the wait interval increased in steps of 1/2 minute un' counts are less than 100 ft^3.
- Repeat for all grid areas, recording recovery time for each grid are

3. Documentation

Make entries in attachment no. 2100.80.

4. Acceptance Criteria

The recovery time should be not more than 2 minutes or as specified individual work zone in the validation master plan.

SOP No.: Val. 2100.80 **Effective date: mm/dd/yyyy**

Approved by:

REASONS FOR REVISION

Effective date: mm/dd/yyyy

■ First time issued for your company, affiliates, and contract manufacturers.

SOP No.: Val. 2100.80   Effective date: mm/dd/yyyy

Approved by:

Attachment No. 2100.80

RECOVERY TEST PROCEDURE
Company Name

Issue Date: mm/dd/yyyy
Revision No.: New

Equipment type: _____

Type of smoke stick: _____

Equipment calibrated on: _____

Next calibration due on: _____

Test performed date: _____

Grid Area	Smoke Generation Time	Particulate Count	Meeting Acceptance Criteria (Recovery Time)

Acceptance criteria:

The recovery time should be not more than 2 min or as specified in the validation master plan.

Remarks:

_____

_____

Conclusion:

_____

_____

Performed by: _____ Date: _____

Checked by: _____ Date: _____

YOUR COMPANY
VALIDATION STANDARD OPERATING PROCEDURE

SOP No.: Val. 2100.90   Effective date: mm/dd/yyyy
                                         Approved by:

TITLE: Particulate Dispersion Test Procedure

AUTHOR: _____
                    Name/Title/Department

        _____
                    Signature/Date

CHECKED BY: _____
                    Name/Title/Department

        _____
                    Signature/Date

APPROVED BY: _____
                    Name/Title/Department

        _____
                    Signature/Date

REVISIONS:

No.	Section	Pages	Initials/Date

SOP No.: Val. 2100.90   Effective date: mm/dd/yyyy
Approved by:

SUBJECT: Particulate Dispersion Test Procedure

PURPOSE

The purpose is to verify the parallelism of airflow throughout the work zone and the capability of the cleanroom to limit the dispersion.

RESPONSIBILITY

It is the responsibility of validation team members to follow the procedures. The quality assurance (QA) manager is responsible for SOP compliance.

PROCEDURE

1. Equipment

- Visual smoke generator
- Particle counter
- Anemometer

2. Method

- Complete the air velocity uniformity tests.
- Divide the work zone into 2 × 2 ft equal areas.
- Set up the smoke generator, with outlet tube pointing in the directio of airflow and located at the center of a grid area at the wo zone entrance plane.
- If smoke is introduced with air pressure, adjust it to provide smoke outlet velocity equal to the room air velocity at that poin
- Operate the particle counter with the sample tube at the norm work level and at a point remote from the smoke source. Verify t the counter indicates particle concentrations less than 100 partic of 0.5 μm or greater.
- Move the sample tube in toward the smoke source from directions at this level to the point where particle counts sho sudden and rapid rise to high levels (10^6 per cubic foot). T defines the envelope of dispersion away from the smoke sou and demonstrates the airflow parallelism control of the room.
- Repeat for all grid areas. Prepare a diagram showing grid ar and corresponding dispersion envelopes.

SOP No.: Val. 2100.90   **Effective date: mm/dd/yyyy**

                                         Approved by:

3. Documentation

Make entries in attachment no. 2100.90.

4. Acceptance Criteria

The dispersion should not extend beyond 2 ft radically from the point of smoke source; i.e., at 2 ft from the generation point, the particle count should be less than $100/ft^3$ of the 0.5-μm size and larger.

REASONS FOR REVISION

Effective date: mm/dd/yyyy

■ First time issued for your company, affiliates, and contract manufacturers.

SOP No.: Val. 2100.90

Effective date: mm/dd/yyyy

Approved by:

Attachment No. 2100.90

Particulate Dispersion Test Procedure

Issue Date: mm/dd/yyyy

Revision No.: New

Equipment name/type: _____

Type of smoke stick: _____

Equipment calibration on: _____

Next calibration due on: _____

Test performed date: _____

Grid Area No.	Work Zone Entrance Plane	Smoke Generation Time	Acceptance Criteria	Actual Particulate Count	Particulate Count at 2 ft from the Smoke Generation Point

Remarks:

_____

_____

Conclusion:

_____

_____

Performed by: _____   Date: _____

Checked by: _____   Date: _____

YOUR COMPANY
VALIDATION STANDARD OPERATING PROCEDURE

SOP No.: Val. 2100.100

Effective date: mm/dd/yyyy

Approved by:

TITLE: Airflow Pattern Test Procedure

AUTHOR: _____

Name/Title/Department

_____

Signature/Date

CHECKED BY: _____

Name/Title/Department

_____

Signature/Date

APPROVED BY: _____

Name/Title/Department

_____

Signature/Date

REVISIONS:

No.	Section	Pages	Initials/Date

SOP No.: Val. 2100.100

Effective date: mm/dd/yyyy

Approved by:

SUBJECT: Airflow Pattern Test Procedure

PURPOSE

The purpose is to perform the smoke pattern studies of class 100 areas to demonstrate the airflow is not turbulent over the surfaces where the product will be exposed and is unidirectional, and the final movement of air is away from the critical surfaces.

RESPONSIBILITY

It is the responsibility of validation team members to follow the procedures. The quality assurance (QA) manager is responsible for SOP compliance.

1. Equipment

- ■ Smoke sticks
- ■ Smoke stick holder
- ■ Video camera

2. Method

- ■ Verify that the laminar airflow units in class 100 areas are operation.
- ■ Verify that the ventilation and air conditioning systems are operating in balance.
- ■ If the system operates according to the specified operating parameters, generate white visible smoke at the critical locations. *critical location is defined as any area where sterilized product material is exposed to the working environment.*
- ■ Generate white smoke inside and over each component that form part of the line 1 ft just over the work surface. Film or take photographs of the smoke as it travels through each critical area of the machine.
- ■ Smoke should flow through these critical areas. If the air does return (backflows) due to turbulence, the system cannot be accepted and must be rebalanced or adjusted. Slight turbulence due equipment configuration is not significant as long as the air does not return to the critical areas.

SOP No.: Val. 2100.100   Effective date: mm/dd/yyyy

Approved by:

- If the system passes the test in the preceding step, continue to film or take photographs while the smoke is generated; the operator should open the curtained area to perform routine interventions. If the smoke backflows permanently to the critical working area at any point during this operation, procedures must be established to prevent cross contamination and re-entry into these areas. If the smoke resumes direction away from the critical surface, the unit passes. Proceed to the next step.
- Determine whether the generated turbulence can carry contaminants from other areas to critical points of the line. If so, adjust the airflow to ensure a minimum of turbulence and rapid cleaning (covers and diffusers can be used over the filling equipment). If turbulence carries contaminants from any area to the critical areas, the system should be re-evaluated and analyzed in terms of the filling, capping, and laminar-flow equipment.
- If the results of the preceding three steps are unsatisfactory, the laminar-airflow system cannot be validated and the rest of the validation tests should not be carried out until a satisfactory operation has been reached. Otherwise, the system is valid and can be certified.

Acceptance Criteria

- Airflow should be unidirectional over the critical areas.
- The smoke should resume direction away from the critical surfaces and it should not backflow permanently to the critical area at any point during routine interventions.
- The movement of air is always away from the critical work surfaces.

Documentation

- Video tape or CD
- Smoke studies at dynamic conditions; see attachment no. 2100.120
- Validation report

REASONS FOR REVISION

ective date: mm/dd/yyyy

- First time issued for your company, affiliates, and contract manufacturers.

Attachment No. 2100.100

SMOKE STUDIES AT DYNAMIC CONDITIONS

Issue Date: mm/dd/yyyy
Revision No.: New

Area/ Activities	Smoke Pattern just below HEPA Filter				Smoke Pattern at Work Surface						
	Laminar		Turbulent		Turbulence		Smoke Travels away from Critical Work Surface		Cascade Airflow		
	Laminar				Significant	Insignificant					
	Yes	No					Yes	No	Yes	No	

YOUR COMPANY
VALIDATION STANDARD OPERATING PROCEDURE

SOP No.: Val. 2100.110   Effective date: mm/dd/yyyy
                                          Approved by:

TITLE: Critical Sampling Point Determination
           in Cleanroom (Viable Count)

AUTHOR: _____
                          Name/Title/Department

           _____
                          Signature/Date

CHECKED BY: _____
                          Name/Title/Department

             _____
                          Signature/Date

APPROVED BY: _____
                          Name/Title/Department

              _____
                          Signature/Date

REVISIONS:

No.	Section	Pages	Initials/Date

SOP No.: Val. 2100.110

Effective date: mm/dd/yyyy

Approved by:

SUBJECT: Critical Sampling Point Determination in Cleanroom (Viable Count)

PURPOSE

The purpose is to establish that confidence sampling locations of environmental samples are capable of detecting all possible contamination sources within monitored areas.

RESPONSIBILITY

It is the responsibility of the microbiologist concerned to follow the procedure. The quality control (QC) manager is responsible for SOP compliance.

References

- ISO 14644-1 1999 (F)
- Federal Standard 209E
- Guidelines on Environmental Monitoring for Aseptic Facilities. working group of the Scottish, QA, 2004

PROCEDURE

- Choose locations, considering the airflow, criticality, and configuration distributions (determine on the area drawing). The number of sampling point locations shall be less than N_L derived from the following equation:

$$N_L = A,$$

where N_L is the minimum number of sampling locations and A the area in square meters.
- Monitor each location at working height by:
  - Settling plate (for passive air monitoring)
  - RCS air samplers (for dynamic air monitoring)
  - Particulate counter (for total particle count)

The monitoring should cover at least 3 working weeks of different shifts.

SOP No.: Val. 2100.110   **Effective date: mm/dd/yyyy**
                                              Approved by:

■ Statistically treat the data to determine the optimal sampling location that represents the air quality of the tested area.

■ Perform survey of tools, equipment, and other surfaces (wall, floor, etc.) used in the area.

■ Choose locations covering the critical and configuration distributions (determine on the area drawing).

■ Check the cleanliness of cleanroom surfaces' bright light or ultraviolet light for sampling location guiding.

■ Monitor by the sterile swab and contact place for each sampling location.

■ The monitoring should cover at least 3 working weeks of different shifts.

SOP No.: Val. 2100.110

Effective date: mm/dd/yyyy

Approved by:

1. Documentation

Filling Room

Room No.	Conductivity	Total Machine Surface Area (m³)	Samp. Location Reference in Drawing	Ref. Drawing No.	Product Risk		Reference Floor Plan	Ref. to Microbial Count Correlation to Particle Count	No. Shifts		Results of Air Sample (CFU)	Correlated to Particulate Count		Reference to Graphic Presentation Air Sampling Point for 3 Weeks	Final Air Monitoring	
					Yes	No			1	2		C	NC		A	NA

Freeze Dryer Room

Room No.	Conductivity	Total Machine Surface Area (m³)	Samp. Location Reference in Drawing	Ref. Drawing No.	Product Risk		Reference Floor Plan	Ref. to Microbial Count Correlation to Particle Count	No. Shifts		Results of Air Sample (CFU)	Correlated to Particulate Count		Reference to Graphic Presentation Air Sampling Point for 3 Weeks	Final Air Monitoring	
					Yes	No			1	2		C	NC		A	NA

SOP No.: Val. 2100.110

Effective date: mm/dd/yyyy

Approved by:

Room No.	Conductivity	Total Machine Surface Area (m³)	Samp. Location Reference in Drawing	Ref. Drawing No.	Product Risk		Reference Floor Plan	Ref. to Microbial Count Correlation to Particle Count	No. Shifts		Results of Air Sample (CFU)	Correlated to Particulate Count		Reference to Graphic Presentation Air Sampling Point for 3 Weeks	Final Air Monitoring	
					Yes	No			1	2		C	NC		A	NA

Solution Preparation Room

Room No.	Conductivity	Total Machine Surface Area (m³)	Samp. Location Reference in Drawing	Ref. Drawing No.	Product Risk		Reference Floor Plan	Ref. to Microbial Count Correlation to Particle Count	No. Shifts		Results of Air Sample (CFU)	Correlated to Particulate Count		Reference to Graphic Presentation Air Sampling Point for 3 Weeks	Final Air Monitoring	
					Yes	No			1	2		C	NC		A	NA

Notes: Yes: product liable to risk; No: product not liable to risk; 1: First shift; 2: Second shift; C: Correlated; NC: Not correlated; A: Acceptable; NA: Not acceptable.

SOP No.: Val. 2100.110   Effective date: mm/dd/yyyy

Approved by:

REASONS FOR REVISION

Effective date: mm/dd/yyyy

- First time issued for your company, affiliates, and contract manufacturers.

YOUR COMPANY
VALIDATION STANDARD OPERATING PROCEDURE

SOP No.: Val. 2100.120

Effective date: mm/dd/yyyy
Approved by:

TITLE: Critical Sampling Point Determination in Cleanroom (Nonviable Count) by Grid Method

AUTHOR: _____
Name/Title/Department

_____
Signature/Date

CHECKED BY: _____
Name/Title/Department

_____
Signature/Date

APPROVED BY: _____
Name/Title/Department

_____
Signature/Date

REVISIONS:

No.	Section	Pages	Initials/Date

SOP No.: Val. 2100.120

Effective date: mm/dd/yyyy

Approved by:

SUBJECT: Critical Sampling Point Determination in Cleanroom (Nonviable Count) by Grid Method

PURPOSE

The purpose is to provide a guideline to identify sampling points for monitoring particulate count.

RESPONSIBILITY

It is the responsibility of validation team members to follow the procedure. The quality assurance (QA) manager is responsible for SOP compliance.

PROCEDURE

1. Objective

To identify sampling points for the monitoring particulate count (nonviable) by generating data for nonviable particulate matter in class 100 filling and lyophilization room by grid technique.

2. Tests

Checks to be performed as a prerequisite: air velocity. Smoke test study for dead end (if any) or air sweeping efficiency should also be reviewed in concluding the studies for critical points for routine monitoring.

3. Equipment

- Alnor thermo anemometer 8575
- CI-500-003A laser particulate counter

4. Procedure

4.1 Air velocity check

- Make sure that HVAC system and laminar flow units are in operation
- Perform air velocity check in unidirectional area and nonunidirectional area in the following rooms per plant SOP.

SOP No.: Val. 2100.120

Effective date: mm/dd/yyyy

Approved by:

Acceptance Criteria

Room No.	Activity	Unidirectional and Nonunidirectional Airflow Velocity Limit
	Filling	0.35–0.55 m/s
	Lyophilization	0.35–0.55 m/s

4.2 Particulate counts

Divide the unidirectional and nonunidirectional areas into grids according to the federal military standard 209E (see attached drawing). Calculate the number of samples according to the area size under the HEPA filters. Identify sampling points on the drawing with grids.

4.3 Number of samples calculation

4.3.1 Unidirectional laminar flow

The minimum number of sample locations required for verification in a clean zone with unidirectional airflow shall not be less than the lesser of (a) or (b):

(a) Minimum number of samples = $\dfrac{A}{2.32}$

Where A is the area of the entrance plane in square meters.

(b) Maximum number of samples == $A \times \dfrac{64}{\left(10^M\right)^{0.5}}$. For class 100, $M = 3.5$

Where A is the area of the entrance plane in square meters and M is the numerical designation of the class.

Laminar Flow Location	Area (m²)	Minimum No. Samples (a)	Maximum No. Samples (b)	For Study No. Samples	Samples ID on Grid
Ampoule filling					
Vial filling					
Conveyor (in)					
Conveyor (out)					
Lyophilizer					

Notes: LFA = laminar flow unit over ampoule filling machine; LFV = laminar flow unit over vial filling machine; LFCI = laminar flow unit over conveyor (in); LFCO = laminar flow unit over conveyor (out); LFF = laminar flow unit over freeze drying unit.

4.3.2 Nonunidirectional laminar flow

■ The minimum number of sample locations required for verification in a clean zone shall be equal to:

$$== A \times \frac{64}{\left(10^M\right)^{0.5}} \quad M = 3.5 \text{ (class 100)}$$

where A is the floor area of the clean zone in square meters and M is the numerical designation of the class.

Room No.	Activity	Area (m²)	Minimum Sample Points	For Study No. Samples	Sample ID on Grid
	Filling				
	Lyophilization				

■ Place the particulate counter in the center of each grid. Open the particulate counter according to the plant SOP and perform three measurements at each location for 3 consecutive days.
■ The particulate counter should be placed at the height of at least 1 m above the ground level, i.e., working level.

SOP No.: Val. 2100.120   Effective date: mm/dd/yyyy
Approved by:

- Tabulate the data and calculate the average particulate count for each grid.
- Identify the grids where the particulate count is within the acceptance criteria, but with value near the higher side of the acceptance criteria.
- Review the existing data, identify the critical locations in unidirectional and nonunidirectional areas, and prepare the nonviable particulate count monitoring SOP for routine monitoring.
- Provide attachment for data tabulation and analysis, review, and recommendation.

5. Documentation

- Air velocity check record, attachment no. 2100.120(A)
- Unidirectional laminar flow particulate matter check record, attachment no. 2100.120(B)
- Nonunidirectional laminar flow particulate matter check record, attachment no. 2100.120(C)

REASONS FOR REVISION

Effective date: mm/dd/yyyy

- First time issued for your company, affiliates, and contract manufacturers.

Attachment No. 2100.120(A)

AIR VELOCITY CHECK RECORD

-m .	Activity	Acceptance	Result Unidirectional	Non-unidirectional	Result Acceptable (Yes/No)	Performed by/Date	Checked by/Date
	Filling	0.35–0.55 m/s					
	Lyophilization	0.35–0.55 m/s					

SOP No.: Val. 2100.120

Effective date: mm/dd/yyyy

Approved by:

Attachment No. 2100.120(B)

UNIDIRECTIONAL LAMINAR FLOW PARTICULATE MATTER CHECK RECORD

Laminar Flow Location	Samples ID on Grid	Condition at Rest/ at Work	Actual Counts No. Counts 0.5 μm			Actual Counts No. Counts 5.0 μm			Most Critical Location	Performed by/Date	Checked by/Date
			Day 1	Day 2	Day 3	3 Day Avg.	Day 1	Day 2	Day 3	3-Day Avg.			
Ampoule filling												
Vial filling												
Conveyor (in)												
Conveyor (out)												
Lyophilizer												

Abbreviations used: LFA = Laminar Flow Unit over Ampoule Filling Machine; LFCI = Laminar Flow Unit over Conveyor (IN); LFF = Laminar Flow Unit over Freeze Drying Unit; LFV = Laminar Flow Unit over Vial Filling Machine; LFCO = Laminar Flow Unit over Conveyor (OUT)

Acceptance Criteria

	Maximum Allowable Number of Particles (Per Cubic Foot of Air)			
	Action Limit		Alert Limit	
Class	0.5 μm	5.0 μm	0.5 μm	5.0 μm
Class 100	100	0	50	0

SOP No.: Val. 2100.120   **Effective date: mm/dd/yyyy**
                                              Approved by:

Attachment No. 2100.120(C)

NONUNIDIRECTIONAL LAMINAR FLOW PARTICULATE MATTER CHECK RECORD

Laminar Flow Location	Samples ID on Grid	Condition at Rest/ at Work	Actual Counts								Most Critical Location	Performed by/Date	Checked by/Date
			No. Counts 0.5 µm				No. Counts 5.0 µm						
			Day 1	Day 2	Day 3	3-Day Avg.	Day 1	Day 2	Day 3	3-Day Avg.			
Ampoule filling													
Vial filling													
Lyophilization													

Acceptance Criteria

Class	Maximum Allowable Number of Particles (Per Cubic Foot of Air)			
	Action Limit		Alert Limit	
	0.5 µm	5.0 µm	0.5 µm	5.0 µm
Class 1000	1000	7	500	4

SECTION

VAL 2200.00

YOUR COMPANY
VALIDATION STANDARD OPERATING PROCEDURE

SOP No.: Val. 2200.10   Effective date: mm/dd/yyyy

Approved by:

TITLE: Cleaning Validation Protocol of Solution
            Preparation Tank

AUTHOR: _____

            Name/Title/Department

            _____

            Signature/Date

CHECKED BY: _____

            Name/Title/Department

            _____

            Signature/Date

APPROVED BY: _____

            Name/Title/Department

            _____

            Signature/Date

REVISIONS:

No.	Section	Pages	Initials/Date

SOP No.: Val. 2200.10

Effective date: mm/dd/yyyy

Approved by:

SUBJECT: Cleaning Validation Protocol of Solution Preparation Tank

PURPOSE

The purpose is to validate the cleaning procedure and ensure that residues of previous product are removed in accordance with the maximum allowable carryover limit calculated, as well as other tests, i.e., visual inspection, pH, conductivity, and TOC.

RESPONSIBILITY

Quality assurance inspector/machine operator and QC analyst are responsible to follow the procedure. The quality assurance (QA) manager is responsible for SOP compliance.

PROCEDURE

The cleaning validation protocols are applicable to:

- Preparation tank
- Nitrogen-purging dip stick
- Triclamps
- Gaskets
- Propeller blades
- Sprinkle

Verify following documentation requirements

- Availability of equipment cleaning procedure
- Verification of equipment cleaning log book records
- Verification of staff training record

Perform following test functions

- Visual inspection of equipment _____ (provide SOP numb
- pH of the final rinse/swab per method _____ (provide meth number)
- Conductivity of the final rinse/swab per procedure _____ (prov method number)

SOP No.: Val. 2200.10

Effective date: mm/dd/yyyy

Approved by:

- TOC of the final rinse/swab per procedure _____ (provide method number)
- Maximum allowable carryover (MAC) _____ (provide method number)

Sampling criteria for swabs and rinses:

- Samples of the internal surfaces are taken by moistening the swab with a suitable solvent, sampling a 2.5-cm^2 area, and then placing the swab in a test tube containing 10 ml of the solvent (specified for each active material from the analytical test method available in the laboratory or from pharmacoepia).
- It is important to take a representative sample of the area because the results will be calculated for the entire surface area at a later stage.
- Samples of the vessel rinses are collected in 500-ml volumetric flasks after final rinsing of the vessels with purified water as described in the individual equipment cleaning SOP.

Types of samples:

- Collect two rinsing samples from the preparation tank drainage. Sampling will be performed two times, one before the start of cleaning procedure and another at the end of final rinsing water, after completion of cleaning procedure.
- Collect two surface samples from each part of preparation tank per attached figures (provide your machine figures).
- Sampling will be performed two times, one before the start of cleaning procedure and another after completion of cleaning procedure.

Handling of samples:

- After collecting, rinse/swab samples are kept in the refrigerator.
- Analyze samples immediately for pH, conductivity, and TOC.
- Samples of rinses/swabs for the HPLC analysis collected at the time of manufacturing analysis are to be completed within 24 hours of collection.
- HPLC samples should be kept at room temperature for at least 2 hours before testing.

SOP No.: Val. 2200.10 **Effective date: mm/dd/yyyy**

Approved by:

Analysis and documentation:

■ The validated HPLC test method will be used for the determination of chemical residues. The validation of test method will be performed by addition and recovery tests.
■ All analysis results will be recorded in the analysis log book. Printouts and chromatograms will also be attached along with analytical log book for reference.
■ All analysis and data are to be verified by second analyst.

1. Documentation Requirements

■ Availability of cleaning SOP: confirm the availability of respective cleaning SOP in the area.
■ Verification of log book record: check the log book and verify the cleaning time and date and product details. Cleaning procedure and date and time are to be verified from log book.
■ Verification of staff training record: verify the staff trained on cleaning SOP. The training of staff responsible for the equipment cleaning is to be verified from the training record and SOP. Refer to attachment no. 2200.10(D).

2. Test Function and Acceptance Criteria

■ Visual inspection of equipment:
  ■ Test function: after cleaning completion, perform visual inspection of equipment — particularly, difficult to clean areas — and record in the analytical log book.
  ■ Acceptance criteria: the visible internal equipment surface and all critical and difficult-to-clean parts are to be optically free from residue and the color of the final rinse should be comparable to WFI.
■ pH determination:
  ■ Test function: after cleaning, perform pH. Refer to attachment no. 2200.10(D).
  ■ Acceptance criteria: is comparable to WFI pH value of 5 to

SOP No.: Val. 2200.10   **Effective date: mm/dd/yyyy**

                                               Approved by:

- Conductivity:
  - Test function: after final cleaning, perform conductivity at room temperature, i.e., 25°C ± 2. Refer to attachment no. 2200.10(D).
  - Acceptance criteria (NMT 1.0 μS/cm): the conductivity of the final rinse is comparable to WFI and NMT 1.0 μS/cm.
- Total organic carbon (TOC):
  - Test function: after cleaning, perform TOC of the final rinse. Refer to attachment no. 2200.10(D).
  - Acceptance criteria (NMT 500 ppb): the TOC of the final rinse is comparable to WFI and NMT 500 ppb.
- Determination of MAC (maximum allowable carryover):
  - Test function: determine MAC of active material, compare with cleaning-rinse results, and record in the analytical log book.
  - Acceptance criteria: the active ingredient in the final rinse is not detected, equal to, or less than the maximum allowable carryover limit determined.
  - For preparation vessel, MAC of active material in the washing rinses is to be checked by HPLC using suitable validated test method for each active material.

Documentation

- Cleaning SOP
- Log book entries
- Chromatograms
- Final report
- Training records

Attachments list:

- Attachment No. 2200.10(A): Equipment Description Figure and Sampling Locations
- Attachment No. 2200.10(B): Cleaning and Testing Responsibilities
- Attachment No. 2200.10(C): Equipment Cleaning Procedure Validation Report
- Attachment No. 2200.10(D): Sampling Technique
- Attachment No. 2200.10(E): Calculations for Surface Swabs
- Attachment No. 2200.10(F): Calculation for Final Rinse Sample

SOP No.: Val. 2200.10 Effective date: mm/dd/yyyy

Approved by:

REASONS FOR REVISION

Effective date: mm/dd/yyyy

- First time issued for your company, affiliates, and contract manu-facturers.

SOP No.: Val. 2200.10   Effective date: mm/dd/yyyy

Approved by:

Attachment No. 2200.10(A)

PROVIDE EQUIPMENT DESCRIPTION FIGURE

Sampling Locations

Equipment Figure No. 1

Sampling location:

R1 (vessel outlet)

Equipment Figure No. 2

Sampling location:

S1 (nitrogen-purging tube)
S2 (inner surface)
S3 (inner surface)

Equipment Figure No. 3

Sampling location:

S4 (outer surface)

Equipment Figure No. 4

Sampling location (blades):

S5   S6   S7   S8

Preparation tank propeller blades

Equipment Figure No. 5

Sampling location (sprinkle):

S9   S10

Water sprinkle point

Attachment No. 2200.10(B)

CLEANING/TESTING RESPONSIBILITIES

Cleaning/ Testing	Done by	Recorded on	Checked by
Equipment cleaning	Machine operator	Equipment usage/ cleaning log book	Production supervisor
Visual inspection	Cleaning validation officer		
Rinse sample	Machine operator/ cleaning validation officer	Sampling sheets 3, 4	Asst. manager QA
Swab sample	Machine operator/ cleaning validation officer	Sampling sheets 3, 4	Asst. manager QA
pH	Cleaning validation officer/QC analyst	Analytical log book	QA/QC office
Conductivity	Cleaning validation officer/QC analyst	Analytical log book	QA/QC office
TOC	Cleaning validation officer/QC analyst	Analytical log book	QA/QC office
MAC	Cleaning validation officer/QC analyst	Analytical log book	Development manager

Notes: TOC = total organic carbon; MAC = maximum allowable carryover.

SOP No.: Val. 2200.10   **Effective date: mm/dd/yyyy**

                                            Approved by:

Attachment No. 2200.10(C)

EQUIPMENT CLEANING PROCEDURE

VALIDATION REPORT

Equipment name: _____

Calibrated/validated on: _____

Location: _____

Room no.: _____

Last product: _____

B. no. of last prod.: _____

Manufacturing date: _____

Active ingredient: _____

Therapeutic group: _____

Cleaning date: _____

Cleaning SOP no.: _____ Revision no.: _____

Sampling technique: _____

Cleaning sample analysis date: _____ Assay result: _____

Test method reference: _____ Ref. analytical log book: _____

Limit of detection: _____

Next product to be manufactured on same equipment: _____

Safety factor: _____

Attachment No. 2200.10(D)

SAMPLING TECHNIQUE

Process description: _____

Process involved: _____

Final Rinsing Volume

Sampling Location	Sampling Property		Type of Sample		Sample Quantity		Sampling Time		Surface Area in Contact with Product Solution	Final Volume of Rinse	
	D	N	S	R	cm²	500 ml	EC	AC		EC	AC
R1		—		—							
S1		—	—								
S2		—	—								
S3		—	—								
S4		—	—								
S5	—		—								
S6	—		—								
S7	—		—								
S8	—		—								
S9		—	—								
S10		—	—								

Notes: EC: sample collected at the end of cleaning before draining the rinse; A◄ sample collected at the end of final rinsing water; S: swab; R: rinse; ▮ difficult to clean; N: normal.

Training Record Verification

Following staff found trained on cleaning of equipment:

Using SOP no. (provide number): _____

Name: _____ ID no. _____ Sign _____ Date _____

Name: _____ ID no. _____ Sign _____ Date _____

Performed by: _____ Checked by: _____

Date: _____ Date: _____

SOP No.: Val. 2200.10

Effective date: mm/dd/yyyy
Approved by:

Attachment No. 2200.10(E)

CALCULATION FOR SURFACE SWABS

Formula

$$MAC = \frac{TD \times BS \times SF}{LDD}$$

where:
MAC = maximum allowable carryover
TD = therapeutic dose (single) = _____
BS = batch size (next product) = _____
SF = safety factor = _____
LDD = largest daily dose (for next product) _____

MAC _____

MAC = _____ mg

Calculation

Total active recovered from the 2.5-cm^2 surface area of the equipment directly in contact with product solution by swabs (A) _____ (mg)

Total active present over total surface area in direct contact with product solution (B)

_____(A) × _____ (B) = _____(C)

Acceptance Criteria

The total quantity of active (C) determined over the total surface area by using validated test method should be less than or equal to the MAC limit.

Recovery (Challenge) Test

No.	Name of Active Material	Concentration of Standard Solution	Method Used for Testing	% Recovery of Active Ingredient	% Recovery per Limit (95–105%)	
					Y	N
		80%				
		100%				
		120%				

SOP No.: Val. 2200.10

Effective date: mm/dd/yyyy

Approved by:

Results

MAC	Actual Results (C)

Swab Analysis Results

Sampling Location	Visual Inspection	Blank WFI	PH (5–7)	TOC (NMT 1.0 µS/cm)	Conductivity (NMT 500 ppb)
S1					
S2					
S3					
S4					
S5					
S6					
S7					
S8					
S9					
S10					

Deviations: _____

_____

_____

Corrective actions/preventive measures: _____

_____

_____

Conclusion: _____

_____

_____

_____   _____

Validation Officer/Date   Asst. QA Manager/Date

SOP No.: Val. 2200.10   Effective date: mm/dd/yyyy
                                         Approved by:

Attachment No. 2200.10(F)

CALCULATION FOR FINAL RINSE SAMPLE

Formula

$$MAC = \frac{TD \times BS \times SF}{LDD}$$

where:

MAC = maximum allowable carryover
TD = therapeutic dose (single) = _____
BS = batch size (next product) = _____
SF = safety factor = _____
LDD = largest daily dose (for next product) _____

MAC _____

MAC = _____ mg

Calculation

00 ml of the rinse contains active (A) = _____ mg

_____(B) ml of the final rinse contains an active (C) = $\dfrac{A\,(mg) \times B\,(ml)}{500\,ml}$

Acceptance Criteria

The total amount of active (C) determined in the total volume of final rinse should be less than or equal to the MAC limit.

Recovery (Challenge) Test

No.	Name of Active Material	Concentration of Standard Solution	Method Used for Testing	% Recovery of Active Ingredient	% Recovery per Limit (95–105%)	
					Y	N
		80%				
		100%				
		120%				

SOP No.: Val. 2200.10

Effective date: mm/dd/yyyy

Approved by:

Results

MAC	Actual Results (C)

Rinses Analysis Results

Sampling Location	Visual Inspection	Blank WFI	PH (5–7)	TOC (NMT 1.0 µS/cm)	Conductivity (NMT 500 ppb)

Deviations: _____

_____

_____

Corrective actions/preventive measures: _____

_____

_____

Conclusion: _____

_____

_____

_____   _____

Validation Officer/Date   QA Manager/Date

| YOUR COMPANY |
| VALIDATION STANDARD OPERATING PROCEDURE |

SOP No.: Val. 2200.20

Effective date: mm/dd/yyyy

Approved by:

TITLE: Cleaning Validation Protocol of Mobile Tank

AUTHOR: _____

Name/Title/Department

_____

Signature/Date

CHECKED BY: _____

Name/Title/Department

_____

Signature/Date

APPROVED BY: _____

Name/Title/Department

_____

Signature/Date

REVISIONS:

No.	Section	Pages	Initials/Date

SOP No.: Val. 2200.20

Effective date: mm/dd/yyyy

Approved by:

SUBJECT: Cleaning Validation Protocol of Mobile Tank

PURPOSE

The purpose is to validate the cleaning procedure and ensure that residues of previous product are removed in accordance with the maximum allowable carryover limit calculated, as well as other tests, i.e., visual inspection, pH, conductivity, and TOC.

RESPONSIBILITY

Quality assurance inspector/machine operator and QC analyst are responsible to follow the procedure. QA manager is responsible for SOP compliance

PROCEDURE

The cleaning validation protocols are applicable to:

- Mobile tank
- Dip tube
- Triclamps
- Gaskets

Verify following documentation requirements

- Availability of equipment cleaning procedure
- Verification of equipment cleaning log book records
- Verification of staff training record

Perform following test functions

- Visual inspection of equipment _____ (provide SOP number)
- pH of the final rinse/swab per method _____ (provide meth number)
- Conductivity of the final rinse/swab per procedure _____ (provi method number)
- TOC of the final rinse/swab per procedure _____ (provide meth number)
- Maximum allowable carryover (MAC) _____ (prov method number)

SOP No.: Val. 2200.20 Effective date: mm/dd/yyyy

Approved by:

Sampling criteria for swabs and rinses:

■ Samples of the internal surfaces are taken by moistening the swab with a suitable solvent, sampling a 2.5-cm² area, and then placing the swab in a test tube containing 10 ml of the solvent (specified for each active material from the analytical test method available in the laboratory or from pharmacopoeia).

■ It is important to take a representative sample of the area because the results will be calculated for the entire surface area at a later stage.

■ Samples of the vessel rinses are collected in 500-ml volumetric flasks after final rinsing of the vessels with purified water as described in the individual equipment cleaning SOP.

Types of samples:

■ Collect two rinsing samples from the preparation tank drainage. Sampling will be performed two times, one before the start of cleaning procedure and another at the end of final rinsing water, after completion of cleaning procedure.

■ Collect two surface samples from each part of preparation tank per attached figures (provide your mobile tank figures).

■ Sampling will be performed two times, one before the start of cleaning procedure and another after completion of cleaning procedure.

Handling of samples:

■ After collecting, rinse/swab samples are kept in the refrigerator.

■ Analyze samples immediately for pH, conductivity, and TOC.

■ Samples of rinses/swabs for the HPLC analysis collected at the time of manufacturing analysis are to be completed within 24 hours of collection.

■ HPLC samples should be kept at room temperature for at least 2 hours before testing.

Analysis and documentation:

■ The validated HPLC test method will be used for the determination of chemical residues. The validation of test method will be performed by addition and recovery tests.

SOP No.: Val. 2200.20 **Effective date: mm/dd/yyyy**

Approved by:

▪ All analysis results will be recorded in the analysis log book. Printouts and chromatograms will also be attached along with analytical log book for reference.
▪ All analysis and data are to be verified by second analyst.

1. Documentation Requirements

▪ Availability of cleaning SOP: confirm the availability of respective cleaning SOP in the area.
▪ Verification of log book record: check the log book and verify the cleaning time and date and product details. Cleaning procedure and date and time are to be verified from log book.
▪ Verification of staff training record: verify the staff trained on cleaning SOP. The training of staff responsible for the equipment cleaning is to be verified from the training record and SOP. Refer to attachment no. 2200.20(D).

2. Test Function and Acceptance Criteria

▪ Visual inspection of equipment:
  ▪ Test function: after cleaning completion, perform visual inspection of equipment — particularly, difficult to clean areas — and record in the analytical log book.
  ▪ Acceptance criteria: the visible internal equipment surface and all critical and difficult to clean parts are to be optically free from residue and the color of the final rinse should be comparable to WFI.
▪ pH determination:
  ▪ Test function: after cleaning, perform pH. Refer to attachment no. 2200.20(D).
  ▪ Acceptance criteria: is comparable to WFI pH value of 5 to
▪ Conductivity:
  ▪ Test function: after final cleaning, perform conductivity at room temperature, i.e., 25°C ± 2. Refer to attachment no. 2200.20(
  ▪ Acceptance criteria (NMT 1.0 µS/cm): the conductivity of final rinse is comparable to WFI and NMT 1.0 µS/cm.
▪ Total organic carbon (TOC):

SOP No.: Val. 2200.20   Effective date: mm/dd/yyyy

Approved by:

- Test function: after cleaning, perform TOC of the final rinse. Refer to attachment no. 2200.20(D).
- Acceptance criteria (NMT 500 ppb): the TOC of the final rinse is comparable to WFI and NMT 500 ppb.
- Determination of MAC (maximum allowable carryover):
  - Test function: determine MAC of active material, compare with cleaning-rinse results, and record in the analytical log book.
  - Acceptance criteria: the active ingredient in the final rinse is not detected, equal to, or less than the maximum allowable carryover limit determined.
  - For preparation vessel, MAC of active material in the washing rinses is to be checked by HPLC using suitable validated test method for each active material.

Documentation

- Cleaning SOP
- Log book entries
- Chromatograms
- Final report
- Training records

Attachments list:

- Attachment No. 2200.20(A): Equipment Description Figure and Sampling Locations
- Attachment No. 2200.20(B): Cleaning and Testing Responsibilities
- Attachment No. 2200.20(C): Equipment Cleaning Procedure Validation Report
- Attachment No. 2200.20(D): Sampling Technique
- Attachment No. 2200.20(E): Calculations for Surface Swabs
- Attachment No. 2200.20(F): Calculation for Final Rinse Sample

REASONS FOR REVISION

Effective date: mm/dd/yyyy

- First time issued for your company, affiliates, and contract manufacturers.

SOP No.: Val. 2200.20

Effective date: mm/dd/yyyy

Approved by:

Attachment No. 2200.20(A)

PROVIDE EQUIPMENT DESCRIPTION FIGURE

Sampling Locations

Equipment Figure No. 1

Sampling location:

R1 (vessel outlet)

Equipment Figure No. 2

Sampling location:

S1 (inner surface)
S2 (nitrogen-purging tube)
S3 (inner surface)
S4 (inner surface)

SOP No.: Val. 2200.20

Effective date: mm/dd/yyyy
Approved by:

Attachment No. 2200.20(B)

CLEANING/TESTING RESPONSIBILITIES

Cleaning/ Testing	Done by	Recorded on	Checked by
Equipment cleaning	Machine operator	Equipment usage/ cleaning log book	Production supervisor
Visual inspection	Cleaning validation officer		
Rinse sample	Machine operator/ cleaning validation officer	Sampling sheets 3, 4	Asst. manager QA
Swab sample	Machine operator/ cleaning validation officer	Sampling sheets 3, 4	Asst. manager QA
pH	Cleaning validation officer/QC analyst	Analytical log book	QA/QC officer
Conductivity	Cleaning validation officer/QC analyst	Analytical log book	QA/QC officer
TOC	Cleaning validation officer/QC analyst	Analytical log book	QA/QC officer
MAC	Cleaning validation officer/QC analyst	Analytical log book	Development manager

Notes: TOC = total organic carbon; MAC = maximum allowable carryover.

Effective date: mm/dd/yyyy

Approved by:

Attachment No. 2200.20(C)

EQUIPMENT CLEANING PROCEDURE

VALIDATION REPORT

Equipment name: _____

Calibrated/validated on: _____

Location: _____

Room no.: _____

Last product: _____

B. no. of last prod.: _____

Manufacturing date: _____

Active ingredient: _____

Therapeutic group: _____

Cleaning date: _____

Cleaning SOP no.: _____ Revision no.: _____

Sampling technique: _____

Cleaning sample analysis date: _____ Assay result:_____

Test method reference: _____ Ref. analytical log book: _____

Limit of detection: _____

Next product to be manufactured on same equipment: _____

Safety factor: _____

SOP No.: Val. 2200.20

Effective date: mm/dd/yyyy
Approved by:

Attachment No. 2200.20(D)

SAMPLING TECHNIQUE

Process description: _____

Process involved: _____

Final Rinsing Volume

Sampling Location	Sampling Property		Type of Sample		Sample Quantity		Sampling Time		Surface Area in Contact with Product Solution	Final Volume of Rinse	
	D	*N*	*S*	*R*	*cm²*	*500 ml*	*EC*	*AC*		*EC*	*AC*
R1		—		—							
S1		—	—								
S2		—	—								
S3		—	—								
S4		—	—								

Notes: EC: sample collected at the end of cleaning before draining the rinse; AC: sample collected at the end of final rinsing water; S: swab; R: rinse; D: difficult to clean; N: normal.

Training Record Verification

llowing staff found trained on cleaning of equipment:

ing SOP no. (provide number): _____

me: _____ ID no._____ Sign _____Date _____

me: _____ ID no._____ Sign _____Date _____

formed by: _____ Checked by_____

te: _____ Date: _____

Attachment No. 2200.20(E)

CALCULATION FOR SURFACE SWABS

Formula

$$MAC = \frac{TD \times BS \times SF}{LDD}$$

where:
MAC = maximum allowable carryover
TD = therapeutic dose (single) = _____
BS = batch size (next product) = _____
SF = safety factor = _____
LDD = largest daily dose (for next product) _____

MAC _____

MAC = _____ mg

Calculation
Total active recovered from the 2.5-cm² surface area of the equipment directl
in contact with product solution by swabs (A) _____ (mg)

Total active present over total surface area in direct contact with product
solution (B)

_____(A) × _____ (B) = _____(C)

Acceptance Criteria
The total quantity of active (C) determined over the total surface area by usir
a validated test method should be less than or equal to the MAC limit.

Recovery (Challenge) Test

S. No.	Name of Active Material	Concentration of Standard Solution	Method Used for Testing	% Recovery of Active Ingredient	% Recover per Limit (95–105%	
					Y	N
1		80%				
2		100%				
3		120%				

SOP No.: Val. 2200.20

Effective date: mm/dd/yyyy

Approved by:

Results

MAC	Actual Results (C)

Swab Analysis Results

Sampling Location	Visual Inspection	Blank WFI	PH (5–7)	TOC (NMT 1.0 μS/cm)	Conductivity (NMT 500 ppb)
S1					
S2					
S3					
S4					
S5					
S6					
S7					
S8					
S9					
S10					

eviations: _____

_____

_____

orrective actions/preventive measures: _____

_____

nclusion: _____

_____

_____

_____   _____
idation Officer/Date   Asst. Manager/Date

Attachment No. 2200.20(F)

CALCULATION FOR FINAL RINSE SAMPLE

Formula

$$MAC = \frac{TD \times BS \times SF}{LDD}$$

where:
MAC = maximum allowable carryover
TD = therapeutic dose (single) = _____
BS = batch size (next product) = _____
SF = safety factor = _____
LDD = largest daily dose (for next product) _____

MAC _____

MAC = _____ mg

Calculation
500 ml of the rinse contains active (A) = _____ mg

_____(B) ml of the final rinse contains an active (C) = $\dfrac{A\,(mg) \times B\,(ml)}{500\,ml}$

Acceptance criteria
The total amount of active (C) determined in the total volume of final rins
should be less than or equal to the MAC limit.

Recovery (Challenge) Test

S. No.	Name of Active Material	Concentration of Standard Solution	Method Used for Testing	% Recovery of Active Ingredient	% Recove per Limi (95–105% Y	N
1		80%				
2		100%				
3		120%				

SOP No.: Val. 2200.20

Effective date: mm/dd/yyyy

Approved by:

Results

MAC	Actual Results (C)

Rinses Analysis Results

Sampling Location	Visual Inspection	Blank WFI	PH (5–7)	TOC (NMT 1.0 μS/cm)	Conductivity (NMT 500 ppb)

Deviations: _____

_____

_____

Corrective actions/preventive measures: _____

_____

_____

Conclusion: _____

_____

_____

_____   _____

Validation Officer/Date   QA Manager/Date

| YOUR COMPANY |
| VALIDATION STANDARD OPERATING PROCEDURE |

SOP No.: Val. 2200.30

Effective date: mm/dd/yyyy

Approved by:

TITLE: Cleaning Validation Protocol of Filtration Assembly

AUTHOR: _____

Name/Title/Department

_____

Signature/Date

CHECKED BY: _____

Name/Title/Department

_____

Signature/Date

APPROVED BY: _____

Name/Title/Department

_____

Signature/Date

REVISIONS:

No.	Section	Pages	Initials/Date

SOP No.: Val. 2200.30

Effective date: mm/dd/yyyy

Approved by:

SUBJECT: Cleaning Validation Protocol of Filtration Assembly

PURPOSE

The purpose is to validate the cleaning procedure and ensure that residues of previous product are removed in accordance with the maximum allowable carryover limit calculated, as well as other tests, i.e., visual inspection, pH, conductivity, and TOC.

RESPONSIBILITY

Quality assurance inspector/machine operator and QC analyst are responsible to follow the procedure. The quality assurance (QA) manager is responsible for SOP compliance.

PROCEDURE

The cleaning validation protocols are applicable to filtration assembly, including:

- Prefiltration
- Final filtration
- Upstream silicon hose
- Downstream silicon hose

Verify following documentation requirements

- Availability of equipment cleaning procedure
- Verification of equipment cleaning log book records
- Verification of staff training record

Perform following test functions

- Visual inspection of equipment _____ (provide SOP number)
- pH of the final rinse/swab per method _____ (provide method number)
- Conductivity of the final rinse/swab per procedure _____ (provide method number)

SOP No.: Val. 2200.30　　　　　　　　　　**Effective date: mm/dd/yyyy**

　　　　　　　　　　　　　　　　　　　　　　　　　Approved by:

- TOC of the final rinse/swab per procedure _____ (provide method number)
- Maximum allowable carryover (MAC) _____ (provide method number)

Sampling criteria for swabs and rinses:

- Samples of the internal surfaces are taken by moistening the swab with a suitable solvent, sampling a 2.5-cm^2 area, and then placing the swab in a test tube containing 10 ml of the solvent (specified for each active material from the analytical test method available in the laboratory or from pharmacopoeia).
- It is important to take a representative sample of the area because the results will be calculated for the entire surface area at a later stage
- Samples of the vessel rinses are collected in 500-ml volumetric flasks after final rinsing of the vessels with purified water as described in the individual equipment cleaning SOP.

Types of samples:

- Collect two rinsing samples from the preparation tank drainage. Sampling will be performed two times, one before the start cleaning procedure and another at the end of final rinsing water after completion of cleaning procedure.
- Collect two surface samples from each part of preparation tank per attached figures (provide your company's filtration assembly photographs).
- Sampling will be performed two times, one before the start of cleaning procedure and another after completion of cleaning procedure

Handling of samples:

- After collecting, rinse/swab samples are kept in the refrigerator
- Analyze samples immediately for pH, conductivity, and TOC.
- Samples of rinses/swabs for the HPLC analysis collected at the time of manufacturing analysis are to be completed within 24 hours collection.
- HPLC samples should be kept at room temperature for at least 2 hours before testing.

SOP No.: Val. 2200.30

Effective date: mm/dd/yyyy

Approved by:

Analysis and documentation:

- The validated HPLC test method will be used for the determination of chemical residues. The validation of test method will be performed by addition and recovery tests.
- All analysis results will be recorded in the analysis log book. Printouts and chromatograms will also be attached along with analytical log book for reference.
- All analysis and data are to be verified by second analyst.

◀. Documentation Requirements

- Availability of cleaning SOP: confirm the availability of respective cleaning SOP in the area.
- Verification of log book record: check the log book and verify the cleaning time and date and product details. Cleaning procedure and date and time are to be verified from log book.
- Verification of staff training record: verify the staff trained on cleaning SOP. The training of staff responsible for the equipment cleaning is to be verified from the training record and SOP. Refer to attachment no. 2200.30(D).

Test Function and Acceptance Criteria

- Visual inspection of equipment:
  - Test function: after cleaning completion, perform visual inspection of equipment — particularly, difficult-to-clean areas — and record in the analytical log book.
  - Acceptance criteria: the visible internal equipment surface and all critical and difficult-to-clean parts are to be optically free from residue and the color of the final rinse should be comparable to WFI.
- pH determination:
  - Test function: after cleaning, perform pH. Refer to attachment no. 2200.30(D).
  - Acceptance criteria: is comparable to WFI pH value of 5 to 7.

SOP No.: Val. 2200.30 **Effective date: mm/dd/yyyy**

Approved by:

- Conductivity:
  - Test function: after final cleaning, perform conductivity at room temperature, i.e., 25°C ± 2. Refer to attachment no. 2200.30(D).
  - Acceptance criteria (NMT 1.0 µS/cm): the conductivity of the final rinse is comparable to WFI and NMT 1.0 µS/cm.
- Total organic carbon (TOC):
  - Test function: after cleaning, perform TOC of the final rinse Refer to attachment no. 2200.30(D).
  - Acceptance criteria (NMT 500 ppb): the TOC of the final rinse is comparable to WFI and NMT 500 ppb.
- Determination of MAC (maximum allowable carryover):
  - Test function: determine MAC of active material, compare wit cleaning-rinse results, and record in the analytical log book.
  - Acceptance criteria: the active ingredient in the final rinse is no detected, equal to, or less than the maximum allowable carryove limit determined.
  - For preparation vessel, MAC of active material in the washin rinses is to be checked by HPLC using suitable validated te method for each active material.

3. Documentation

- Cleaning SOP
- Log book entries
- Chromatograms
- Final report
- Training records

Attachments list:

- Attachment No. 2200.30(A): Equipment Description Figure Sampling Locations
- Attachment No. 2200.30(B): Cleaning and Testing Responsibili
- Attachment No. 2200.30(C): Equipment Cleaning Procedure \ dation Report
- Attachment No. 2200.30(D): Sampling Technique
- Attachment No. 2200.30(E): Calculation for Final Rinse Sample

SOP No.: Val. 2200.30   **Effective date: mm/dd/yyyy**
                                         Approved by:

REASONS FOR REVISION

Effective date: mm/dd/yyyy

- First time issued for your company, affiliates, and contract manufacturers.

Effective date: mm/dd/yyyy
                                                      Approved by:

Attachment No. 2200.30(A)

PROVIDE EQUIPMENT DESCRIPTION FIGURE

Sampling Locations

Equipment Figure No. 1

Sampling location: prefiltration unit

R1 (vessel outlet)

Equipment Figure No. 2

Sampling location: postfiltration unit

R2 (vessel outlet)

Equipment Figure No. 3

Sampling location: upstream silicon hose

R3

Equipment Figure No. 4

Sampling location: downstream silicon hose

R4

SOP No.: Val. 2200.30

Effective date: mm/dd/yyyy
Approved by:

Attachment No. 2200.30(B)

CLEANING/TESTING RESPONSIBILITIES

Cleaning/ Testing	Done by	Recorded on	Checked by
Equipment cleaning	Machine operator	Equipment usage/ cleaning log book	Production supervisor
Visual inspection	Cleaning validation officer		
Rinse sample	Machine operator/ cleaning validation officer	Sampling sheets 3, 4	Asst. manager QA
Swab sample	Machine operator/ cleaning validation officer	Sampling sheets 3, 4	Asst. manager QA
H	Cleaning validation officer/QC analyst	Analytical log book	QA/QC officer
onductivity	Cleaning validation officer/QC analyst	Analytical log book	QA/QC officer
OC	Cleaning validation officer/QC analyst	Analytical log book	QA/QC officer
AC	Cleaning validation officer/QC analyst	Analytical log book	Development manager

otes: TOC = total organic carbon; MAC = maximum allowable carryover.

SOP No.: Val. 2200.30   Effective date: mm/dd/yyyy

Approved by:

Attachment No. 2200.30(C)

EQUIPMENT CLEANING PROCEDURE

VALIDATION REPORT

Equipment name: _____

Calibrated/validated on: _____

Location: _____

Room no.: _____

Last product: _____

B. no. of last prod.: _____

Manufacturing date: _____

Active ingredient: _____

Therapeutic group: _____

Cleaning date: _____

Cleaning SOP no.: _____ Revision no.: _____

Sampling technique: _____

Cleaning sample analysis date: _____ Assay result:_____

Test method reference: _____ Ref. analytical log book:_____

Limit of detection: _____

Next product to be manufactured on same equipment: _____

Safety factor: _____

SOP No.: Val. 2200.30

Effective date: mm/dd/yyyy

Approved by:

Attachment No. 2200.30(D)

SAMPLING TECHNIQUE

Process description: _____

Process involved: _____

Final Rinsing Volume

Sampling Location	Sampling Property		Type of Sample		Sample Quantity		Sampling Time		Surface Area in Contact with Product Solution	Final Volume of Rinse	
	D	N	S	R	cm²	500 ml	EC	AC		EC	AC
R1	—		—								
R2	—		—								
R3	—		—								
R4	—		—								

Notes: EC: sample collected at the end of cleaning before draining the rinse; AC: sample collected at the end of final rinsing water; S: swab; R: rinse; D: difficult to clean; N: normal.

Training Record Verification

Following staff found trained on cleaning of equipment:

Using SOP no. (provide number): _____

Name: _____ ID no. _____ Sign_____Date _____

Name: _____ ID no. _____ Sign_____Date _____

Performed by: _____ Checked by_____

Date: _____ Date: _____

SOP No.: Val. 2200.30

Effective date: mm/dd/yyyy
Approved by:

Attachment No. 2200.30(E)

CALCULATION FOR FINAL RINSE SAMPLE

Formula

$$MAC = \frac{TD \times BS \times SF}{LDD}$$

where:

MAC = maximum allowable carryover
TD = therapeutic dose (single) = _____
BS = batch size (next product) = _____
SF = safety factor = _____
LDD = largest daily dose (for next product) _____

MAC _____

MAC = _____ mg

Calculation
500 ml of the rinse contains active (A) = _____ mg

_____(B) ml of the final rinse contains an active (C) = $\dfrac{A\,(mg) \times B\,(ml)}{500\,ml}$

Acceptance Criteria
The total amount of active (C) determined in the total volume of final rins
should be less than or equal to the MAC limit.

Recovery (Challenge) Test

S. No.	Name of Active Material	Concentration of Standard Solution	Method Used for Testing	% Recovery of Active Ingredient	% Recove per Limi (95–105%	
					Y	
1		80%				
2		100%				
3		120%				

SOP No.: Val. 2200.30

Effective date: mm/dd/yyyy

Approved by:

Results

MAC	Actual Results (C)

Rinses Analysis Results

Sampling Location	Visual Inspection	Blank WFI	PH (5–7)	TOC (NMT 1.0 μS/cm)	Conductivity (NMT 500 ppb)

Deviations: _____

_____

_____

Corrective actions/preventive measures: _____

_____

_____

Conclusion: _____

_____

_____

_____   _____

Validation Officer/Date   QA Manager/Date

YOUR COMPANY
VALIDATION STANDARD OPERATING PROCEDURE

SOP No.: Val. 2200.40

Effective date: mm/dd/yyyy

Approved by:

TITLE: Cleaning Validation Protocol of Freeze Dryer

AUTHOR: _____

Name/Title/Department

_____

Signature/Date

CHECKED BY: _____

Name/Title/Department

_____

Signature/Date

APPROVED BY: _____

Name/Title/Department

_____

Signature/Date

REVISIONS:

No.	Section	Pages	Initials/Date

SOP No.: Val. 2200.40

Effective date: mm/dd/yyyy

Approved by:

SUBJECT: Cleaning Validation Protocol of Freeze Dryer

PURPOSE

The purpose is to validate the cleaning procedure and ensure that residues of previous product are removed in accordance with the maximum allowable carryover limit calculated, as well as other tests, i.e., visual inspection, pH, conductivity, and TOC.

RESPONSIBILITY

Quality assurance inspector/machine operator and QC analyst are responsible to follow the procedure. The quality assurance (QA) manager is responsible for SOP compliance.

PROCEDURE

The cleaning validation protocols are applicable to the freeze dryer chamber, including:

- Walls
- Shelves (provide number)
- Bottom

Verify following documentation requirements

- Availability of equipment cleaning procedure
- Verification of equipment cleaning log book records
- Verification of staff training record

Perform following test functions

- Visual inspection of equipment _____ (provide SOP number)
- pH of the final rinse/swab per method _____ (provide method number)
- Conductivity of the final rinse/swab per procedure _____ (provide method number)

SOP No.: Val. 2200.40   **Effective date: mm/dd/yyyy**
                                                       Approved by:

- TOC of the final rinse/swab per procedure _____ (provide method number)
- Maximum allowable carryover (MAC) _____ (provide method number)

Sampling criteria for swabs and rinses:

- Samples of the internal surfaces are taken by moistening the swab with a suitable solvent, sampling a 2.5-cm^2 area, and then placing the swab in a test tube containing 10 ml of the solvent (specified for each active material from the analytical test method available in the laboratory or from pharmacopoeia).
- It is important to take a representative sample of the area because the results will be calculated for the entire surface area at a later stage
- Samples of the vessel rinses are collected in 500-ml volumetric flask after final rinsing of the vessels with purified water as described in the individual equipment cleaning SOP.

Types of samples:

- Collect two rinsing samples from the preparation tank drainage Sampling will be performed two times, one before the start cleaning procedure and another at the end of final rinsing water after completion of cleaning procedure.
- Collect two surface samples from each part of preparation tank per attached figures (provide photographs of your company freeze dryer).
- Sampling will be performed two times, one before the start of clean procedure and another after completion of cleaning procedure

Handling of samples:

- After collecting, rinse/swab samples are kept in the refrigerato
- Analyze samples immediately for pH, conductivity, and TOC.
- Samples of rinses/swabs for the HPLC analysis collected at the t of manufacturing analysis are to be completed within 24 hour collection.
- HPLC samples should be kept at room temperature for at l 2 hours before testing.

SOP No.: Val. 2200.40   Effective date: mm/dd/yyyy

Approved by:

Analysis and documentation:

- The validated HPLC test method will be used for the determination of chemical residues. The validation of test method will be performed by addition and recovery tests.
- All analysis results will be recorded in the analysis log book. Printouts and chromatograms will also be attached along with analytical log book for reference.
- All analysis and data are to be verified by second analyst.

1. Documentation Requirements

- Availability of cleaning SOP: confirm the availability of respective cleaning SOP in the area.
- Verification of log book record: check the log book and verify the cleaning time and date and product details. Cleaning procedure and date and time are to be verified from log book.
- Verification of staff training record: verify the staff trained on cleaning SOP. The training of staff responsible for the equipment cleaning is to be verified from the training record and SOP. Refer to attachment no. 2200.40(D).

Test Function and Acceptance Criteria

- Visual inspection of equipment:
  - Test function: after cleaning completion, perform visual inspection of equipment — particularly, difficult-to-clean areas — and record in the analytical log book.
  - Acceptance criteria: the visible internal equipment surface and all critical and difficult-to-clean parts are to be optically free from residue and the color of the final rinse should be comparable to WFI.
- pH determination:
  - Test function: after cleaning, perform pH. Refer to attachment no. 2200.40(D).
  - Acceptance criteria: is comparable to WFI pH value of 5 to 7.

SOP No.: Val. 2200.40   Effective date: mm/dd/yyyy

Approved by:

- Conductivity:
  - Test function: after final cleaning, perform conductivity at room temperature, i.e., 25°C ± 2. Refer to attachment no. 2200.40(D).
  - Acceptance criteria (NMT 1.0 µS/cm): the conductivity of the final rinse is comparable to WFI and NMT 1.0 µS/cm.
- Total organic carbon (TOC):
  - Test function: after cleaning, perform TOC of the final rinse. Refer to attachment no. 2200.40(D).
  - Acceptance criteria (NMT 500 ppb): the TOC of the final rinse is comparable to WFI and NMT 500 ppb.
- Determination of MAC (maximum allowable carryover):
  - Test function: determine MAC of active material, compare with cleaning rinse results, and record in the analytical log book.
  - Acceptance criteria: the active ingredient in the final rinse is not detected, equal to, or less than the maximum allowable carryover limit determined.
  - For preparation vessel, MAC of active material in the washing rinses is to be checked by HPLC using suitable validated test method for each active material.

3. Documentation

- Cleaning SOP
- Log book entries
- Chromatograms
- Final report
- Training records

Attachments list:

- Attachment No. 2200.40(A): Equipment Description Figure and Sampling Locations
- Attachment No. 2200.40(B): Cleaning and Testing Responsibility
- Attachment No. 2200.40(C): Equipment Cleaning Procedure Validation Report
- Attachment No. 2200.40(D): Sampling Technique
- Attachment No. 2200.40(E): Calculations for Surface Swabs

SOP No.: Val. 2200.40   Effective date: mm/dd/yyyy
                                         Approved by:

REASONS FOR REVISION

Effective date: mm/dd/yyyy

- First time issued for your company, affiliates, and contract manufacturers.

Effective date: mm/dd/yyyy

Approved by:

Attachment No. 2200.40(A)

PROVIDE EQUIPMENT DESCRIPTION FIGURE

Sampling Locations

Equipment Figure No. 1

Sampling location: shelves

S1	S1
S1	S1
S1	S1
S1	S1
S1	S1
S1	S1
S1	S1

SOP No.: Val. 2200.40   Effective date: mm/dd/yyyy
                                         Approved by:

Attachment No. 2200.40(B)

CLEANING/TESTING RESPONSIBILITIES

Cleaning/ Testing	Done by	Recorded on	Checked by
Equipment cleaning	Machine operator	Equipment usage/ cleaning log book	Production supervisor
Visual inspection	Cleaning validation officer		
Rinse sample	Machine operator/ cleaning validation officer	Sampling sheets 3, 4	Asst. manager QA
Swab sample	Machine operator/ cleaning validation officer	Sampling sheets 3, 4	Asst. manager QA
pH	Cleaning validation officer/QC analyst	Analytical log book	QA/QC officer
Conductivity	Cleaning validation officer/QC analyst	Analytical log book	QA/QC officer
TOC	Cleaning validation officer/QC analyst	Analytical log book	QA/QC officer
MAC	Cleaning validation officer/QC analyst	Analytical log book	Development manager

Notes: TOC = total organic carbon; MAC = maximum allowable carryover.

Attachment No. 2200.40(C)

EQUIPMENT CLEANING PROCEDURE

VALIDATION REPORT

Equipment name: _____

Calibrated/validated on: _____

Location: _____

Room no.: _____

Last product: _____

B. no. of last prod.: _____

Manufacturing date: _____

Active ingredient: _____

Therapeutic group: _____

Cleaning date: _____

Cleaning SOP no.: _____ Revision no.: _____

Sampling technique: _____

Cleaning sample analysis date: _____ Assay result: _____

Test method reference: _____ Ref. analytical log book: ____

Limit of detection: _____

Next product to be manufactured on same equipment: _____

Safety factor: _____

SOP No.: Val. 2200.40

Effective date: mm/dd/yyyy

Approved by:

Attachment No. 2200.40(D)

SAMPLING TECHNIQUE

Process description: _____

Process involved: _____

Final Rinsing Volume

Sampling Location	Sampling Property		Type of Sample		Sample Quantity		Sampling Time		Surface Area in Contact with Product Solution	Final Volume of Rinse	
	D	N	S	R	cm²	500 ml	EC	AC		EC	AC
S1	—	—									
S2	—	—									
S3	—	—									
S4	—	—									
S5	—	—									
S6	—	—									
S7	—	—									
S8	—	—									
S9	—	—									
S10	—	—									
S11	—	—									
S12	—	—									
S13	—	—									
S14	—	—									

Notes: EC: sample collected at the end of cleaning before draining the rinse; AC: sample collected at the end of final rinsing water; S: swab; R: rinse; D: difficult to clean; N: normal.

SOP No.: Val. 2200.40   **Effective date: mm/dd/yyyy**

Approved by:

Training Record Verification

Following staff found trained on cleaning of equipment:

Using SOP no. (provide number): _____

Name: _____ ID no. _____ Sign _____ Date _____

Name: _____ ID no. _____ Sign _____ Date _____

Performed by: _____   Checked by: _____

Date: _____   Date: _____

SOP No.: Val. 2200.40

Effective date: mm/dd/yyyy

Approved by:

Attachment No. 2200.40(E)

CALCULATION FOR SURFACE SWABS

Formula

$$MAC = \frac{TD \times BS \times SF}{LDD}$$

where:
MAC = maximum allowable carryover
TD = therapeutic dose (single) = _____
BS = batch size (next product) = _____
SF = safety factor = _____
LDD = largest daily dose (for next product) _____

MAC _____

MAC = _____ mg

Calculation

Total active recovered from the 2.5-cm^2 surface area of the equipment directly in contact with product solution by swabs (A) _____ (mg)

Total active present over total surface area in direct contact with product solution (B)

_____(A) × _____ (B) = _____(C)

Acceptance Criteria

The total quantity of active (C) determined over the total surface area by using validated test method should be less than or equal to the MAC limit.

Recovery (Challenge) Test

S. No.	Name of Active Material	Concentration of Standard Solution	Method Used for Testing	% Recovery of Active Ingredient	% Recovery per Limit (95–105%)	
					Y	N
1		80%				
2		100%				
3		120%				

SOP No.: Val. 2200.40

Effective date: mm/dd/yyyy

Approved by:

Results

MAC	Actual Results (C)

Swab Analysis Results

Sampling Location	Visual Inspection	Blank WFI	PH (5–7)	TOC (NMT 1.0 μS/cm)	Conductivity (NMT 500 ppb)
S1					
S2					
S3					
S4					
S5					
S6					
S7					
S8					
S9					
S10					
S11					
S12					
S13					
S14					

Deviations: _____

_____

_____

Corrective actions/preventive measures: _____

_____

_____

Conclusion: _____

_____

_____

_____   _____
Validation Officer/Date   QA Manager/Date

YOUR COMPANY
VALIDATION STANDARD OPERATING PROCEDURE

SOP No.: Val. 2200.50 Effective date: mm/dd/yyyy

Approved by:

TITLE: Cleaning Validation Protocol of Vial-Filling
Machine Parts

AUTHOR: _____

Name/Title/Department

_____

Signature/Date

CHECKED BY: _____

Name/Title/Department

_____

Signature/Date

APPROVED BY: _____

Name/Title/Department

_____

Signature/Date

REVISIONS:

No.	Section	Pages	Initials/Date

SOP No.: Val. 2200.50   Effective date: mm/dd/yyyy

Approved by:

SUBJECT: Cleaning Validation Protocol of Vial-Filling Machine Parts

PURPOSE

The purpose is to validate the cleaning procedure and ensure that residue of previous product are removed in accordance with the maximum allow able carryover limit calculated, as well as other tests, i.e., visual inspection pH, conductivity, and TOC.

RESPONSIBILITY

Quality assurance inspector/machine operator and QC analyst are respor sible to follow the procedure. The quality assurance (QA) manager responsible for SOP compliance.

PROCEDURE

The cleaning validation protocols are applicable to:

- Buffer tank
- Pipe to manifold
- Manifold
- Prepump silicon hose
- Pump
- Postpump silicon hose
- Filling needles
- Vibratory sorter for stopper
- Stopper feed track

Verify following documentation requirements

- Availability of equipment cleaning procedure
- Verification of equipment cleaning log book records
- Verification of staff training record

Perform following test functions

- Visual inspection of equipment _____ (provide SOP number
- pH of the final rinse/swab per method _____ (provide met number)

SOP No.: Val. 2200.50   **Effective date: mm/dd/yyyy**

                                             Approved by:

- Conductivity of the final rinse/swab per procedure _____ (provide method number)
- TOC of the final rinse/swab per procedure _____ (provide method number)
- Maximum allowable carryover (MAC) _____ (provide method number)

Sampling criteria for swabs and rinses:

- Samples of the internal surfaces are taken by moistening the swab with a suitable solvent, sampling a 2.5-cm^2 area, and then placing the swab in a test tube containing 10 ml of the solvent (specified for each active material from the analytical test method available in the laboratory or from pharmacopoeia).
- It is important to take a representative sample of the area because the results will be calculated for the entire surface area at a later stage.
- Samples of the vessel rinses are collected in 500-ml volumetric flasks after final rinsing of the vessels with purified water as described in the individual equipment cleaning SOP.

Types of samples:

- Collect two rinsing samples from the preparation tank drainage. Sampling will be performed two times, one before the start of cleaning procedure and another at the end of final rinsing water, after completion of cleaning procedure.
- Collect two surface samples from each part of preparation tank per attached (provide photographs of vial-filing machine parts with sample identifications).
- Sampling will be performed two times, one before the start of cleaning procedure and another after completion of cleaning procedure.

Handling of samples:

- After collecting, rinse/swab samples are kept in the refrigerator.
- Analyze samples immediately for pH, conductivity, and TOC.
- Samples of rinses/swabs for the HPLC analysis collected at the time of manufacturing analysis are to be completed within 24 hours of collection.
- HPLC samples should be kept at room temperature for at least 2 hours before testing.

SOP No.: Val. 2200.50

Effective date: mm/dd/yyyy

Approved by:

Analysis and documentation:

- The validated HPLC test method will be used for the determination of chemical residues. The validation of test method will be performed by addition and recovery tests.
- All analysis results will be recorded in the analysis log book Printouts and chromatograms will also be attached along with analytical log book for reference.
- All analysis and data are to be verified by second analyst.

1. Documentation Requirements

- Availability of cleaning SOP: confirm the availability of respectiv cleaning SOP in the area.
- Verification of log book record: check the log book and verify th cleaning time and date and product details. Cleaning procedu and date and time are to be verified from log book.
- Verification of staff training record: verify the staff trained c cleaning SOP. The training of staff responsible for the equipme cleaning is to be verified from the training record and SOP. Ref to attachment no. 2200.50(D).

2. Test Function and Acceptance Criteria

- Visual inspection of equipment:
  - Test function: after cleaning completion, perform visual insp tion of equipment — particularly, difficult-to-clean areas — a record in the analytical log book.
  - Acceptance criteria: the visible internal equipment surface a all critical and difficult-to-clean parts are to be optically f from residue and the color of the final rinse should be com rable to WFI.
- pH determination:
  - Test function: after cleaning, perform pH. Refer to attachn no. 2200.50(D).
  - Acceptance criteria: is comparable to WFI pH value of 5 tc

SOP No.: Val. 2200.50   **Effective date: mm/dd/yyyy**
                                            Approved by:

- Conductivity:
  - Test function: after final cleaning, perform conductivity at room temperature, i.e., 25°C ± 2. Refer to attachment no. 2200.50(D).
  - Acceptance criteria (NMT 1.0 μS/cm): the conductivity of the final rinse is comparable to WFI and NMT 1.0 μS/cm.
- Total organic carbon (TOC):
  - Test function: after cleaning, perform TOC of the final rinse. Refer to attachment no. 2200.50(D).
  - Acceptance criteria (NMT 500 ppb): the TOC of the final rinse is comparable to WFI and NMT 500 ppb.
- Determination of MAC (maximum allowable carryover):
  - Test function: determine MAC of active material, compare with cleaning-rinse results, and record in the analytical log book.
  - Acceptance criteria: the active ingredient in the final rinse is not detected, equal to, or less than the maximum allowable carryover limit determined.
  - For preparation vessel, MAC of active material in the washing rinses is to be checked by HPLC using suitable validated test method for each active material.

8. Documentation

- Cleaning SOP
- Log book entries
- Chromatograms
- Final report
- Training records

Attachments list:

- Attachment No. 2200.50(A): Equipment Description Figure and Sampling Locations
- Attachment No. 2200.50(B): Cleaning and Testing Responsibilities
- Attachment No. 2200.50(C): Equipment Cleaning Procedure Validation Report
- Attachment No. 2200.50(D): Sampling Technique
- Attachment No. 2200.50(E): Calculations for Surface Swabs
- Attachment No. 2200.50(F): Calculations for Final Rinse Sample

SOP No.: Val. 2200.50

Effective date: mm/dd/yyyy

Approved by:

REASONS FOR REVISION

Effective date: mm/dd/yyyy

■ First time issued for your company, affiliates, and contract manufacturers.

SOP No.: Val. 2200.50   **Effective date: mm/dd/yyyy**

Approved by:

Attachment No. 2200.50(A)

PROVIDE EQUIPMENT DESCRIPTION FIGURE

Sampling Locations

Equipment Figure No. 1

ampling location: needles, manifold, pipe to manifold

1 S2 S3 S4 S5 S6 S7 S8 S9 S10 S11 S12 S13 S14 S15 S16 S17 S18 S19 S20

Equipment Figure No. 2

ampling location: vibratory sorters for stoppers, stopper feed track

1   S2   S3   S4   S5   S6

Equipment Figure No. 3

ampling location: vibratory pumps

Effective date: mm/dd/yyyy
                                                    Approved by:

Attachment No. 2200.50(B)

CLEANING/TESTING RESPONSIBILITIES

Cleaning/ Testing	Done by	Recorded on	Checked by
Equipment cleaning	Machine operator	Equipment usage/ cleaning log book	Production supervisor
Visual inspection	Cleaning validation officer		
Rinse sample	Machine operator/ cleaning validation officer	Sampling sheets 3, 4	Asst. manager QA
Swab sample	Machine operator/ cleaning validation officer	Sampling sheets 3, 4	Asst. manager QA
pH	Cleaning validation officer/QC analyst	Analytical log book	QA/QC officer
Conductivity	Cleaning validation officer/QC analyst	Analytical log book	QA/QC officer
TOC	Cleaning validation officer/QC analyst	Analytical log book	QA/QC officer
MAC	Cleaning validation officer/QC analyst	Analytical log book	Developmer manager

Notes: TOC = total organic carbon; MAC = maximum allowable carryover.

Effective date: mm/dd/yyyy

Approved by:

Attachment No. 2200.50(C)

EQUIPMENT CLEANING PROCEDURE

VALIDATION REPORT

Equipment name: _____

Calibrated/validated on: _____

Location: _____

Room no.: _____

Last product: _____

B. no. of last prod.: _____

Manufacturing date: _____

Active ingredient: _____

Therapeutic group: _____

Cleaning date: _____

Cleaning SOP no.: _____ Revision no.: _____

Sampling technique: _____

Cleaning sample analysis date: _____ Assay result: _____

Test method reference: _____ Ref. analytical log book: _____

Limit of detection: _____

Next product to be manufactured on same equipment: _____

Safety factor: _____

SOP No.: Val. 2200.50

Effective date: mm/dd/yyyy

Approved by:

Attachment No. 2200.50(D)

SAMPLING TECHNIQUE

Process description: _____

Process involved: _____

Final Rinsing Volume

Sampling Location	Sampling Property		Type of Sample		Sample Quantity		Sampling Time		Surface Area in Contact with Product Solution	Final Volume of Rinse	
	D	N	S	R	cm²	500 ml	EC	AC		EC	A
R1		—		—							
R2		—		—							
S1		—	—								
S2		—	—								
S3		—	—								
S4		—	—								
S5		—	—								
S6		—	—								
S7		—	—								
S8		—	—								
S9		—	—								
S10		—	—								
S11		—	—								
S12		—	—								
S13		—	—								
S14		—	—								
S15		—	—								
S16		—	—								
S17		—	—								
S18		—	—								

SOP No.: Val. 2200.50   **Effective date: mm/dd/yyyy**

                                                 Approved by:

S19		—	—								
S20		—	—								
S21		—	—								
S22		—	—								
S23		—	—								
S240		—	—								
S25		—	—								
S26		—	—								
S27		—	—								

Notes: EC: sample collected at the end of cleaning before draining the rinse; AC: sample collected at the end of final rinsing water; S: swab; R: rinse; D: difficult to clean; N: normal.

Training Record Verification

ollowing staff found trained on cleaning of equipment:

Jsing SOP no. (provide number): _____

Jame: _____ ID no. _____ Sign _____ Date _____

Jame: _____ ID no. _____ Sign _____ Date _____

erformed by: _____ Checked by: _____

ate: _____ Date: _____

Attachment No. 2200.50(E)

CALCULATION FOR SURFACE SWABS

Formula

$$MAC = \frac{TD \times BS \times SF}{LDD}$$

where:

MAC = maximum allowable carryover

TD = therapeutic dose (single) = _____

BS = batch size (next product) = _____

SF = safety factor = _____

LDD = largest daily dose (for next product) _____

MAC _____

MAC = _____ mg

Calculation

Total active recovered from the 2.5-cm² surface area of the equipment directly in contact with product solution by swabs (A) _____ (mg)

Total active present over total surface area in direct contact with product solution (B)

_____(A) × _____ (B) = _____(C)

Acceptance Criteria

The total quantity of active (C) determined over the total surface area by using a validated test method should be less than or equal to the MAC limit.

Recovery (Challenge) Test

S. No.	Name of Active Material	Concentration of Standard Solution	Method Used for Testing	% Recovery of Active Ingredient	% Recovery per Limit (95–105%)	
					Y	N
1		80%				
2		100%				
3		120%				

SOP No.: Val. 2200.50

Effective date: mm/dd/yyyy

Approved by:

Results

MAC	Actual Results (C)

Swab Analysis Results

Sampling Location	Visual Inspection	Blank WFI	PH (5–7)	TOC (NMT 1.0 µS/cm)	Conductivity (NMT 500 ppb)
S1					
S2					
S3					
S4					
S5					
S6					
S7					
S8					
S9					
S10					
S11					
S12					
S13					
S14					
S15					
S16					
S17					
S18					
S19					
S20					
S21					
S22					
S23					

SOP No.: Val. 2200.50
Effective date: mm/dd/yyyy
Approved by:

S24					
S25					
S26					
S27					

Deviations: _____

_____

_____

Corrective actions/preventive measures: _____

_____

_____

Conclusion: _____

_____

_____

_____   _____
Validation Officer/Date   Asst. QA Manager/Date

SOP No.: Val. 2200.50   Effective date: mm/dd/yyyy
                                         Approved by:

Attachment No. 2200.50(F)

CALCULATION FOR FINAL RINSE SAMPLE

Formula

$$MAC = \frac{TD \times BS \times SF}{LDD}$$

where:
MAC = maximum allowable carryover
TD = therapeutic dose (single) = _____
BS = batch size (next product) = _____
SF = safety factor = _____
LDD = largest daily dose (for next product) _____

MAC _____

MAC = _____ mg

alculation
00 ml of the rinse contains active (A) = _____ mg

_____(B) ml of the final rinse contains an active (C) = $\frac{A\,(mg) \times B\,(ml)}{500\,ml}$

cceptance Criteria
he total amount of active (C) determined in the total volume of final rinse
ould be less than or equal to the MAC limit.

Recovery (Challenge) Test

	Name of Active Material	Concentration of Standard Solution	Method Used for Testing	% Recovery of Active Ingredient	% Recovery per Limit (95–105%)	
					Y	N
		80%				
		100%				
		120%				

SOP No.: Val. 2200.50

Effective date: mm/dd/yyyy

Approved by:

Results

MAC	Actual Results (C)

Rinses Analysis Results

Sampling Location	Visual Inspection	Blank WFI	PH (5–7)	TOC (NMT 1.0 μS/cm)	Conductivity (NMT 500 ppb)

Deviations: _____

_____

_____

Corrective actions/preventive measures: _____

_____

_____

Conclusion: _____

_____

_____

_____   _____
Validation Officer/Date   QA Manager/Date

SECTION

VAL 2300.00

| YOUR COMPANY |
| VALIDATION STANDARD OPERATING PROCEDURE |

SOP No.: Val. 2300.10

Effective date: mm/dd/yyyy

Approved by:

TITLE: Recommended Reading

AUTHOR: _____
Name/Title/Department

_____
Signature/Date

CHECKED BY: _____
Name/Title/Department

_____
Signature/Date

APPROVED BY: _____
Name/Title/Department

_____
Signature/Date

REVISIONS:

No.	Section	Pages	Initials/Date

SOP No.: Val. 2300.10

Effective date: mm/dd/yyyy
Approved by:

SUBJECT: Recommended Reading

PURPOSE

The purpose is to provide guidelines for keeping the record of literature cited in validation documents.

RESPONSIBILITY

The validation officer and validation team members are responsible to follow the procedure. The quality assurance (QA) manager is responsible for SOP compliance.

PROCEDURE

1. Citation of Literature

- The validation officer is responsible to keep himself or herself updated with the current FDA and GMP guidelines and maintain an index with issue dates.
- The references used in the development of abbreviated new drug application (ANDA) file should be cited at the end of the sterility assurance report.
- Some recent literature is provided as examples:

RECOMMENDED READING

Baker, G.S. and Rhodes, C.T., *Modern Pharmaceutics*, Marcel Dekker, Inc., New York, 1996.

Berry, I.R. and Harpaz, D., *Validation of Bulk Pharmaceutical Chemicals*, Interpharm Press, Inc., Buffalo Grove, IL, 1997.

Berry, I.R. and Nash, R.A., *Pharmaceutical Process Validation*, Marcel Dekker, Inc., New York, 1993.

British Pharmacopoeia, 1993.

Bulk Pharmaceutical Chemicals, ISPE, 1996.

Carleton, F.J. and Agalloco, J.P., *Validation of Aseptic Pharmaceutical Processes*, Marcel Dekker, Inc., New York, 1986.

Carleton, F.J. and Agalloco, J.P., *Validation of Pharmaceutical Processes*, Marcel Dekker, Inc., New York, 1986.

SOP No.: Val. 2300.10   Effective date: mm/dd/yyyy

Approved by:

Cloud, P., *How to Develop and Manage Qualification Protocols for FDA Compliance*, Interpharm Press, Buffalo Grove, IL, 1999.

Cloud P., *Pharmaceutical Equipment Validation Qualification*, Interpharm Press, Buffalo Grove, IL, 1998.

Commission of the European Communities, The rules governing medicinal products in the European Community, Guide to II/2244/87-EN, Rev. 3, 1989.

DeSain, C., *Documentation Basics That Support Good Manufacturing Practices*, Advanstar Communications, Cleveland, 1993.

Development Pharmaceutics and Process Validation, CPMP, 1988.

Gibson, W. and Evans, K.P., *Validation Fundamentals How to, What to, When to Validate*, Interpharm Press, Buffalo Grove, IL, 1998.

Good Manufacturing Practice for Medical Products, European Community, 1992.

Guide to Inspection of High Purity Water Systems, Interpharm Press, Buffalo Grove, IL, 1993.

Guide to Inspection of Microbiological Pharmaceutical Quality Control Laboratories, Interpharm Press, Buffalo Grove, IL, 1993.

Guideline of the Commission for Principles of Good Manufacturing Practice for Medical Products, Directive 91/356/EEC, 1991.

Guideline on General Principles of Process Validation. May 1987. Center for Drugs and Biologics, Food and Drug Administration, Rockville, MD.

Guideline on PDA (Parenteral Drug Association Inc.) of Validation of Pharmaceuticals and Biopharmaceuticals, February 1993.

Guidelines for Good Validation Practice, FIP, 1988.

Guidelines on the Validation of Manufacturing Process, WHO, 1996.

Huber, L., *Validation and Qualification in Analytical Laboratories*, Interpharm Press, Buffalo Grove, IL, 1998.

Huber, L., *Validation of Computerized Analytical Systems*, Interpharm Press, Buffalo Grove, IL, 2000.

International Standard of ISO 13408-1 for aseptic processing of health care products.

Isaacs, A., Validating machinery with electronic control systems, *Manuf. Chemist*, 19, 1992.

ISO 14644-1 Cleanrooms and Associated Controlled Environments, Classification of Air Cleanliness.

Knapp, J.Z. and Kushner, H.K., Generalized methodology for evaluation of parenteral inspection, *J. Parenteral Drug Assoc.*, 34(I), 1980.

Loftus, B.T. and Nash, R.A., *Pharmaceutical Process Validation*, Marcel Dekker, Inc., New York, 1984.

Oral Solid Dosage Forms, ISPE, 1997.

Parenteral Drug Association, Inc., 2002.

Parenteral Drug Association, Technical Report No. 3, Validation of dry heat process used for sterilization and depyrogenation, 1981.

Part 211 — Current Good Manufacturing Practice for Fisher Pharmaceuticals.

SOP No.: Val. 2300.10   Effective date: mm/dd/yyyy
                                         Approved by:

PMA's Computer Systems Validation Committee, Validation concepts for computer systems used in the manufacture of drug products, *Pharm. Technol.*, 10(5), 24, 1986.

PMA's Computer Systems Validation Committee, Computer system validation — staying current: introduction, *Pharm. Technol.*, 13(5), 60, 1989.

Principles of Qualification and Validation in Pharmaceutical Manufacture, PIC, 1996.

Stokes, T. et al., *Good Computer Validation Practices: Common Sense Implementation*, Interpharm Press, Inc., Buffalo Grove, IL, 1994.

Technical Report No. 26, Sterilization Filtration of Liquids, Parenteral Drug Association, Inc., 1998.

Technical Report No. 36, Current Practices in the Validation of Aseptic Processing.

United States Pharmacopia, USP-23, 1995.

United States Pharmacopia 24, 2000.

Validation Documentation Inspection Guide, Interpharm Press, Buffalo Grove, IL, 1993.

Wingate, G., *Validating Automated Manufacturing and Laboratory Applications: Putting Principles into Practice*, Interpharm Press, Inc., Buffalo Grove, IL, 1997.

REASONS FOR REVISION

Effective date: mm/dd/yyyy

■ First time issued for your company, affiliates, and contract manufacturers.

T - #0122 - 071024 - C0 - 234/156/61 - PB - 9780367390778 - Gloss Lamination